# HYDRODYNAMICS, MASS AND HEAT TRANSFER IN CHEMICAL ENGINEERING

T0199546

# Topics in Chemical Engineering
A series edited by R. Hughes, University of Salford, UK

Please see the back of this book for other titles in the Topics in Chemical Engineering series.

# HYDRODYNAMICS, MASS AND HEAT TRANSFER IN CHEMICAL ENGINEERING

## A.D. Polyanin
*Institute for Problems in Mechanics, Moscow, Russia*

## A.M. Kutepov
*Moscow State Academy of Chemical Engineering, Moscow, Russia*

## A.V. Vyazmin
*Karpov Institute of Physical Chemistry, Moscow, Russia*

## D.A. Kazenin
*Moscow State Academy of Chemical Engineering, Moscow, Russia*

CRC Press
Taylor & Francis Group
Boca Raton London New York

CRC Press is an imprint of the
Taylor & Francis Group, an **informa** business

CRC Press
Taylor & Francis Group
6000 Broken Sound Parkway NW, Suite 300
Boca Raton, FL 33487-2742

First issued in paperback 2019

© 2002 by Taylor & Francis Group, LLC
CRC Press is an imprint of Taylor & Francis Group, an Informa business

This book has been produced from camera-ready copy
supplied by the authors

No claim to original U.S. Government works

ISBN-13: 978-0-415-27237-7 (hbk)
ISBN-13: 978-0-367-39690-9 (pbk)

*British Library Cataloguing in Publication Data*
A catalogue record for this book is available from the British Library

*Library of Congress Cataloging in Publication Data*
A catalog record for this book has been requested

**Visit the Taylor & Francis Web site at**
**http://www.taylorandfrancis.com**

**and the CRC Press Web site at**
**http://www.crcpress.com**

# CONTENTS

# Introduction to the Series

The subject matter of chemical engineering covers a very wide spectrum of learning and the number of subject areas encompassed in both undergraduate and graduate courses is inevitably increasing each year. This wide variety of subjects makes it difficult to cover the whole subject matter of chemical engineering in a single book. The present series is therefore planned as a number of books covering areas of chemical engineering which, although important, are not treated at any length in graduate and postgraduate standard texts. Additionally, the series will incorporate recent research material which has reached the stage where an overall survey is appropriate, and where sufficient information is available to merit publication in book form for the benefit of the profession as a whole.

Inevitably, with a series such as this, constant revision is necessary if the value of the texts for both teaching and research purposes is to be maintained. I would be grateful to individuals for criticisms and for suggestions for future editions.

R. HUGHES

# Preface

The book contains a concise and systematic exposition of fundamental problems of hydrodynamics, heat and mass transfer, and physicochemical hydrodynamics, which constitute the theoretical basis of chemical engineering science.

In the selection of the material, the authors have given preference to simple exact, approximate, and empirical formulas that can be used in a wide range of practical applications. A number of new formulas are presented. Special attention has been paid to universal formulas that can be used to describe entire classes of problems (that differ in geometric or other factors). Such formulas provide a lot of information in compact form.

Each section of the book usually begins with a brief physical and mathematical statement of the problem considered. Then final results are usually given for the desired variables in the form of final relationships and tables (as a rule, the solution method is not presented, only some explanations and necessary references are given). This approach simplifies the understanding of the text for a wider readership.

Only the most important problems that admit exact analytical solution are discussed in more detail. Such solutions play an important role in the proper understanding of qualitative features of many phenomena and processes in various areas of natural and engineering sciences. The corresponding sections of the book may be used by college and university lecturers of courses in chemical engineering science, hydrodynamics, heat and mass transfer, and physicochemical hydrodynamics for graduate and postgraduate students.

In Chapters 1 and 2 we study fluid flows, which underlie numerous processes of chemical engineering science. We present up-to-date results about translational and shear flows past particles, drops, and bubbles of various shapes at a wide range of Reynolds numbers. Single particles and systems of particles are considered. Film and jet flows, fluid flows through tubes and channels of various shapes, and flow past plates, cylinders, and disks are examined.

In Chapters 3 and 4 we analyze mass and heat transfer in plane channels, tubes, and fluid films. We consider the mass and heat exchange between particles, drops, or bubbles and uniform or shear flows at various Peclet and Reynolds numbers. The results presented are of great importance in obtaining scientifically justified methods for a number of technological processes such as dissolution, drying, adsorption, aerosol and colloid sedimentation, heterogeneous catalytic reactions, absorption, extraction, and rectification.

In Chapter 5 some problems of mass and heat transfer with various complicating factors are discussed. Mass transfer problems are investigated for various kinetics of volume and surface chemical reactions. Nonlinear problems of convective mass and heat exchange are considered taking into account the dependence of the transfer coefficients on concentration (temperature). Nonisothermal flows through tubes and channels accompanied by dissipative heating of liquid are also studied. Qualitative features of heat transfer in liquids with temperature-dependent viscosity are discussed, and various thermohydrodynamic phenomena related to the fact that the surface tension coefficient is temperature dependent are analyzed.

In Chapter 6 we consider problems of hydrodynamics and mass and heat transfer in non-Newtonian fluids and describe the basic models for rheologically complicated fluids, which are used in chemical technology. Namely, we consider the motion and mass exchange of power-law and viscoplastic fluids through tubes, channels, and films. The flow past particles, drops, and bubbles in non-Newtonian fluid are also analyzed.

Chapter 7 deals with the basic concepts and properties of very specific technological media, namely, foam systems. Important processes such as surfactant interface accumulation, syneresis, and foam rupture are considered.

The supplements contain tables with exact solutions of the heat equation. In addition, the equation of convective diffusion, the continuity equation, equations of motion in some curvilinear orthogonal coordinate systems, and some other reference materials are given.

The topics in the present book are arranged in increasing order of difficulty, which substantially simplifies understanding the material. A detailed table of contents readily allows the reader to find the desired information. A lot of material and its compact presentation permit the book to be used as a concise handbook in chemical engineering science and related fields in hydrodynamics, heat and mass transfer, etc.

The authors are grateful to A. E. Rednikov and Yu. S. Ryazantsev, who wrote Sections 5.8–5.10, Z. D. Zapryanov, who contributed to Sections 1.2, 1.3, and 2.9, and A. G. Petrov, who contributed to Subsections 2.4-3, 2.8-2, and 2.8-4. We express our deep gratitude to V. E. Nazaikinskii and A. I. Zhurov for fruitful discussions and valuable remarks.

The work on this book was supported in part by the Russian Foundation for Basic Research.

The authors hope that the book will be useful for researchers and engineers, as well as postgraduate and graduate students, in chemical engineering science, hydrodynamics, heat and mass transfer, mechanics of disperse systems, physico-chemical hydrodynamics, power engineering, meteorology, and biomechanics.

# Basic Notation

## Latin Symbols

$a$    characteristic scale of length; radius of spherical particle or circular cylinder

$a_e$    radius of volume-equivalent sphere

$a_p$    radius of perimeter-equivalent sphere (for body of revolution)

$C$    concentration

$C_i$    nonperturbed concentration (in incoming flow remote from particle)

$C_s$    concentration at surface of particle (or tube)

$c$    dimensionless concentration (introduced differently in various problems, see Table 3.1 in Section 3.1)

$c_f$    drag coefficient (for particles, drops, and bubbles); local coefficient of friction

$c_p$    specific heat at constant pressure

$D$    diffusion coefficient

$D_t$    turbulent diffusion coefficient

$d$    diameter of circular tube, spherical particle, or circular cylinder

$d_e$    equivalent diameter

$e_{ij}$    shear rate tensor components

$F$    viscous drag forces acting on particle, drop or bubbles

$F_\parallel, F_\perp$    drag forces of body of revolution for its parallel and perpendicular positions in translational flow

$F_T$    thermocapillary force acting on drop

$F_s$    kinetic function of surface reaction, $F_s = F_s(C)$

$F_v$    kinetic function of volume reaction, $F_v = F_v(C)$

Fr    Froude number

$f_s$    dimensionless kinetic function of surface reaction, $f_s = f_s(c)$

$f_v$    dimensionless kinetic function of volume reaction, $f_v = f_v(c)$

$\langle f_v \rangle$    mean value of dimensionless kinetic function, $\langle f_v \rangle = \int_0^1 f_v(c)\,dc$

$G_{km}$    shear matrix coefficients

Gr    Grashof number

$g$    acceleration due to gravity

$g_{ij}$    metric tensor components

$h$    film thickness; half-width of plane channel

$I$    dimensionless total diffusion flux

$I_*$    total diffusion flux

$I_T$    dimensionless total heat flux

$i_X, i_Y, i_Z$    unit vectors of Cartesian coordinate system

$J_0$    momentum of jet

$j$    dimensionless diffusion flux

$j_*$    diffusion flux

$j_T$     dimensionless heat flux

$K_s$     rate constant for surface chemical reaction

$K_v$     rate constant for volume chemical reaction

Ku     Kutateladze number

$k$     consistence factor of power-law fluid

$k_s$     dimensionless rate constant for surface chemical reaction

$k_v$     dimensionless rate constant for volume chemical reaction

Le     Lewis number

Ma     Marangoni number

Mo     Morton number

Nu     Nusselt number; mean Nusselt number

$\text{Nu}_X$     local Nusselt number

$\text{Nu}_\infty$     limit Nusselt number

$n$     rate order of chemical reaction (surface or volume) or rheological parameter of power-law fluid

$P$     pressure

$P_i$     nonperturbed pressure remote from particle (drop or bubble)

Pe     diffusion Peclet number, $\text{Pe} = aU/D$

$\text{Pe}_T$     heat Peclet number

$\text{Pr}_t$     turbulent Prandtl number

$Q$     volume rate of flow (through tube cross-section)

$R, \theta, \varphi$     spherical coordinate system, $R = \sqrt{X^2 + Y^2 + Z^2}$

$\mathcal{R}, Z, \varphi$     cylindrical coordinate system, $\mathcal{R} = \sqrt{X^2 + Y^2}$

Re     Reynolds number, $\text{Re} = aU/\nu$

$\text{Re}_d$     Reynolds number based on diameter, $\text{Re}_d = dU/\nu$

$\text{Re}_X$     local Reynolds number, $\text{Re}_X = XU/\nu$

$r$     dimensionless radial spherical coordinate, $r = R/a$

$S$     dimensionless area of surface, $S = S_*/a^2$

$S_*$     dimensional area of surface

Sc     Schmidt number, $\text{Sc} = \nu/D$

Sh     mean Sherwood number, $\text{Sh} = I/S$

$\text{Sh}_b$     mean Sherwood number for bubble

$\text{Sh}_p$     mean Sherwood number for particle

$\text{Sh}_\beta$     mean Sherwood number for drop

$\text{Sh}_0$     asymptotic value of mean Sherwood number at small values of characteristic parameter of problem

$\text{Sh}_\infty$     asymptotic value of mean Sherwood number at large values of characteristic parameter of problem

St     Strouhal number

$T$     dimensionless temperature

$T_*$     temperature

$T_i$     nonperturbed temperature (in incoming flow remote from particle)

$T_s$     temperature at surface of particle (or tube)

$\overline{T}$     average component of temperature for turbulent flow

$\langle T \rangle$     bulk body temperature

$\langle T \rangle_m$     mean flow rate temperature

$t$     time

| | |
|---|---|
| $U$ | characteristic flow velocity |
| $U_i$ | nonperturbed fluid velocity (in incoming flow remote from particle) |
| $U_{max}$ | maximum fluid velocity at surface of film or on tube axis |
| $U_T$ | thermocapillary drift velocity of drop |
| $U_*$ | friction velocity (for turbulent flows), $U_* = \sqrt{\tau_s/\rho}$ |
| $\mathbf{V}$ | fluid velocity vector |
| $\langle V \rangle$ | mean flow rate velocity, $\langle V \rangle = Q/S_*$ |
| $V_X, V_Y, V_Z$ | fluid velocity components in Cartesian coordinate system |
| $\overline{V}_X$ | average component of velocity in turbulent flow |
| $V_R, V_\theta, V_\varphi$ | fluid velocity components in spherical coordinate system |
| $V_{\mathcal{R}}, V_Z, V_\varphi$ | fluid velocity components in cylindrical coordinate system |
| $V_R^{(1)}, V_\theta^{(1)}$ | fluid velocity components in continuous phase (outside drops) in axisymmetric case |
| $V_R^{(2)}, V_\theta^{(2)}$ | fluid velocity components in disperse phase (inside drops) in axisymmetric case |
| $\mathbf{v}$ | dimensionless fluid velocity vector |
| $v_x, v_y, v_z$ | dimensionless fluid velocity components in Cartesian coordinate system |
| $v_r, v_\theta, v_\varphi$ | dimensionless fluid velocity components in spherical coordinate system |
| $\mathsf{We}$ | Weber number, $\mathsf{We} = aU_i^2\rho_1/\sigma$ ($\sigma$ is surface tension) |
| $X, Y, Z$ | Cartesian coordinate system |
| $X_1, X_2, X_3$ | Cartesian coordinate system, $X_1 = X$, $X_2 = Y$, $X_3 = Z$ |
| $x, y, z$ | dimensionless Cartesian coordinate system |

# Greek Symbols

| | |
|---|---|
| $\beta$ | viscosity ratio, $\beta = \mu_2/\mu_1$ |
| $\Delta$ | Laplace operator |
| $\Delta P$ | total pressure drop along a tube part of length $L$, $\Delta P > 0$ |
| $\delta$ | thickness of hydrodynamic boundary layer |
| $\delta_T$ | thickness of thermal boundary layer |
| $\delta_{km}$ | Kronecker delta |
| $\Theta_*$ | friction temperature (for turbulent flows) |
| $\theta$ | angular coordinate |
| $\varkappa$ | thermal conductivity coefficient |
| $\kappa$ | von Karman constant |
| $\lambda$ | drag coefficient (for tubes and channels) |
| $\lambda_m$ | eigenvalues |
| $\mu$ | viscosity |
| $\mu_p$ | plastic viscosity for Shvedov–Bingham fluid |
| $\mu_1$ | viscosity of continuous phase |
| $\mu_2$ | viscosity of disperse phase |
| $\nu$ | kinematic viscosity, $\nu = \mu/\rho$ |
| $\nu_t$ | turbulent viscosity coefficient |
| $\Pi$ | shape factor, $\Pi = \mathsf{Sh}\, S_*/a$; disjoining pressure |
| $\rho$ | density |
| $\rho_1$ | density of continuous phase |
| $\rho_2$ | density of disperse phase |
| $\varrho$ | dimensionless cylindrical coordinate, $\varrho = \mathcal{R}/a$ |

| | |
|---|---|
| $\Sigma$ | perimeter-equivalent factor, $S_*/(4\pi a_\mathrm{p}^2)$ |
| $\tau_s$ | friction stress on wall |
| $\tau_{ij}$ | shear stress tensor components |
| $\tau_0$ | yield stress (for Shvedov–Bingham fluid) |
| $\varphi$ | angular coordinate (polar angle) |
| $\phi$ | volume fraction of disperse phase |
| $\chi$ | thermal diffusivity |
| $\Psi$ | stream function |
| $\Psi^{(1)}$ | stream function in continuous phase |
| $\Psi^{(2)}$ | stream function in disperse phase |
| $\psi$ | dimensionless stream function |

# Chapter 1
# Fluid Flows in Films, Jets, Tubes, and Boundary Layers

The information on velocity and pressure fields necessary for studying the distribution and transformation of reactants in reaction equipment can often be obtained from purely hydrodynamic considerations. The same hydrodynamic equations describe a vast variety of actual fluid flows depending on numerous geometric, physical, and mode factors that determine the flow region, type, and structure. There are various classifications of flows according to their specific properties, for example, the widely used classification based on the Reynolds number Re, which is the most significant state-geometric parameter.* This classification distinguishes flows at low Re [179], at high Re (boundary layers [427]), and at supercritical Re (turbulent flows [188]) and is methodologically important in that it introduces a small parameter (Re or $Re^{-1}$), which permits one to solve nonlinear hydrodynamic problems reliably by using expansions with respect to that parameter. Although this classification is undoubtedly fruitful and convenient for those studying hydrodynamic problems mathematically and numerically, in the present book we focus our attention on the practical needs of industrial engineers who deal with specific units of equipment where the type of flow of the reactive medium is virtually predetermined by the design. Accordingly, our treatment of hydrodynamics consists of two chapters. Chapter 1 deals with flows of extended fluid media interacting with each other or with containing walls (flows in films, tubes, channels, jets, and boundary layers near a solid surface). In Chapter 2 we consider the hydrodynamic interaction of particles of various nature (solid, liquid, or gaseous) with the ambient continuous phase.

## 1.1. Hydrodynamic Equations and Boundary Conditions

In this section we present equations and boundary conditions used in solving hydrodynamic problems. Their detailed derivation, as well as an analysis of

---

* There are some other flow classifications, for example, with respect to specific properties of the boundary of the flow region: fluid flow with free boundaries [385], fluid flow with interface [226, 501], and flow along a permeable boundary [524]. This classification also allows one to describe properties of various flows and suggest methods for studying these flows.

scope, various physical statements and solutions of related problems, and applied issues can be found, e.g., in the books [26, 126, 260, 276, 427, 440, 502]. We consider fluids with constant density $\rho$ and dynamic viscosity $\mu$.

### 1.1-1. Laminar Flows. Navier–Stokes Equations

First, laminar flows of fluids are considered. For brevity, in what follows we often refer to "laminar flows" simply as "flows."

*Navier–Stokes equations.* The closed system of equations of motion for a viscous incompressible Newtonian fluid consists of the continuity equation

$$\frac{\partial V_X}{\partial X} + \frac{\partial V_Y}{\partial Y} + \frac{\partial V_Z}{\partial Z} = 0 \tag{1.1.1}$$

and the three Navier–Stokes equations [326, 477]

$$\frac{\partial V_X}{\partial t} + V_X \frac{\partial V_X}{\partial X} + V_Y \frac{\partial V_X}{\partial Y} + V_Z \frac{\partial V_X}{\partial Z}$$
$$= -\frac{1}{\rho} \frac{\partial P}{\partial X} + \nu \left( \frac{\partial^2 V_X}{\partial X^2} + \frac{\partial^2 V_X}{\partial Y^2} + \frac{\partial^2 V_X}{\partial Z^2} \right) + g_X,$$

$$\frac{\partial V_Y}{\partial t} + V_X \frac{\partial V_Y}{\partial X} + V_Y \frac{\partial V_Y}{\partial Y} + V_Z \frac{\partial V_Y}{\partial Z} \tag{1.1.2}$$
$$= -\frac{1}{\rho} \frac{\partial P}{\partial Y} + \nu \left( \frac{\partial^2 V_Y}{\partial X^2} + \frac{\partial^2 V_Y}{\partial Y^2} + \frac{\partial^2 V_Y}{\partial Z^2} \right) + g_Y,$$

$$\frac{\partial V_Z}{\partial t} + V_X \frac{\partial V_Z}{\partial X} + V_Y \frac{\partial V_Z}{\partial Y} + V_Z \frac{\partial V_Z}{\partial Z}$$
$$= -\frac{1}{\rho} \frac{\partial P}{\partial Z} + \nu \left( \frac{\partial^2 V_Z}{\partial X^2} + \frac{\partial^2 V_Z}{\partial Y^2} + \frac{\partial^2 V_Z}{\partial Z^2} \right) + g_Z,$$

Equations (1.1.1) and (1.1.2) are written in an orthogonal Cartesian system $X, Y,$ and $Z$ in physical space; $t$ is time; $g_X$, $g_Y$, and $g_Z$ are the mass force (e.g., the gravity force) density components; $\nu = \mu/\rho$ is the kinematic viscosity of the fluid. The three components of the fluid velocity $V_X, V_Y, V_Z$, and the pressure $P$ are the unknowns.

By introducing the fluid velocity vector $\mathbf{V} = \mathbf{i}_X V_X + \mathbf{i}_Y V_Y + \mathbf{i}_Z V_Z$, where $\mathbf{i}_X$, $\mathbf{i}_Y$, and $\mathbf{i}_Z$ are the unit vectors of the Cartesian coordinate system, and by using the symbolic differential operators

$$\nabla \equiv \mathbf{i}_X \frac{\partial}{\partial X} + \mathbf{i}_Y \frac{\partial}{\partial Y} + \mathbf{i}_Z \frac{\partial}{\partial Z}, \qquad \Delta \equiv \frac{\partial^2}{\partial X^2} + \frac{\partial^2}{\partial Y^2} + \frac{\partial^2}{\partial Z^2},$$

one can rewrite system (1.1.1), (1.1.2) in the compact vector form

$$\nabla \cdot \mathbf{V} = 0, \tag{1.1.3}$$

$$\frac{\partial \mathbf{V}}{\partial t} + (\mathbf{V} \cdot \nabla)\mathbf{V} = -\frac{1}{\rho}\nabla P + \nu \Delta \mathbf{V} + \mathbf{g}. \tag{1.1.4}$$

The continuity and Navier–Stokes equations in cylindrical and spherical coordinate systems are given in Supplement 5.

***Stream function.*** Most of the problems considered in the first two chapters possess some symmetry properties. In these cases, instead of the fluid velocity components, it is often convenient to introduce a stream function $\Psi$ on the basis of the continuity equation (1.1.3). Then (1.1.3) is satisfied automatically. Usually, the stream function is introduced in the following three cases.

1. In plane problems, all variables are independent of the coordinate $Z$, and the continuity equation (1.1.3) becomes

$$\frac{\partial V_X}{\partial X} + \frac{\partial V_Y}{\partial Y} = 0. \tag{1.1.5}$$

The stream function $\Psi(X, Y)$ is introduced by the relations

$$V_X = \frac{\partial \Psi}{\partial Y}, \qquad V_Y = -\frac{\partial \Psi}{\partial X}. \tag{1.1.6}$$

The continuity equation is satisfied identically.

2. In axisymmetric problems, all variables are independent of the axial coordinate $Z$ in the cylindrical coordinates $\mathcal{R}, \theta, Z$. The continuity equation has the form (both sides are multiplied by $\mathcal{R}$)

$$\frac{\partial}{\partial \mathcal{R}}(\mathcal{R} V_\mathcal{R}) + \frac{\partial V_\theta}{\partial \theta} = 0, \tag{1.1.7}$$

and the stream function is introduced by

$$V_\mathcal{R} = \frac{1}{\mathcal{R}} \frac{\partial \Psi}{\partial \theta}, \qquad V_\theta = -\frac{\partial \Psi}{\partial \mathcal{R}}. \tag{1.1.8}$$

3. In axisymmetric problems, all variables are independent of the coordinate $\varphi$ in the spherical coordinates $R, \theta, \varphi$. The continuity equation has the form (both sides are multiplied by $R$)

$$\frac{1}{R} \frac{\partial}{\partial R}(R^2 V_R) + \frac{1}{\sin \theta} \frac{\partial}{\partial \theta}(V_\theta \sin \theta) = 0, \tag{1.1.9}$$

and the stream function is introduced by

$$V_R = \frac{1}{R^2 \sin \theta} \frac{\partial \Psi}{\partial \theta}, \qquad V_\theta = -\frac{1}{R \sin \theta} \frac{\partial \Psi}{\partial R}. \tag{1.1.10}$$

In all these cases, the stream function depends only on two orthogonal coordinates. The streamlines are determined by the equation $\Psi = $ const. To each line there corresponds a constant value of the stream function. The fluid velocity vector is tangent to the streamline. (Note that the streamlines coincide with the trajectories of fluid particles only in the stationary case.)

Table 1.1 presents equations for the stream function, obtained from the Navier–Stokes equations (1.1.1), (1.1.2) in various coordinate systems.

*Dimensionless form of equations.* To analyze the hydrodynamic equations (1.1.3), (1.1.4), it is convenient to introduce dimensionless variables and un-known functions as follows:

$$\tau = \frac{Ut}{a}, \quad x = \frac{X}{a}, \quad y = \frac{Y}{a}, \quad z = \frac{Z}{a}, \quad \mathbf{v} = \frac{\mathbf{V}}{U}, \quad p = \frac{P}{\rho U^2},$$

where $a$ and $U$ are the characteristic length and the characteristic velocity, respectively. As a result, we obtain

$$\nabla \cdot \mathbf{v} = 0, \tag{1.1.11}$$

$$\frac{\partial \mathbf{v}}{\partial t} + (\mathbf{v} \cdot \nabla)\mathbf{v} = -\nabla p + \frac{1}{\mathsf{Re}}\Delta \mathbf{v} + \frac{1}{\mathsf{Fr}}\frac{\mathbf{g}}{g}. \tag{1.1.12}$$

In Eq. (1.1.12), the following basic dimensionless state-geometric parameters of the flow are used:

$$\mathsf{Re} = \frac{aU}{\nu} \text{ is the Reynolds number,} \qquad \mathsf{Fr} = \frac{U^2}{ga} \text{ is the Froude number.}$$

Small values of Reynolds numbers correspond to slow ("creeping") flows and high Reynolds numbers, to rapid flows. Since these limit cases contain a small or large dimensionless parameter, one can efficiently use various modifications of the perturbation method [224, 258, 485].

---

### 1.1-2. Initial Conditions and the Simplest Boundary Conditions

For the solution of system (1.1.1), (1.1.2) to determine the velocity and pressure fields uniquely, we must impose initial and boundary conditions.

In nonstationary problems, where the terms with partial derivatives with respect to time are retained in the equation of motion, the initial velocity field must be given in the entire flow region and satisfy the continuity equation (1.1.1) there. The initial pressure field need not be given, since the equations do not contain the derivative of pressure with respect to time.*

As a rule, the region occupied by a moving reactive mixture is not the entire space but only a part bounded by some surfaces. According to whether the point at infinity belongs to the flow region or not, the problem of finding the unknown functions is called the exterior or interior problem of hydrodynamics, respectively.

On the surface $S$ of a solid body moving in a flow of a viscous fluid, the no-slip condition is imposed. This condition says that the vector $\mathbf{V}|_S$ of the fluid

---

* Obviously, if an arbitrary initial pressure field is given, it may happen that the velocity fields obtained from the equations of motion do not satisfy the continuity equation for $t > 0$ [404]. No such problems arise in the stationary case.

TABLE 1.1.
Stream function equations equivalent to the Navier–Stokes equations

| Coordinate system | Velocity components | Equation for the stream function | Notation |
|---|---|---|---|
| **Two-dimensional plane flow** | | | |
| Cartesian; the velocity is independent of $Z$; $V_Z = 0$ | $V_X = \dfrac{\partial \Psi}{\partial Y}$, $\;V_Y = -\dfrac{\partial \Psi}{\partial X}$ | $\dfrac{\partial}{\partial t}(\Delta\Psi) + \dfrac{\partial(\Psi, \Delta\Psi)}{\partial(X,Y)} = \nu \Delta^2 \Psi$ | $\Delta \equiv \dfrac{\partial^2}{\partial X^2} + \dfrac{\partial^2}{\partial Y^2}$, $\;\Delta^2 \equiv \Delta(\Delta)$ |
| Cylindrical; the velocity is independent of $Z$; $V_Z = 0$ | $V_{\mathcal{R}} = \dfrac{1}{\mathcal{R}}\dfrac{\partial \Psi}{\partial \theta}$, $\;V_\theta = -\dfrac{\partial \Psi}{\partial \mathcal{R}}$ | $\dfrac{\partial}{\partial t}(\Delta\Psi) + \dfrac{1}{\mathcal{R}}\dfrac{\partial(\Psi, \Delta\Psi)}{\partial(\mathcal{R},\theta)} = \nu \Delta^2 \Psi$ | $\Delta \equiv \dfrac{1}{\mathcal{R}}\dfrac{\partial}{\partial \mathcal{R}}\left(\mathcal{R}\dfrac{\partial}{\partial \mathcal{R}}\right) + \dfrac{1}{\mathcal{R}^2}\dfrac{\partial^2}{\partial \theta^2}$, $\;\Delta^2 \equiv \Delta(\Delta)$ |
| **Axisymmetric flow** | | | |
| Cylindrical; the velocity is independent of $\theta$; $V_\theta = 0$ | $V_{\mathcal{R}} = \dfrac{1}{\mathcal{R}}\dfrac{\partial \Psi}{\partial Z}$, $\;V_Z = -\dfrac{1}{\mathcal{R}}\dfrac{\partial \Psi}{\partial \mathcal{R}}$ | $\dfrac{\partial}{\partial t}(\mathcal{G}\Psi) - \dfrac{1}{\mathcal{R}}\dfrac{\partial(\Psi, \mathcal{G}\Psi)}{\partial(\mathcal{R},Z)} - \dfrac{2}{\mathcal{R}^2}\dfrac{\partial \Psi}{\partial Z}\mathcal{G}^2 = \nu \mathcal{G}^2 \Psi$ | $\mathcal{G} \equiv \dfrac{\partial^2}{\partial \mathcal{R}^2} - \dfrac{1}{\mathcal{R}}\dfrac{\partial}{\partial \mathcal{R}} + \dfrac{\partial^2}{\partial Z^2}$, $\;\mathcal{G}^2 \equiv \mathcal{G}(\mathcal{G})$ |
| Spherical; independent of $\varphi$; $V_\varphi = 0$ | $V_R = \dfrac{1}{R^2 \sin\theta}\dfrac{\partial \Psi}{\partial \theta}$, $\;V_\theta = -\dfrac{1}{R\sin\theta}\dfrac{\partial \Psi}{\partial R}$ | $\dfrac{\partial}{\partial t}(\mathcal{G}\Psi) + \dfrac{1}{R^2 \sin\theta}\dfrac{\partial(\Psi, \mathcal{G}\Psi)}{\partial(R,\theta)} - \dfrac{2\mathcal{G}\Psi}{R^2 \sin^2\theta} \times \left(\cos\theta \dfrac{\partial \Psi}{\partial R} - \dfrac{\sin\theta}{R}\dfrac{\partial \Psi}{\partial \theta}\right) = \nu \mathcal{G}^2 \Psi$ | $\mathcal{G} \equiv \dfrac{\partial^2}{\partial R^2} + \dfrac{\sin\theta}{R^2}\dfrac{\partial}{\partial \theta}\left(\dfrac{1}{\sin\theta}\dfrac{\partial}{\partial \theta}\right)$, $\;\mathcal{G}^2 \equiv \mathcal{G}(\mathcal{G})$ |

*Remark.* The Jacobians are calculated by the formula $\dfrac{\partial(f,g)}{\partial(x,y)} = \dfrac{\partial f}{\partial x}\dfrac{\partial g}{\partial y} - \dfrac{\partial f}{\partial y}\dfrac{\partial g}{\partial x}$.

velocity on the surface of the solid is equal to the vector $\mathbf{V}_0$ of the solid velocity. If the solid is at rest, then $\mathbf{V}|_S = 0$. In the projections on the normal $\mathbf{n}$ and the tangent $\tau$ to the surface $S$, this condition reads

$$V_n\big|_S = 0, \qquad V_\tau\big|_S = 0. \tag{1.1.13}$$

More complicated boundary conditions are posed on an interface between two fluids (e.g., see 2.2 and 5.9).

To solve the exterior hydrodynamic problem, one must impose a condition at infinity (that is, remote from the body, the drop, or the bubble).

---

### 1.1-3. Translational and Shear Flows

*Translational flow.* For uniform translational flow with velocity $\mathbf{U}_i$ around a finite body, the boundary condition remote from the body has the form

$$\mathbf{V} \to \mathbf{U}_i \quad \text{as} \quad R \to \infty, \tag{1.1.14}$$

where $R = \sqrt{X^2 + Y^2 + Z^2}$.

Let us consider more complicated situations, which are typical of gradient flows of inhomogeneous structure.

*Shear flows.* An arbitrary stationary velocity field $\mathbf{V}(\mathbf{R})$ in an incompressible medium can be approximated near the point $\mathbf{R} = 0$ by two terms of the Taylor series:

$$V_k(\mathbf{R}) = V_k(0) + G_{km} X_m,$$
$$G_{km} \equiv (\partial V_k / \partial X_m)_{\mathbf{R}=0}, \quad G_{11} + G_{22} + G_{33} = 0. \tag{1.1.15}$$

Here $V_k$ and $G_{km}$ are the fluid velocity and the shear tensor components in the Cartesian coordinates $X_1$, $X_2$, $X_3$. The sum is taken over the repeated index $m$; since the fluid is incompressible, it follows that the sum of the diagonal entries $G_{mm}$ is zero.

For viscous flows around particles whose size is much less than the characteristic size of flow inhomogeneities, the velocity distribution (1.1.15) can be viewed as the velocity field remote from the particle. The special case $G_{km} = 0$ corresponds to uniform translational flow. For $V_k(0) = 0$, Eq. (1.1.15) describes the velocity field in an arbitrary linear shear flow.

Any tensor $\mathbf{G} = [G_{km}]$ can be represented as the sum of a symmetric and an antisymmetric tensor, $\mathbf{G} = \mathbf{E} + \mathbf{\Omega}$, or

$$[G_{km}] = [E_{km}] + [\Omega_{km}],$$
$$E_{km} = E_{mk} = \tfrac{1}{2}(G_{km} + G_{mk}), \quad \Omega_{km} = -\Omega_{mk} = \tfrac{1}{2}(G_{km} - G_{mk}). \tag{1.1.16}$$

By rotating the coordinate system, one can reduce the symmetric tensor $\mathbf{E} = [E_{km}]$ to a diagonal form with diagonal entries $E_1$, $E_2$, $E_3$ being the roots of the cubic equation $\det[E_{km} - \lambda\delta_{km}] = 0$ for $\lambda$; here $\delta_{km}$ is the Kronecker delta. The diagonal entries $E_1$, $E_2$, $E_3$ of the tensor $[E_{km}]$ reduced to the principal axes determine the intensity of tensile (or compressive) motion along the coordinate axes. Since the fluid is incompressible, only two of these entries are independent; namely, $E_1 + E_2 + E_3 = 0$.

The decomposition of the tensor $[G_{km}]$ into the symmetric and antisymmetric parts corresponds to the representation of the velocity field of a linear shear fluid flow as the superposition of linear straining flow with extension coefficients $E_1$, $E_2$, $E_3$ along the principal axes and the rotation of the fluid as a solid at the angular velocity $\omega = (\Omega_{32}, \Omega_{13}, \Omega_{21})$.

For a uniform translational flow, the velocity of the nonperturbed flow is independent of the coordinates; therefore, all $G_{km} = 0$. In this case we have the simplest flow around a body with the boundary condition (1.1.14) at infinity.

**Examples of shear flows.** Now let us consider the most frequently encountered types of linear shear flows [518].

1°. Simple shear (Couette) flow:

$$V_X = GY, \qquad V_Y = 0, \qquad V_Z = 0,$$

$$[G_{km}] = \begin{bmatrix} 0 & G & 0 \\ 0 & 0 & 0 \\ 0 & 0 & 0 \end{bmatrix}, \quad [E_{km}] = \begin{bmatrix} 0 & \frac{1}{2}G & 0 \\ \frac{1}{2}G & 0 & 0 \\ 0 & 0 & 0 \end{bmatrix}, \quad [\Omega_{km}] = \begin{bmatrix} 0 & \frac{1}{2}G & 0 \\ -\frac{1}{2}G & 0 & 0 \\ 0 & 0 & 0 \end{bmatrix}.$$

In this case, $G$ is called the gradient of the flow rate or the shear rate. The Couette flow occurs between two parallel moving planes or in the gap between coaxial cylinders rotating at different angular velocities.

2°. Plane irrotational flow:

$$V_X = \tfrac{1}{2}GY, \qquad V_Y = \tfrac{1}{2}GX, \qquad V_Z = 0,$$

$$[G_{km}] = \begin{bmatrix} 0 & \frac{1}{2}G & 0 \\ \frac{1}{2}G & 0 & 0 \\ 0 & 0 & 0 \end{bmatrix}, \quad [E_{km}] = \begin{bmatrix} 0 & \frac{1}{2}G & 0 \\ \frac{1}{2}G & 0 & 0 \\ 0 & 0 & 0 \end{bmatrix}, \quad [\Omega_{km}] = \begin{bmatrix} 0 & 0 & 0 \\ 0 & 0 & 0 \\ 0 & 0 & 0 \end{bmatrix}.$$

This flow has the same extension component as the simple shear flow but has no rotational component.

3°. Plane straining flow:

$$V_X = \tfrac{1}{2}GX, \qquad V_Y = -\tfrac{1}{2}GY, \qquad V_Z = 0,$$

$$[G_{km}] = \begin{bmatrix} \frac{1}{2}G & 0 & 0 \\ 0 & -\frac{1}{2}G & 0 \\ 0 & 0 & 0 \end{bmatrix}, \quad [E_{km}] = \begin{bmatrix} \frac{1}{2}G & 0 & 0 \\ 0 & -\frac{1}{2}G & 0 \\ 0 & 0 & 0 \end{bmatrix}, \quad [\Omega_{km}] = \begin{bmatrix} 0 & 0 & 0 \\ 0 & 0 & 0 \\ 0 & 0 & 0 \end{bmatrix}.$$

This flow can be obtained in the Taylor device, consisting of four rotating cylinders [474, 475]. Note that flow 2° is the same as flow 3° but in a different coordinate system (rotated about the $Z$-axis by 45° counterclockwise).

4°. Plane solid-body rotation:

$$V_X = GY, \qquad V_Y = -GX, \qquad V_Z = 0,$$

$$[G_{km}] = \begin{bmatrix} 0 & G & 0 \\ -G & 0 & 0 \\ 0 & 0 & 0 \end{bmatrix}, \quad [E_{km}] = \begin{bmatrix} 0 & 0 & 0 \\ 0 & 0 & 0 \\ 0 & 0 & 0 \end{bmatrix}, \quad [\Omega_{km}] = \begin{bmatrix} 0 & G & 0 \\ -G & 0 & 0 \\ 0 & 0 & 0 \end{bmatrix}.$$

The fluid rotates around the $Z$-axis at the angular velocity $G$.

5°. Axisymmetric shear (axisymmetric straining flow):

$$V_X = -\tfrac{1}{2}GX, \qquad V_Y = -\tfrac{1}{2}GY, \qquad V_Z = GZ,$$

$$[G_{km}] = \begin{bmatrix} -\tfrac{1}{2}G & 0 & 0 \\ 0 & -\tfrac{1}{2}G & 0 \\ 0 & 0 & G \end{bmatrix}, \quad [E_{km}] = \begin{bmatrix} -\tfrac{1}{2}G & 0 & 0 \\ 0 & -\tfrac{1}{2}G & 0 \\ 0 & 0 & G \end{bmatrix}, \quad [\Omega_{km}] = \begin{bmatrix} 0 & 0 & 0 \\ 0 & 0 & 0 \\ 0 & 0 & 0 \end{bmatrix}.$$

This flow can be implemented by elongating a cylindrical deformable thread or by using a device similar to the Taylor device [475] with two toroidal shafts rotating in opposite directions.

6°. Extensiometric flow:

$$V_X = G_1 X, \quad V_Y = G_2 Y, \quad V_Z = G_3 Z, \qquad G_1 + G_2 + G_3 = 0;$$

$$[G_{km}] = \begin{bmatrix} G_1 & 0 & 0 \\ 0 & G_2 & 0 \\ 0 & 0 & G_3 \end{bmatrix}, \quad [E_{km}] = \begin{bmatrix} G_1 & 0 & 0 \\ 0 & G_2 & 0 \\ 0 & 0 & G_3 \end{bmatrix}, \quad [\Omega_{km}] = \begin{bmatrix} 0 & 0 & 0 \\ 0 & 0 & 0 \\ 0 & 0 & 0 \end{bmatrix}.$$

This flow is a generalization of flow 5° to the nonaxisymmetric case.

7°. Orthogonal rheometric flow:

$$V_X = GY - HZ, \qquad V_Y = 0, \qquad V_Z = HX,$$

$$[G_{km}] = \begin{bmatrix} 0 & G & -H \\ 0 & 0 & 0 \\ H & 0 & 0 \end{bmatrix}, \quad [E_{km}] = \begin{bmatrix} 0 & \tfrac{1}{2}G & 0 \\ \tfrac{1}{2}G & 0 & 0 \\ 0 & 0 & 0 \end{bmatrix}, \quad [\Omega_{km}] = \begin{bmatrix} 0 & \tfrac{1}{2}G & -H \\ -\tfrac{1}{2}G & 0 & 0 \\ H & 0 & 0 \end{bmatrix}.$$

This flow combines shear along the $X$-axis with rotation around the $Y$- and $Z$-axes.

When modeling gradient nonperturbed flow around a body, the boundary conditions at infinity (remote from the body) must be taken in the following form: the fluid velocity components tend to the corresponding components of the above gradient flows as $R \to \infty$.

## 1.1-4. Turbulent Flows

*Reynolds equations.* Formally, stationary solutions of the Navier–Stokes equations are possible for any Reynolds numbers [477]. But practically, only stable flows with respect to small perturbations, always present in the flow, can exist. For sufficiently high Reynolds numbers, the stationary solutions become unstable, i.e., the amplitude of small perturbations increases with time. For this reason, stationary solutions can only describe real flows at not too high Reynolds numbers.

The flow in the boundary layer on a flat plate is laminar up to $\mathsf{Re}_X = U_i X/\nu \approx 3.5 \times 10^5$, and that in a circular smooth tube for $\mathsf{Re} = a\langle V \rangle/\nu < 1500$ [427]. For higher Reynolds numbers, the laminar flow loses its stability and a transient regime of development of unstable modes takes place. For $\mathsf{Re}_X > 10^7$ and $\mathsf{Re} > 2500$, a fully developed regime of turbulent flow is established which is characterized by chaotic variations in the basic macroscopic flow parameters in time and space.

When mathematically describing a fully developed turbulent motion of fluid, it is common to represent the velocity components and pressure in the form

$$V_i = \overline{V}_i + V_i', \quad P = \overline{P} + P', \tag{1.1.17}$$

where the bar and prime denote the time-average and fluctuating components, respectively. The averages of the fluctuations are zero, $\overline{V_i'} = \overline{P'} = 0$.

The representation (1.1.17) of the hydrodynamic parameters of turbulent flow as the sum of the average and fluctuating components followed by the averaging process made it possible, based on the continuity equation (1.1.3) and the Navier–Stokes equations (1.1.4), to obtain (under some assumptions) the Reynolds equations

$$\nabla \cdot \overline{\mathbf{V}} = 0,$$

$$\frac{\partial \overline{\mathbf{V}}}{\partial t} + (\overline{\mathbf{V}} \cdot \nabla)\overline{\mathbf{V}} = -\frac{1}{\rho}\nabla \overline{P} + \nu \Delta \overline{\mathbf{V}} + \mathbf{g} + \frac{1}{\rho}\nabla \cdot \sigma^t \tag{1.1.18}$$

for the averaged pressure and velocity fields. These equation contain the Reynolds turbulent shear stress tensor $\sigma^t$ whose components are defined as

$$\sigma_{ij}^t = -\rho\overline{V_i'V_j'}. \tag{1.1.19}$$

The variable $\rho\overline{V_i'V_j'}$ is the average rate at which the turbulent fluctuations transfer the $j$th momentum component along the $i$th axis.

*The closure problem. Turbulent viscosity.* Unlike the Navier–Stokes equations completed by the continuity equation, the Reynolds equation form an unclosed system of equations, since these contain the *a priori* unknown turbulent stress tensor $\sigma^t$ with components (1.1.19). Additional hypotheses must be invoked to close system (1.1.18). These hypotheses are of much greater significance compared with those used for the derivation of the Navier–Stokes equations [430].

So far the closure problem for the system of Reynolds equations has not been theoretically solved in a conclusive way. In engineering calculations, various assumptions that the Reynolds stresses depend on the average turbulent flow parameters are often adopted as closure conditions. These conditions are usually formulated on the basis of experimental data, dimensional considerations, analogies with molecular rheological models, etc.

Two traditional approaches to the closure of the Reynolds equation are outlined below. These approaches are based on Boussinesq's model of turbulent viscosity completed by Prandtl's or von Karman's hypotheses [276, 427]. For simplicity, we confine our consideration to the case of simple shear flow, where the transverse coordinate $Y = X_2$ is measured from the wall (the results are also applicable to turbulent boundary layers). According to Boussinesq's model, the only nonzero component of the Reynolds turbulent shear stress tensor and the divergence of this tensor are defined as

$$\sigma_{1,2}^t = \rho\nu_t\frac{\partial \overline{V}}{\partial Y}, \qquad \nabla \cdot \sigma^t = \rho\frac{\partial}{\partial Y}\left(\nu_t\frac{\partial \overline{V}}{\partial Y}\right), \tag{1.1.20}$$

where $\overline{V}$ stands for the longitudinal average velocity component. Formulas (1.1.20) contain the turbulent ("eddy") viscosity $\nu_t$, which is not a physical constant but is a function of geometric and kinematic flow parameters. It is necessary to specify this function to close the Reynolds equations.

Following Prandtl, we have

$$\nu_t = \kappa^2 Y^2 \left| \frac{\partial \overline{V}}{\partial Y} \right|, \tag{1.1.21}$$

where $\kappa = 0.4$ is the von Karman empirical constant.*

Von Karman suggested a more complicated expression for the turbulent viscosity, namely,

$$\nu_t = \kappa^2 \left| \frac{\partial \overline{V}}{\partial Y} \right|^3 \bigg/ \left| \frac{\partial^2 \overline{V}}{\partial Y^2} \right|^2. \tag{1.1.22}$$

Some justification and the scope of relations (1.1.21) and (1.1.22) can be found, for example, in the books [276, 427]. There are a number of other ways of closing the Reynolds equations also based on the notion of turbulent viscosity [41, 80, 163, 223].

*Other models and methods of turbulence theory.* Dimensional and similarity methods are widely used in turbulence theory [23, 65, 135, 161, 162, 230, 432]. Under some assumptions, these methods permit relations (1.1.21) and (1.1.22) and their generalizations to be obtained [276, 427]. In this approach, the experimental data are used for the statistical estimation of the parameters and coefficients occurring in the relationships obtained and for the selection of simple and sufficiently accurate approximate formulas. A comprehensive presentation of the results obtained in turbulence theory by the dimensional and similarity methods can be found in [188, 211, 212, 483].

The turbulence models based on the Reynolds equations (1.1.18) and relations like (1.1.21) and (1.1.22) pertain to first-order closure models. These model only permit fairly simple turbulent flows to be described. The necessity of investigation of complex flows and their fluctuating properties has led researchers to the construction of more complicated, second-order closure models,** which contain a lot of empirical constants. Apart from the average components of velocity $\overline{V}_i$ and average pressure $\overline{P}$, the kinetic energy of turbulent fluctuations, $\mathcal{K}$, and the dissipation rate of the energy of these fluctuations, $\mathcal{E}$ are usually taken to be basic dynamical parameters of turbulence in second-order closure models. The scalar quantities $\mathcal{K}$ and $\mathcal{E}$ are governed by special differential equations of transfer, which must be solved together with the Reynolds equations. However, various additional hypotheses and rheological relations must be used in this approach. There are quite a few second-order closure models [178, 409, 453, 456, 458]. The so-called $\mathcal{K}$-$\mathcal{E}$ model is the most widespread [57, 458].

More generally, the problem of closure of the Reynolds equations is treated as the problem of establishing mathematical relationships between two-point correlation moments of various order [41, 260, 290, 492]. Keller and Fridman [221]

---

  \* In the literature [57, 80, 276, 289, 398], the von Karman constant is most frequently taken to be $\kappa = 0.40$ or 0.41, although other values can sometimes be encountered [41, 316].

  \*\* These models are often referred to as multiparameter or differentiable models.

suggested a procedure for obtaining a chain of additional equations in which the second correlation moments are expressed via the third, the third via the fourth, and so on. The solvability of the infinite chain of moment equations is discussed in [492]. In practice, the chain is truncated at equations of sufficiently high order and various hypotheses for the relationships between the higher order moments are used.

Statistical methods also are applied in turbulence theory. These methods use the averaging over the ensemble of possible realizations of the process and take into account the probability distribution density for each quantity. In this case, various hypotheses and experimental data for the probability distribution must be used to obtain specific results. The statistical approach is related to a fairly high level of complexity of describing turbulent flows [290, 460, 492].

Another group of methods relies on straightforward numerical simulation of turbulent flows. Numerical analysis is based directly on the Navier–Stokes equations [228, 229, 288, 314–316, 376] or equivalent variational principles [160, 310]. The computations are carried out until statistically steady-state flow regimes characterized by steady values of average quantities are attained. This approach involves a lot of computation but does not require the use of physical hypotheses and empirical constants. Note that no rigorous mathematical estimates of the accuracy of the numerical method for the simulation of turbulent flows have been available so far.

A variety of other methods for the investigation and mathematical modeling of turbulence are known, which are based on various arguments and hypotheses, e.g., see [36, 192, 426, 435, 451, 459].

## 1.2. Flows Caused by a Rotating Disk

### 1.2-1. Infinite Plane Disk

*Statement of the problem.* In this section we describe one of the few cases in which a nonlinear boundary value problem for the Navier–Stokes equations admits an exact closed-form solution.

Let us consider the flow caused by an infinite plane disk rotating at a constant angular velocity $\omega$. The no-slip condition on the disk surface results in a rather complicated three-dimensional motion of the fluid, which is drawn in from the bulk along the rotation axis and thrown away to the periphery near the disk surface. This flow is quite a good model of the hydrodynamics of disk agitators, widely used in chemical technology, as well as disk electrodes, used as sensors in electrochemistry [270].

Let us use the cylindrical coordinate system $\mathcal{R}$, $\varphi$, $Z$, where the coordinate $Z$ is measured from the disk surface along the rotation axis. Taking account of the problem symmetry (the unknown variables are independent of the angular coordinate $\varphi$), we rewrite the continuity and the Navier–Stokes equations in the form

$$
\frac{\partial V_{\mathcal{R}}}{\partial \mathcal{R}} + \frac{\partial V_Z}{\partial Z} + \frac{V_{\mathcal{R}}}{\mathcal{R}} = 0,
$$
$$
V_{\mathcal{R}}\frac{\partial V_{\mathcal{R}}}{\partial \mathcal{R}} + V_Z\frac{\partial V_{\mathcal{R}}}{\partial Z} - \frac{V_\varphi^2}{\mathcal{R}} = -\frac{1}{\rho}\frac{\partial P}{\partial \mathcal{R}} + \nu\left(\Delta V_{\mathcal{R}} - \frac{V_{\mathcal{R}}}{\mathcal{R}^2}\right),
$$
$$
V_{\mathcal{R}}\frac{\partial V_\varphi}{\partial \mathcal{R}} + V_Z\frac{\partial V_\varphi}{\partial Z} + \frac{V_{\mathcal{R}}V_\varphi}{\mathcal{R}} = \nu\left(\Delta V_\varphi - \frac{V_\varphi}{\mathcal{R}^2}\right), \tag{1.2.1}
$$
$$
V_{\mathcal{R}}\frac{\partial V_Z}{\partial \mathcal{R}} + V_Z\frac{\partial V_Z}{\partial Z} = -\frac{1}{\rho}\frac{\partial P}{\partial Z} + \nu\Delta V_Z,
$$

where $\Delta$ is the Laplace operator in the cylindrical coordinates:

$$
\Delta \equiv \frac{1}{\mathcal{R}}\frac{\partial}{\partial \mathcal{R}}\left(\mathcal{R}\frac{\partial}{\partial \mathcal{R}}\right) + \frac{\partial^2}{\partial Z^2}. \tag{1.2.2}
$$

To complete the mathematical statement of the problem, we supplement the hydrodynamic equations (1.2.1) by some boundary conditions, namely, the no-slip condition on the disk surface and the conditions of nonperturbed radial and angular motions and pressure remote from the disk:

$$
\begin{array}{llll}
V_{\mathcal{R}} = 0, & V_\varphi = \mathcal{R}\omega, & V_Z = 0 & \text{at } Z = 0, \\
V_{\mathcal{R}} \to 0, & V_\varphi \to 0, & P \to P_{\mathrm{i}} & \text{as } Z \to \infty.
\end{array} \tag{1.2.3}
$$

***Solution of the problem.*** Following Karman, we seek the solution of problem (1.2.1)–(1.2.3) in the form

$$
V_{\mathcal{R}} = \omega\mathcal{R}u_1(z), \quad V_\varphi = \omega\mathcal{R}u_2(z), \quad V_Z = \sqrt{\nu\omega}\,v(z),
$$
$$
P = P_{\mathrm{i}} + \rho\nu\omega p(z), \quad \text{where} \quad z = \sqrt{\omega/\nu}\,Z. \tag{1.2.4}
$$

Substituting these expressions into (1.2.1)–(1.2.3) and performing some transformations, we arrive at the system of ordinary differential equations (the primes stand for derivatives with respect to $z$)

$$
\begin{aligned}
u_1'' &= v u_1' + u_1^2 - u_2^2, \\
u_2'' &= v u_2' + 2u_1 u_2, \\
v'' &= v v' + p', \\
v' &= -2u_1
\end{aligned} \tag{1.2.5}
$$

with the boundary conditions

$$
\begin{array}{llll}
u_1 = 0, & u_2 = 1, & v = 0 & \text{at } z = 0, \\
u_1 \to 0, & u_2 \to 0, & p \to 0 & \text{as } z \to \infty.
\end{array} \tag{1.2.6}
$$

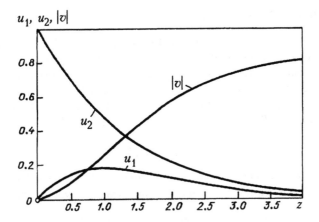

**Figure 1.1.** The distribution of velocity components near a rotating disk

Note that the axial distribution of pressure can be found from the third equation in (1.2.5) after the first two equations have been solved. The pressure is expressed via the transverse velocity by

$$p = v'(z) - \tfrac{1}{2}v^2(z) - v'(\infty) + \tfrac{1}{2}v(\infty). \tag{1.2.7}$$

Numerical results for problem (1.2.5), (1.2.6) can be found in [95, 427]. The corresponding plots of $u_1$, $u_2$, and $|v|$ against $z$ are shown in Figure 1.1.

The following expansions of the unknown functions are valid near and remote from the disk surface, respectively [276]:

$$u_1(z) \simeq 0.51\,z - 0.5\,z^2, \quad u_2(z) \simeq 1 - 0.616\,z,$$
$$v(z) \simeq -0.51\,z^2 + 0.333\,z^3, \quad p(z) \simeq 0.393 - 1.02\,z \tag{1.2.8}$$

as $z \to 0$;

$$u_1(z) \simeq 0.934\,\exp(-0.886\,z), \quad u_2(z) \simeq 1.208\,\exp(-0.886\,z),$$
$$v(z) \simeq -0.886, \quad p(z) \simeq 0.393 \tag{1.2.9}$$

as $z \to \infty$.

Using formulas (1.2.9), one can estimate the perturbations caused by the rotating disk in the fluid remote from the disk surface. It follows from the boundary conditions (1.2.3) that the pressure, as well as the radial and the angular velocity, is not perturbed as $z \to \infty$. However, the remote dimensionless axial velocity is not zero, $v(\infty) = -0.886$. This is the rate at which the disk draws the ambient fluid. Figure 1.1 shows that the pressure and the radial and angular velocities are perturbed only near the disk surface, in the so-called dynamic boundary layer. The thickness of this layer is independent of the radial coordinate* and is approximately equal to $\delta = 3\sqrt{\nu/\omega}$.

---

\* In Section 3.2 it will be shown that the diffusion boundary layer near a rotating disk is also of constant thickness. This allows one to assume that the surface of a rotating disk, used as an electrode in electrochemical experiments, is uniformly approachable.

## 1.2-2. Disk of Finite Radius

*Laminar flow.* All the above considerations apply to a disk of infinite radius. However, for a circular disk of finite radius $a$ that is much greater than the thickness of the boundary layer ($a \gg 3\sqrt{\nu/\omega}$), these statements hold approximately, and so we can obtain some important practical estimates.

Using the capture rate of the fluid by the disk, $V_Z(\infty) = -0.886\sqrt{\nu\omega}$, one can find the rate of flow of the fluid captured by the disk of radius $a$ and thrown away:

$$q = 0.886\,\pi a^2 \sqrt{\nu\omega}. \qquad (1.2.10)$$

Since the disk is two-sided, the total rate of flow is in fact twice as large, $Q = 2q$. It is convenient to express the total rate of flow via the Reynolds number:

$$Q = 1.77\,\pi a^3 \omega\, \mathrm{Re}^{-1/2}, \qquad \mathrm{Re} = a^2\omega/\nu. \qquad (1.2.11)$$

In a similar way, one can estimate the frictional torque exerted by the fluid on the disk. It is given by the integral

$$m = -2\pi\mu \int_0^a \mathcal{R}^2 \left(\frac{\partial V_\varphi}{\partial Z}\right)_{Z=0} d\mathcal{R}.$$

For the two-sided torque $M = 2m$, we have the estimate

$$M = 0.616\,\pi\rho a^4 \sqrt{\nu\omega^3}. \qquad (1.2.12)$$

The dimensionless frictional torque coefficient is

$$c_M \equiv \frac{M}{\frac{1}{2}\rho a^5 \omega^2} = 3.87\,\mathrm{Re}^{-1/2}. \qquad (1.2.13)$$

The theoretical estimate (1.2.13) is corroborated by experiments for the Reynolds number less than the critical value $\mathrm{Re}_* \approx 3 \times 10^5$, at which the flow becomes unstable and a transition to the turbulent flow starts.

*Turbulent flow.* Approximate computations based on the integral boundary layer method lead to the following estimates for a disk of radius $a$ in a turbulent flow ($\mathrm{Re} > 3 \times 10^5$) [276]:

the two-sided rate of flow is

$$Q = 0.438\,a^3 \omega\, \mathrm{Re}^{-1/5}; \qquad (1.2.14)$$

the two-sided frictional torque coefficient is

$$c_M = 0.146\,\mathrm{Re}^{-1/5}. \qquad (1.2.15)$$

The thickness of the turbulent dynamical boundary layer over the disk can be estimated by the formula $\delta = 0.5\,a\,\mathrm{Re}^{-1/5}$.

**Figure 1.2.** The definition of the wetting angle

# 1.3. Hydrodynamics of Thin Films

## 1.3-1. Preliminary Remarks

Film type flows are widely used in chemical technology (in contact devices of absorption, chemosorption, and rectification columns as well as evaporators, dryers, heat exchangers, film chemical reactors, extractors, and condensers.

As a rule, the liquid and the gas phase are simultaneously fed into an apparatus where the fluids undergo physical and chemical treatment. Therefore, generally speaking, there is a dynamic interaction between the phases until the flooding mode sets in the countercurrent flows of gas and liquid. However, for small values of gas flow rate one can neglect the dynamic interaction and assume that the liquid flow in a film is due to the gravity force alone.

The value of the Reynolds number $\mathsf{Re} = Q/\nu$, where $Q$ is the volume rate of flow per unit film width, determines whether the flow in the gravitational film is laminar, wave, or turbulent. It is well known [11, 54, 226] that laminar flow becomes unstable at the critical value $\mathsf{Re}_* = 2$ to 6. However, the point starting from which the waves actually occur is noticeably shifted downstream [54]. Even in the range $6 \leq \mathsf{Re} \leq 400$, corresponding to wave flows [11], a considerable part of the film remains wave-free. Since this part is much larger than the initial part where the velocity profile and the film width reach their steady-state values, we see that for films in which viscous and gravity forces are in balance, the hydrodynamic laws of steady-state laminar flow virtually determine the rate of mass exchange in various apparatuses, like packed absorbing and fractionating columns, widely used in chemical and petroleum industry. In these columns, the films flow over the packing surface whose linear dimensions do not exceed a few centimeters (Raschig rings, Palle rings, Birle seats, etc. [226]).

Paradoxically, the range of flow rates (or Reynolds numbers) for which the assumption of laminar flow can be used in practice is bounded below (rather than above). Indeed, there is [500] a threshold value $Q_{min}$ of the volume rate of flow per unit film width such that for $Q < Q_{min}$ the flow in separate jets is energetically favorable. It was theoretically established in [191] that

$$Q_{min} = 2.15 \left(\frac{\nu\sigma^3}{\rho^3 g}\right)^{1/5} (1 - \cos\theta)^{3/5},$$

where $\sigma$ is the surface tension for the liquid and $\theta$ is the wetting angle for the wall material and the liquid (see Figure 1.2), determined by Young's fundamental relation [26]

$$\sigma_{gw} = \sigma \cos\theta + \sigma_{fw},$$

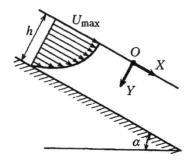

**Figure 1.3.** Steady waveless laminar flow in thin film on an inclined plane

where $\sigma_{gw}$ and $\sigma_{fw}$ are the specific excess surface energies for the gas–wall and liquid–wall interfaces.

Recently, the criterion of nonbreaking film flow was thermodynamically substantiated with the aid of Prigogine's principle of minimum entropy production including the case of a double film flow [88].

In practice, $Q_{min}$ can be reduced by wall hydrophilization [54], that is, by treating the surface by alcohol, which decreases the wetting angle.

### 1.3-2. Film on an Inclined Plane

Let us consider a thin liquid film flowing by gravity on a solid plane surface (Figure 1.3). Let $\alpha$ be the angle of inclination. We assume that the motion is sufficiently slow, so that we can neglect inertial forces (that is, convective terms) compared with the viscous friction and the gravity force. Let the film thickness $h$ (which is assumed to be constant) be much less than the film length. In this case, in the first approximation, the normal component of the liquid velocity is small compared with the longitudinal component, and we can neglect the derivatives along the film surface compared with the normal derivatives.

These assumptions result in the one-dimensional velocity and pressure profiles $V = V(Y)$ and $P = P(Y)$, where $Y$ is the coordinate measured along the normal to the film surface. The corresponding hydrodynamic equations of thin films express the balance of viscous and gravity forces [41, 441]:

$$\mu \frac{d^2V}{dY^2} + \rho g \sin \alpha = 0,$$
$$\frac{dP}{dY} - \rho g \cos \alpha = 0.$$

(1.3.1)

To these equations one must add the boundary conditions

$$\frac{dV}{dY} = 0, \quad P = P_0 \quad \text{at} \quad Y = 0,$$
$$V = 0 \quad\quad\quad \text{at} \quad Y = h,$$

(1.3.2)

which show that the tangent stress is zero, the pressure is equal to the atmosphere pressure at the free surface, and the no-slip condition is satisfied at the surface of the plane.

The solution of problem (1.3.1), (1.3.2) has the form

$$V = U_{max}(1 - y^2),$$
$$P = P_0 + \rho g h \cos \alpha \, y,$$

(1.3.3)

where $U_{max} = \frac{1}{2}(g/\nu)h^2 \sin \alpha$ is the maximum flow velocity (the velocity at the free boundary) and $y = Y/h$ is the dimensionless transverse coordinate.

The volume rate of flow per unit width is given by the formula

$$Q = \int_0^h V(Y)\, dY = \frac{g h^3 \sin \alpha}{3\nu} = \frac{2}{3} U_{max} h.$$

(1.3.4)

The mean flow rate velocity $\langle V \rangle$ is equal to 2/3 of the maximum velocity,

$$\langle V \rangle = \frac{2}{3} U_{max}.$$

Let us find the Reynolds number for the film flow:

$$\mathsf{Re} = \frac{Q}{\nu} = \frac{g h^3 \sin \alpha}{3 \nu^2}.$$

This allows us to express the film thickness via the Reynolds number and the volume rate of flow per unit width:

$$h = \left( \frac{3 \nu^2}{g \sin \alpha} \mathsf{Re} \right)^{1/3} = \left( \frac{3 \nu}{g \sin \alpha} Q \right)^{1/3}.$$

## 1.3-3. Film on a Cylindrical Surface

Let us consider a thin liquid film of thickness $h$ flowing by gravity on the surface of a vertical circular cylinder of radius $a$. In the cylindrical coordinates $\mathcal{R}, \varphi, Z$, the only nonzero component of the liquid velocity satisfies the equation

$$\mu \left( \frac{d^2 V_Z}{d\mathcal{R}^2} + \frac{1}{\mathcal{R}} \frac{\partial V_Z}{\partial \mathcal{R}} \right) + \rho g = 0.$$

(1.3.5)

The boundary conditions on the wall and on the free surface can be written as

$$V_Z = 0 \quad \text{at} \quad \mathcal{R} = a, \qquad \frac{dV_Z}{d\mathcal{R}} = 0 \quad \text{at} \quad \mathcal{R} = a + h.$$

(1.3.6)

The solution of problem (1.3.5), (1.3.6) is given by the formula [41]

$$V_Z(\mathcal{R}) = \frac{\rho g}{4\mu} \left\{ a^2 - \mathcal{R}^2 + \left[ (a + h)^2 - a^2 \right] \frac{\ln(\mathcal{R}/a)}{\ln(1 + h/a)} \right\}.$$

(1.3.7)

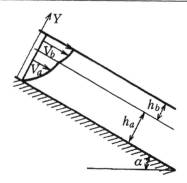

**Figure 1.4.** A double-film flow

## 1.3-4. Two-Layer Film

It is convenient to manage some processes of chemical technology (like liquid-phase extraction, as well as nitration and sulfonation of liquid hydrocarbons) in double hydrodynamic films.

Figure 1.4 shows the scheme of a double-film flow and the coordinate system used. The boundary problem for the $X$-components $V_a(Y)$ and $V_b(Y)$ of film the velocities consists of the equations

$$\mu_a \frac{d^2 V_a}{dY^2} - \rho_a g \sin \alpha = 0,$$

$$\mu_b \frac{d^2 V_b}{dY^2} - \rho_b g \sin \alpha = 0$$

and the boundary conditions

$$
\begin{array}{lll}
V_a = 0 & \text{at} & Y = 0, \\
V_a = V_b & \text{at} & Y = h_a, \\
\mu_a \dfrac{dV_a}{dY} = \mu_b \dfrac{dV_b}{dY} & \text{at} & Y = h_a, \\
\dfrac{dV_b}{dY} = 0 & \text{at} & Y = h_a + h_b.
\end{array}
$$

The solution of the problem on the laminar flow of two immiscible liquid films is given by the formulas [470]

$$V_a = \frac{\rho_a g \sin \alpha}{2\mu_a} \left[ 2\left( h_a + h_b \frac{\rho_b}{\rho_a} \right) Y - Y^2 \right] \qquad \text{for} \quad 0 \le Y \le h_a,$$

$$V_b = \frac{\rho_b g \sin \alpha}{2\mu_b} \left[ \left( \frac{\rho_a \mu_b}{\rho_b \mu_a} - 1 \right) h_a^2 + 2 h_a h_b \left( \frac{\mu_b}{\mu_a} - 1 \right) + 2(h_a + h_b) Y - Y^2 \right]$$

$$\text{for} \quad h_a \le Y \le h_a + h_b.$$

For the volume rate of flow per unit width in each film, we have the expressions

$$Q_a = \frac{\rho_a^2 h_a^3 g \sin\alpha}{3\mu_a}\left(1 + \frac{3}{2}\frac{\rho_b h_b}{\rho_a h_a}\right),$$

$$Q_b = \frac{\rho_b^2 h_b^3 g \sin\alpha}{3\mu_b}\left(1 + 3\frac{h_a \mu_b}{h_b \mu_a} + 3\frac{h_a^2 \rho_a \mu_b}{h_b^2 \rho_b \mu_a}\right).$$

For a given ratio $Q_a/Q_b$, the corresponding ratio of film widths $\lambda = h_a/h_b$ satisfies the cubic equation

$$\lambda^3 + \frac{3}{2}\left[\left(\frac{\mu_a}{\mu_b}\frac{Q_a}{Q_b}\right)^{2/3}\left(\frac{\rho_b}{\rho_a}\right)^{7/3} - \frac{\rho_b}{\rho_a}\frac{Q_a}{Q_b}\right]\lambda^2 - 3\left(\frac{\rho_b}{\rho_a}\right)^2 \lambda - \frac{\mu_a \rho_b^2}{\mu_b \rho_a^2}\frac{Q_a}{Q_b} = 0.$$

This equation is solved graphically in [470].

# 1.4. Jet Flows

Jet flows form a wide class of frequently encountered motions of viscous fluids. In this section we restrict our consideration to steady jet flows of an incompressible liquid in the space filled with a liquid with the same physical properties (such flows are known as "submerged" jets). We consider the problem about a jet-source in infinite space [26, 260] and give some practically important information about the wake structure past moving bodies [3, 427, 501].

### 1.4-1. Axisymmetric Jets

***Statement of the problem. Exact solution.*** In infinite space, we consider the flow caused by a liquid jet discharging from a thin tube. We treat the jet source as a point source, since the size and shape of the nozzle section are unessential remote from the source. The jet is axisymmetric about the flow direction. If there is no rotation of the fluid, then the motion considered in the spherical coordinates $(R, \theta, \varphi)$ is independent of the azimuth coordinate $\varphi$, and moreover, the condition $V_\varphi = 0$ must be satisfied.

The corresponding hydrodynamic problem is described by the equations of motion

$$V_R \frac{\partial V_R}{\partial R} + \frac{V_\theta}{R}\frac{\partial V_R}{\partial\theta} - \frac{V_\theta^2}{R} = -\frac{1}{\rho}\frac{\partial P}{\partial R}$$

$$+ \nu\left(\Delta V_R - \frac{2V_R}{R^2} - \frac{2}{R^2}\frac{\partial V_\theta}{\partial\theta} - \frac{2V_\theta\cot\theta}{R^2}\right),$$

$$V_R \frac{\partial V_\theta}{\partial R} + \frac{V_\theta}{R}\frac{\partial V_\theta}{\partial\theta} + \frac{V_R V_\theta}{R} = -\frac{1}{\rho R}\frac{\partial P}{\partial\theta}$$

$$+ \nu\left(\Delta V_\theta + \frac{2}{R^2}\frac{\partial V_R}{\partial\theta} - \frac{V_\theta}{R^2 \sin^2\theta}\right),$$

(1.4.1)

where

$$\Delta \equiv \frac{1}{R^2}\frac{\partial}{\partial R}\left(R^2\frac{\partial}{\partial R}\right) + \frac{1}{R^2\sin\theta}\frac{\partial}{\partial\theta}\left(\sin\theta\frac{\partial}{\partial\theta}\right),$$

and the continuity equation, which will be identically satisfied if we introduce the stream function $\Psi$ according to (1.1.10).

We seek the stream function and the pressure in the form

$$\Psi(R,\theta) = \nu R f(\xi), \quad P = P_{\mathrm{i}} + \frac{\rho\nu^2}{R^2}g(\xi), \qquad \xi = \cos\theta. \qquad (1.4.2)$$

Let us first express the fluid velocity components in (1.4.1) via the stream function (1.1.10) and then substitute the expressions (1.4.2). As a result, we obtain the following system of ordinary differential equations for the unknown functions $f$ and $g$:

$$
\begin{aligned}
g &= -\frac{f^2}{2(1-\xi^2)} - \frac{1}{2}\frac{d}{d\xi}[ff' - (1-\xi^2)f''], \\
g' &= -f'' - \frac{1}{2}\frac{d}{d\xi}\frac{f^2}{1-\xi^2}.
\end{aligned}
\qquad (1.4.3)
$$

By eliminating $g$ from system (1.4.3) and by integrating three times, we obtain the equation

$$f^2 - 2(1-\xi^2)f' - 4\xi f = C_1\xi^2 + C_2\xi + C_3 \qquad (1.4.4)$$

for $f$, where $C_1$, $C_2$, and $C_3$ are arbitrary constants of integration.

These constants must be determined with regard to the flow singularities on the symmetry axis [26]. For $C_1 = C_2 = C_3 = 0$, we obtain the particular solution describing the simplest flow with minimum number of singularities. In this case, the equation for $f$ is simplified dramatically, and the substitution

$$f(\xi) = (1 - \xi^2)h(\xi)$$

yields the separable equation $2h' - h^2 = 0$. The solution is $h(\xi) = 2(A - \xi)^{-1}$, where $A$ is another constant of integration.

Finally, we obtain the following formulas for $f$ and $g$:

$$f(\xi) = \frac{2(1-\xi^2)}{A-\xi}, \quad g(\xi) = \frac{4(A\xi - 1)}{(A-\xi)^2}. \qquad (1.4.5)$$

To find the value of $A$, we need to know only one quantitative characteristic of the jet-source, namely, the momentum

$$J_0 = \int_S \rho V^2\, dS, \qquad (1.4.6)$$

where the integral is taken over the nozzle cross-section $S$; here $V$ is the local velocity at an arbitrary point of the section.

It may seem that the mass flow rate $G_0 = \int_S \rho V \, dS$ is an equally important quantitative characteristic that influences the flow picture, but this is not the case. In fact, the stream function has no singularities on the flow axis. This function does not undergo a jump at the origin and is zero on both rays $\theta = 0$ (i.e., for $\xi = 1$) and $\theta = \pi$ (i.e., for $\xi = -1$). This means that the jet-source generating the flow in question is a source of momentum rather than mass [26]. Thus, the value of $G_0$ is unessential for this flow.

To obtain a relation between $A$ and $J_0$, we equate the jet momentum with the axial projection of the total flux of momentum through an arbitrary sphere centered at the origin. Then from the solution (1.4.5) we eventually obtain [260]

$$J_0 = 16\pi\nu^2\rho A \left[ 1 + \frac{4}{3(A^2 - 1)} - \frac{A}{2} \ln \frac{A+1}{A-1} \right]; \qquad (1.4.7)$$

the graph of (1.4.7) can be found in [501].

As the jet momentum $J_0$ varies from 0 to $\infty$, the value of $A$ varies from $\infty$ to 1. Since the solution is valid only for laminar flows, it follows that only the case of small $J_0$ (weak jets) is of practical use. In this case, one can find $A$ by the approximate formula

$$A = \frac{16\pi\rho\nu^2}{J_0}. \qquad (1.4.8)$$

Sometimes it is convenient to represent $A$ via the Reynolds number $\mathsf{Re} = Ud/\nu$, where $d$ is the nozzle diameter and $U$ is the characteristic velocity. By setting $J_0 = \frac{1}{4}\pi d^2 \rho U^2$, we obtain

$$A = \frac{64}{\mathsf{Re}^2}. \qquad (1.4.9)$$

Since, according to [3], laminar flows become unstable for $\mathsf{Re} > 5$, we see that the minimum value of $A$ for which one can use the above relations for laminar flows is about 2.5.

Although a jet-source involves the fluid in the entire space in motion, the streamline picture (described, for example, in the monographs [26, 260, 501]) suggests that it is meaningful to speak about the jet boundaries and the law of jet expansion. The point is that on each streamline there is a pronounced turning point that lies at the minimum distance from the jet axis. It is reasonable to define the jet boundary as the set of these points. Then the jet boundary is determined by the conditional minimum of the function $R \sin \theta$ under the constraint $\dfrac{R \sin^2 \theta}{A - \cos \theta} = \text{const}$ and is a conical surface with vertex at the origin (Figure 1.5) and vertex half-angle

$$\theta_0 = \arccos\frac{1}{A}. \qquad (1.4.10)$$

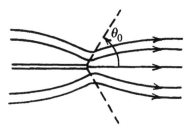

**Figure 1.5.** Streamlines near a laminar jet-source and the conventional jet width

The estimates (1.4.8) and (1.4.9) show that the stronger a jet, the narrower it is. The values $A \approx 2.5$ and $\theta_0 \approx 65°$ correspond to the narrowest possible laminar jet.

In practice, the jet range is also of interest. The maximum velocity which is attained at the jet axis ($\theta = 0$) and can be calculated according to the above relations is

$$V_{max} = \frac{\nu}{R} \frac{2}{A - 1}. \tag{1.4.11}$$

Thus, this velocity is greater for stronger jets (lower values of $A$) and is inversely proportional to the distance $R$ along the axis. We point out that all these characteristics are independent of the mass flow rate in the jet and are determined by the momentum alone.

The pressure distribution in the jet is determined by (1.4.2) and (1.4.5). Along the jet axis (for $\xi = 1$), the pressure varies according to the formula

$$P = P_i + \frac{4\rho\nu^2}{R^2(A - 1)}, \tag{1.4.12}$$

and hence virtually coincides with the nonperturbed pressure in the ambient medium even at small distances from the source.

***Boundary layer approximation.*** The Landau problem, which was described above, is an example of an exact solution of the Navier–Stokes equations. Schlichting [427] proposed another approach to the jet-source problem, which gives an approximate solution and is based on the boundary layer theory (see Section 1.7). The main idea of this method is to neglect the gradients of normal stresses in the equations of motion. In the cylindrical coordinates ($R$, $\varphi$, $Z$), with regard to the axial symmetry ($V_\varphi = 0$) and in the absence of rotational motion in the flow ($\partial/\partial\varphi = 0$), the system of boundary layer equations has the form

$$V_Z \frac{\partial V_Z}{\partial Z} + V_R \frac{\partial V_Z}{\partial R} = \frac{\nu}{R} \frac{\partial}{\partial R}\left(R \frac{\partial V_Z}{\partial R}\right),$$

$$\frac{\partial V_Z}{\partial Z} + \frac{\partial V_R}{\partial R} + \frac{V_R}{R} = 0 \tag{1.4.13}$$

with the boundary conditions

$$V_R = 0, \quad \frac{\partial V_Z}{\partial R} = 0 \quad \text{at} \quad R = 0,$$

$$V_Z \to 0 \quad\quad\quad \text{as} \quad R \to \infty. \tag{1.4.14}$$

We seek the stream function $\Psi$, determined by

$$V_Z = \frac{1}{R}\frac{\partial \Psi}{\partial R}, \qquad V_R = -\frac{1}{R}\frac{\partial \Psi}{\partial Z}, \qquad (1.4.15)$$

in the form

$$\Psi = \nu Z F(\eta), \qquad \eta = \frac{R}{\sqrt{K}Z}, \qquad (1.4.16)$$

where $\eta$ is a self-similar variable.

As a result, we obtain the following boundary value problem for $F$:

$$\left(F'' - \frac{F'}{\eta}\right)' + \left(\frac{FF'}{\eta}\right)' = 0,$$

$$\frac{F'}{\eta} = 1, \quad \frac{F}{\eta} = 0 \quad \text{at} \quad \eta = 0, \qquad (1.4.17)$$

$$F' \to 0 \qquad \qquad \text{as} \quad \eta \to \infty.$$

The constant $K$ in the definition of $\eta$ is determined by the jet momentum $J_0$ as follows:

$$K = \frac{16\pi}{3}\frac{\rho\nu^2}{J_0}. \qquad (1.4.18)$$

Problem (1.4.17) has an exact closed-form solution. The final expression for the flow field has the form [427]

$$V_Z = \frac{3}{8\pi}\frac{J_0}{\rho\nu}\frac{1}{Z}\left(1 + \frac{\eta^2}{4}\right)^{-2},$$

$$V_R = \frac{1}{4}\sqrt{\frac{3}{\pi}\frac{J_0}{\rho}}\frac{1}{Z}\left(\eta - \frac{\eta^3}{4}\right)\left(1 + \frac{\eta^2}{4}\right)^{-2}. \qquad (1.4.19)$$

Obviously, the solution (1.4.19) differs from the Landau solution, but many characteristics of flow remain the same. For example, the dependence between the flow field and the jet momentum is the same, and the velocity at the jet axis decreases inversely proportional to the distance from the source.

*Turbulent jet.* As was already pointed out, for laminar jets the solution is of limited practical use (it is valid only for Re < 5). However, it was shown in [427] that a similar approach can be applied to the case of turbulent jets. It turns out that the apparent kinematic turbulent viscosity $\nu_t$ is constant for turbulent jet flows. This constant can be obtained only empirically, since it depends on the nozzle geometry. The flow field in the jet is still determined by formulas (1.4.19) with the only difference that the physical constant $\nu$ of the medium must be replaced by the empirical constant $\nu_t$. Finding this constant experimentally is a separate problem. A coarse estimate can be obtained from the following relation for $K$ [3]:

$$K = \frac{16\pi}{3}\frac{\rho\nu_t^2}{J_0} \qquad (K \approx 0.002 \text{ to } 0.005).$$

### 1.4-2. Plane Jets

*Laminar jet.* The problem of a jet emerging from a long narrow slit and mixing with the surrounding fluid was studied by Schlichting [427]. In the boundary layer approximation, Schlichting obtained the expressions

$$V_X = 0.454 \left( \frac{J^2}{\rho^2 \nu X} \right)^{1/3} (1 - \tanh^2 \xi), \qquad \xi = 0.275 \left( \frac{J}{\rho \nu^2} \right)^{1/3} \frac{Y}{X^{2/3}};$$

$$V_Y = 0.550 \left( \frac{J\nu}{\rho X^2} \right)^{1/3} [2\xi(1 - \tanh^2 \xi) - \tanh \xi].$$

Here the Cartesian coordinate system $X$, $Y$ is used, where the coordinate $X$ is measured along the jet axis. The quantity $J = \rho \int_{-\infty}^{\infty} V_X^2 \, dY = \text{const}$ is the momentum of the jet per unit width.

The rate of flow of the jet per unit width is given by the formula

$$G = \rho \int_{-\infty}^{\infty} V_X \, dY = 3.302 \left( \frac{\nu}{\rho} JX \right)^{1/3}.$$

It is apparent that the rate of flow increases with the distance from the slit. This is due to the fact that the jet drags the surrounding fluid.

The jet remains laminar until $\text{Re} = 30$, where $\text{Re}$ is the Reynolds number based on the slit width [427].

*Turbulent jet.* Now consider a turbulent plane jet. Estimates show that the turbulent viscosity in this case is constant within every cross-section of the jet (but varies along the jet axis, just as was the case for an axisymmetric jet). In the boundary layer approximation, the solution of the corresponding hydrodynamic problem results in the following fluid velocity components [395, 427]:

$$V_X = \frac{1}{2} \sqrt{\frac{3\sigma J}{\rho X}} (1 - \tanh^2 \eta), \quad V_Y = \frac{1}{4} \sqrt{\frac{3J}{\sigma \rho X}} [2\eta(1 - \tanh^2 \eta) - \tanh \eta],$$

where $\eta = \sigma Y / X$ and $\sigma \approx 7.67$ is an empirical constant.

### 1.4-3. Structure of Wakes Behind Moving Bodies

The flow in the wake behind a body moving in an unbounded liquid possesses all the properties of free jet flows and can be calculated by methods of the boundary layer theory [427]. Note that the wake behind a moving body is almost invariably turbulent even if the boundary layer on the body surface is laminar. This is due to the fact that there are points of inflection on all velocity profiles of the wake; such velocity distributions are known to be particularly unstable.

For practical estimates, we present the distribution of the axial velocity component in the plane wake behind the moving body [427]:

$$\frac{V_X}{U_i} = 1 - 0.976 \left(\frac{c_f d}{X}\right)^{1/2} \left[1 - \left(\frac{Y}{b(X)}\right)^{3/2}\right]^2. \tag{1.4.20}$$

Here $c_f$ is the drag coefficient of the body, $d$ is the center section diameter, and $b(X)$ is the local half-width of the wake. The coordinate $X$ is measured from the rear point of the body and $Y$ is the transverse coordinate. The definition of the local half-width $b(X)$ of the wake is a matter of convention; it can be estimated as

$$b(X) = 0.569 \, (c_f X d)^{1/2}. \tag{1.4.21}$$

Formulas (1.4.20), (1.4.21) hold for $X > 50 \, c_f d$, that is, describe only the so-called "remote" wake.

The similar relation for the wake behind a body of revolution has the form

$$1 - \frac{V_X}{U_i} \sim \left(\frac{c_f F}{X^2}\right)^{1/3}, \qquad b(X) \sim (c_f F X)^{1/3}.$$

Here $F$ is the midsection area of the body.

## 1.5. Laminar Flows in Tubes

Laminar steady-state fluid flows in tubes of various cross-sections were studied by many authors (e.g., see [179, 276, 427]). Such flows are often encountered in practice (water-, gas- and oil pipelines, heat exchangers, etc.). It is worth noting that in these cases the corresponding hydrodynamic equations admit an exact closed-form solution. In what follows we describe the most important results in that direction.

### 1.5-1. Statement of the Problem

Let us consider a laminar steady-state fluid flow in a rectilinear tube of constant cross-section. The fluid streamlines in such systems are strictly parallel (we neglect the influence of the tube endpoints on the flow). We shall use the Cartesian coordinates $X, Y, Z$ with $Z$-axis directed along the flow. Let us take into account the fact that the transverse velocity components of the fluid are zero and the longitudinal component depends only on the transverse coordinates. In this case, the continuity equation (1.1.1) and the first two Navier–Stokes equations in (1.1.2) are satisfied automatically, and it follows from the third equation in (1.1.2) that

$$\frac{\partial^2 V}{\partial X^2} + \frac{\partial^2 V}{\partial Y^2} = \frac{1}{\mu} \frac{dP}{dZ}, \tag{1.5.1}$$

where $V \equiv V_Z$ is the longitudinal velocity component.

Equation (1.5.1) must be supplemented by the no-slip condition

$$V = 0 \qquad \text{on the tube surface.} \tag{1.5.2}$$

The pressure gradient $dP/dZ$ in the steady state is constant along the tube and can be represented in the form

$$\frac{dP}{dZ} = -\frac{\Delta P}{L}, \tag{1.5.3}$$

where $\Delta P > 0$ is the total pressure drop along a tube part of length $L$.

The main flow characteristics are the volume rate of flow

$$Q = \int_S V \, dS \tag{1.5.4}$$

and the mean flow rate velocity

$$\langle V \rangle = \frac{Q}{S_*}, \tag{1.5.5}$$

where $S_*$ is the area of the tube cross-section.

### 1.5-2. Plane Channel

First, let us consider the flow between two infinite parallel planes at a distance $2h$ from each other. The coordinate $X$ is measured from one of the planes along the normal. Since the fluid velocity is independent of the coordinate $Y$, we can rewrite (1.5.1) in the form

$$\frac{d^2V}{dX^2} = -\frac{\Delta P}{\mu L}.$$

The solution of this equation under the no-slip boundary conditions on the planes ($V = 0$ for $X = 0$ and $X = 2h$) has the form

$$V = \frac{\Delta P}{2\mu L} X(2h - X). \tag{1.5.6}$$

Formula (1.5.6) describes the parabolic velocity field in a plane Poiseuille flow symmetric with respect to the midplane $X = h$ of the channel.

The volume rate of flow per unit width of the channel can be found by integrating (1.5.6) over the cross-section:

$$Q = \frac{2h^3 \Delta P}{3\mu L}. \tag{1.5.7}$$

The mean flow rate velocity is

$$\langle V \rangle = \frac{h^2 \Delta P}{3\mu L}. \tag{1.5.8}$$

The maximum velocity is attained on the midplane of the channel:

$$U_{\max} = \frac{h^2 \Delta P}{2\mu L} \qquad (\text{at} \quad X = h).$$

## 1.5-3. Circular Tube

In the case of a circular tube, Eq. (1.5.1), with regard to (1.5.3), acquires the form

$$\frac{1}{\mathcal{R}}\frac{\partial}{\partial \mathcal{R}}\left(\mathcal{R}\frac{\partial V}{\partial \mathcal{R}}\right) = -\frac{\Delta P}{\mu L}, \qquad \mathcal{R} = \sqrt{X^2 + Y^2}. \tag{1.5.9}$$

The solution of this equation under the no-slip condition on the surface of a tube of radius $a$ ($V = 0$ for $\mathcal{R} = a$) describes an axisymmetric Poiseuille flow with parabolic velocity profile:

$$V = \frac{\Delta P}{4\mu L}\left(a^2 - \mathcal{R}^2\right). \tag{1.5.10}$$

The volume rate of flow can be obtained by integrating over the cross-section:

$$Q = 2\pi \int_0^a \mathcal{R}V \, d\mathcal{R} = \frac{\pi a^4 \Delta P}{8\mu L}. \tag{1.5.11}$$

By using (1.5.5), we obtain the mean flow rate velocity

$$\langle V \rangle = \frac{a^2 \Delta P}{8\mu L}. \tag{1.5.12}$$

The maximum fluid velocity is attained at the tube axis:

$$U_{\max} = \frac{a^2 \Delta P}{4\mu L} \qquad \text{(at } \mathcal{R} = 0). \tag{1.5.13}$$

Now let us consider the flow in an annular channel between two coaxial circular cylinders of radii $a_1$ and $a_2$ ($a_1 < a_2$). In this case, Eq. (1.5.9) remains valid. The solution of this equation satisfying the no-slip conditions on the cylinder surfaces,

$$V = 0 \quad \text{at} \quad \mathcal{R} = a_1, \qquad V = 0 \quad \text{at} \quad \mathcal{R} = a_2,$$

has the form

$$V = \frac{\Delta P}{4\mu L}\left[a_2^2 - \mathcal{R}^2 + \frac{a_2^2 - a_1^2}{\ln(a_2/a_1)}\ln\frac{\mathcal{R}}{a_2}\right]. \tag{1.5.14}$$

The volume rate of flow is given by the formula

$$Q = \frac{\pi \Delta P}{8\mu L}\left[a_2^4 - a_1^4 - \frac{(a_2^2 - a_1^2)^2}{\ln(a_2/a_1)}\right]. \tag{1.5.15}$$

## 1.5-4. Tube of Elliptic Cross-Section

Now let us consider a tube whose cross-section is an ellipse with semiaxes $a$ and $b$. The surface of this tube is given by the equation

$$\left(\frac{X}{a}\right)^2 + \left(\frac{Y}{b}\right)^2 = 1. \tag{1.5.16}$$

The solution of Eq. (1.5.1) under the no-slip condition on the surface (1.5.16) has the form [179]

$$V = \frac{a^2 b^2 \Delta P}{2\mu L(a^2 + b^2)}\left(1 - \frac{X^2}{a^2} - \frac{Y^2}{b^2}\right). \tag{1.5.17}$$

The volume rate of flow is

$$Q = \frac{\pi \Delta P}{4\mu L}\frac{a^3 b^3}{a^2 + b^2}. \tag{1.5.18}$$

By using formula (1.5.5), we find the mean flow rate velocity

$$\langle V \rangle = \frac{\Delta P}{4\mu L}\frac{a^2 b^2}{a^2 + b^2}. \tag{1.5.19}$$

The maximum velocity is attained at the tube axis

$$U_{\max} = \frac{a^2 b^2 \Delta P}{2\mu L(a^2 + b^2)} \qquad (\text{at } X = Y = 0). \tag{1.5.20}$$

In the special case $a = b$, formulas (1.5.17)–(1.5.20) are reduced to the corresponding formulas (1.5.10)–(1.5.13) for a circular tube.

## 1.5-5. Tube of Rectangular Cross-Section

Now let us consider a tube of rectangular cross-section with sides $a$ and $b$. We assume that the flow region is described by the inequalities $0 \le X \le a$ and $0 \le Y \le b$. The solution of Eq. (1.5.1) under the no-slip conditions on the tube surface has the form [179]

$$V = \frac{\Delta P}{2\mu L}X(a-X) + \sum_{m=1}^{\infty}\sin\left(\frac{\pi m X}{a}\right)\left(A_m \cosh\frac{\pi m Y}{a} + B_m \sinh\frac{\pi m Y}{a}\right),$$

$$A_m = \frac{a^2 \Delta P}{\pi^3 m^3 \mu L}[\cos(\pi m) - 1], \quad B_m = -A_m\frac{\cosh(\pi m k) - 1}{\sinh(\pi m k)}, \quad k = \frac{b}{a}. \tag{1.5.21}$$

By integrating the expression for $V$, we obtain the volume rate of flow

$$Q = \frac{\Delta P}{24\mu L} ab(a^2 + b^2)$$
$$- \frac{8\Delta P}{\pi^5 \mu L} \sum_{m=1}^{\infty} \frac{1}{(2m-1)^5} \left[ a^4 \tanh\left( \pi b \frac{2m-1}{2a} \right) + b^4 \tanh\left( \pi a \frac{2m-1}{2b} \right) \right].$$

$$(1.5.22)$$

For a tube of square section with side $a$, this formula acquires the form

$$Q = \frac{a^4 \Delta P}{12 \mu L} \left[ 1 - \frac{192}{\pi^5} \sum_{m=1}^{\infty} \frac{1}{(2m-1)^5} \tanh\left( \pi \frac{2m-1}{2} \right) \right], \qquad (1.5.23)$$

or, after summing the series,

$$Q = 0.0351 \frac{a^4 \Delta P}{\mu L}. \qquad (1.5.23a)$$

It is useful to rewrite the last expression as

$$\frac{Q}{Q_0} = 0.883,$$

where $Q_0$ is the volume rate of flow for a circular tube with the same cross-section area. The volume rate of flow for a square tube is smaller, because the cross-section has corners near which the velocity of a viscous fluid decreases noticeably.

## 1.5-6. Tube of Triangular Cross-Section

Now we suppose that the cross-section of the tube is an equilateral triangle with side $b$. We place the origin at the center of the cross-section and measure the coordinate $X$ along one of the sides of the triangle. In this case, the solution of Eq. (1.5.1) under the boundary condition (1.5.2) has the form

$$V = \frac{\sqrt{3}\,\Delta P}{6\mu b L} \left( Y - \frac{b}{2\sqrt{3}} \right) \left( Y + \sqrt{3}\,X - \frac{b}{\sqrt{3}} \right) \left( Y - \sqrt{3}\,X - \frac{b}{\sqrt{3}} \right).$$

The volume rate of flow of this flow is given by the formula

$$Q = \frac{\sqrt{3}}{320} \frac{b^4 \Delta P}{\mu L}.$$

It is useful to compare $Q$ with the volume rate of flow for a circular tube of the same cross-section area:

$$\frac{Q}{Q_0} = 0.726.$$

This expression shows that the volume rate of flow for a tube whose cross-section is an equilateral triangle is substantially lower than for tubes of square or circular cross-section of the same area.

## 1.5-7. Tube of Arbitrary Cross-Section

By using dimensional considerations, one can obtain the following formulas for the volume rate of flow and the maximum velocity in a rectilinear tube of constant cross-section of arbitrary shape:

$$Q = K \frac{S_*^2 \Delta P}{\mu L}, \qquad U_{max} = K_v \frac{S_* \Delta P}{\mu L}, \qquad (1.5.24)$$

where $S_*$ is the tube cross-section area and $K$ and $K_v$ are dimensionless coefficients depending on the shape of the cross-section. The coefficients $K$ and $K_v$ can be obtained either experimentally or theoretically.

The most important dimensionless geometric parameter characterizing the cross-section shape is the ratio $\sqrt{S_*}/P$, where $P$ is the cross-section perimeter. In calculations it is convenient to use the shape parameter

$$\xi = 2\sqrt{\pi} \frac{\sqrt{S_*}}{P}, \qquad (1.5.25)$$

which always lies in the range $0 \leq \xi \leq 1$. The value $\xi = 1$ corresponds to a circular tube. This condition is ensured by choosing the proportionality factor $2\sqrt{\pi}$ in formula (1.5.25).

For tubes of convex cross-section that is nearly circular or at least does not deviate very much from the circular shape, it is natural to assume that the coefficients $K$ and $K_v$ in (1.5.24) depend only on $\xi$:

$$K = K(\xi), \qquad K_v = K_v(\xi). \qquad (1.5.26)$$

Then in many cases the function $K = K(\xi)$ (which determines the volume rate of flow $Q$ in (1.5.24)) is well approximated by the linear function

$$K = K_0 \xi, \qquad K_0 = \frac{1}{8\pi} \approx 0.0398. \qquad (1.5.27)$$

For $\xi = 1$, the approximate formula (1.5.27) gives the exact value of $K_0$ corresponding to a circular tube. For example, formula (1.5.27) can be used for tubes whose cross-section is a regular $N$-gon ($N = 4, 5, \ldots$). In particular, for a tube of square cross-section, Eq. (1.5.25) gives $\xi = \frac{1}{2}\sqrt{\pi} \approx 0.886$. Substituting this into (1.5.27), we obtain $K = 0.0353$. This differs from the exact value $K = 0.0351$ only by 0.6% (see (1.5.23a)). For a tube of elliptic cross-section with axial ratio $a/b = 1.5$, the error in (1.5.27) is about 5%.

The hydrodynamic drag for the laminar flow of a fluid in tubes of various shape is considered in [80]. The drag coefficient $\lambda$ between the pressure drop and the characteristic pressure head is introduced by the relation

$$\lambda = \frac{d_e}{\frac{1}{2}\rho \langle V \rangle^2} \frac{\Delta P}{L}, \qquad (1.5.28)$$

## TABLE 1.2
Values of the drag coefficients for laminar flow in tubes of various shape

| Tube profile | | $\lambda\,\mathsf{Re}_d$ | Equivalent diameter $d_e$ |
|---|---|---|---|
| Circular tube of diameter $d$ | | 64.000 | $d$ |
| Flat tube of width $2h$ | | 96.000 | $4h$ |
| Elliptic tube with semiaxes $a$ and $b$ | $b/a =$ 1.00 | 64.000 | $\dfrac{\pi b}{E\left(\sqrt{1-b^2/a^2}\,\right)}$, where $E(\vartheta)$ is the complete elliptic integral of the second kind |
| | 0.80 | 64.392 | |
| | 0.50 | 67.292 | |
| | 0.25 | 72.960 | |
| | 0.125 | 76.584 | |
| | 0.0625 | 78.144 | |
| | 0 | 78.956 | |
| Tube of rectangular cross-section with sides $a$ and $b$ | $b/a =$ 1.00 | 58.008 | $\dfrac{2ab}{a+b}$ |
| | 0.714 | 58.260 | |
| | 0.50 | 62.192 | |
| | 0.25 | 72.932 | |
| | 0.125 | 82.336 | |
| | 0.05 | 89.908 | |
| | 0 | 96.000 | |
| Equilateral triangle with side $a$ | | 53.348 | $\dfrac{a\sqrt{3}}{3}$ |
| Regular hexagon with side $a$ | | 60.216 | $a\sqrt{3}$ |
| Semi-circle of diameter $d$ | | 63.068 | $\dfrac{\pi d}{\pi+2}$ |

where $d_e$ is equivalent (or "hydraulic") diameter and $\langle V\rangle$ is the mean flow rate velocity.

Let us introduce the equivalent (or "hydraulic") diameter $d_e$ by the formula $d_e = 4S_*/\mathcal{P}$, where $S_*$ is the area of the tube cross-section and $\mathcal{P}$ is the cross-section perimeter. For tubes of circular cross-section, $d_e$ coincides with the diameter, and for a plane channel, $d_e$ is twice the height of the channel.

Table 1.2 presents values of the drag coefficients for tubes with various shapes of the cross-section (according to [80]). The Reynolds number $\mathsf{Re}_d = \langle d_e V\rangle/\nu$ can be calculated from the mean flow rate velocity and the equivalent diameter.

For a tube whose cross-section is a regular $N$-gon, the value $\lambda\,\mathsf{Re}_d$ is given by the approximate formula

$$\lambda \operatorname{Re}_d = \frac{64\,N - 82}{N - 0.95}. \qquad (1.5.29)$$

The comparison with Table 1.2 shows that the maximum error in (1.5.29) is less than 0.6% at $N = 3,\ 4,\ 6,\ \infty$.

# 1.6. Turbulent Flows in Tubes

## 1.6-1. Tangential Stress. Turbulent Viscosity

A flow of fluid through a smooth tube of circular cross-section remains laminar while $\operatorname{Re} = a\langle V\rangle/\nu < 1500$ [427], where $a$ is the tube radius and $\langle V\rangle$ the mean flow rate velocity of the fluid. For higher Reynolds numbers, the loss of stability of the laminar flow is observed and an intermediate regime occurs. For $\operatorname{Re} > 2500$, a fully developed regime of turbulent flow is established which is characterized by a chaotic variation of the velocity and pressure in time and space.

In a turbulent flow in a tube, there are two significantly different regions of flow. In the first, entry region, the average velocity profile $\overline{V}$ changes dramatically with the distance from the entry cross-section. In the second, stabilized flow region, the average velocity profile is the same at each cross-section. The length of the entry (stabilization) region depends on the Reynolds number and the roughness of the walls and occupies a few dozen diameters (from 25 to 100, according to [427]). For rough estimates, this length is frequently taken to be 50 tube diameters.

In the stabilized region of flow, the average fluid velocity $\overline{V}$ is directed along the tube axis and depends only on the distance $Y$ from the tube wall. The integration of the Reynolds equations (1.1.18) yields the following expression for the shear stress [289, 427]:

$$\tau = \tau_s(1 - Y/a), \qquad (1.6.1)$$

where $\tau_s$ is the friction stress at the wall. Near the tube axis, as $Y/a \to 1$, it follows from (1.6.1) that $\tau/\tau_s \to 0$. Near the wall, as $Y/a \to 0$, we have $\tau/\tau_s \to 1$.

The friction stress at the wall for a circular tube is calculated by the formula $\tau_s = \frac{1}{2}a(\Delta P/L)$, where $\Delta P > 0$ is the total pressure drop along a tube part of length $L$.

In accordance with Boussinesq's model (1.1.20), the shear stress can be represented as

$$\tau = \rho(\nu + \nu_t)\frac{\partial \overline{V}}{\partial Y}, \qquad (1.6.2)$$

where $\nu$ is the kinematic viscosity and $\nu_t$ is the turbulent viscosity.

Formulas (1.6.1) and (1.6.2) have formed the basis of most theoretical investigations on the determination of the average fluid velocity and the drag coefficient in the stabilized region of turbulent flow in a circular tube (and a plane channel of width $2a$). The corresponding results obtained on the basis of Prandtl's relation (1.1.21) and von Karman's relation (1.1.22) for the turbulent viscosity can be found in [276, 427]. In what follows, major attention will be paid to empirical and semiempirical formulas that approximate numerous experimental data quite well.

---

#### 1.6-2. Structure of the Flow. Velocity Profile in a Circular Tube

---

The experiments show that in the region of stabilized flow, two characteristic subregions can be singled out [56, 398], namely,

$$0 \leq Y/a \leq \sigma \quad \text{(wall region)},$$
$$\sigma \leq Y/a \leq 1 \quad \text{(core of turbulent flow)},$$

where $Y$ is the transverse coordinate measured from the wall and $\sigma = 0.1$ to $0.2$.

To describe the turbulent flow in the wall region, one introduces the so-called friction velocity $U_*$ and the dimensionless internal coordinate $y^+$ according to the formulas

$$U_* = \sqrt{\tau_s/\rho}, \quad y^+ = YU_*/\nu, \tag{1.6.3}$$

where $\tau_s$ is the shear stress at the wall and $\rho$ is the fluid density. According to von Karman [394] (see also [276, 398]), it is convenient to single out three subdomains in the wall region:

$$\text{wall region} = \begin{cases} \text{viscous sublayer } (\nu_t \ll \nu), \\ \text{buffer layer } (\nu_t \sim \nu), \\ \text{logarithmic layer } (\nu_t \gg \nu). \end{cases}$$

In the viscous sublayer, the turbulent viscosity tends to zero near the wall ($\nu_t$ is proportional to $Y^3$, see [289]). In the logarithmic layer, the turbulent viscosity depends on the transverse coordinate linearly, $\nu_t = 0.4U_*Y$.

The approximate ranges of the above subdomains in terms of $y^+$ are as follows [276, 289]:

viscous sublayer    $0 \leq y^+ \leq 5$,
buffer layer         $5 \leq y^+ \leq 30$,
logarithmic layer   $30 \leq y^+$.

In the viscous sublayer the distribution of the average velocity is linear [138, 268, 276]:

$$\frac{\overline{V}(Y)}{U_*} = y^+ \quad \text{for} \quad 0 \leq y^+ \leq 5. \tag{1.6.4}$$

Note that turbulent fluctuations can penetrate into the viscous sublayer, although the turbulent friction is small there.

The average velocity profile in the thickest, logarithmic layer can be described by the formula

$$\frac{\overline{V}(Y)}{U_*} = 2.5 \ln y^+ + 5 \qquad \text{for} \qquad 30 \le y^+ \qquad (1.6.5)$$

quite well. This relation is referred to as Prandtl's law of the wall. It is worth noting that the factor 2.5 occurring in Eq. (1.6.5) comes from $1/\kappa$, where $\kappa = 0.4$ is the von Karman constant [80, 276, 289, 398]. The value 5 of the constant term is recommended in [289] on the basis of numerous experimental data; other values can also be found in the literature [254, 276].

It should be pointed out that formulas (1.6.4) and (1.6.5) not only are supported by experimental data quite well, but also have certain theoretical justification [138, 276, 289, 427].

Various approximate formulas for the velocity distribution in the buffer layer can be found, for example, in [138, 276, 398]. Choosing not to focus on these formulas (since these do not have sufficient theoretical justification), we only indicate a simple interpolation formula with a wider scope:

$$\frac{\overline{V}(Y)}{U_*} = y^+ \cos^{3.24} \xi + (2.5 \ln y^+ + 5) \sin^{2.24} \xi, \quad \xi = \frac{\pi}{2} \frac{y^+}{18 + y^+}. \qquad (1.6.6)$$

This formula well agrees with the experimental data [223, 289] and defines a continuous and smooth profile of the average velocity over the entire wall domain (including the viscous sublayer, buffer layer, and logarithmic layer).

The numerous experimental data provide the evidence that the average velocity $\overline{V}$ in the core of turbulent flow can be approximated as [217, 276, 427]

$$\frac{U_{max} - \overline{V}(Y)}{U_*} = 2.5 \ln \frac{a}{Y}, \qquad (1.6.7)$$

where $U_{max}$ is the velocity at the flow axis. Equation (1.6.7) is universal and is referred to as von Karman's velocity defect law;* it is applicable for smooth and rough tubes and any Reynolds numbers corresponding to turbulent flows. Formula (1.6.7) can be somewhat refined by adding the term $0.6(1 - Y/a)^2$ to the right-hand side, thus extending the formula to the range $0.01 \le Y/a < 1$.

Darcy's formula [105]

$$\frac{U_{max} - \overline{V}(Y)}{U_*} = 5.08 \left(1 - \frac{Y}{a}\right)^{3/2} \qquad (1.6.8)$$

is worth mentioning. It provides a more accurate prediction of the turbulent flow near the flow axis ($Y/a \approx 1$) compared with Eq. (1.6.7) but has a narrower scope, $0.25 \le Y/a \le 1$.

---

* It is remarkable that Eq. (1.6.7) was first suggested for the logarithmic layer, but it turned out that it can well be extrapolated to almost the entire domain of turbulent core.

In engineering calculations, it is not uncommon to approximate the average fluid velocity in a turbulent flow by the power law [276, 427]

$$\frac{\overline{V}(Y)}{U_{\max}} = \left(\frac{Y}{a}\right)^{1/n}, \tag{1.6.9}$$

where the parameter $n$ slowly increases with the Reynolds number. The value $n = 7$ suggested by Blasius is most frequently used. In this case, formula (1.6.9) agrees well with experimental data within the range $3 \times 10^3 \leq \mathsf{Re}_d \leq 10^5$, where $\mathsf{Re}_d = d\langle V\rangle/\nu$ is the Reynolds number for a tube of diameter $d = 2a$.

The average velocity profile for the entire cross-section of a circular tube can be calculated using the unified interpolation formula

$$\frac{\overline{V}(Y)}{U_*} = 2.5\ln(1 + 4\zeta) + 7.5\left(1 - e^{-\zeta} - \zeta e^{-3\zeta}\right) + 2.5\ln\frac{1.5\,(2 - \eta)}{1 + 2(1 - \eta)^2}, \tag{1.6.10}$$

where $\zeta = 0.1\,YU_*/\nu$ and $\eta = Y/a$. This formula is a bit simpler than Reichardt's formula [394]. The former is obtained from the latter by a slight change in numerical coefficients, which provides a better agreement with the experimental data in the logarithmic layer, with the same accuracy in the other domains.

Other interpolation formulas for the velocity distribution over the entire cross-section of a circular tube can be found, for example, in [56, 80, 212, 289].

### 1.6-3. Drag Coefficient of a Circular Tube

The drag coefficient $\lambda$ is expressed via other hydrodynamic parameters as follows:

$$\lambda = \frac{\Delta P}{L}\frac{4a}{\rho\langle V\rangle^2} = \frac{8\tau_s}{\rho\langle V\rangle^2} = 8\left(\frac{U_*}{\langle V\rangle}\right)^2. \tag{1.6.11}$$

In the region of stabilized turbulent flow in a smooth circular tube, the drag coefficient can be estimated from the Prandtl–Nikuradze implicit formula [318]

$$\frac{1}{\sqrt{\lambda}} = 0.88\ln\left(\mathsf{Re}_d\sqrt{\lambda}\right) - 0.82, \tag{1.6.12}$$

where $\mathsf{Re}_d$ is the Reynolds number determined by the diameter. Within the range $3 \times 10^3 \leq \mathsf{Re}_d \leq 3 \times 10^6$, the maximum deviation of the result predicted by Eq. (1.6.12) from experimental data is about 2%. In practice, it is more convenient to use simpler explicit formulas of Blasius and Nikuradze:

$$\lambda = \begin{cases} 0.3164\,\mathsf{Re}_d^{-0.25} & \text{for } 3 \times 10^3 \leq \mathsf{Re}_d \leq 10^5, \\ 0.0032 + 0.221\,\mathsf{Re}_d^{-0.237} & \text{for } 10^5 \leq \mathsf{Re}_d. \end{cases} \tag{1.6.13}$$

These formulas are also accurate within 2% [276, 427]. The first line in Eq. (1.6.13) follows from the assumption that the average velocity profile is given by the power law (1.6.9) with $n = 7$.

In the transient zone, it is recommended to calculate the drag coefficient by the formula [254]

$$\lambda = 6.3 \times 10^{-4}\sqrt{\mathsf{Re}_d} \qquad (2200 \leq \mathsf{Re}_d \leq 4000).$$

## 1.6-4. Turbulent Flow in a Plane Channel

Qualitatively, the picture of stabilized turbulent flow in a plane channel is similar to that in a circular tube. Indeed, in the viscous sublayer adjacent to the channel walls, the velocity distribution increases linearly with the distance from the wall: $\overline{V}(Y)/U_* = y^+$. In the logarithmic layer, the average velocity profile can be described by the expression [289]

$$\frac{\overline{V}(Y)}{U_*} = 2.5 \ln y^+ + 5.2.$$

In the flow core, the average velocity distribution in a plane channel of width $2h$ can be approximately described by formulas of the form [212, 289]

$$\frac{U_{\max} - \overline{V}(Y)}{U_*} = A\left(1 - \frac{Y}{h}\right)^m.$$

According to the book [398], $A = 6.5$ and $m = 1.9$.

In the region of stabilized turbulent flow, the drag coefficient can be determined from the implicit relation [289]

$$\frac{1}{\sqrt{\lambda}} = 0.86 \ln\left(\text{Re } \sqrt{\lambda}\right) - 0.35,$$

where $\lambda = 8\left(U_*/\langle \overline{V}\rangle\right)^2$ and $\text{Re} = 2h\langle\overline{V}\rangle/\nu$. This relation is in good agreement with experimental data for all $\text{Re} \leq 10^4$.

## 1.6-5. Drag Coefficient for Tubes of Other Shape

The drag coefficient $\lambda$ for turbulent flows in rectilinear tubes of noncircular cross-section can also be computed using relations (1.6.12) and (1.6.13), where the equivalent diameter

$$d_e = \frac{4S_*}{\mathcal{P}} \tag{1.6.14}$$

should be regarded as the characteristic length used to calculate the Reynolds number. In Eq. (1.6.14), $S_*$ is the cross-section area of the tube and $\mathcal{P}$ the cross-section perimeter. The values of the drag coefficient predicted by this approach fairly well agree with experimental data for tubes of rectangular and triangular cross-section [267, 427].

More detailed information about the structure of turbulent flows in a circular (noncircular) tube and a plane channel, as well as various relations for determining the average velocity profile and the drag coefficient, can be found in the books [138, 198, 268, 276, 289], which contain extensive literature surveys. Systematic data for rough tubes and results of studying fluctuating parameters of turbulent flow can also be found in the cited references.

# 1.7. Hydrodynamic Boundary Layer on a Flat Plate

## 1.7-1. Preliminary Remarks

In practice, one often deals with outside flows around stationary extended equipment elements such as plates, guiding elements, or tubes. In this case the action of external mass forces can often be neglected, and the hydrodynamic laws are determined by the relation between pressure, viscous, and inertial forces. Then the system of dimensionless steady-state hydrodynamic equations becomes

$$\nabla \cdot \mathbf{v} = 0,$$

$$(\mathbf{v} \cdot \nabla)\mathbf{v} = -\nabla p + \frac{1}{\mathsf{Re}} \Delta \mathbf{v}. \tag{1.7.1}$$

The system contains a single parameter, the Reynolds number. The solution in the general case is very complicated (the system is nonlinear), but it can be simplified if we consider the passage to the limit as $\mathsf{Re} \to 0$ or $\mathsf{Re} \to \infty$. In this section, we solve the problem on longitudinal flat-plate flow assuming that $\mathsf{Re} \to \infty$, that is, for a liquid with "vanishing viscosity." The solution is not straightforward: one cannot just disregard the term $\mathsf{Re}^{-1} \Delta \mathbf{v}$, thus obtaining the equation of an ideal fluid. The mathematical difficulty is that the small parameter $\mathsf{Re}^{-1}$ occurs in the term with higher derivatives. By neglecting this term, we change the order and the type of the equation, and the solution of the original system need not converge as $\mathsf{Re}^{-1} \to 0$ to the solution of the limit system with $\mathsf{Re}^{-1} = 0$. Here we deal with a singular perturbation [485]. Moreover, it is clear from physical considerations that an ideal fluid flow past a body cannot satisfy the no-slip condition on the surface. Actually, the tangential velocity varies from zero on the surface of the body to the velocity of the undisturbed flow remote from the body.

For fluids with low viscosity, this change of velocity occurs in a thin fluid layer adjacent to the surface of the body. Prandtl termed this layer a boundary layer. The magnitude of $\Delta \mathbf{v}$ is very large in this layer. Thus, although $\mathsf{Re}^{-1}$ is a small parameter, one cannot disregard the term $\mathsf{Re}^{-1} \Delta \mathbf{v}$ in the boundary layer. Nevertheless, since the longitudinal and the transverse coordinates play different roles in the boundary layer, we can simplify the equations. The corresponding formal estimate of terms in the second equation in (1.7.1) is described in the monographs [123, 270, 276, 427].

## 1.7-2. Laminar Boundary Layer

***Statement of the Blasius problem.*** We consider the steady-state problem on the longitudinal zero-pressure-gradient flow ($\nabla P \equiv 0$) past a half-infinite flat plate ($0 \le X < \infty$). We assume that the coordinates $X$ and $Y$ are directed along the plate and transverse to the plate, respectively, and the origin is placed at the front edge of the plate. The velocity of the incoming flow is $U_i$.

Let us write out the final system of boundary layer equations for an incompressible fluid

$$V_X \frac{\partial V_X}{\partial X} + V_Y \frac{\partial V_X}{\partial Y} = \nu \frac{\partial^2 V_X}{\partial Y^2},$$

$$\frac{\partial V_X}{\partial X} + \frac{\partial V_Y}{\partial Y} = 0. \tag{1.7.2}$$

Equations (1.7.2) are written in a dimensional form, since there is no natural characteristic length in the problem.

The natural boundary conditions read

$$\begin{aligned} Y &= 0, & V_X &= V_Y = 0, \\ Y &\to \infty, & V_X &\to U_i. \end{aligned} \tag{1.7.3}$$

**Exact solution. Friction coefficient.** Following Blasius [43], we express the fluid velocity components via the stream function $\Psi$ according to (1.1.6) and substitute them into (1.7.2). Then we seek the stream function in the form

$$\Psi(X, Y) = \sqrt{\nu X U_i}\, f(\eta), \qquad \eta = Y\sqrt{\frac{U_i}{\nu X}}, \tag{1.7.4}$$

where $\eta$ is a self-similar variable.

We obtain the following boundary value problem for the function $f(\eta)$:

$$\begin{aligned} 2f''' + ff'' &= 0; \\ \eta = 0, \qquad & f = f' = 0; \\ \eta \to \infty, \qquad & f' \to 1. \end{aligned} \tag{1.7.5}$$

Detailed tables containing the numerical solution of this problem can be found, for example, in [427].

By using (1.1.6), we can calculate the fluid velocity components as follows:

$$V_X = U_i f'(\eta), \qquad V_Y = \frac{1}{2}\sqrt{\frac{\nu U_i}{X}}\left[\eta f'(\eta) - f(\eta)\right]. \tag{1.7.6}$$

The obtained solution also allows us to calculate some variables that are of practical interest. For example, the local frictional stress on the wall is

$$\tau_w(X) = \mu \left(\frac{\partial V_X}{\partial Y}\right)_{Y=0} = \mu U_i \sqrt{\frac{U_i}{\nu X}} f''(0) = 0.332\,\mu U_i \sqrt{\frac{U_i}{\nu X}}, \tag{1.7.7}$$

and the local friction coefficient is given by

$$c_f(X) = \frac{\tau_w(X)}{\frac{1}{2}\rho U_i^2} = 0.664\sqrt{\frac{\nu}{U_i X}}. \tag{1.7.8}$$

The total friction coefficient for a plate of length $L$ is given by the formula

$$\langle c_f \rangle = \frac{1}{L} \int_0^L c_f(X)\, dX = \frac{1.328}{\sqrt{\mathsf{Re}_L}}, \tag{1.7.9}$$

where $\mathsf{Re}_L = U_i L / \nu$ is the Reynolds number for the plate in the flow.

Formula (1.7.8) is known as the Blasius law for the drag in longitudinal flat-plate flow. This formula can be used in laminar flow, that is, for $\mathsf{Re}_L < 3.5 \times 10^5$.

Although the boundary layer in this statement of the problem is asymptotic, that is, extends infinitely along the coordinate $Y$, one can approximately estimate its thickness if we adopt the convention that the velocity on the boundary of the layer differs from the undisturbed flow velocity at most by 1%.* Then the boundary layer thickness is

$$\delta(X) \approx 5\sqrt{\nu X / U_i}. \tag{1.7.10}$$

The Blasius solution shows that the longitudinal velocity profiles are affinely similar to each other for all cross-sections of the boundary layer.

Nikuradze [427] carried out thorough experiments to verify the Blasius theory. His results confirm that these conclusions are valid both for velocity profiles and for the friction coefficient.

***Reversed statement of the Blasius problem.*** In applications [72], one often deals with the "reversed" statement of the Blasius problem, in which a half-infinite plate moves in its plane at a velocity $U_i$. In this case, the boundary value problem (1.7.5) is replaced by

$$2f''' + ff'' = 0;$$
$$\eta = 0, \qquad f = 0, \quad f' = 1; \tag{1.7.11}$$
$$\eta \to \infty, \qquad f' \to 0,$$

This problem has also been solved numerically, and the function $f(\eta)$ is tabulated in [424]. We point out that in this case the solution differs from the corresponding Blasius solution. Thus, although physical consideration suggests that the "inversion" of flow is possible, the solution shows that it is impossible from the mathematical viewpoint. This is due to the fact that problems (1.7.5) and (1.7.11) are nonlinear.

In this case, in contrast with (1.7.7), the local friction stress on the wall is given by

$$\tau_w(X) = 0.444\, \mu U_i \sqrt{\frac{U_i}{\nu X}}. \tag{1.7.12}$$

---

* Along with the conventional thickness of the boundary layer, the following variables are also used: the plug flow depth $\delta_* = \int_0^\infty \left(1 - V_X / U_i\right)\, dY$ and the momentum loss depth $\delta_{**} = \int_0^\infty (V_X / U_i)\left(1 - V_X / U_i\right)\, dY$. The Blasius solutions implies $\delta_* = 1.7208\sqrt{\nu X / U_i}$ and $\delta_{**} = 0.664\sqrt{\nu X / U_i}$.

## 1.7-3. Turbulent Boundary Layer

***Structure of the flow. Velocity profile.*** The flow in the boundary layer on a flat plate is laminar until $\mathsf{Re}_X = U_i X / \nu \approx 3 \times 10^5$. On a longer plate, the boundary layer becomes turbulent, that is, its thickness increases sharply and the longitudinal velocity profile alters.

In accordance with Boussinesq's model the turbulence boundary layer on a flat plate is described by the equations

$$\overline{V}_X \frac{\partial \overline{V}_X}{\partial X} + \overline{V}_Y \frac{\partial \overline{V}_X}{\partial Y} = \frac{\partial}{\partial Y}\left[(\nu + \nu_t)\frac{\partial \overline{V}_X}{\partial Y}\right],$$

$$\frac{\partial \overline{V}_X}{\partial X} + \frac{\partial \overline{V}_Y}{\partial Y} = 0,$$

where $\nu_t$ is the turbulent viscosity.

These equations have formed the basis of most theoretical investigations on the determination of the average fluid velocity and the drag coefficient in the stabilized region of turbulent flow on a flat plate. In what follows, major attention will be paid to empirical and semiempirical formulas that approximate numerous experimental data quite well.

Experiments show [56, 212, 289, 427] that the turbulent boundary layer on a flat plate includes two qualitatively different regions, namely, the wall region (adjacent to the plate surface) and the outer region (bordering the unperturbed stream). By analogy with the flow through a circular tube, it is common to subdivide the thin wall region into three subdomains (von Karman's scheme):

| wall region | = | viscous sublayer | + | buffer layer | + | logarithmic layer |

The friction velocity $U_*$ and the dimensionless internal coordinate $y^+$ defined by relations (1.6.3) are introduced to describe the turbulent flow of fluid in the wall region. The transverse coordinate $Y$ in Eq. (1.6.3) is measures from the plate surface.

The velocity profile in the viscous sublayer is linear, $\overline{V}_X(Y)/U_* = y^+$.

In the logarithmic layer and the outer region, the average velocity profile is quite well described by the relations [289]

$$\frac{U_i - \overline{V}_X(Y)}{U_*} = \begin{cases} -2.5 \ln(Y/\delta) + 2.2 & \text{for } Y/\delta \leq 0.15, \\ 9.6\,[1 - (Y/\delta)]^2 & \text{for } 0.15 \leq Y/\delta \leq 1, \end{cases}$$

where $\delta = \delta(X)$ is the boundary layer thickness.

Note the Spalding implicit interpolation formula [455]

$$y^+ = 2.5\,V^+ + 0.135\left[\exp(-V^+) - 1 - V^+ - \tfrac{1}{2}(V^+)^2 - \tfrac{1}{6}(V^+)^3\right],$$

where $y^+ = Y U_*/\nu$ and $V^+ = 0.4\,\overline{V}(Y)/U_*$. This formula quite well describes the average velocity profile within the entire wall region $0 \leq Y/\delta \leq 0.15$.

According to [212, 260], the turbulent boundary layer thickness can be calculated from the formula

$$\delta = 0.33\, X \sqrt{\tfrac{1}{2} c_f}. \tag{1.7.13}$$

***The local coefficient of friction.*** In the case of two-sided flow past a flat plate, the local coefficient of friction $c_f = c_f(X)$ is expressed via other hydrodynamic parameters as

$$c_f = \frac{\tau_s}{\tfrac{1}{2}\rho U_i^2} = 2\left(\frac{U_*}{U_i}\right)^2.$$

For the turbulent boundary layer on a flat plate, von Karman's friction law with modified numerical coefficients [212, 289],

$$\frac{1}{\sqrt{c_f}} = 1.77 \ln\left(c_f \, \mathsf{Re}_X\right) + 2.4, \tag{1.7.14}$$

is typically used, which quite well agrees with experimental data. In Eq. (1.7.14), $\mathsf{Re}_X = U_i X/\nu$ is the local Reynolds number. Here it is assumed that the flow turbulization in the boundary layer starts from the front edge of the plate.

Relation (1.7.14) defines the friction coefficient versus the Reynolds number implicitly. For $\mathsf{Re}_X > 10^9$, it is more convenient to use the simpler explicit formula

$$c_f = 0.0262 \, \mathsf{Re}_X^{-1/7}, \tag{1.7.15}$$

suggested by Falkner [127, 289]. The corresponding mean coefficient of friction for a plate of length $L$ is given by

$$\langle c_f \rangle = \frac{1}{L} \int_0^L c_f \, dX = 0.0306 \, \mathsf{Re}_L^{-1/7}, \tag{1.7.16}$$

where $\mathsf{Re}_L = U_i L/\nu$.

The comparison of Eq. (1.7.16) with Eq. (1.7.9) reveals that the resistance of a flat plate to turbulent flow is much greater than to laminar flow and decreases with increasing Reynolds number considerably slower.

Note also Schlichting formulas [427]

$$c_f = \left(2 \log_{10} \mathsf{Re}_X - 0.65\right)^{-2.3}, \qquad \langle c_f \rangle = 0.455 \left(\log_{10} \mathsf{Re}_L\right)^{-2.58},$$

which are accurate within few percent for $10^5 \le \mathsf{Re}_L \le 10^9$.

More detailed information about the structure of turbulent flows on a flat plate, as well as various relations for determining the average velocity profile and the local coefficient of friction, can be found in the references [135, 138, 268, 276, 289, 427], which contain extensive literature surveys.

# 1.8. Gradient Boundary Layers

## 1.8-1. Equations and Boundary Conditions

The Blasius problem on longitudinal flow past a plate, considered in Subsection 1.7-2, may serve as an example in which the boundary layer equations are used under the assumption that the pressure is constant in the entire flow region. In the general case, this condition is not satisfied. For flow past bodies of arbitrary shape at high Reynolds numbers, one can only assume that the transverse pressure gradient in the boundary layer is zero. The longitudinal pressure gradient is determined either experimentally or by using relations that describe "effectively inviscid" flow (the term introduced by Batchelor [26]) outside the boundary layer. Since the thickness of the boundary layer is small, we assume in this case that the distortion of the shape of the body in flow and hence of the flow hydrodynamics is small, that is, the influence of the boundary layer on the outside flow is negligible.

The steady-state flow in the plane laminar boundary layer near the surface of a body of arbitrary shape is described by the system of equations [276, 427]

$$V_X \frac{\partial V_X}{\partial X} + V_Y \frac{\partial V_X}{\partial Y} = U \frac{dU}{dX} + \nu \frac{\partial^2 V_X}{\partial Y^2},$$
$$\frac{\partial V_X}{\partial X} + \frac{\partial V_Y}{\partial Y} = 0. \tag{1.8.1}$$

Here the $Y$-axis is directed along the normal to the surface $Y = 0$ of the body, and the $X$-axis is directed along the surface; $V_X$ and $V_Y$ are, respectively, the longitudinal and transverse fluid velocity components. The longitudinal component $U = U(X)$ of the outer velocity near the surface of the body is determined by the solution of a simpler auxiliary problem for a flow of an ideal fluid past the body (the model of an ideal fluid is used for the description of the flow outside the boundary layer at Re $\gg 1$).

To complete the statement of the problem, Eqs. (1.8.1) must be supplemented by the no-slip boundary conditions for the fluid velocity at the surface of the body,

$$V_X = V_Y = 0 \quad \text{at} \quad Y = 0, \tag{1.8.2}$$

and also by the condition

$$V_X \to U(X) \quad \text{as} \quad Y \to \infty \tag{1.8.3}$$

for the asymptotic matching of the longitudinal velocity component at the outer edge of the boundary layer with the fluid velocity in the flow core.

## 1.8-2. Boundary Layer on a V-Shaped Body

We shall now investigate a plane problem involving laminar flow past a V-shaped body (a wedge). In a potential flow of an ideal fluid past the front critical point

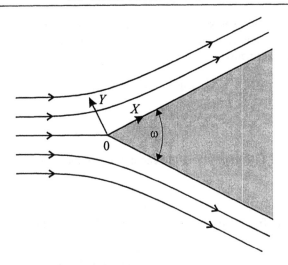

**Figure 1.6.** Schematic of a flow over a wedge with an angle of taper $\omega$

of the V-shaped body with an angle $\omega$ of taper (see Figure 1.6), the velocity close to the vertex is

$$U(X) = AX^m. \tag{1.8.4}$$

Here the $X$-axis is directed along the wedge surface, $A$ is a constant, and the exponent $m$ and angle $\omega$ are related by [427]

$$m = \frac{\omega}{2\pi - \omega}.$$

The steady-state flow in the plane boundary layer near the surface of a V-shaped body is described by system (1.8.1), where the $Y$-axis is directed along the normal to the wedge surface (given by $Y = 0$), $\rho$ is the fluid density, $V_X$ and $V_Y$ are, respectively, the longitudinal and transverse components of the fluid velocity, and $U = U(X)$ is defined by Eq. (1.8.4).

In the special case $m = 0$, problem (1.8.1)–(1.8.3) is reduced to problem (1.7.2)–(1.7.3) for the steady-state flow past a flat plate. The value $m = 1$, which corresponds to the angle $\omega = \pi$, characterizes the plane flow near the critical line.

The solution of problem (1.8.1)–(1.8.3) was found by Falkner and Skan [128] and is described in detail, for example, in [80, 276, 427]. The solution has the form

$$V_X = AX^m \Phi'(\xi), \qquad \xi = \sqrt{\tfrac{1}{2}(m + 1)A/\nu}\, X^{\frac{m-1}{2}} Y;$$

$$V_Y = -\sqrt{\tfrac{1}{2}(m + 1)A\nu}\, X^{\frac{m-1}{2}} \left[ \Phi(\xi) + \frac{m-1}{m+1}\xi\Phi'(\xi) \right], \tag{1.8.5}$$

where the primes denote the derivatives with respect to $\xi$.

Substituting the expressions (1.8.5) into system (1.8.1) and the boundary conditions (1.8.2) and (1.8.3) and using (1.8.4), we obtain the following problem for the unknown function $\Phi = \Phi(\xi)$:

$$\Phi''' + \Phi\Phi'' = b\left[(\Phi')^2 - 1\right],$$
$$\Phi = \Phi' = 0 \quad \text{at} \quad \xi = 0, \qquad (1.8.6)$$
$$\Phi' \to 1 \quad \text{as} \quad \xi \to \infty,$$

where the constant $b$ depends on the geometric parameter $m$ of the problem as follows:

$$b = \frac{2m}{m+1}. \qquad (1.8.7)$$

The viscous friction on the wedge surface is determined by

$$\tau_s = \mu \left.\frac{\partial V_X}{\partial Y}\right|_{Y=0} = \sqrt{\tfrac{1}{2}(m+1)A^3\nu\rho^2}\ \Phi''(0)X^{\frac{3m-1}{2}}, \qquad (1.8.8)$$

and the friction coefficient is

$$c_f = \frac{\tau_s}{\tfrac{1}{2}\rho U^2} = \sqrt{2(m+1)\nu/A}\ \Phi''(0)X^{-\frac{m+1}{2}}. \qquad (1.8.9)$$

To employ expressions (1.8.8) and (1.8.9), one must know the value of the second derivative of $\Phi$ on the surface $\xi = 0$.

To estimate the value of the second derivative, one can use the approximate formula [359]

$$\Phi''(0) = \sqrt{0.22 + 1.3b}, \quad b = 2m/(m+1). \qquad (1.8.10)$$

A comparison of the results based on formula (1.8.10) with the data of [276] obtained using a numerical solution of problem (1.8.6) was made in [359]. The error of formula (1.8.10) does not exceed 0.9 percent for $0 \le b < \infty$.

## 1.8-3. Qualitative Features of Boundary Layer Separation

*Preliminary remarks. Separation point.* Note that in the hydrodynamics of inviscid flow past a body of nonzero thickness it is assumed that there are regions near the body in which the flow accelerates from the front stagnation point to the midsection and decelerates behind the midsection. According to the Bernoulli theorem, a pressure counter-gradient arises in the deceleration region, which acts both in the outer flow and in the boundary layer. For the inviscid flow, the fluid particles store sufficiently much kinetic energy in the acceleration region to overcome this barrier, but in the frictional flow, the fluid particles that remain in the boundary layer cannot reach the region of higher pressure. They are pushed away from the wall, and an opposite flow arises downstream. This phenomenon

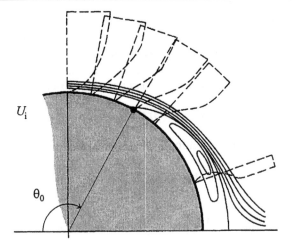

**Figure 1.7.** Flow separation and the onset of a recirculation region in flow past a cylinder

is known as the boundary layer separation. The separation point is determined by the following mathematical condition:

$$\frac{\partial V_X}{\partial Y} = 0 \qquad \text{at} \quad Y = 0. \qquad (1.8.11)$$

Here $X$ is the curvilinear coordinate along the surface of the body and $Y$ is the coordinate normal to the surface.

The boundary layer equations are valid only in the region between the front stagnation point and the separation point. Behind the separation point there is a wake region with absolutely different hydrodynamic laws. The position of the separation point can be determined either experimentally or by using numerical or approximate analytical methods.

The position of the separation point depends on the Reynolds number. At low Reynolds numbers, we have a flow without separation (see Figure 1.7). In a cross flow around a circular cylinder, the separation occurs if the Reynolds number (the cylinder diameter is taken as the characteristic length) is greater than 5 [486]. For $5 \leq \text{Re} \leq 40$, a separation region with steady-state symmetric adjacent vortices is formed (there is no boundary layer yet).

As Re increases, the adjacent vortices are elongated and shedding of vortices occurs (Karman's vortices are formed). Finally, for Re > 1000, the remote wake becomes completely turbulent [117]. At the same time, the separation point moves toward the midsection and even a bit farther upstream. For such values of Re, we can speak about a pronounced boundary layer. In a large part of the boundary layer, the flow remains laminar [486]. Strong turbulence within the boundary layer occurs for considerably higher Reynolds numbers ($\text{Re} \approx 2 \times 10^5$), at which the cylinder drag drops rapidly [117]. This phenomenon is called the drag crisis.

In what follows we consider the laminar boundary layer in a cross flow past a circular cylinder. This problem is also of practical importance, since tube elements are widely used in industrial equipment.

**Cross flow past a cylinder.** The problem on a cross flow past a cylinder comprises Eqs. (1.8.1) and boundary conditions (1.8.2)–(1.8.3). For a circular cylinder, we have [427]

$$U(X) = 2U_i \sin(X/a), \qquad (1.8.12)$$

where $U_i$ is the velocity of uniform flow remote from the body and $a$ is the cylinder radius.

Near the front point of the body ($X = 0$) one can use the power series expansion

$$U(X) = 2U_i\left(x - \tfrac{1}{3!}x^3 + \tfrac{1}{5!}x^5 - \tfrac{1}{7!}x^7 + \cdots\right), \qquad x = X/a. \qquad (1.8.13)$$

In the range $10^3 < \mathrm{Re} < 10^5$ (here the radius $a$ of the circular cylinder is taken as the characteristic length, $\mathrm{Re} = aU_i/\nu$), the laminar boundary layer approximation is valid, and the separation point $\theta_0$ moves from 109° to 85° [427]. In this case, retaining only the first three terms in the expansion (1.8.13) provides fairly good accuracy.

To integrate system (1.8.1) with the boundary conditions (1.8.2)–(1.8.3), we use the expansion (1.8.13) and represent the fluid velocity via the stream function (1.1.6) in the form of a series (the Blasius method),

$$\Psi = \sqrt{2\nu aU_i}\left[x f_1(y) + x^3 f_2(y) + x^5 f_3(y) + \cdots\right], \qquad y = \sqrt{2U_i/(\nu a)}\,Y, \qquad (1.8.14)$$

where $f_1(y)$, $f_2(y)$, $f_3(y)$, ... are unknown functions.

To find the unknown functions, one writes out a system of linked boundary value problems for ordinary differential equations. The numerical solution of these problems for the first six terms of the expansion (1.8.14) is tabulated in detail in [427]. The corresponding analytical expressions for the velocity components in the boundary layer can be calculated by formulas (1.1.6).

This permits the local friction coefficient at the cylinder wall to be calculated as

$$c_f(x) = \sqrt{\frac{\nu}{aU_i}}\left(6.973\,x - 2.732\,x^3 + 0.292\,x^5 - 0.0183\,x^7 + \cdots\right),$$

which is valid until the separation point.

The position of the separation point can be obtained theoretically by using condition (1.8.11). The calculation with regard to the first six terms of the expansions (1.8.13) and (1.8.14) implies $\theta_0 = 108.8°$. This is an estimate of the limit position of the separation point, since the influence of the unsteady wake behind the cylinder on the separation point is not taken into account. The experimental values of $\theta_0$ obtained are smaller and depend on the Reynolds number.

Numerous data about the hydrodynamic characteristics for bodies of unfavorable streamline shape are given in the monograph [117]. Note that the results given by the theory of laminar boundary layer are valid only for the Reynolds number in the range from several hundreds to $10^5$.

Sufficient information about the drag coefficient of the circular cylinder in a wide range of the Reynolds numbers can be found in Section 2.7.

## 1.9. Transient and Pulsating Flows

### 1.9-1. Transient or Oscillatory Motion of an Infinite Flat Plate

Let us consider a semi-infinite fluid bounded by a by a rigid plane $-\infty < X < \infty$, $Y \geq 0$. Two exact solutions are known for the transient flow near a plate [427]. These solutions correspond to rather simple flows governed by linear equations of motion. However, in these cases the Navier–Stokes equations are linearized because the nonlinear convective terms are identically zero ($V_X \partial V_X / \partial X \equiv 0$) rather than because we neglect these terms. The equation of motion has the form

$$\frac{\partial V_X}{\partial t} - \nu \frac{\partial^2 V_X}{\partial Y^2} = 0. \tag{1.9.1}$$

*Transient motion of a flat plate.* One of these problems (known as Stokes' first problem) describes the flow near an infinite flat plate instantaneously set in motion at a velocity $U_0$ in the plate plane. In this case, the initial and boundary conditions for Eq. (1.9.1) are written as follows:

$$V_X = 0 \text{ at } t = 0, \quad V_X = U_0 \text{ at } Y = 0, \quad V_X = 0 \text{ as } Y \to \infty. \tag{1.9.2}$$

The solution of problem (1.9.1), (1.9.2) is given by the formula

$$V_X(t, Y) = U_0 \operatorname{erfc}\left(\frac{Y}{2\sqrt{\nu t}}\right), \tag{1.9.3}$$

where $\operatorname{erfc} z = 1 - \frac{2}{\pi} \int_0^z \exp(-x^2)\, dx$ is the complementary error function.

The problem admits a generalization to the case in which the plate velocity is a given function of time. Then instead of (1.9.2) the following conditions are used:

$$V_X = 0 \text{ at } t = 0, \quad V_X = U(t) \text{ at } Y = 0, \quad V_X = 0 \text{ as } Y \to \infty. \tag{1.9.4}$$

The solution of problem (1.9.1), (1.9.4) is given by

$$V_X(t, Y) = \frac{Y}{2\sqrt{\pi t}} \int_0^t U(\tau) \exp\left[-\frac{Y^2}{4\nu(t-\tau)}\right] \frac{d\tau}{(t-\tau)^{3/2}}; \tag{1.9.5}$$

it was obtained in [79] with the help of the Duhamel theorem. By introducing the new integration variable

$$\eta = \frac{Y}{2\sqrt{\nu(t-\tau)}}, \tag{1.9.6}$$

one obtains another form of the solution,

$$V_X(t, Y) = \frac{2}{\sqrt{\pi}} \int_{\frac{Y}{2\sqrt{\nu t}}}^{\infty} U\left(t - \frac{Y^2}{4\nu\eta^2}\right) \exp(-\eta^2)\, d\eta. \tag{1.9.7}$$

We readily see that $V_X(t, Y) = U(t)$ for $Y = 0$, and for $U(t) = U_0$ the expressions (1.9.5) and (1.9.7) are reduced to (1.9.3).

*Oscillatory motion of a flat plate.* Another nonstationary problem that admits an exact solution [427] (known as Stokes' second problem) describes the flow near an infinite plate oscillating in its plane. This is a problem without initial data. In this case the boundary conditions are posed as follows:

$$V_X = U_0 \cos \omega t \quad \text{at} \quad Y = 0, \qquad V_X = 0 \quad \text{as} \quad Y \to \infty. \tag{1.9.8}$$

The solution of problem (1.9.1), (1.9.8) has the form

$$V_X(t, Y) = U_0 \exp\left(-Y\sqrt{\frac{\omega}{2\nu}}\right) \cos\left(\omega t - Y\sqrt{\frac{\omega}{2\nu}}\right). \tag{1.9.9}$$

This expression shows that the fluid oscillations decrease in amplitude and gradually lag in phase with increasing distance from the plate.

Let us consider the problem, similar to Stokes' first problem, in which a stagnant viscous incompressible fluid occupying the half-space $Y > 0$ is set in motion for $t > 0$ by a constant tangential stress $\tau_0$ acting on the fluid surface (the simplest model of flow in the near-surface layer of water under the action of wind) [140]. In this case, the initial and boundary conditions for Eq. (1.9.1) are written as follows:

$$V_X = 0 \text{ at } t = 0, \quad \mu\frac{\partial V_X}{\partial Y} = \tau_0 \text{ at } Y = 0, \quad V_X = 0 \text{ as } Y \to \infty. \tag{1.9.10}$$

The solution of problem (1.9.1), (1.9.10) is given by the formula

$$V_X = \frac{2\tau_0}{\mu}\sqrt{\frac{\nu t}{\pi}} [\sqrt{\pi}\, \xi \, \text{erfc}\, \xi + \exp(-\xi^2)], \qquad \xi = \frac{Y}{2\sqrt{\nu t}}. \tag{1.9.11}$$

Note that on the free boundary $(Y = 0)$ the expression in the square brackets in (1.9.11) is equal to unity.

## 1.9-2. Transient or Pulsating Flows in Tubes

*Transient flow in tubes under instantaneously applied pressure gradient.* The problem on a transient laminar viscous flow in an infinite circular tube under a constant pressure gradient instantaneously applied at $t = 0$ is considered in [449]. The equation of motion in the cylindrical coordinates has the form

$$\frac{1}{\nu}\frac{\partial V_Z}{\partial t} = \frac{1}{\mu}f(t) + \frac{\partial^2 V_Z}{\partial R^2} + \frac{1}{R}\frac{\partial V_Z}{\partial R}, \tag{1.9.12}$$

$$f(t) = \begin{cases} 0 & \text{for } t \le 0, \\ \dfrac{\Delta P}{L} & \text{for } t > 0, \end{cases} \tag{1.9.13}$$

where $\Delta P$ is the pressure drop on a part of length $L$ of the tube. In this case the initial and boundary conditions can be written as follows:

$$V_Z = 0 \quad \text{at } t = 0, \qquad V_Z = 0 \quad \text{at } R = a. \tag{1.9.14}$$

The solution has the form [449]

$$V_Z(t, R) = \frac{a^2}{4\mu}\frac{\Delta P}{L}\left[1 - \frac{R^2}{a^2} - 8\sum_{k=1}^{\infty}\frac{J_0\left(\lambda_k\frac{R}{a}\right)}{\lambda_k^3 J_1(\lambda_k)}\exp\left(-\frac{\nu\lambda_k^2 t}{a^2}\right)\right], \tag{1.9.15}$$

where the $\lambda_k$ are the roots of the equation

$$J_0(\lambda_k) = 0. \tag{1.9.16}$$

Here $J_0$ and $J_1$ are the Bessel functions of the first kind.
The volume rate of flow is given by the formula

$$Q(t) = \frac{\pi a^4}{8\mu}\frac{\Delta P}{L}\left[1 - 32\sum_{k=1}^{\infty}\frac{\exp(-\nu\lambda_k^2 t/a^2)}{\lambda_k^4}\right], \tag{1.9.17}$$

which turns as $t \to \infty$ into formula (1.5.11) for the volume rate of flow in a steady-state tube flow.
A similar problem for an annular channel with a homogeneous initial velocity field was solved in [296]. For $t > 0$ and $b \le R \le a$, we consider the equation of motion in cylindrical coordinates

$$\frac{1}{\nu}\frac{\partial V_Z}{\partial t} = \frac{1}{\mu}\frac{\Delta P}{L} + \frac{\partial^2 V_Z}{\partial R^2} + \frac{1}{R}\frac{\partial V_Z}{\partial R} \tag{1.9.18}$$

with the initial and boundary conditions

$$V_Z = U = \text{const} \quad \text{at } t = 0, \tag{1.9.19}$$

$$V_Z = 0 \quad \text{at } R = a, \qquad V_Z = 0 \quad \text{at } R = b. \tag{1.9.20}$$

The solution is

$$V_Z(t, \mathcal{R}) = \frac{2u}{a^2 \varepsilon} \left[ a^2 - \mathcal{R}^2 - \frac{(a^2 - b^2) \ln \frac{a}{\mathcal{R}}}{\ln \frac{a}{b}} \right]$$

$$+ 2 \sum_{k=1}^{\infty} \left( U - \frac{8u}{\lambda_k^2 \varepsilon} \right) \frac{\mathcal{Z}_0 \left( \lambda_k \frac{\mathcal{R}}{a} \right) \exp \left( -t \lambda_k^2 \frac{\nu}{a^2} \right)}{\lambda_k \mathcal{Z}_1(\lambda_k) + q \lambda_k \mathcal{Z}_1(q \lambda_k)}, \qquad (1.9.21)$$

where

$$q = \frac{b}{a}, \quad \varepsilon = 1 + q^2 + \frac{1 - q^2}{\ln q}, \quad u = \frac{a^2}{8\mu} \frac{\Delta P}{L} \varepsilon,$$

$$\mathcal{Z}_n(\lambda) = \frac{J_n \left( \lambda \frac{\mathcal{R}}{a} \right) Y_0(q\lambda) - Y_n \left( \lambda \frac{\mathcal{R}}{a} \right) J_0(q\lambda)}{J_0(q\lambda) Y_0(q\lambda)}. \qquad (1.9.22)$$

Here $J_n(\lambda)$ and $Y_n(\lambda)$ are the Bessel functions of the first and second kind, respectively, [261], and the $\lambda_k$ are the roots of the transcendental equation

$$J_0(\lambda) Y_0(q\lambda) - J_0(q\lambda) Y_0(\lambda) = 0. \qquad (1.9.23)$$

If $b = 0$ and $U = 0$, then the solution (1.9.21) turns into the expression (1.9.15) for a circular tube, note also that the solution tends to the steady-state solution (1.5.14) (with $a_1 = b$ and $a_2 = a$) for annular tubes as $t \to \infty$.

*Pulsating laminar flow in a circular tube.* Let us give an exact solution of yet another problem without initial conditions. Consider the problem about a laminar viscous flow in an infinite circular tube with the axial pressure gradient varying according to a harmonic law. Since the problem is axisymmetric, in the cylindrical coordinates $\mathcal{R}$, $Z$ it can be represented in the form

$$\frac{\partial V_Z}{\partial t} - \nu \left( \frac{\partial^2 V_Z}{\partial \mathcal{R}^2} + \frac{1}{\mathcal{R}} \frac{\partial V_Z}{\partial \mathcal{R}} \right) = A \cos \omega t \qquad (1.9.24)$$

with the boundary condition

$$V_Z = 0 \quad \text{at} \quad \mathcal{R} = a, \qquad (1.9.25)$$

where $A$ is the amplitude and $\omega$ is the frequency of oscillations of $-\frac{1}{\rho} \frac{\partial P}{\partial Z}$.

The asymptotic solutions of problem (1.9.24), (1.9.25) at small and large frequencies are presented in [427]. For $a\sqrt{\omega/\nu} \ll 1$ we have

$$V_Z(t, \mathcal{R}) = \frac{A}{4\nu} (a^2 - \mathcal{R}^2) \cos \omega t. \qquad (1.9.26)$$

Hence, at small frequencies, the pressure and the velocity oscillate in phase, and the velocity amplitude is distributed along the tube radius according to the same parabolic law as the velocity in steady-state flow.

For $a\sqrt{\omega/\nu} \gg 1$ we have

$$V_Z(t, \mathcal{R}) = \frac{A}{\omega}\left[\sin\omega t - \sqrt{\frac{a}{\mathcal{R}}}e^{-\zeta}\sin(\omega t - \zeta)\right], \quad \zeta = \sqrt{\frac{\omega}{2\nu}}\,(a - \mathcal{R}). \quad (1.9.27)$$

It follows from this expression that near the tube axis the fluid and pressure oscillations are opposite in phase, and the viscosity effects are noticeable only in the near-wall boundary layer with characteristic size $\sqrt{\nu/\omega}$. By calculating the time-averaged square of the velocity $\overline{V_Z^2}$, one can see that this variable attains its maximum value at the distance $3.22\sqrt{\nu/\omega}$ from the wall. This is just the annular effect experimentally discovered in [400].

The exact solution of problem (1.9.24), (1.9.25) has the form [276]

$$V_Z(t, \mathcal{R}) = \frac{A}{\omega}\left\{\left[1 - \frac{\mathrm{bei}(a/b)\mathrm{bei}(\mathcal{R}/b) - \mathrm{ber}(a/b)\mathrm{ber}(\mathcal{R}/b)}{\mathrm{bei}^2(a/b) + \mathrm{ber}^2(a/b)}\right]\sin\omega t \right.$$
$$\left. + \frac{\mathrm{bei}(a/b)\mathrm{ber}(\mathcal{R}/b) - \mathrm{ber}(a/b)\mathrm{bei}(\mathcal{R}/b)}{\mathrm{bei}^2(a/b) + \mathrm{ber}^2(a/b)}\cos\omega t\right\}, \quad b = \sqrt{\frac{\nu}{\omega}}.$$
$$(1.9.28)$$

The Kelvin functions $\mathrm{ber}\,(x) = \mathrm{Re}\,J_0(x\sqrt{i}\,)$ and $\mathrm{bei}\,(x) = \mathrm{Im}\,J_0(x\sqrt{i}\,)$, where $J_0$ is the Bessel function of the first kind and order zero, are tabulated, for example, in [202].

### 1.9-3. Transient Rotational Fluid Motion

***Statement of the problem.*** Two problems on the transient rotational motion of an incompressible viscous fluid around a cylindrical surface which suddenly begins to rotate are also considered in [449]. In this case only the velocity component $V_\varphi$ is nonzero. This variable satisfies the equation

$$\frac{\partial V_\varphi}{\partial t} = \nu\left(\frac{\partial^2 V_\varphi}{\partial \mathcal{R}^2} + \frac{1}{\mathcal{R}}\frac{\partial V_\varphi}{\partial \mathcal{R}} - \frac{V_\varphi}{\mathcal{R}^2}\right) \qquad (1.9.29)$$

with the initial and boundary conditions

$$V_\varphi = 0 \qquad \text{at} \quad t = 0, \qquad (1.9.30)$$
$$V_\varphi = \omega a \qquad \text{at} \quad \mathcal{R} = a. \qquad (1.9.31)$$

***Fluid motion inside a hollow rotating cylinder.*** The first problem corresponds to the motion of a fluid inside the cylinder (in the region $\mathcal{R} \leq a$). In this case the solution has the form

$$V_\varphi(t, \mathcal{R}) = \omega a\left[\frac{\mathcal{R}}{a} + 2\sum_{k=1}^{\infty}\frac{J_1(\lambda_k \mathcal{R}/a)}{\lambda_k J_1'(\lambda_k)}\exp\left(-t\lambda_k^2\frac{\nu}{a^2}\right)\right], \qquad (1.9.32)$$

where the $\lambda_k$ are the real roots of the transcendental equation

$$J_1(\lambda) = 0. \tag{1.9.33}$$

Here $J_1(\lambda)$ is the Bessel functions of the first kind and order one.

Using (1.9.32), one can calculate the tangential stresses

$$\tau_w = \mu \left( \frac{\partial V_\varphi}{\partial \mathcal{R}} - \frac{V_\varphi}{\mathcal{R}} \right)_{\mathcal{R}=a} = 2\mu\omega \sum_{k=1}^{\infty} \exp\left( -t\lambda_k^2 \frac{\nu}{a^2} \right) \tag{1.9.34}$$

on the surface of the rotating cylinder and the angular momentum per unit length needed to maintain rotation at a constant angular velocity $\omega$:

$$M = 4\pi\mu\omega a^2 \sum_{k=1}^{\infty} \exp\left( -t\lambda_k^2 \frac{\nu}{a^2} \right). \tag{1.9.35}$$

***Fluid motion outside a rotating cylinder.*** The second problem corresponds to the rotation of a cylinder in an infinite fluid. In this case Eq. (1.9.29) must be considered in the region $\mathcal{R} \geq a$. The boundary condition (1.9.31) is supplemented by the condition that the velocity decays at infinity,

$$V_\varphi \to 0 \qquad \text{as} \quad \mathcal{R} \to \infty. \tag{1.9.36}$$

The solution of the problem has the form

$$V_\varphi(t, \mathcal{R}) = \frac{\omega a^2}{\mathcal{R}} + \frac{2\omega a}{\pi} \int_0^\infty \exp(-\nu\beta^2 t)F(\mathcal{R}, \beta)\,d\beta,$$
$$F(\mathcal{R}, \beta) = \frac{J_1(\mathcal{R}\beta)Y_1(a\beta) - J_1(a\beta)Y_1(\mathcal{R}\beta)}{\beta[J_1^2(a\beta) + Y_1^2(a\beta)]}. \tag{1.9.37}$$

It should be noted that the spectrum of eigenvalues of this problem is continuous, $0 \leq \beta < \infty$.

The tangential stresses on the cylinder wall are

$$\tau_w = -2\mu\omega \left[ 1 - \frac{a}{\pi} \int_0^\infty \exp(-\nu\beta^2 t)G(\beta)\,d\beta \right],$$
$$G(\beta) = \frac{J_1'(a\beta)Y_1(a\beta) - J_1(a\beta)Y_1'(a\beta)}{J_1^2(a\beta) + Y_1^2(a\beta)}. \tag{1.9.38}$$

To maintain the rotation of the cylinder at a constant angular velocity $\omega$ in infinite fluid, it is necessary to apply the following time-dependent angular momentum per unit length of the cylinder:

$$M = 4\pi\mu\omega a^2 \left[ 1 + \frac{2}{\pi^2} \int_0^\infty \frac{\exp(-\nu\beta^2 t)}{J_1^2(a\beta) + Y_1^2(a\beta)} \frac{d\beta}{\beta} \right]. \tag{1.9.39}$$

***Fluid motion between rotating cylinders.*** The problem about the transient motion of an initially stagnant fluid in the gap between two coaxial cylinders of radii $b$ and $a$ $(b < a)$ that suddenly begin to rotate at angular velocities $\omega_b$ and $\omega_a$ is considered in [40]. The only nonzero velocity component $V_\varphi$ satisfies Eq. (1.9.29) with the initial condition (1.9.30) and the following boundary conditions:

$$V_\varphi = b\omega_b \quad \text{at} \quad \mathcal{R} = b, \qquad V_\varphi = a\omega_a \quad \text{at} \quad \mathcal{R} = a. \qquad (1.9.40)$$

By introducing the notation

$$q = \frac{b}{a}, \qquad \omega = \frac{\omega_b}{\omega_b - \omega_a}, \qquad (1.9.41)$$

and solving problem (1.9.29), (1.9.30), (1.9.40) by separation of variables, we obtain the velocity field

$$\frac{V_\varphi(t, \mathcal{R})}{a(\omega_a - \omega_b)} = \frac{1 - \omega(1 - q^2)}{1 - q^2} \frac{\mathcal{R}}{a} - \frac{q^2}{1 - q^2} \frac{a}{\mathcal{R}}$$

$$- \pi \sum_{k=1}^{\infty} A_k \exp\left(-\frac{\nu \lambda_k^2 t}{a^2}\right) \left[ Y_1(q\lambda_k) J_1\left(\lambda_k \frac{\mathcal{R}}{a}\right) - J_1(q\lambda_k) Y_1\left(\lambda_k \frac{\mathcal{R}}{a}\right) \right],$$

$$A_k = J_1(\lambda_k) \frac{(1 - \omega) J_1(q\lambda_k) + q\omega J_1(\lambda_k)}{J_1^2(q\lambda_k) - J_1^2(\lambda_k)}, \qquad (1.9.42)$$

where $J_1(\lambda)$ and $Y_1(\lambda)$ are respectively the Bessel functions of the first and second kind [261] and the $\lambda_k$ are the real roots of the transcendental equation

$$J_1(\lambda) Y_1(q\lambda) - J_1(q\lambda) Y_1(\lambda) = 0; \qquad (1.9.43)$$

these roots are tabulated in [40].

For small values of $t$ we have the asymptotic expression

$$V_\varphi(t, \mathcal{R}) = \sqrt{\frac{a}{\mathcal{R}}} \left\{ b\omega_b \sqrt{q} \left[ \text{erfc}\left(\frac{\mathcal{R} - aq}{2\sqrt{\nu t}}\right) - \text{erfc}\left(\frac{2a - aq - \mathcal{R}}{2\sqrt{\nu t}}\right) \right] \right.$$

$$\left. + a\omega_a \left[ \text{erfc}\left(\frac{a - \mathcal{R}}{2\sqrt{\nu t}}\right) - \text{erfc}\left(\frac{a - aq + \mathcal{R}}{2\sqrt{\nu t}}\right) \right] \right\}. \qquad (1.9.44)$$

One can see that the conditions $V_\varphi(0, b) = b\omega_b$ and $V_\varphi(0, a) = a\omega_a$ are satisfied as $t \to 0$.

The first two terms on the right-hand side in (1.9.42) determine the steady-state solution obtained by passing to the limit as $t \to \infty$. This solution has the form [140]

$$V_\varphi = \frac{\omega_a a^2 - \omega_b b^2}{a^2 - b^2} \mathcal{R} + \frac{a^2 b^2 (\omega_b - \omega_a)}{a^2 - b^2} \frac{1}{\mathcal{R}}. \qquad (1.9.45)$$

The moments $M_1$ and $M_2$ per unit length of the inner and outer cylinders are

$$M_1 = -M_2 = -4\pi\mu\frac{a^2b^2(\omega_b - \omega_a)}{a^2 - b^2}. \tag{1.9.46}$$

In the limit cases, relation (1.9.45) is simplified drastically. For example, as $a \to \infty$ and $\omega_a = 0$, we have potential flow with velocity profile

$$V_\varphi = \frac{b^2\omega_b}{\mathcal{R}}.$$

For $b = 0$ and $\omega_b = 0$, the velocity field is described by the expression

$$V_\varphi = \omega_a\mathcal{R}.$$

The above steady-state problem can be generalized to the case in which the pressure gradient $\frac{\partial P}{\partial Z} = -\frac{\Delta P}{L}$ is constant along the flow axis [140]. In this case, the velocity $V_\varphi$ is given by (1.9.45). However, we also have the tangential velocity component

$$V_Z = -\frac{\Delta P}{4\mu L}\left[(a^2 - b^2)\frac{\ln(\mathcal{R}/b)}{\ln(a/b)} - \mathcal{R}^2 + b^2\right]. \tag{1.9.47}$$

If the gap between the cylinders is small, $\frac{a-b}{b} = \epsilon \ll 1$, then it is convenient to introduce a new variable $y$ by setting

$$\mathcal{R} = b(1 + \epsilon y).$$

In this case, the velocity components satisfy the relations

$$V_\varphi = b[\omega_b + (\omega_a - \omega_b)y], \qquad V_Z \approx -\frac{\epsilon^2 b^2 \Delta P}{2\mu L}y(1 - y), \tag{1.9.48}$$

the first of which coincides with the corresponding relation for plane-parallel Couette flow and the second, with that for plane-parallel Poiseuille flow.

# Chapter 2
# Motion of Particles, Drops, and Bubbles in Fluid

Solving the problem on the interaction of a solid particle, drop, or bubble with the surrounding continuous phase underlies the design and analysis of many technological processes. The industrial applications of such interaction include classification of suspensions in hydrocyclones, sedimentation of colloids, pneumatic conveyers, fluidization, heterogeneous catalysis in suspension, dissolving solid particles, extraction from drops, absorption, and evaporation into bubbles [69, 107, 111, 122, 137, 478, 505].

The description of a large variety of meteorological phenomena is also based on the analysis of motion of a collection of drops in air. The recent increase in atmosphere pollution is a serious problem, which requires understanding the transfer of mechanical, chemical, and radioactive particles in the atmosphere.

In rarefied systems of particles, drops, or bubbles, the particle-particle interaction can be neglected in the first approximation; then one deals with the behavior of a single particle moving in fluid. In this case, the streamline pattern depends on the particle shape, the flow type (translational or shear), and a number of other geometric factors.

One of the main methods for obtaining approximate closed-form solutions of the corresponding hydrodynamic problems is to linearize the Navier–Stokes equations for low Reynolds numbers. This method is often used in this chapter when we study the motion of small particles, drops, and bubbles in a fluid.

A number of physical and mathematical statements of problems on motion of particles, drops, and bubbles in fluid, as well as various formulas and experimental data, can be found, for example, in [94, 179, 183, 255, 270, 287, 517].

## 2.1. Exact Solutions of the Stokes Equations

### 2.1-1. Stokes Equations

One of the main approaches to the analysis and simplification of the Navier–Stokes equations is as follows. One assumes that the nonlinear inertia term $(\mathbf{V} \cdot \nabla)\mathbf{V}$ is small compared with the linear viscous term $\nu \Delta \mathbf{V}$ and hence can be neglected altogether or taken into account in some special way. This method is well-founded for $\mathrm{Re} = LU/\nu \ll 1$ and is widely used for studying the motion of particles, drops, and bubbles in fluids. Low Reynolds numbers are typical of the following three cases: slow (creeping) flows, highly viscous fluids, and small dimensions of particles.

For steady-state flows of viscous incompressible fluid, by neglecting the inertia terms in (1.1.4) and by including all conservative mass forces in the pressure $P$, we arrive at the Stokes equations [464]

$$\nabla \cdot \mathbf{V} = 0,$$
$$\mu \Delta \mathbf{V} = \nabla P. \tag{2.1.1}$$

The Stokes equations (2.1.1) are linear and hence much simpler than the nonlinear Navier–Stokes equations. For any two solutions $\{\mathbf{V}_1, P_1\}$ and $\{\mathbf{V}_2, P_2\}$ of (2.1.1), the sum $\{\alpha \mathbf{V}_1 + \beta \mathbf{V}_2, \alpha P_1 + \beta P_2\}$ satisfies the same equations for any constants $\alpha$ and $\beta$.

In axisymmetric problems, all the variables in the spherical coordinates $R$, $\theta$, $\varphi$ are independent of $\varphi$, and the third component of the fluid velocity is zero, $V_\varphi = 0$. The Stokes equations (2.1.1) in the spherical coordinates have the form

$$\frac{1}{R}\frac{\partial}{\partial R}(R^2 V_R) + \frac{1}{\sin\theta}\frac{\partial}{\partial \theta}(V_\theta \sin\theta) = 0,$$

$$\mu\left(\Delta V_R - \frac{2V_R}{R^2} - \frac{2}{R^2}\frac{\partial V_\theta}{\partial \theta} - \frac{2V_\theta \cot\theta}{R^2}\right) = \frac{\partial P}{\partial R}, \tag{2.1.2}$$

$$\mu\left(\Delta V_\theta + \frac{2}{R^2}\frac{\partial V_R}{\partial \theta} - \frac{V_\theta}{R^2\sin^2\theta}\right) = \frac{1}{R}\frac{\partial P}{\partial \theta},$$

where

$$\Delta \equiv \frac{1}{R^2}\frac{\partial}{\partial R}\left(R^2\frac{\partial}{\partial R}\right) + \frac{1}{R^2\sin\theta}\frac{\partial}{\partial \theta}\left(\sin\theta\frac{\partial}{\partial \theta}\right).$$

The fluid velocity components can be expressed via the stream function $\Psi$ as follows:

$$V_R = \frac{1}{R^2\sin\theta}\frac{\partial \Psi}{\partial \theta}, \quad V_\theta = -\frac{1}{R\sin\theta}\frac{\partial \Psi}{\partial R}. \tag{2.1.3}$$

Then the first equation in (2.1.2) (the continuity equation) is satisfied automatically. Let us substitute the expressions (2.1.3) into the second and the third equations in (2.1.2). On eliminating the terms containing the pressure, we obtain the following equation for the stream function:

$$E^2(E^2\Psi) = 0, \quad E^2 \equiv \frac{\partial^2}{\partial R^2} + \frac{\sin\theta}{R^2}\frac{\partial}{\partial \theta}\left(\frac{1}{\sin\theta}\frac{\partial}{\partial \theta}\right). \tag{2.1.4}$$

## 2.1-2. General Solution for the Axisymmetric Case

The general solution of Eq. (2.1.4) has the form [179]

$$\Psi(R,\theta) = \sum_{n=0}^{\infty}\left(A_n R^n + B_n R^{1-n} + C_n R^{n+2} + D_n R^{3-n}\right)\mathcal{J}_n(\cos\theta)$$

$$+ \sum_{n=2}^{\infty}\left(\tilde{A}_n R^n + \tilde{B}_n R^{1-n} + \tilde{C}_n R^{n+2} + \tilde{D}_n R^{3-n}\right)\mathcal{H}_n(\cos\theta), \tag{2.1.5}$$

where $A_n$, $B_n$, $C_n$, $D_n$, $\tilde{A}_n$, $\tilde{B}_n$, $\tilde{C}_n$, and $\tilde{D}_n$ are arbitrary constants and $\mathcal{J}_n(\zeta)$ and $\mathcal{H}_n(\zeta)$ are the Gegenbauer functions of the first and the second kind, respectively, which can be linearly expressed via the Legendre functions $P_n(\zeta)$ and $Q_n(\zeta)$ as follows:

$$\mathcal{J}_n(\zeta) = \frac{P_{n-2}(\zeta) - P_n(\zeta)}{2n - 1}, \quad \mathcal{H}_n(\zeta) = \frac{Q_{n-2}(\zeta) - Q_n(\zeta)}{2n - 1} \quad (n \geq 2).$$

The Gegenbauer functions of the first kind are polynomials and can be represented in the form

$$\mathcal{J}_n(\zeta) = -\frac{1}{(n-1)!}\left(\frac{d}{d\zeta}\right)^{n-2}\left(\frac{\zeta^2 - 1}{2}\right)^{n-1}$$

$$= \frac{1 \cdot 3 \ldots (2n-3)}{1 \cdot 2 \ldots n}\left[\zeta^n - \frac{n(n-1)}{2(2n-3)}\zeta^{n-2} + \frac{n(n-1)(n-2)(n-3)}{2 \cdot 4(2n-3)(2n-5)}\zeta^{n-4} - \cdots\right].$$

In particular,

$$\mathcal{J}_0(\zeta) = 1, \quad \mathcal{J}_1(\zeta) = -\zeta, \quad \mathcal{J}_2(\zeta) = \tfrac{1}{2}(1 - \zeta^2), \quad \mathcal{J}_3(\zeta) = \tfrac{1}{2}\zeta(1 - \zeta^2),$$
$$\mathcal{J}_4(\zeta) = \tfrac{1}{8}(1 - \zeta^2)(5\zeta^2 - 1), \quad \mathcal{J}_5(\zeta) = \tfrac{1}{8}\zeta(1 - \zeta^2)(7\zeta^2 - 3).$$

The Gegenbauer functions of the second kind are given by the formulas

$$\mathcal{H}_0(\zeta) = -\zeta, \quad \mathcal{H}_1(\zeta) = -1,$$
$$\mathcal{H}_n(\zeta) = \frac{1}{2}\mathcal{J}_n(\zeta)\ln\frac{1+\zeta}{1-\zeta} + \mathcal{K}_n(\zeta), \quad n \geq 2,$$

where $\mathcal{K}_n(\zeta)$ has the following expression via the Gegenbauer functions of the first kind:

$$\mathcal{K}_n(\zeta) = -\sum_{k}^{\frac{1}{2}n \leq k \leq \frac{1}{2}n + \frac{1}{2}} \frac{(2n - 4k + 1)}{(2k - 1)(n - k)}\left[1 - \frac{(2k - 1)(n - k)}{n(n - 1)}\right]\mathcal{J}_{n-2k+1}(\zeta);$$

here the series starts from $\mathcal{J}_0$ or $\mathcal{J}_1$ depending on whether $n$ is odd or even. In particular,

$$\mathcal{K}_2(\zeta) = \tfrac{1}{2}\zeta, \quad \mathcal{K}_3(\zeta) = \tfrac{1}{6}(3\zeta^2 - 2),$$
$$\mathcal{K}_4(\zeta) = \tfrac{1}{24}\zeta(15\zeta^2 - 13), \quad \mathcal{K}_5(\zeta) = \tfrac{1}{120}(105\zeta^4 - 115\zeta^2 + 16).$$

For $n \geq 2$, the Gegenbauer functions of the second kind are infinite at the points $\zeta = \pm 1$, which correspond to $\theta = 0$ and $\theta = \pi$. Therefore, if there are no singularities in the physical statement of the problem, then all the constants with tildes in (2.1.5) must be zero. Moreover, for $n = 0$ and $n = 1$, the remaining constants, if nonzero, result in infinite values of the tangential velocity $V_\theta$ on the flow axis. Therefore, in a majority of problems on Stokes flow past particles, drops, and bubbles, the stream function in the spherical coordinates has the form

$$\Psi(R, \theta) = \sum_{n=2}^{\infty} \left( A_n R^n + B_n R^{1-n} + C_n R^{n+2} + D_n R^{3-n} \right) \mathcal{J}_n(\cos \theta). \quad (2.1.6)$$

The corresponding fluid velocity components and pressure are given by the formulas

$$V_R = -\sum_{n=2}^{\infty} \left( A_n R^{n-2} + B_n R^{-n-1} + C_n R^n + D_n R^{1-n} \right) P_{n-1}(\cos \theta),$$

$$V_\theta = \sum_{n=2}^{\infty} \left[ n A_n R^{n-2} - (n-1) B_n R^{-n-1} \right.$$

$$\left. + (n+2) C_n R^n - (n-3) D_n R^{1-n} \right] \frac{\mathcal{J}_n(\cos \theta)}{\sin \theta},$$

$$P = -2\mu \sum_{n=2}^{\infty} \left( \frac{2n+1}{n-1} C_n R^{n-1} + \frac{2n-3}{n} D_n R^{-n} \right) P_{n-1}(\cos \theta) + \text{const}.$$

$$(2.1.7)$$

In this case, the force exerted by the fluid on any spherical boundary described by the equation $R = \text{const}$ is given by

$$F = 4\pi\mu D_2. \quad (2.1.8)$$

It is of interest that this force is completely determined by only one coefficient of the series (2.1.6).

Formulas (2.1.6)–(2.1.8) provide a basis for solving a wide class of problems in chemical hydrodynamics.

## 2.2. Spherical Particles, Drops, and Bubbles in Translational Stokes Flow

### 2.2-1. Flow Past a Spherical Particle

Let us consider a solid spherical particle of radius $a$ in a translational Stokes flow with velocity $U_i$ and dynamic viscosity $\mu$ (Figure 2.1). We assume that the fluid has a dynamic viscosity $\mu$. We use the spherical coordinate system $R$, $\theta$, $\varphi$ with origin at the center of the particle and with angle $\theta$ measured from the direction of the incoming flow (that is, from the rear stagnation point on the particle surface). In view of the axial symmetry, only two components of the fluid velocity, namely, $V_R$ and $V_\theta$, are nonzero, and all the unknowns are independent of the third coordinate $\varphi$.

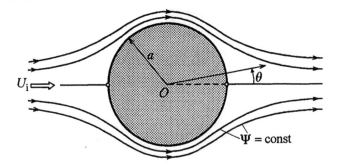

**Figure 2.1.** Translational Stokes flow past a spherical particle

The fluid velocity distribution is given by the Stokes equations (2.1.1) with the no-slip boundary conditions on the surface of the solid sphere,

$$V_R = V_\theta = 0 \qquad \text{at} \quad R = a, \tag{2.2.1}$$

and the boundary conditions at infinity of the form

$$V_R \to U_i \cos \theta, \quad V_\theta \to -U_i \sin \theta \qquad \text{as} \quad R \to \infty, \tag{2.2.2}$$

which correspond to condition (1.1.14) of translational nonperturbed flow remote from the particle.

By passing from the fluid velocity components $V_R$, $V_\theta$ to the stream function $\Psi$ according to formulas (2.1.3), we arrive at Eq. (2.1.4). It follows from the remote boundary conditions (2.2.2) that in the general solution (2.1.5) it suffices to retain only the first term (corresponding to the case $n = 2$). The no-slip conditions (2.2.1) allow us to find the unknown constants $A_2$, $B_2$, $C_2$, and $D_2$. The resulting expression for the stream function,

$$\Psi = \frac{1}{2} U_i R^2 \left( 1 - \frac{3}{2} \frac{a}{R} + \frac{1}{2} \frac{a^3}{R^3} \right) \sin^2 \theta, \tag{2.2.3}$$

allows us to find the fluid velocity and pressure in the form

$$V_R = U_i \left( 1 - \frac{3}{2} \frac{a}{R} + \frac{1}{2} \frac{a^3}{R^3} \right) \cos \theta,$$

$$V_\theta = -U_i \left( 1 - \frac{3}{4} \frac{a}{R} - \frac{1}{4} \frac{a^3}{R^3} \right) \sin \theta, \tag{2.2.4}$$

$$P = P_i - \frac{3\mu U_i a \cos \theta}{2R^2},$$

where $P_i$ is the nonperturbed pressure remote from the particle.

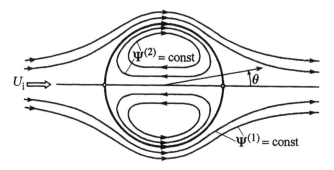

**Figure 2.2.** Translational Stokes flow past a spherical drop

The dynamic particle-fluid interaction is characterized by the drag force, which is defined as the projection of the resultant of all hydrodynamic forces on the flow direction:

$$F = \int_S (\tau_{RR} \cos\theta - \tau_{R\theta} \sin\theta) \, dS,$$

where $S$ is the particle surface.

The stresses on the spherical particle surface are given by the relations

$$\tau_{RR} = \left(-P + 2\mu \frac{\partial V_R}{\partial R}\right)_{R=a}, \qquad \tau_{R\theta} = \mu \left(\frac{\partial V_\theta}{\partial R} - \frac{V_\theta}{R} + \frac{1}{R} \frac{\partial V_R}{\partial \theta}\right)_{R=a}.$$

Using the expressions (2.2.4) and carrying out the integration, we find that the viscous drag acting on the spherical particle is given by

$$F = 6\pi\mu a U_{\mathrm{i}}. \tag{2.2.5}$$

This expression is known as the Stokes formula.

## 2.2-2. Flow Past a Spherical Drop or Bubble

Now let us consider a spherical drop of radius $a$ in a translational Stokes flow of another fluid with velocity $U_{\mathrm{i}}$ (Figure 2.2). We assume that the dynamic viscosities of the outer and inner fluids are equal to $\mu_1$ and $\mu_2$, respectively. The unknown variables outside and inside the drop are indicated by the superscripts (1) and (2), respectively.

To obtain the velocity and pressure profiles for the fluid in each phase, we shall use the Stokes equations (2.1.1). As previously, the condition that the flow is uniform remote from the drop has the form (2.2.2).

Let us write out four conditions that must be satisfied on the boundary of a spherical drop.

There is no flow across the interface:

$$V_R^{(1)} = V_R^{(2)} = 0 \qquad \text{at} \quad R = a. \tag{2.2.6}$$

The tangential velocity is continuous across the interface:

$$V_\theta^{(1)} = V_\theta^{(2)} \qquad \text{at} \quad R = a. \tag{2.2.7}$$

The jump of the normal stress across the interface is equal to the pressure increment due to interfacial tension:

$$P^{(1)} - 2\mu_1 \frac{\partial V_R^{(1)}}{\partial R} + \frac{2\sigma}{a} = P^{(2)} - 2\mu_2 \frac{\partial V_R^{(2)}}{\partial R} \qquad \text{at} \quad R = a, \tag{2.2.8}$$

where $\sigma$ is the interfacial tension.

The tangential stress is continuous across the interface:

$$\mu_1 \left( \frac{\partial V_\theta^{(1)}}{\partial R} - \frac{V_\theta^{(1)}}{R} \right) = \mu_2 \left( \frac{\partial V_\theta^{(2)}}{\partial R} - \frac{V_\theta^{(2)}}{R} \right) \qquad \text{at} \quad R = a. \tag{2.2.9}$$

We also use the boundedness of the solution at the drop center:

$$V_R^{(2)} < \infty, \quad V_\theta^{(2)} < \infty \qquad \text{at} \quad R = 0. \tag{2.2.10}$$

We introduce the stream function $\Psi^{(m)}$ in each of the phases ($m = 1, 2$) according to formulas (2.1.3). Conditions (2.2.6)–(2.2.10) allow us to determine the constants in the general solutions (2.1.5) inside and outside the drop. As a result, we obtain the Hadamard–Rybczynski solution [94, 179]

$$\Psi^{(1)} = \frac{1}{4} U_i R^2 \left( 2 - \frac{2 + 3\beta}{1 + \beta} \frac{a}{R} + \frac{\beta}{1 + \beta} \frac{a^3}{R^3} \right) \sin^2 \theta,$$

$$\Psi^{(2)} = -\frac{U_i}{4(1 + \beta)} R^2 \left( 1 - \frac{R^2}{a^2} \right) \sin^2 \theta, \tag{2.2.11}$$

where $\beta = \mu_2 / \mu_1$.

By using formulas (2.1.3), we calculate the velocity and the pressure outside the drop:

$$V_R^{(1)} = U_i \left[ 1 - \frac{2 + 3\beta}{2(1 + \beta)} \frac{a}{R} + \frac{\beta}{2(1 + \beta)} \frac{a^3}{R^3} \right] \cos \theta,$$

$$V_\theta^{(1)} = -U_i \left[ 1 - \frac{2 + 3\beta}{4(1 + \beta)} \frac{a}{R} - \frac{\beta}{4(1 + \beta)} \frac{a^3}{R^3} \right] \sin \theta, \tag{2.2.12}$$

$$P^{(1)} = P_0^{(1)} - \frac{\mu_1 U_i a (2 + 3\beta)}{2(1 + \beta)} \frac{\cos \theta}{R^2}.$$

The velocity and the pressure inside the drop are given by

$$V_R^{(2)} = -\frac{U_i}{2(1 + \beta)} \left( 1 - \frac{R^2}{a^2} \right) \cos \theta,$$

$$V_\theta^{(2)} = \frac{U_i}{2(1 + \beta)} \left( 1 - 2\frac{R^2}{a^2} \right) \sin \theta, \tag{2.2.13}$$

$$P^{(2)} = P_0^{(2)} + \frac{5\mu_2 U_i R \cos \theta}{a^2 (1 + \beta)}.$$

The constants $P_0^{(1)}$ and $P_0^{(2)}$ in the expressions (2.2.12) and (2.2.13) for the pressure fields are related by

$$P_0^{(2)} - P_0^{(1)} = \frac{2\sigma}{a}. \tag{2.2.14}$$

The drag force acting on the spherical drop is

$$F = 2\pi a U_i \frac{2\mu_1 + 3\mu_2}{\mu_1 + \mu_2}. \tag{2.2.15}$$

As $\beta = \mu_2/\mu_1 \to \infty$, Eq. (2.2.15) becomes the Stokes formula (2.2.5) for a solid particle. The passage to the limit as $\beta \to 0$ corresponds to a gas bubble.

## 2.2-3. Steady-State Motion of Particles and Drops in a Fluid

In chemical technology one often meets the problem of a steady-state motion of a spherical particle, drop, or bubble with velocity $U_i$ in a stagnant fluid. Since the Stokes equations are linear, the solution of this problem can be obtained from formulas (2.2.12) and (2.2.13) by adding the terms $\bar{V}_R = -U_i \cos\theta$ and $\bar{V}_\theta = U_i \sin\theta$, which describe a translational flow with velocity $U_i$ in the direction opposite to the incoming flow. Although the dynamic characteristics of flow remain the same, the streamline pattern looks different in the reference frame fixed to the stagnant fluid. In particular, the streamlines inside the sphere are not closed.

The drag coefficient for a drop is calculated by the formula

$$c_f = \frac{F}{\frac{1}{2}\rho_1 U_i^2 \pi a^2} = \frac{4}{\text{Re}} \left( \frac{2 + 3\beta}{1 + \beta} \right), \qquad \text{where} \quad \text{Re} = \frac{\rho_1 U_i a}{\mu_1}. \tag{2.2.16}$$

By equating the drag force $F$ of the sphere with the difference $\frac{4}{3}\pi a^3 g \Delta\rho$ between the gravity and buoyancy forces, one can estimate the steady-state velocity of relative motion of phases (the velocity at which a spherical drop falls or rises) as

$$U = \frac{2}{3} \frac{ga^2 \Delta\rho}{\mu_1} \left( \frac{1 + \beta}{2 + 3\beta} \right), \tag{2.2.17}$$

where $\Delta\rho$ is the difference between the densities of the outer and the inner fluid and $g$ is the free fall acceleration.

Relations (2.2.16) and (2.2.17) cover the entire range $0 \le \beta \le \infty$ of the phase viscosity ratio. In the limit cases $\beta = 0$ (a gas bubble in a highly viscous liquid) and $\beta \to \infty$ (a solid particle in a fluid), these formulas become

$$c_f = \frac{8}{\text{Re}}, \quad U = \frac{1}{3} \frac{ga^2 \Delta\rho}{\mu_1} \qquad \text{(gas bubble)}, \tag{2.2.18}$$

$$c_f = \frac{12}{\text{Re}}, \quad U = \frac{2}{9} \frac{ga^2 \Delta\rho}{\mu_1} \qquad \text{(solid particle)}. \tag{2.2.19}$$

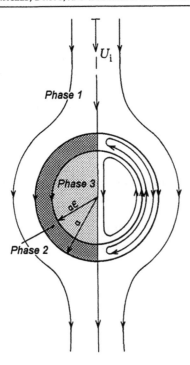

**Figure 2.3.** Flow past a drop with a membrane phase

The last expression for $c_f$ is known as the Stokes law for the drag coefficients of solid spherical particles. This law is confirmed by experiments for Re $< 0.1$. The drag law (2.2.18) for spherical bubbles holds only for extremely pure liquids without any surfactants.

According to [270], even very small quantities of surfactants are adsorbed on the bubble surface and lead to its "solidification." This results in eliminating internal circulation, and hence the bubble rises according to the Stokes law (2.2.19) for solid particles.

### 2.2-4. Flow Past Drops With a Membrane Phase

In chemical technology, compound drops are often used, where the continuous phase (phase 1) and the liquid core of the drop (phase 3) are separated by a liquid buffer or membrane phase (phase 2). The streamlines of steady-state flow in immiscible phases 2 and 3 are closed (Figure 2.3). If the membrane layer is thin, then the flow in this layer is very retarded and nearly stagnant.

Let $a$ be the outer radius of the compound drop, and let $a\varepsilon$ be the radius of the core ($0 \le \varepsilon \le 1$). The exact solution of the problem on the flow past a compound drop in a translational Stokes flow with velocity $U_i$ can be found in [416], where the stream functions in the phases are given. The drag force is also computed there:

$$F = 6\pi\mu a U_i \lambda, \qquad \lambda = \frac{2}{3}\frac{\beta_3 + 6\beta_2^2 F(\varepsilon) + \beta_2(2 + 3\beta_3)G(\varepsilon)}{\beta_3 + 4\beta_2^2 F(\varepsilon) + 2\beta_2(1 + \beta_3)G(\varepsilon)}, \qquad (2.2.20)$$

where $\beta_2 = \mu_2/\mu_1$ and $\beta_3 = \mu_3/\mu_1$ are the phase viscosity ratios and

$$F(\varepsilon) = \frac{1 - \varepsilon^5}{(1 - \varepsilon)^3(4\varepsilon^2 + 7\varepsilon + 4)}, \qquad G(\varepsilon) = \frac{(1 + \varepsilon)(2\varepsilon^2 + \varepsilon + 2)}{(1 - \varepsilon)(4\varepsilon^2 + 7\varepsilon + 4)}.$$

Let us note three important limit cases of formula (2.2.20).
1. If the drop core is small, we have

$$\lambda \to \frac{2 + 3\beta_2}{3(1 + \beta_2)} \qquad \text{as} \quad \varepsilon \to 0, \qquad (2.2.21)$$

and formula (2.2.20) turns into Eq. (2.2.15) related to the Hadamard–Rybczynski solution with $\beta = \beta_2$.
2. If the membrane layer is thin, we have

$$\lambda \to 1 \qquad \text{as} \quad \varepsilon \to 1. \qquad (2.2.22)$$

Thus the motion in a thin membrane layer is strongly retarded and the drop in a flow behaves as a solid particle. This result can be interpreted as a purely hydrodynamic alternative to the explanation [270] of the solidification effect (the surface of a bubble rising in a liquid with traces of surfactants becomes "solid").
3. With increasing core viscosity, we obtain the drag force of a solid spherical particle of radius $a\varepsilon$ covered by a liquid film of thickness $a(1 - \varepsilon)$:

$$F = 6\pi\mu a U_i \lambda, \qquad \lambda = \frac{2}{3}\frac{1 + 3\beta_2 G(\varepsilon)}{1 + 2\beta_2 G(\varepsilon)}. \qquad (2.2.23)$$

For small core radius ($\varepsilon \to 0$), this formula tends to the Hadamard–Rybczynski formula (2.2.15) for a drop; if the membrane is thin ($\varepsilon \to 1$), then we obtain the Stokes formula (2.2.5) for a solid sphere.

### 2.2-5. Flow Past a Porous Spherical Particle

Let us consider the problem about a porous particle of radius $a$ in a translational fluid flow with velocity $U_i$. We assume that the flow outside the particle obeys the Stokes equations (2.1.1) with viscosity $\mu$. We also assume that the percolation flow of an incompressible fluid inside the particle satisfies Darcy's law [97, 346, 423]:

$$\mathbf{V}^{(2)} = -\frac{K}{\mu}\nabla P^{(2)}, \qquad \nabla \cdot \mathbf{V}^{(2)} = 0, \qquad (2.2.24)$$

where $K$ is the permeability coefficient.

To complete the statement of the problem, along with condition (2.2.2) of remote flow uniformity and condition (2.2.10) of boundedness of the solution, we need some boundary conditions on the interface. One of these conditions is obtained by substituting $\mu_2 = 0$ and $\sigma = 0$ into (2.2.8) and means that the normal stress is equal to the internal pressure. Furthermore, the normal velocity must be continuous across the interface,

$$V_R^{(1)} = V_R^{(2)} \qquad \text{at} \quad R = a, \tag{2.2.25}$$

and the jump of the tangential velocity across the interface must be proportional to the normal derivative,

$$\lambda\sqrt{K}\frac{\partial V_\theta^{(1)}}{\partial R} = \left(V_\theta^{(1)} - V_\theta^{(2)}\right) \qquad \text{at} \quad R = a. \tag{2.2.26}$$

The last condition was obtained experimentally and justified theoretically in [33, 34, 422]. The parameter $\lambda$ is a dimensionless empirical constant depending on the material and the intrinsic geometry of the porous medium interface ($0.25 \le \lambda \le 10$). For example, $\lambda = 10$ for aloxite ($K = 1.6 \times 10^{-9}\,\text{m}^2$) and $\lambda = 0.25, 0.69$, and $1.28$ for some foametals ($K = 8.2 \times 10^{-8}, 3.9 \times 10^{-8}$, and $0.97 \times 10^{-8}\,\text{m}^2$, respectively).

By solving the stated hydrodynamic problem, we arrive at the following expressions for the velocity components outside and inside a porous particle [372]:

$$V_R^{(1)} = U_i\left(1 - \frac{3}{2}B_1\frac{a}{R} + \frac{1}{2}B_2\frac{a^3}{R^3}\right)\cos\theta,$$

$$V_\theta^{(1)} = -U_i\left(1 - \frac{3}{4}B_1\frac{a}{R} - \frac{1}{4}B_2\frac{a^3}{R^3}\right)\sin\theta; \tag{2.2.27}$$

$$V_R^{(2)} = \frac{3}{2}U_i\frac{K}{a^2}B_3\cos\theta, \qquad V_\theta^{(2)} = -\frac{3}{2}U_i\frac{K}{a^2}B_3\sin\theta,$$

where the dimensionless constants $B_1$, $B_2$, and $B_3$ are given by

$$B_1 = \frac{1 + \lambda k^{1/2}}{\Lambda}, \qquad B_2 = \frac{1 - \lambda k^{1/2}}{\Lambda}, \qquad B_3 = \frac{1 + 5\lambda k^{1/2}}{\Lambda},$$

$$k = \frac{K}{a^2}, \qquad \Lambda = 1 + 2\lambda k^{1/2} + \frac{3}{2}k + \frac{15}{2}\lambda k^{3/2}. \tag{2.2.28}$$

For the drag force of the porous particle, we have

$$F = 6\pi\mu a U_i B_1, \tag{2.2.29}$$

where the parameter $B_1$ is defined in (2.2.28).

In the limit cases $K \to 0$ and $\lambda \to 0$, formula (2.2.29) is reduced to (2.2.5) and corresponds to a Stokes flow around a solid sphere.

## 2.3. Spherical Particles in Translational Flow at Various Reynolds Numbers

### 2.3-1. Oseen's and Higher Approximations as Re → 0

Neglecting the inertia term for the flow past a sphere is adequate to experiments only in the limit case Re → 0. According to [94], even for Re = 0.05 the error in estimating the drag by formula (2.2.19) is 1.5 to 2% and for Re = 0.5 the error becomes 10.5 to 11%. Therefore, one can use the estimate $c_f = 12/\text{Re}$ for the drag coefficient only for Re < 0.2 (in this case, the maximum error does not exceed 5%). An attempt to improve the Stokes approximation by taking account of the convective terms iteratively leads to an equation for which it is impossible to construct the solution satisfying the condition at infinity. This is known as the Whitehead paradox and is due to the fact that the solution has a singularity at infinity.

A method for overcoming this paradox was proposed by Oseen [325] who suggested to approximate the inertia term by the expression $(\mathbf{U_i} \cdot \nabla)\mathbf{V}$, since the difference between the flow velocity $\mathbf{V}$ and the incoming flow velocity $\mathbf{U_i}$ is small remote from the sphere. The system of Oseen's equations has the form

$$(\mathbf{U_i} \cdot \nabla)\mathbf{V} = -\frac{1}{\rho}\nabla P + \nu\Delta\mathbf{V},$$
$$\nabla \cdot \mathbf{V} = 0. \tag{2.3.1}$$

This system is also linear but gives a better approximation (remote from the particle) than the Stokes system.

By using the formula [325]

$$\Psi = \frac{U_i R^2 \sin^2\theta}{2}\left(1 + \frac{a^3}{2R^3}\right)$$
$$- \frac{3}{2\,\text{Re}}U_i a^2(1 + \cos\theta)\left[1 - \exp\left(-\text{Re}\,\frac{1 - \cos\theta}{2}\frac{R}{a}\right)\right] \tag{2.3.2}$$

one can express the solution of (2.3.1) that satisfies the no-slip conditions (2.2.1) at the solid sphere surface and the boundary conditions (2.2.2) at infinity via the stream function (2.1.3).

As a result, we obtain the drag coefficient

$$c_f = \frac{12}{\text{Re}}\left(1 + \frac{3}{8}\text{Re}\right); \tag{2.3.3}$$

this expression improves the Stokes law (2.2.19).

The difference between Oseen's approximation and experimental data is 0 to 1.0% for Re ≤ 0.05 and 4 to 6% for Re = 0.5.

Proudman and Pearson [382] tried to obtain analytic solutions in a wider range of Reynolds numbers. They solved the system of Navier–Stokes equations by the method of matched asymptotic expansions [258, 485] in domains near the sphere and at some distance from it. As a result, they found the first three terms of the asymptotic expansion as $\mathsf{Re} \to 0$ for the drag coefficient in the form

$$c_f = \frac{12}{\mathsf{Re}} \left[ 1 + \frac{3}{8}\mathsf{Re} + \frac{9}{40}\mathsf{Re}^2 \ln \mathsf{Re} + O(\mathsf{Re}^2) \right]. \qquad (2.3.4)$$

For $\mathsf{Re} < 0.35$, the difference between the results predicted by formula (2.3.4) and the experimental data does not exceed 1.5%, but divergence is rapid at higher $\mathsf{Re}$.

Chester and Breach (see [94]) found the next two terms of the expansion of the drag coefficient with respect to the Reynolds number:

$$c_f = \frac{12}{\mathsf{Re}} \left[ 1 + \frac{3}{8}\mathsf{Re} + \frac{9}{40}\mathsf{Re}^2 \left( \ln \mathsf{Re} + \gamma + \frac{5}{3}\ln 2 - \frac{323}{360} \right) \right.$$
$$\left. + \frac{27}{80}\mathsf{Re}^3 \ln \mathsf{Re} + O(\mathsf{Re}^3) \right], \qquad (2.3.5)$$

where $\gamma \approx 0.5772$ is the Euler constant.

---

### 2.3-2. Flow Past Spherical Particles in a Wide Range of Re

For $\mathsf{Re} > 0.5$, asymptotic solutions no longer give an adequate description of translational flow of a viscous fluid past a spherical particle.

Numerous available numerical solutions the Navier–Stokes equations, as well as experimental data (see a review in [94]), provide a detailed analysis of the flow pattern for increasing Reynolds numbers. For $0.5 < \mathsf{Re} < 10$, there is no flow separation, although the fore-and-aft symmetry typical of inertia-free Stokes flow past a sphere is more and more distorted. Finally, at $\mathsf{Re} \approx 10$, flow separation occurs at the rear of the particle.

The range $10 < \mathsf{Re} < 65$ is characterized by the existence of a closed stable area at the rear, in which there is an axisymmetric recirculating wake (Figure 2.4). As $\mathsf{Re}$ increases, the wake lengthens, and the separation ring $\theta_s$ moves forward from the rear point ($\theta_s = 0°$ at $\mathsf{Re} = 10$) to $\theta_s = 72°$ at $\mathsf{Re} = 200$ according to the law [94]

$$\theta_s = 42.5 \left( \ln \frac{\mathsf{Re}}{10} \right)^{0.483} \qquad \text{for} \quad 10 < \mathsf{Re} < 200. \qquad (2.3.6)$$

Here, as well as in (2.3.7), $\theta_s$ is measured in degrees.

At $\mathsf{Re} > 65$, the vorticity region in the rear area ceases to be stable and becomes unsteady. At $65 < \mathsf{Re} < 200$, a long oscillating wake is formed behind the particle, which gradually becomes turbulent for $200 < \mathsf{Re} < 1.5 \times 10^5$. Simultaneously, the separation point moves upstream according to the law [94]

$$\theta_s = 102 - 213 \, (\mathsf{Re})^{-0.37} \qquad \text{for} \quad 200 < \mathsf{Re} < 1.5 \times 10^5. \qquad (2.3.7)$$

For $\mathsf{Re} > 1500$, the corresponding hydrodynamic problems can be solved by methods of the boundary layer theory [427]. However, because of the flow

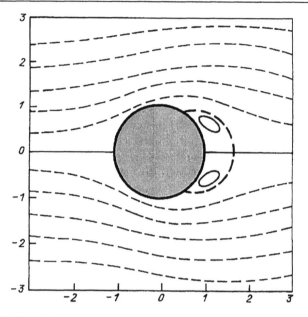

**Figure 2.4.** Qualitative flow pattern past a sphere with stable separation area $(10 < \text{Re} < 65)$

separation at the rear of the sphere, the law of potential flow past a sphere outside its boundary layer is valid only on some part of the front hemisphere (for $\theta > 150°$) [94]. This does not allow us to obtain appropriate estimates for the longitudinal pressure gradient on a large part of the boundary layer. Another approximate method of solution for the problem on flow past a sphere at high Reynolds number was used in [462].

---

### 2.3-3. Formulas for Drag Coefficient in a Wide Range of Re

We give two simple approximate formulas for the drag coefficient of a spherical particle [46, 94]

$$c_f = \frac{12}{\text{Re}}\left(1 + 0.241\,\text{Re}^{0.687}\right), \qquad 0 \le \text{Re} \le 400,$$
$$c_f = \frac{12}{\text{Re}}\left(1 + 0.0811\,\text{Re}^{0.879}\right), \qquad 200 \le \text{Re} \le 2500, \qquad (2.3.8)$$

where the Reynolds number is determined with respect to the radius. In formulas (2.3.8) the maximum error does not exceed 5% in the given ranges.

In a wide range of Reynolds numbers, one can use the following more complicated approximation for the drag coefficient [94]:

$$c_f = \frac{12}{\text{Re}}\left(1 + 0.241\,\text{Re}^{0.687}\right) + 0.42\left(1 + 1.902 \times 10^4\,\text{Re}^{-1.16}\right)^{-1},$$

whose maximum error does not exceed 6% for $\text{Re} < 1.5 \times 10^5$.

At $Re \approx 1.5 \times 10^5$, one can observe the "drag crisis" characterized by a sharp decrease in the drag coefficient; the boundary layer becomes more and more turbulent; the separation point shifts abruptly to the aft area.

For $Re \geq 1.7 \times 10^5$, the drag coefficient can be calculated by the formulas

$$c_f = \begin{cases} 28.18 - 5.3 \log_{10} Re & \text{for } 1.7 \times 10^5 \leq Re \leq 2 \times 10^5, \\ 0.1 \log_{10} Re - 0.46 & \text{for } 2 \times 10^5 < Re \leq 5 \times 10^5, \\ 0.19 - 4 \times 10^4 Re^{-1} & \text{for } 5 \times 10^5 < Re, \end{cases}$$

presented in [94].

# 2.4. Spherical Drops and Bubbles in Translational Flow at Various Reynolds Numbers

## 2.4-1. Bubble in a Translational Flow

In the case of a spherical bubble in a translational flow at small Reynolds numbers, the solution of Oseen's equation (2.3.1) results is a two-term asymptotic expansion for the drag coefficient [476]:

$$c_f = \frac{8}{Re} + 1 \qquad \text{(as } Re \to 0), \tag{2.4.1}$$

which refines formula (2.2.18).

The drag coefficient monotonically decreases as the Reynolds number increases. For high Reynolds numbers, one can use the approximation of ideal fluid for solving the problem on the flow past a bubble. In this case, the leading term of the asymptotic expansion of the drag coefficient has the form [291]

$$c_f = \frac{24}{Re} \qquad \text{(as } Re \to \infty). \tag{2.4.2}$$

In [359] the interpolation formula

$$c_f = \frac{8}{Re} + \frac{16}{Re + 16}, \tag{2.4.3}$$

was proposed, which gives the drag coefficient of a spherical bubble in the entire range of Reynolds numbers. In the limit cases $Re \to 0$ and $Re \to \infty$, this formula gives correct asymptotic results (2.4.1) and (2.4.2); its maximum error for intermediate Reynolds numbers is less than 4.5%.

In [283], a number of experimental data are presented on the rise of air bubbles in clean mixtures of distilled water and pure, reagent grade, glycerin covering a wide range of the Morton numbers.

The motion of air bubbles in free-surface turbulent shear flows is considered in [83] in detail.

## 2.4-2. Drop in a Translational Liquid Flow

In [476] the following asymptotic expansion was obtained for the drag coefficient of a drop in a translational flow at small Reynolds numbers:

$$c_f = \frac{3\beta + 2}{\beta + 1}\left(\frac{4}{\text{Re}} + \frac{\text{Re}}{2} + \frac{1}{40}\,\text{Re}^2 \ln \text{Re}\right), \quad \text{Re} = \frac{aU_i}{\nu}; \qquad (2.4.4)$$

this is a more precise version of (2.2.16).

The spherical form of a drop or a bubble in Stokes flow follows from the fact that the flow is inertia-free. However, even for the case in which the inertia forces dominate viscous forces and the Reynolds number cannot be considered small, the drop remains undeformed if the inertia forces are small compared with the capillary forces. The ratio of inertial to capillary forces is measured by the Weber number $\text{We} = \rho_1 U_i^2 a/\sigma$, where $\sigma$ is the surface tension at the drop boundary. For small $\text{We}$, a deformable drop will conserve the spherical form.

In Section 2.2 it was already noted that even a small amount of surfactants in any of the adjacent phases may lead to the "solidification" of the interface, so that the laws of flow around a drop become close to those for a solid particle. This effect often occurs in practice. However, if both phases are carefully purified (do not contain contaminants), the flow around a drop possesses some special features.

Flow separation in the case of a drop is delayed compared with the case of a solid particle, and the vorticity region (wake) is considerably narrower. While in the case of a solid sphere, the flow separates and the rear wake occurs at $\text{Re} \approx 10$ (the number $\text{Re}$ is determined by the sphere radius), in the case of a drop there may be no separation until $\text{Re} \approx 50$. For $1 \leq \text{Re} \leq 50$, numerical methods are widely used. The results of numerical calculations are discussed in [94]. For these Reynolds numbers, the internal circulation is more intensive than is predicted by the Hadamard–Rybczynski solution. The velocity at the drop boundary increases rapidly with the Reynolds number even for highly viscous drops. In the limit case of small viscosity of the disperse phase, $\beta \to 0$ (this corresponds to the case of a gas bubble), one can use the approximation of ideal fluid for the outer flow at $\text{Re} \gg 1$.

According to the data presented in [94], to estimate the drag of a spherical drop with high accuracy, one can use the following formula, which approximates numerical results obtain by the Galerkin method:

$$c_f = \frac{1.83\,(783\,\beta^2 + 2142\,\beta + 1080)}{(60 + 29\,\beta)(4 + 3\,\beta)}\,\text{Re}^{-0.74} \quad \text{for} \ \ 2 < \text{Re} < 50, \qquad (2.4.5)$$

where $\beta$ is the ratio of dynamic viscosities of the drop and the outer liquid.

The drag coefficient of a spherical drop can also be determined by the formula [359]

$$c_f(\beta, \text{Re}) = \frac{1}{\beta + 1}c_f(0, \text{Re}) + \frac{\beta}{\beta + 1}c_f(\infty, \text{Re}). \qquad (2.4.6)$$

Here $c_f(0, \text{Re})$ is the drag coefficient of the spherical bubble, which can be calculated by the formula (2.4.3), and $c_f(\infty, \text{Re})$ is the drag coefficient of a solid spherical particle, which can be calculated by (2.3.8). The approximate expression (2.4.6) gives three correct terms of the expansion for small Reynolds numbers; for $0 \le \text{Re} \le 50$, the maximum error is less than 5%.

The drop may conserve its spherical form until $\text{Re} \approx 300$ [94]. Since usually the boundary layer on a drop or a bubble is considerably thinner than on a solid sphere, one can use methods based on the boundary layer theory even for $50 < \text{Re} < 300$. By using these methods, the following formula was obtained in [94] for the drag coefficient for $\text{Re} \gg 1$:

$$c_f = \frac{24}{\text{Re}}\left[1 + \frac{3}{2}\beta + \frac{(2+3\beta)^2}{\text{Re}^{1/2}}(B_1 + B_2 \ln \text{Re})\right]. \qquad (2.4.7)$$

The constants $B_1$ and $B_2$ are given in the following table ($\beta = \mu_2/\mu_1$ is the ratio of viscosities of the inner and outer phases, and $\gamma = \rho_2/\rho_1$ is the corresponding phase density ratio).

| $\beta\gamma$ | 25 | 4.0 | 1.0 | 0.25 | 0.04 | 0 |
|---|---|---|---|---|---|---|
| $B_1$ | −0.429 | −0.457 | −0.460 | −0.446 | −0.434 | −0.391 |
| $B_2$ | 0.00202 | 0.00620 | 0.0100 | 0.0113 | 0.00842 | 0 |

### 2.4-3. Drop in a Translational Gas Flow

In [340], an approximate solution was obtained for the problem about a translational gas flow (of small viscosity) past a viscous spherical drop. The solution uses expansions with respect to the small parameter

$$\varepsilon = \frac{\mu_1}{\mu_2}\sqrt{\text{Re}_1} \ll 1.$$

The fluid motion outside and inside the drop is characterized by different Reynolds numbers (inside the drop, $\varepsilon U_i$ is used as the characteristic velocity):

$$\text{Re}_1 = \frac{aU_i\rho_1}{\mu_1}, \qquad \text{Re}_2 = \frac{a\varepsilon U_i\rho_2}{\mu_2},$$

and no special restrictions are imposed on these values. This situation is typical of rain drops moving in air; in this case $\mu_1/\mu_2 = 1.8 \times 10^{-2}$ and the parameter $\varepsilon$ is small in the wide range $0 < \text{Re}_1 < 10^3$.

We seek the solution of the problem in the form of asymptotic expansions with respect to the small parameter $\varepsilon$. The leading term of the expansion outside the drop is determined by the solution of the problem about the flow past a solid sphere. The leading term inside the drop corresponds to the viscous fluid flow caused by the tangential stress on the interface (the tangential stress depends only on the outer Reynolds number $Re_1$ and is taken from well-known numerical solutions [113, 402]).

The flow field inside the drop depends only on the two parameters $Re_1$ and $Re_2$, and the dependence on $Re_2$ proves unessential. The maximum dimensionless velocity of the fluid inside the drop $v_{max} = v_{max}(Re_1, Re_2)$ is attained at the drop boundary. For $Re_1 \geq 2.5$ [338], the following estimates hold:

$$v_{max}(Re_1, \infty) \leq v_{max}(Re_1, Re_2) \leq v_{max}(Re_1, 0) \qquad (2.4.8)$$

where

$$v_{max}(Re_1, \infty) = 0.15 + 0.42\, Re_1^{-0.32},$$
$$v_{max}(Re_1, 0) = \left(0.15 + 0.42\, Re_1^{-0.32}\right)\left(1 + \frac{Re_1}{50 + 2\, Re_1}\right). \qquad (2.4.9)$$

The difference between the upper and the lower bounds is not very large, which shows that the dependence of the internal flow on the parameter $Re_2$ is weak; the above estimates are in a good agreement with known numerical results [262, 498].

The study of the internal flow shows that the toroidal vortex deforms as $Re_1$ increases and separates from the boundary near the rear point at $Re_1 = 150$ (for $\theta_s \approx 30°$). In this case, the second vortex is formed near the internal separation point, and the velocity in the region of the second vortex is much less (approximately, 30 times) than the maximum velocity in the region of the first vortex.

The dimensionless velocity components in the spherical coordinates are approximately described by the following formulas for the Hill vortex [26]:

$$v_r = v_{max}(r^2 - 1)\cos\theta, \quad v_\theta = v_{max}(1 - 2r^2)\sin\theta, \qquad r = R/a,$$

where the maximum velocity $v_{max} = v_{max}(Re_1, Re_2)$ is characterized by the estimates (2.4.8) and (2.4.9).

For $Re_2 \gg 1$, the Hill vortex occupies the entire drop except for a thin boundary layer adjacent to the surface, in which there is a convective-diffusion transfer of vorticity [496].

As $Re_2$ decreases, the streamlines are slightly deformed and the singular streamline (on which the fluid velocity is zero), lying in the meridian plane inside the drop, slightly moves toward the front surface.

For $Re_1 < 150$, there is no flow separation inside the drop. For $Re \geq 150$, near the rear point, a second vortex is formed, whose velocity is by an order of magnitude less than $v_{max}$ [336, 340].

For $Re_1 \leq 2.5$, the difference between $v_{max}(Re_1, \infty)$ and $v_{max}(Re_1, 0)$ is practically zero, so that $v_{max}$ is independent of $Re_2$ [339] and can be obtained from the solution for small Reynolds numbers [476].

## 2.4-4. Dynamics of an Extending (Contracting) Spherical Bubble

In chemical technology, one often meets the problem about a spherically symmetric deformation (contraction or extension) of a gas bubble in an infinite viscous fluid. In the homobaric approximation (the pressure is homogeneous inside the bubble) [306, 312], only the motion of the outer fluid is of interest. The Navier–Stokes equations describing this motion in the spherical coordinates have the form

$$\rho\left(\frac{\partial V_R}{\partial t} + V_R\frac{\partial V_R}{\partial R}\right) = -\frac{\partial P}{\partial R} + 2\mu\frac{\partial^2 V_R}{\partial R^2}, \qquad (2.4.10)$$

$$\frac{\partial}{\partial R}(R^2 V_R) = 0. \qquad (2.4.11)$$

If there is no mass transfer across the bubble boundary, the fluid velocity on the boundary is equal to the velocity of the boundary,

$$V_R = \dot{a} \qquad \text{at} \quad R = a, \qquad (2.4.12)$$

where $\dot{a} = da/dt$.

The solution of Eq. (2.4.11) with regard to (2.4.12) and the condition that the fluid velocity is zero at infinity has the form

$$V_R = \frac{a^2\dot{a}}{R^2}. \qquad (2.4.13)$$

By substituting (2.4.13) into (2.4.10) and then integrating with respect to $R$ from $a$ to $\infty$, we arrive at the relations

$$\rho(a\ddot{a} + \tfrac{3}{2}\dot{a}^2) = P|_{R=a} - P_\infty(t), \qquad (2.4.14)$$

where $P_\infty(t)$ is the fluid pressure at infinity; the bubble pulsations are caused just by the variations of $P_\infty(t)$.

The fluid pressure $P|_{R=a}$ at the bubble boundary can be determined from the condition on the jump of normal stresses on the discontinuity surface, that is, the bubble boundary [24, 430]. Under the homobaric assumptions, the gas in the bubble does not move, which implies that

$$P = P_b - \frac{2\sigma}{a} + 2\mu\frac{\partial V_R}{\partial R} \qquad \text{at} \quad R = a, \qquad (2.4.15)$$

where $\sigma$ is the surface tension coefficient at the bubble boundary and $P_b$ is the pressure inside the bubble.

Taking into account (2.4.13) and substituting (2.4.15) into (2.4.14), we obtain the Rayleigh equation

$$\rho\left(a\ddot{a} + \frac{3}{2}\dot{a}^2\right) + 4\mu\frac{\dot{a}}{a} + \frac{2\sigma}{a} = -P_\infty(t) + P_b, \qquad (2.4.16)$$

which describes the dynamics of the bubble radius variation under the action of pressure that varies at infinity. The initial conditions for this equation are usually posed as

$$a = a_0, \quad \dot{a} = 0 \quad \text{at} \quad t = 0. \tag{2.4.17}$$

If the gas in the bubble extends and contracts adiabatically, then the gas pressure in the bubble is related to the initial pressure $P_0$ by the adiabatic equation

$$P_b = P_0 \left(\frac{a_0}{a}\right)^{3\gamma}, \tag{2.4.18}$$

where $\gamma$ is the adiabatic exponent.

For $\gamma = 1$, the solution of problem (2.4.16)–(2.4.18) can be written in the parametric form [513, 515]

$$a = \frac{a_0}{\left[H(\tau)\right]^2}, \qquad t = \int_s^\tau \frac{d\tau}{\left[H(\tau)\right]^5},$$

where the function $H(\tau)$ and the coefficient $s$ are determined by

$$H(\tau) = \exp(-\tau^2)\left[2s \int_s^\tau \exp(\tau^2)\,d\tau + \exp(s^2)\right], \qquad s = \frac{4\mu}{\rho a_0^2},$$

and the parameter $\tau$ varies in the range $s \le \tau < \infty$.

In [513, 515], it is shown that problem (2.4.16)–(2.4.18) can be solved in quadratures also for $\gamma = \frac{2}{3}, \frac{5}{6}$ and can be reduced to the Bessel equation for $\gamma = \frac{11}{12}, \frac{7}{6}$.

In the books [312, 313], the problems of dynamics and heat and mass transfer of a pulsating gas bubble (with allowance for various complicating factors) are considered in detail.

# 2.5. Spherical Particles, Drops, and Bubbles in Shear Flows

### 2.5-1. Axisymmetric Straining Shear Flow

*Statement of the problem.* Let us consider a linear shear flow at low Reynolds numbers past a solid spherical particle of radius $a$. In the general case, the Stokes equation (2.1.1) must be completed by the no-slip condition (2.2.1) on the particle boundary and the following boundary conditions remote from the particle (see Section 1.1):

$$V_k \to G_{kj} X_j \quad \text{as} \quad R \to \infty, \tag{2.5.1}$$

where $X_1$, $X_2$, $X_3$ are the Cartesian coordinates, the $V_k$ are the fluid velocity components, $G_{kj}$ are the shear tensor components, $k, j = 1, 2, 3$, and $j$ is the summation index.

In the problem of linear shear flow past a spherical drop (bubble), the Stokes equations (2.1.1) and the boundary conditions at infinity (2.5.1) must be completed by the boundary conditions on the interface and the condition that the solution is bounded inside the drop. In particular, in the axisymmetric case, the boundary conditions (2.2.6)–(2.2.10) are used.

In the sequel we consider some special cases of shear flows described in Section 1.1.

**Solid particles.** In the case of axisymmetric straining shear flow, the boundary conditions (2.5.1) remote from the particle have the form

$$V_X \to -\tfrac{1}{2}GX, \quad V_Y \to -\tfrac{1}{2}GY, \quad V_Z \to GZ \qquad \text{as} \quad R \to \infty. \qquad (2.5.2)$$

To solve the corresponding hydrodynamic problem, it is convenient to use the spherical coordinates and introduce the stream function according to (2.1.3). Condition (2.5.2) acquires the form

$$\Psi \to \tfrac{1}{2}GR^3 \sin^2\theta \cos\theta \qquad \text{as} \quad R \to \infty. \qquad (2.5.3)$$

It follows that in the general solution (2.1.5) of the Stokes equations one must retain only the terms with $n = 3$. The unknown constants $A_3$, $B_3$, $C_3$, $D_3$ are determined by the boundary no-slip conditions (2.2.1). As a result, we obtain the stream function [474, 475]

$$\Psi = \frac{1}{2}Ga^3 \left( \frac{R^3}{a^3} - \frac{5}{2} + \frac{3}{2}\frac{a^2}{R^2} \right) \sin^2\theta \cos\theta. \qquad (2.5.4)$$

By substituting (2.5.4) into (2.1.3), we find the velocity and pressure in the form

$$V_R = \frac{1}{2}Ga \left( \frac{R}{a} - \frac{5}{2}\frac{a^2}{R^2} + \frac{3}{2}\frac{a^4}{R^4} \right)(3\cos^2\theta - 1),$$

$$V_\theta = -\frac{3}{2}Ga \left( \frac{R}{a} - \frac{a^4}{R^4} \right) \sin\theta \cos\theta, \qquad (2.5.5)$$

$$P = P_i - \frac{5}{2}G\mu \frac{a^3}{R^3}(3\cos^2\theta - 1),$$

where $P_i$ is the pressure remote from the particle.

**Drops and bubbles.** Axisymmetric shear flow past a drop was studied in [474, 475]. We denote the dynamic viscosities of the fluid outside and inside the drop by $\mu_1$ and $\mu_2$. Far from the drop, the stream function satisfies (2.5.3) just as in the case of a solid particle. Therefore, we must retain only the terms with $n = 3$ in the general solution (2.1.5). We find the unknown constants from the boundary conditions (2.2.6)–(2.2.10) and obtain

$$\Psi^{(1)} = \frac{1}{2}Ga^3 \left( \frac{R^3}{a^3} - \frac{1}{2}\frac{5\beta+2}{\beta+1} + \frac{3}{2}\frac{\beta}{\beta+1}\frac{a^2}{R^2} \right) \sin^2\theta \cos\theta,$$

$$\Psi^{(2)} = \frac{3}{4}\frac{Ga^3}{\beta+1}\frac{R^3}{a^3} \left( \frac{R^2}{a^2} - 1 \right) \sin^2\theta \cos\theta, \qquad (2.5.6)$$

where $\Psi^{(1)}$ and $\Psi^{(2)}$ are the stream functions outside and inside the drop and $\beta = \mu_2/\mu_1$.

The fluid velocity outside the drop is given by

$$V_R^{(1)} = \frac{1}{2} Ga \left( \frac{R}{a} - \frac{1}{2} \frac{5\beta+2}{\beta+1} \frac{a^2}{R^2} + \frac{3}{2} \frac{\beta}{\beta+1} \frac{a^4}{R^4} \right) (3\cos^2\theta - 1),$$

$$V_\theta^{(1)} = -\frac{3}{2} Ga \left( \frac{R}{a} - \frac{\beta}{\beta+1} \frac{a^4}{R^4} \right) \sin\theta \cos\theta.$$

$$(2.5.7)$$

The fluid velocity inside the drop is given by

$$V_R^{(2)} = \frac{3}{4} \frac{Ga}{\beta+1} \frac{R}{a} \left( \frac{R^2}{a^2} - 1 \right) (3\cos^2\theta - 1),$$

$$V_\theta^{(2)} = -\frac{9}{4} \frac{Ga}{\beta+1} \frac{R}{a} \left( \frac{5}{3} \frac{R^2}{a^2} - 1 \right) \sin\theta \cos\theta.$$

$$(2.5.8)$$

The limit case as $\beta \to 0$ corresponds to a gas bubble.

Analyzing (2.5.7) and (2.5.8), we see that this flow has a symmetry axis ($Z$-axis) and a symmetry plane ($XY$-plane). On the sphere surface there are two critical points ($\theta = 0$ and $\theta = \pi$) and a critical line ($\theta = \pi/2$).

### 2.5-2. Three-Dimensional Shear Flows

*Arbitrary three-dimensional straining shear flow.* Such flow is characterized by the boundary condition (2.5.1) remote from the drop with symmetric matrix of shear coefficients, $G_{kj} = G_{jk}$. The solution of the problem on an arbitrary three-dimensional straining shear flow past a drop leads to the following expressions for the velocities outside and inside the drop [26, 475]:

$$V_k^{(1)} = G_{kj} X_j \left( 1 - \frac{\beta}{\beta+1} \frac{a^5}{R^5} \right)$$

$$- \frac{1}{2a^2} G_{jl} X_k X_j X_l \left( \frac{5\beta+2}{\beta+1} \frac{a^5}{R^5} - \frac{5\beta}{\beta+1} \frac{a^7}{R^7} \right),$$

$$(2.5.9)$$

$$V_k^{(2)} = \frac{1}{2} \frac{1}{\beta+1} \left( 5\frac{R^2}{a^2} - 3 \right) G_{kj} X_j - \frac{1}{\beta+1} \frac{G_{jl} X_k X_j X_l}{a^2},$$

In these formulas $k$, $j$, $l = 1, 2, 3$ and the summation is carried over $j$ and $l$.

The value $\beta = 0$ corresponds to the case of a gas bubble, and the limit case $\beta \to \infty$ to a solid particle.

A plane straining shear flow around a spherical drop (see Section 1.1) is described by the expressions (2.5.9) with $G_{11} = -G_{22}$, $G_{33} = 0$, and $G_{ij} = 0$ for $i \ne j$.

Note that since the problem of Stokes flow is linear, one can find the velocity and pressure fields in translational-shear flows as the superposition of solutions describing the translational flow considered in Section 2.2 and shear flows considered in the present section.

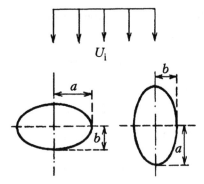

**Figure 2.5.** Flow past oblate and prolate ellipsoids of revolution

*Other results about shears flow past spherical particles.* The motion of a freely floating solid spherical particle in a simple shear flow was considered in [100]. In this case, all the coefficients $G_{ij}$ except for $G_{12}$ in the boundary conditions (2.5.1) are zero. The fact that the shear tensor has an antisymmetric component (see Section 1.1) results in the rotation of the particle because of the fluid no-slip condition on the particle boundary. The corresponding three-dimensional hydrodynamic problem was solved in the Stokes approximation. It was discovered that near the particle there is an area in which all streamlines are closed and outside this area, all streamlines are nonclosed.

The motion of a free spherical particle in an arbitrary plane shear flow was studied in [342, 343].

Arbitrary three-dimensional straining shear flows past a porous particle were considered in [524]. The flow outside the particle was described by using the Stokes equations (2.1.1). It was assumed that the percolation of the outer liquid into the particle obeys Darcy's law (2.2.24). The boundary conditions (2.5.1) remote from the particles and the conditions at the boundary of the particle described in Section 2.2 were satisfied. An exact closed solution for the fluid velocities and pressure inside and outside the porous particle was obtained.

# 2.6. Flow Past Nonspherical Particles

## 2.6-1. Translational Stokes Flow Past Ellipsoidal Particles

The axisymmetric problem about a translational Stokes flow past an ellipsoidal particle admits an exact closed-form solution. Here we restrict our consideration to a brief summary of the corresponding results presented in [179].

*Oblate ellipsoid of revolution.* Let us consider an oblate ellipsoid of revolution (on the left in Figure 2.5) with semiaxes $a$ and $b$ ($a > b$) in a translational Stokes flow with velocity $U_i$. We assume that the fluid viscosity is equal to $\mu$.

We pass from the Cartesian coordinates $X$, $Y$, $Z$ to the reference frame

$\sigma$, $\tau$, $\varphi$ fixed to the oblate ellipsoid of revolution by using the transformations

$$X^2 = m^2(1 + \sigma^2)(1 - \tau^2)\cos^2 \varphi,$$
$$Y^2 = m^2(1 + \sigma^2)(1 - \tau^2)\sin^2 \varphi,$$
$$Z = m\sigma\tau, \quad \text{where } m = \sqrt{a^2 - b^2}$$
$$(\sigma \geq 0, \ -1 \leq \tau \leq 1).$$

As a result, the ellipsoid surface is given by a constant value of the coordinate $\sigma$:

$$\sigma = \sigma_0, \quad \text{where } \sigma_0 = \left[(a/b)^2 - 1\right]^{-1/2}. \tag{2.6.1}$$

Since the problem is axisymmetric, we introduce the stream function as

$$V_\sigma = \frac{1}{m^2\sqrt{(1 + \sigma^2)(\sigma^2 + \tau^2)}} \frac{\partial \Psi}{\partial \tau}, \quad V_\tau = -\frac{1}{m^2\sqrt{(1 - \tau^2)(\sigma^2 + \tau^2)}} \frac{\partial \Psi}{\partial \sigma}. \tag{2.6.2}$$

Then the Stokes equation (2.1.1) are reduced to one equation for $\Psi$, which can be solved by the separation of variables. By satisfying the boundary condition for a uniform flow remote from the particle and the no-slip conditions at the particle boundary, we obtain

$$\Psi = \frac{1}{2}m^2 U_i(1 - \tau^2)\left[\sigma^2 + 1 - \frac{(\sigma_0^2 + 1)\sigma - (\sigma_0^2 - 1)(\sigma^2 + 1)\operatorname{arccot}\sigma}{\sigma_0 - (\sigma_0^2 - 1)\operatorname{arccot}\sigma_0}\right]. \tag{2.6.3}$$

In the similar problem on the motion of an oblate ellipsoid with velocity $U_i$ in a stagnant fluid, we have

$$\Psi = -\frac{1}{2}m^2 U_i(1 - \tau^2)\frac{(\sigma_0^2 + 1)\sigma - (\sigma_0^2 - 1)(\sigma^2 + 1)\operatorname{arccot}\sigma}{\sigma_0 - (\sigma_0^2 - 1)\operatorname{arccot}\sigma_0}. \tag{2.6.4}$$

The force exerted on the ellipsoid by the fluid is

$$F = \frac{8\pi\mu U_i\sqrt{a^2 - b^2}}{\sigma_0 - (\sigma_0^2 - 1)\operatorname{arccot}\sigma_0}. \tag{2.6.5}$$

As $\sigma_0 \to 0$, an oblate ellipsoid degenerates into an infinite thin disk of radius $a$. By passing to the limit in (2.6.4), we can obtain the following expression for the stream function:

$$\Psi = -\frac{1}{\pi}a^2 U_i(1 - \tau^2)\left[\sigma + (\sigma^2 + 1)\operatorname{arccot}\sigma\right].$$

The disk moving in the direction perpendicular to its plane with velocity $U_i$ in a stagnant fluid experiences the drag force

$$F = 16\mu a U_i, \tag{2.6.6}$$

which is less than the force acting on a sphere of the same radius (for the sphere, we have $F = 6\pi\mu a U_i$). Formula (2.6.6) is confirmed by experimental data.

*Prolate ellipsoid of revolution.* To solve the corresponding problem about an ellipsoidal particle (on the right in Figure 2.5) in a translational Stokes flow, we use the reference frame $\sigma$, $\tau$, $\varphi$ fixed to the prolate ellipsoid of revolution. The transformation to the coordinates $(\sigma, \tau, \varphi)$ is determined by the formulas

$$X^2 = m^2(\sigma^2 - 1)(1 - \tau^2)\cos^2\varphi, \quad Y^2 = m^2(\sigma^2 - 1)(1 - \tau^2)\sin^2\varphi, \quad Z = m\sigma\tau,$$

$$\text{where} \quad m = \sqrt{a^2 - b^2} \quad (\sigma \geq 1 \geq \tau \geq -1).$$

In this case, the ellipsoid surface is given by the equation

$$\sigma = \sigma_0, \quad \text{where} \quad \sigma_0 = \left[1 - (b/a)^2\right]^{-1/2}. \tag{2.6.7}$$

Here, as previously, the larger semiaxis is denoted by $a$.

The fluid velocity is given by

$$V_\sigma = \frac{1}{m^2\sqrt{(\sigma^2 - 1)(\sigma^2 - \tau^2)}} \frac{\partial\Psi}{\partial\tau}, \quad V_\tau = -\frac{1}{m^2\sqrt{(1 - \tau^2)(\sigma^2 - \tau^2)}} \frac{\partial\Psi}{\partial\sigma} \tag{2.6.8}$$

in terms of the stream function

$$\Psi = \frac{1}{2}m^2 U_{\mathrm{i}}(1 - \tau^2)\left[\sigma^2 - 1 - \frac{(\sigma_0^2 + 1)(\sigma^2 - 1)\operatorname{arctanh}\sigma - (\sigma_0^2 - 1)\sigma}{(\sigma_0^2 + 1)\operatorname{arctanh}\sigma_0 - \sigma_0}\right], \tag{2.6.9}$$

where $\operatorname{arctanh}\sigma = \dfrac{1}{2}\ln\dfrac{\sigma + 1}{\sigma - 1}$.

In the problem about a prolate ellipsoid of revolution moving with velocity $U_{\mathrm{i}}$ in a stagnant fluid, the corresponding stream function has the form

$$\Psi = -\frac{1}{2}m^2 U_{\mathrm{i}}(1 - \tau^2)\frac{(\sigma_0^2 + 1)(\sigma^2 - 1)\operatorname{arctanh}\sigma - (\sigma_0^2 - 1)\sigma}{(\sigma_0^2 + 1)\operatorname{arctanh}\sigma_0 - \sigma_0}. \tag{2.6.10}$$

The drag force is calculated as

$$F = \frac{8\pi\mu U_{\mathrm{i}}\sqrt{a^2 - b^2}}{(\sigma_0^2 + 1)\operatorname{arctanh}\sigma_0 - \sigma_0}. \tag{2.6.11}$$

If $a \gg b$, then the prolate ellipsoid degenerates into a needle-like rod. In this case, the force acting on the needle of length $a$ and radius $b$ which moves in the direction of its axis with velocity $U_{\mathrm{i}}$ has the form

$$F = \frac{4\pi\mu a U_{\mathrm{i}}}{\ln(a/b) + 0.193}. \tag{2.6.12}$$

By comparing the expressions for forces acting on oblate and prolate ellipsoids with the corresponding expression for the sphere with equivalent equatorial radius, we can write

$$F_{\mathrm{el}} = 6\pi\mu l U_{\mathrm{i}} K\left(\frac{b}{a}\right), \tag{2.6.13}$$

where $l = a$ for an oblate ellipsoid and $l = b$ for a prolate ellipsoid. Table 2.1 shows numerical values of the correction factor $K$ for various semiaxis ratios $b/a$.

TABLE 2.1
The correction factor $K(b/a)$ in (2.6.13)

| $b/a$ | $K$, oblate ellipsoid | $K$, prolate ellipsoid |
|-------|-----------------------|------------------------|
| 0     | 0.849                 | $\infty$               |
| 0.1   | 0.852                 | 2.647                  |
| 0.2   | 0.861                 | 1.785                  |
| 0.3   | 0.874                 | 1.470                  |
| 0.4   | 0.889                 | 1.305                  |
| 0.5   | 0.905                 | 1.204                  |
| 0.6   | 0.923                 | 1.136                  |
| 0.7   | 0.941                 | 1.087                  |
| 0.8   | 0.961                 | 1.051                  |
| 0.9   | 0.980                 | 1.022                  |
| 1.0   | 1.000                 | 1.000                  |

## 2.6-2. Translational Stokes Flow Past Bodies of Revolution

Let us consider bodies of revolution of any shape with arbitrary orientation in a translational flow at low Reynolds numbers. We assume that the axis of the body of revolution forms an angle $\omega$ with the direction of the fluid velocity at infinity (Figure 2.6). The unit vector $\mathbf{i}$ directed along the flow can be represented as the sum $\mathbf{i} = \boldsymbol{\tau}\cos\omega + \mathbf{n}\sin\omega$, where $\boldsymbol{\tau}$ is the unit vector directed along the body axis and $\mathbf{n}$ the unit vector in the plane of rotation of the body. In the Stokes approximation, the drag force is given by the following expression in the general case [179, 359]:

$$\mathbf{F} = \boldsymbol{\tau} F_{\parallel}\cos\omega + \mathbf{n}F_{\perp}\sin\omega, \qquad (2.6.14)$$

where $F_{\parallel}$ and $F_{\perp}$ are the drag forces of the body of revolution for its parallel ($\omega = 0$) and perpendicular ($\omega = \pi/2$) positions in the translational flow.

The projection of the drag force on the incoming flow direction is equal to the scalar product

$$(\mathbf{F}\cdot\mathbf{i}) = F_{\parallel}\cos^2\omega + F_{\perp}\sin^2\omega. \qquad (2.6.15)$$

It follows from (2.6.14) and (2.6.15) that to calculate the drag force of a body of revolution of any shape with arbitrary orientation in a Stokes flow, it suffices to know the value of this force only for two special positions of the body in space. The "axial" ($F_{\parallel}$) and "transversal" ($F_{\perp}$) drags can be obtained both theoretically

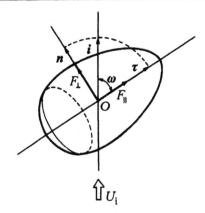

**Figure 2.6.** Body of revolution in translational flow (arbitrary orientation)

and experimentally. In what follows we present the expressions for $F_{\parallel}$ and $F_{\perp}$, given in [179] for some bodies of revolution of nonspherical shape.

For a thin circular disk of radius $a$, one has

$$F_{\parallel} = 16\mu a U_i, \qquad F_{\perp} = \tfrac{32}{3}\mu a U_i. \qquad (2.6.16)$$

For a dumbbell-like particle consisting of two adjacent spheres of equal radius $a$, one has

$$
\begin{aligned}
F_{\parallel} &= 12\pi\mu a U_i \lambda_{\parallel}, & \lambda_{\parallel} &\approx 0.645, \\
F_{\perp} &= 12\pi\mu a U_i \lambda_{\perp}, & \lambda_{\perp} &\approx 0.716.
\end{aligned}
\qquad (2.6.17)
$$

In these formulas, the product $12\pi\mu a U_i$ is equal to the sum of drag forces for two isolated spheres of radius $a$.

For oblate ellipsoids of revolution with semiaxes $a$ and $b$, one has

$$F_{\parallel} = 3.77\,(4a + b), \qquad F_{\perp} = 3.77\,(3a + 2b), \qquad (2.6.18)$$

where $a$ is the equatorial radius $(a > b)$.

For prolate ellipsoids of revolution with semiaxes $a$ and $b$, one has

$$F_{\parallel} = 3.77\,(a + 4b), \qquad F_{\perp} = 3.77\,(2a + 3b), \qquad (2.6.19)$$

where $b$ is the equatorial radius $(b > a)$.

Formulas (2.6.18) and (2.6.19) are approximate. They work well for slightly deformed ellipsoids of revolution. In (2.6.18), the maximum error is less than 6% for any ratio of the semiaxes.

## 2.6-3. Translational Stokes Flow Past Particles of Arbitrary Shape

A particle of an arbitrary shape moving in an infinite fluid that is at rest at infinity is subject to the action of the hydrodynamic force and angular momentum due to its translational motion and rotation, respectively [179]:

$$\mathbf{F} = \mu(\mathbf{K}\,\mathbf{U} + \mathbf{S}\,\omega), \tag{2.6.20}$$

$$\mathbf{M} = \mu(\mathbf{S}\,\mathbf{U} + \mathbf{\Omega}\,\omega), \tag{2.6.21}$$

where $\mathbf{K}$, $\mathbf{S}$, and $\mathbf{\Omega}$ are tensors of rank two depending on the particle geometry.

The symmetric tensor $\mathbf{K} = [K_{ij}]$ is called translational. It characterizes the drag of a body under translational motion and depends only on the size and shape of the body. In the principal axes, the translation tensor is reduced to the diagonal form

$$\mathbf{K} = \begin{bmatrix} K_1 & 0 & 0 \\ 0 & K_2 & 0 \\ 0 & 0 & K_3 \end{bmatrix}, \tag{2.6.22}$$

where $K_1$, $K_2$, $K_3$ are the principal drags acting on the body as it moves along the major axes. For orthotropic bodies (with three symmetry planes orthogonal to each other), the principal axes of the translational tensor are perpendicular to the corresponding symmetry planes. For axisymmetric bodies, one of the axes (say, the first) is a major axis and $K_2 = K_3$. For a sphere of radius $a$, any three pairwise perpendicular axes are major, and $K_1 = K_2 = K_3 = 6\pi a$.

A symmetric tensor $\mathbf{\Omega}$ is called a rotational tensor. It depends both on the shape and size of the particle and on the choice of the origin. The rotational tensor characterizes the drag under rotation of the body and has the diagonal form with entries $\Omega_1, \Omega_2, \Omega_3$ in the principal axes (the positions of the principal axes of the rotational and translational tensors in space are different). For axisymmetric bodies, one of the major axes (for instance, the first) is parallel to the symmetry axis, and in this case $\Omega_2 = \Omega_3$. For a spherical particle, we have $\Omega_1 = \Omega_2 = \Omega_3$.

The tensor $\mathbf{S}$ is symmetric only at a point $O$ unique for each body, this point is called the center of hydrodynamic reaction. This tensor is called the conjugate tensor and characterizes the crossed reaction of the body under translational and rotational motion (the drag moment in the translational motion and the drag force in the rotational motion). For bodies with orthotropic, axial, or spherical symmetry, the conjugate tensor is zero. However, it is necessary to take this tensor into account for bodies with helicoidal symmetry (propeller-like bodies).

In problems of gravity settling of particles, the translational tensor is most important.

The principal drags of some nonspherical bodies are given in [179] and below.

For a thin circular disk of radius $a$, we have

$$K_1 = 16a, \qquad K_2 = \tfrac{32}{3}a, \qquad K_3 = \tfrac{32}{3}a. \tag{2.6.23}$$

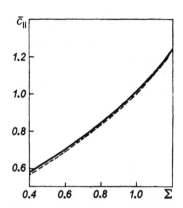

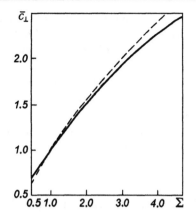

**Figure 2.7.** Relative drag coefficient for axisymmetric particles moving along the axis. Solid line, the approximate formula (2.6.28); dashed line, the exact solution for an ellipsoid of revolution

**Figure 2.8.** Relative drag coefficient for axisymmetric particles moving in a direction perpendicular to the axis. Solid line, the approximate formula (2.6.29); dashed line, the exact solution for an ellipsoid of revolution

For needle-like ellipsoids of length $l$ and radius $a$, one has

$$K_1 = \frac{4\pi l}{2\ln(l/a) - 1}, \quad K_2 = \frac{8\pi l}{2\ln(l/a) + 1}, \quad K_3 = \frac{8\pi l}{2\ln(l/a) + 1}. \qquad (2.6.24)$$

For thin circular cylinders of length $l$ and radius $a$, one has

$$K_1 = \frac{4\pi l}{\ln(l/a) - 0.72}, \quad K_2 = \frac{4\pi l}{\ln(l/a) + 0.5}, \quad K_3 = \frac{4\pi l}{\ln(l/a) + 0.5}. \qquad (2.6.25)$$

For a dumbbell-like particle consisting of two adjacent spheres of equal radius $a$, one has

$$K_1 = 24.3\,a, \qquad K_2 = 27.0\,a, \qquad K_3 = 27.0\,a. \qquad (2.6.26)$$

For an arbitrary ellipsoid with semiaxes $a$, $b$, and $c$, one has

$$K_1 = \frac{16\pi}{\chi + a^2\alpha}, \quad K_2 = \frac{16\pi}{\chi + b^2\beta}, \quad K_3 = \frac{16\pi}{\chi + c^2\gamma}. \qquad (2.6.27)$$

Here the parameters $\alpha$, $\beta$, $\gamma$, and $\chi$ are given by the integrals

$$\alpha = \int_\lambda^\infty \frac{d\lambda}{(a^2+\lambda)\Delta}, \quad \beta = \int_\lambda^\infty \frac{d\lambda}{(b^2+\lambda)\Delta}, \quad \gamma = \int_\lambda^\infty \frac{d\lambda}{(c^2+\lambda)\Delta}, \quad \chi = \int_\lambda^\infty \frac{d\lambda}{\Delta},$$

where $\Delta = \sqrt{(a^2 + \lambda)(b^2 + \lambda)(c^2 + \lambda)}$ and the lower limit of integration is the positive root of the cubic equation

$$\frac{x^2}{a^2 + \lambda} + \frac{y^2}{b^2 + \lambda} + \frac{z^2}{c^2 + \lambda} = 1.$$

For axisymmetric bodies of an arbitrary shape, we introduce the notion of a perimeter-equivalent sphere. To this end, we project all points of the surface of the body on a plane perpendicular to its axis. The projection is a circle of radius $a_\perp$. The perimeter-equivalent sphere has the same radius.

Experimental data and numerical results for principal values of the translational tensor for some axisymmetric and orthotropic bodies (cylinders, doubled cones, parallelepipeds) were discussed in [94]. It was established that the results are well approximated by the following dependence for the relative coefficient of the axial drag:

$$\bar{c}_{\parallel} = 0.244 + 1.035\,\Sigma - 0.712\,\Sigma^2 + 0.441\,\Sigma^3. \tag{2.6.28}$$

This dependence is shown in Figure 2.7 by solid line. The values of the axial drag of an axisymmetric body relative to the drag of the perimeter-equivalent sphere are plotted on the ordinate. The values of the perimeter-equivalent factor $\Sigma$ equal to the ratio of the surface area of the particle to the surface area of the perimeter-equivalent sphere are plotted on the abscissa.

Figure 2.8 illustrates the relative values of the transversal drag against a similar shape factor. The dashed line shows exact results for spheroids. One can calculate the relative coefficients of transversal drag by the formula [94]

$$\bar{c}_{\perp} = 0.392 + 0.621\,\Sigma - 0.04\,\Sigma^2, \tag{2.6.29}$$

which is well consistent with experimental data.

---

### 2.6-4. Sedimentation of Isotropic Particles

The steady-state rate $U_i$ of particle settling in mass force fields, primarily, in the gravitational field, is an important hydrodynamic characteristic of processes in chemical technology such as settling and sedimentation. Any spherically isotropic body homogeneous with respect to density experiences the same drag under translational motion regardless of its orientation. Such a body is also isotropic with respect to any pair of forces that arise when it rotates around an arbitrary axis passing though its center. If at the initial time such a body has some orientation in fluid and can fall without initial rotation, then this body falls vertically without rotation conserving its initial orientation.

It is convenient to describe the free fall of nonspherical isotropic particles by using the sphericity parameter

$$\psi = \frac{S_e}{S_*}, \tag{2.6.30}$$

where $S_*$ is the surface area of the particle and $S_e$ is the surface area of the volume-equivalent sphere. If the motion is slow, one can calculate the settling rate by the empirical formula [179]

$$\mathbf{U}_i = \frac{2}{9}\frac{Q\Delta\rho a_e^2}{\mu}\mathbf{g}, \tag{2.6.31}$$

where $a_e$ is the radius of the volume-equivalent sphere and

$$Q = 0.843 \ln \frac{\psi}{0.065}. \tag{2.6.32}$$

We present some values of the sphericity factor $\psi$ for some particles: sphere, 1.000; octahedron, 0.846; cube, 0.806; tetrahedron, 0.670.

### 2.6-5. Sedimentation of Nonisotropic Particles

If the velocity of a spherical particle in Stokes settling is always codirected with the gravity force, even for homogeneous axisymmetric particles the velocity is directed vertically if and only if the vertical coincides with one of the principal axes of the translational tensor **K**. If the angle between the symmetry axis and the vertical is $\varphi$, then the velocity direction is given by the angle [94]

$$\theta = \pi + \arctan\left(\frac{K_2}{K_1} \tan \varphi\right), \tag{2.6.33}$$

where $K_1$ and $K_2$ are the axial and transversal principal drags to the translational motion. The velocity is given by [94]

$$U_i = \frac{V \Delta \rho g}{\mu} (K_1^2 \cos^2 \theta + K_2^2 \sin^2 \theta)^{-1/2}, \tag{2.6.34}$$

where $V$ is the particle volume.

If the settling direction is not vertical, this means that a falling particle is subject to the action of a transverse force, which leads to its horizontal displacement. An additional complication is that the center of hydrodynamic reaction (including the buoyancy force) does not coincide with the particle center of mass. In this case, in addition to the translational motion, the particle is subject to rotation under the action of the arising moment of forces (e.g., the "somersault" of a bullet with displaced center of mass ). For axisymmetric particles, this rotation stops when the system "the mass center + the reaction center" becomes stable, that is, the mass center is ahead of the reaction center. In this case, the settling trajectory becomes stable and rectilinear.

However, in the more general case of an asymmetric particle, the combined action of the lateral and rotational forces may lead to motion along a spatial, for instance, spiral trajectory. At the same time, a steady-state settling trajectory with helicoidal (propeller-like) symmetry remains rectilinear, notwithstanding the body rotation [179].

Two theorems are useful for estimating the steady-state rate of settling of Stokes nonspherical particles. One theorem, proved by Hill and Power [187], states that the Stokes drag of an arbitrary body moving in viscous fluid is larger than the Stokes drag of any inscribed body. Thus, to determine the upper and lower bounds for the Stokes drag of a body of an exotic shape, one can suggest a reasonable set of inscribed and circumscribed bodies with known drags. The other theorem, proved by Weinberger [507], states that among all particles of different shape but equal volume, the spherical particle has the maximum Stokes settling rate.

---

### 2.6-6. Mean Velocity of Nonisotropic Particles Falling in a Fluid

The mean flow rate velocity $\langle U \rangle$ of a particle, which is obtained in a large number of experiments when an arbitrarily oriented particle falls in a fluid, is determined for the Stokes flow by the formula [179]:

$$\langle U \rangle = \frac{V \Delta \rho}{\mu \overline{K}} \mathbf{g}, \qquad (2.6.35)$$

where $V$ is the particle volume, $\Delta \rho$ is the difference between the densities of the particle and the fluid, $\mathbf{g}$ is the free fall acceleration, and $\overline{K}$ is the average drag expressed via the principal drags as

$$\frac{1}{\overline{K}} = \frac{1}{3} \left( \frac{1}{K_1} + \frac{1}{K_2} + \frac{1}{K_3} \right). \qquad (2.6.36)$$

The average drag force acting on an arbitrarily oriented particle falling in fluid is given by

$$\langle \mathbf{F} \rangle = -\mu \overline{K} \langle \mathbf{U} \rangle. \qquad (2.6.37)$$

Formulas (2.6.35)–(2.6.37) are important in view of some problems of Brownian motion.

In the special case of a spherical particle, one must set $V = \frac{4}{3}\pi a^3$ and $\overline{K} = 6\pi a$ in (2.6.35).

Let us calculate the average drag for a thin disk of radius $a$. To this end, we substitute the principal drags (2.6.23) into (2.6.36). As a result, we obtain

$$\overline{K} = 12a. \qquad (2.6.38)$$

Substituting the coefficients $K_1$, $K_2$, and $K_3$ from (2.6.24)–(2.6.27) into (2.6.36) and using (2.6.35), one obtains the average settling rate for the abovementioned nonspherical bodies.

---

### 2.6-7. Flow Past Nonspherical Particles at Higher Reynolds Numbers

The Stokes flow around particles of any shape is separation-free, that is, the stream lines come from infinity, bend round the body everywhere closely attaining the body surface, and return to infinity. However, for higher Reynolds numbers, separation occurs, which leads to wake formation behind the body [98, 517]. As the Reynolds number increases, the size of the wake region (the wake length) grows differently for different bodies. Figure 2.9 shows experimental and numerical data [94] for the ratio of the wake length $L_W$ to the equatorial section diameter $d$ against the Reynolds number for various axisymmetric bodies,

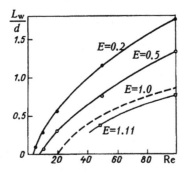

**Figure 2.9.** Relative length of the stern vortex

namely, a sphere and ellipsoids with various ratios $E$ of the axial to the equatorial dimension.

With further increase of the Reynolds number, the wake becomes nonsteady and completely turbulent and goes to infinity. The force action of the flow on the body is closely related to the wake size and state. The limit asymptotic cases of this action are the Stokes (as Re $\to$ 0) and the Newtonian (as Re $\to$ $\infty$) regimes of flow. We have already considered the characteristics of the Stokes flow. The Newtonian regime of a flow is characterized by the fact that the drag coefficient $c_f$ of the body is constant.

In axial flow past disks, which are the limit cases of axisymmetric bodies of small length, the drag coefficients are given in [94] for the entire range of Reynolds numbers calculated with respect to the radius. These formulas approximate numerical results and experimental data:

$$
\begin{aligned}
c_f &= 10.2\,\mathrm{Re}^{-1}(1 + 0.318\,\mathrm{Re}) && \text{for} && \mathrm{Re} \le 0.005, \\
c_f &= 10.2\,\mathrm{Re}^{-1}(1 + 10^s) && \text{for} && 0.005 < \mathrm{Re} \le 0.75, \\
c_f &= 10.2\,\mathrm{Re}^{-1}(1 + 0.239\,\mathrm{Re}^{0.792}) && \text{for} && 0.75 < \mathrm{Re} \le 66.5, \\
c_f &= 1.17 && \text{for} && \mathrm{Re} > 66.5,
\end{aligned}
\qquad (2.6.39)
$$

where $s = -0.61 + 0.906\log_{10}\mathrm{Re} - 0.025\,(\log_{10}\mathrm{Re})^2$.

The steady-state settling rate of a particle of an arbitrary shape (for Newtonian regime of a flow at high Reynolds numbers) can be obtained by the formula [94]

$$
U = 0.69\,\gamma^{1/36}[ga_e(\gamma - 1)(1.08 - \psi)]^{1/2} \quad \text{for} \quad 1.1 < \gamma < 8.6,
\qquad (2.6.40)
$$

where $\gamma$ is the particle-fluid density ratio, $a_e$ is the radius of the volume-equivalent sphere, and $\psi$ is the ratio of the area of the volume-equivalent sphere to the particle surface area.

# 2.7. Flow Past a Cylinder (the Plane Problem)

## 2.7-1. Translational Flow Past a Cylinder

In chemical technology and power engineering, equipment containing heat exchanging pipes and various cylindrical links immersed into moving fluid is often used. The estimation of the hydrodynamic action on these elements is based on the solution of the plane problem on the flow past a cylinder.

*Low Reynolds numbers.* In [216, 382] the problem on a circular cylinder of radius $a$ in translational flow of viscous incompressible fluid with velocity $U_i$ at low Reynolds numbers was solved by the method of matched asymptotic expansions. The study was carried out on the basis of the Navier–Stokes equations* (1.1.4) in the polar coordinates $\mathcal{R}$, $\theta$. Thus, the following expression for the stream function was obtained for $\mathcal{R}/a \sim 1$:

$$\Psi = aU \left( \frac{\mathcal{R}}{a} \ln \frac{\mathcal{R}}{a} - \frac{1}{2} \frac{\mathcal{R}}{a} + \frac{1}{2} \frac{a}{\mathcal{R}} \right) \sin \theta, \qquad (2.7.1)$$

where

$$U = U_i \left( \Delta - 0.87 \Delta^3 \right), \quad \Delta = \left( \ln \frac{3.703}{\mathsf{Re}} \right)^{-1}, \quad \mathsf{Re} = \frac{aU_i \rho}{\mu}.$$

The fluid velocity can be found by using formulas (1.1.8).

By using the stream function (2.7.1), one can calculate the drag coefficient

$$c_f = \frac{F}{aU_i^2 \rho} = \frac{4\pi}{\mathsf{Re}} \left( \Delta - 0.87 \Delta^3 \right), \qquad (2.7.2)$$

where $F$ is the force per unit length of the cylinder.

Comparison with experimental data shows that formula (2.7.2) can be used for $0 < \mathsf{Re} < 0.4$ [485].

*Nonseparating flow past a cylinder at moderate Reynolds numbers.* According to experimental data [486], a nonseparating flow past a circular cylinder is realized at $\mathsf{Re} \leq 2.5$. At such Reynolds numbers, one can use the following approximate formula for calculating the drag coefficient [94]:

$$c_f = 5.65 \, \mathsf{Re}^{-0.78} \left( 1 + 0.26 \, \mathsf{Re}^{0.82} \right) \qquad \text{for} \quad 0.05 \leq \mathsf{Re} \leq 2.5; \qquad (2.7.3)$$

this formula was obtained from experimental and numerical results.

*Separated flow past a cylinder at moderate Reynolds numbers.* If the Reynolds number becomes larger than the critical value $\mathsf{Re} \approx 2.5$, the vortex counterflow with closed streamlines arises near the rear point, that is, separation occurs [486]. As the Reynolds number increases, the separation point gradually

---

\* An attempt to solve the hydrodynamic problem on the flow past a cylinder by using the linear Stokes equations (2.1.1) leads to the Stokes paradox [179, 485].

moves from the axis upward along the cylinder surface. The drag coefficient for a separated flow past a cylinder at moderate Reynolds numbers can be calculated by the empirical formulas [94]

$$
\begin{aligned}
c_f &= 5.65 \times 10^{-0.78}\left(1 + 0.333\,\text{Re}^{0.55}\right) &&\text{for}\quad 2.5 < \text{Re} \leq 20,\\
c_f &= 5.65 \times 10^{-0.78}\left(1 + 0.148\,\text{Re}^{0.82}\right) &&\text{for}\quad 20 < \text{Re} \leq 200.
\end{aligned}
\tag{2.7.4}
$$

*Separated flow past a cylinder at high Reynolds numbers.* With further increase of Re, the rear vortices become longer and then alternative vortex separation occurs (the Karman vortex street is formed). Simultaneously, the separation point moves closer to the equatorial section. The frequency $\nu_f$ of vortex shedding from the rear area is an important characteristic of the flow past a cylinder. It can be determined from the empirical formula [117]

$$
\text{St} = \frac{0.13}{c_f}\left[1 - \exp(-2.38\,c_f)\right],
\tag{2.7.5}
$$

where $\text{St} = a\nu_f/U_i$ is the Strouhal number.

We also present another useful formula for the vortex separation frequency: $\nu_f = 0.08\,U_i/b$, where $b$ is the half-width of the wake at the point of destruction.

Starting from $\text{Re} \approx 0.5 \times 10^3$, one can speak of a developed hydrodynamic boundary layer. The flow remains laminar in a considerable part of this layer [486]. If the Reynolds number varies in the range $0.5 \times 10^3 < \text{Re} < 0.5 \times 10^5$, the separation point $\theta_s$ of the laminar boundary layer gradually moves from $71.2°$ to $95°$ [427, 486].

For $\text{Re} > 2000$, the wake becomes totally turbulent at large distances from the body.

According to [254], the curve $c_f(\text{Re})$ contains two straight-line segments (self-similarity areas), where the drag coefficient is practically constant,

$$
\begin{aligned}
c_f &= 1.0 &&\text{for}\quad 3 \times 10^2 < \text{Re} < 3 \times 10^3,\\
c_f &= 1.1 &&\text{for}\quad 4 \times 10^3 < \text{Re} < 10^5.
\end{aligned}
\tag{2.7.6}
$$

In the intermediate region between these segments, the drag coefficient monotonically increases with Reynolds number.

Extensive information about flow around circular cylinders is presented in [519].

*Developed turbulence in the boundary layer on a cylinder.* Developed turbulence within the boundary layer takes place at higher Reynolds numbers $\text{Re} \approx 10^5$ and is accompanied by the "drag crisis." According to [522], the cylinder drag first decreases sharply to $c_f \approx 0.3$ at $\text{Re} = 3.5 \times 10^5$ and then begins to increase and again enters the self-similar regime characterized by the constant value

$$
c_f = 0.9 \qquad \text{for}\quad \text{Re} > 5 \times 10^5.
\tag{2.7.7}
$$

In the book [117], some data are given on the hydrodynamic characteristics of bodies of various shapes; these data mainly pertain to the region of pre-crisis self-similarity. The influence of roughness of the cylinder surface and the turbulence level of the incoming flow on the drag coefficient is discussed in [522]. In [211], the relationship between hydrodynamic flow characteristics in turbulent boundary layers and the longitudinal pressure gradient is studied. Analysis of the transition to turbulence in the wake of circular cylinders is presented in [333].

Note that in some problems of heat and mass transfer and chemical hydro-dynamics, the velocity fields near the body can be determined by the flow laws of ideal nonviscous fluid. This situation is typical of flows in a porous medium [75, 153, 346] and of interaction between bodies and liquid metals (see Section 4.11, where the solution of heat problem for a translational ideal flow past an elliptical cylinder is given).

---

### 2.7-2. Shear Flow Around a Circular Cylinder

*Fixed cylinder.* Let us consider a fixed circular cylinder in an arbitrary steady-state linear shear flow of viscous incompressible fluid in the plane normal to the cylinder axis. The velocity field of such a flow remote from the cylinder in the Cartesian coordinates $X_1$, $X_2$ can be represented in the general case as follows:

$$\mathbf{V} \to \mathbf{G}\mathcal{R} \qquad \text{as} \quad \mathcal{R} \to \infty. \tag{2.7.8}$$

The shear tensor in (2.7.8) can be represented as the sum of the symmetric and antisymmetric tensors $\mathbf{G} = \mathbf{E} + \mathbf{\Omega}$ that correspond to the straining and rotational components of the fluid motion at infinity,

$$\mathbf{G} = \begin{bmatrix} G_{11} & G_{12} \\ G_{21} & G_{22} \end{bmatrix} = \begin{bmatrix} E_1 & E_2 \\ E_2 & -E_1 \end{bmatrix} + \begin{bmatrix} 0 & -\Omega \\ \Omega & 0 \end{bmatrix}, \quad \mathcal{R} = \begin{bmatrix} X_1 \\ X_2 \end{bmatrix},$$

$$E_1 = G_{11} = -G_{22}, \quad E_2 = \tfrac{1}{2}(G_{12} + G_{21}), \quad \Omega = \tfrac{1}{2}(G_{21} - G_{12}),$$

and are determined by the three independent values $E_1$, $E_2$, and $\Omega$.

In the Stokes approximation (as $\text{Re} \to 0$), the solution of the corresponding hydrodynamic problem with the boundary conditions (2.7.8) at infinity and no-slip conditions at the boundary of the cylinder ($\mathbf{V} = 0$ for $\mathcal{R} = a$) leads to the stream function [93, 166]

$$\Psi = \frac{1}{2}a^2\overline{E}\left(\frac{\mathcal{R}}{a} - \frac{a}{\mathcal{R}}\right)^2 \sin 2\overline{\theta} - \frac{1}{2}a^2\Omega\left(\frac{\mathcal{R}^2}{a^2} - 1 - 2\ln\frac{\mathcal{R}}{a}\right), \tag{2.7.9}$$

where

$$\overline{E} = (E_1^2 + E_2^2)^{1/2}, \quad \overline{\theta} = \theta + \Delta\theta \quad \left(\frac{E_1}{\overline{E}} = \cos(2\Delta\theta), \quad \frac{E_2}{\overline{E}} = -\sin(2\Delta\theta)\right).$$

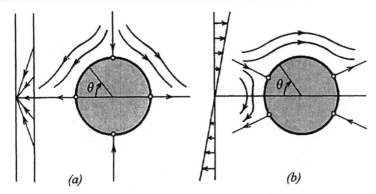

**Figure 2.10.** Linear shear flow past a fixed circular cylinder: (a) straining flow ($E_2 = 0$ and $\Omega = 0$); (b) simple shear flow ($E_1 = 0$ and $E_2 = -\Omega$)

These expressions are written in the coordinates $\mathcal{R}$, $\bar{\theta}$ obtained from the initial point by revolution by the angle $\Delta\theta$ and related to the principal axes of the symmetric tensor $\mathbf{E}$ (in the principal axes, the tensor $\mathbf{E}$ can be reduced to the diagonal form with diagonal entries $\bar{E}$ and $-\bar{E}$). Purely straining shear corresponds to the value $\Omega = 0$, and simple shear is determined by the parameters $E_1 = 0$ and $E_2 = -\Omega$.

The fluid velocity field can be obtained by substituting (2.7.9) into (1.1.8).

The structure of the streamlines $\Psi = \mathrm{const}$ substantially depends on the parameters $\bar{E}$ and $\Omega$. For the qualitative analysis of the flow, it is convenient to introduce the dimensionless rotation velocity of the flow at large distances from the cylinder:

$$\Omega_E = \Omega/\bar{E}.$$

For $0 \leq |\Omega_E| \leq 1$, all streamlines are not closed, and on the surface of the cylinder there are four critical points with the angular coordinates

$$\bar{\theta}_k = \frac{(-1)^{k+1}}{2}\arcsin\frac{\Omega_E}{2} + \frac{\pi}{2}(k-1), \quad \theta_k = \bar{\theta}_k + \frac{1}{2}\arctan\frac{E_2}{E_1}; \quad k = 1, 2, 3, 4.$$

Figure 2.10 presents a qualitative picture of streamlines for straining ($\Omega_E = 0$) and simple shear ($\Omega_E = 1$) flow. As the dimensionless angular flow velocity $\Omega_E$ at infinity increases from zero to unity, the critical separation point $\theta_1$ on the surface of the cylinder moves by $30°$.

For $|\Omega_E| > 1$, the streamlines near the cylinder surface are closed, and the streamlines far from the surface are not closed.

Note that in [523], a similar plane problem on an arbitrary linear shear flow past a porous cylinder was solved. The flow outside the cylinder was described by using the Stokes equations and the percolation of the outer fluid inside the porous cylinder was assumed to obey Darcy's law (2.2.24). The amount of the fluid penetrating into the cylinder per unit time was determined.

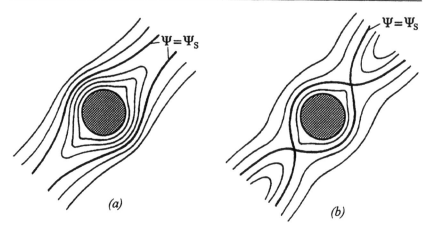

**Figure 2.11.** Linear shear flow past a freely rotating circular cylinder in the $\mathcal{R}\bar{\theta}$-plane (the limit streamlines $\Psi = \Psi_s$ are marked bold): (a) simple shear flow ($|\Omega_E| = 1$); (b) general case of plane shear flow ($0 < |\Omega_E| < 1$)

***Freely rotating cylinder.*** Now let us consider a circular cylinder freely floating in an arbitrary linear shear Stokes flow (Re $\rightarrow$ 0). The velocity distribution for such a flow remote from the cylinder is still given by relations (2.7.8).

In view of the no-slip condition on the surface of the circular cylinder freely floating in a shear flow, this cylinder rotates at the constant angular velocity equal to the rotation flow velocity at infinity. This means that the following boundary conditions for the fluid velocity components must be satisfied on the cylinder surface:

$$V_{\mathcal{R}} = 0, \quad V_\theta = a\Omega \quad \text{at} \quad \mathcal{R} = a. \tag{2.7.10}$$

The solution of the hydrodynamic problem on a freely rotating cylinder in an arbitrary shear Stokes flow with the boundary conditions (2.7.8) and (2.7.10) has the form [93, 166]

$$\Psi = \frac{1}{2}a^2\overline{E}\left(\frac{\mathcal{R}}{a} - \frac{a}{\mathcal{R}}\right)^2 \sin 2\bar{\theta} - \frac{1}{2}a^2\Omega\left(\frac{\mathcal{R}^2}{a^2} - 1\right),$$

$$V_{\mathcal{R}} = \frac{a^2\overline{E}}{\mathcal{R}}\left(\frac{\mathcal{R}}{a} - \frac{a}{\mathcal{R}}\right)^2 \cos 2\bar{\theta}, \quad V_{\bar\theta} = \Omega\mathcal{R} - \overline{E}\mathcal{R}\left(1 - \frac{a^4}{\mathcal{R}^4}\right)\sin 2\bar{\theta}, \tag{2.7.11}$$

where the parameters $E_1$, $E_2$, $\overline{E}$, and $\Omega$ are introduced in the same way as in the problem on the flow past a fixed cylinder.

In this case, for $\Omega_E \neq 0$ there are no critical points on the cylinder surface and there exist two qualitatively different types of flow characterized by the angular velocity $\Omega$. Namely, for $0 < |\Omega_E| \leq 1$, there are both closed and nonclosed streamlines; in this case, near the cylinder surface, there is an area with completely closed streamlines, and at large distances from the cylinder, the streamlines are not closed (Figure 2.11). For $|\Omega_E| > 1$, all streamlines are closed.

It follows from formulas (2.7.11) that for $0 < |\Omega_E| \le 1$, in flow there are two critical points with coordinates

$$\bar{\theta}_1^\circ = \frac{\pi}{4}, \quad \bar{\theta}_2^\circ = \frac{5\pi}{4}, \quad \mathcal{R}_{1,2}^\circ = a\left(\frac{1}{1-\Omega_E}\right)^{1/4}, \qquad (2.7.12)$$

at which the velocity vanishes: $V_R^\circ = V_\theta^\circ = 0$. These singular points are self-intersection points of the limit streamline that separates areas with closed and nonclosed streamlines (Figure 2.11).

The limit streamline is given by the equation

$$\Psi = \Psi_s, \qquad \Psi_s = a^2 E\left[\tfrac{1}{2}\Omega_E - 1 + (1-\Omega_E)^{1/2}\right].$$

For $\Omega_E \to 0$, it follows from formula (2.7.12) that $\mathcal{R}_{1,2}^\circ \to a$, that is, the critical points tend to the surface of the cylinder as the angular flow velocity decreases. In the other limit case, $\Omega_E \to 1$, which corresponds to simple shear, we obtain $\mathcal{R}_{1,2}^\circ \to \infty$ (that is, the critical points go to infinity).

# 2.8. Flow Past Deformed Drops and Bubbles

The dynamic interaction between flow and drops and bubbles floating in the flow may deform or even destroy them. This phenomenon is important for chemical technological processes since it may change the interfacial area and the relative velocity of phases and cause transient effects. In this case, the viscous and inertial forces are perturbing actions, and the capillary forces are obstructing actions. The bubble shape depends on the Reynolds number $\mathsf{Re} = a_e U_i \mu/\rho$ and the Weber number $\mathsf{We} = a_e U_i^2 \rho/\sigma$, where $\mu$ and $\rho$ are the dynamic viscosity and the density of the continuous phase, $\sigma$ is the surface tension coefficient, and $a_e$ is the radius of the sphere volume-equivalent to the bubble.

## 2.8-1. Weak Deformations of Drops at Low Reynolds Numbers

**Translational flow.** At low Reynolds and Weber numbers, the axisymmetric problem on the slow translational motion of a drop with steady-state velocity $U_i$ in a stagnant fluid was studied in [476] under the assumption that $\mathsf{We} = O(\mathsf{Re}^2)$. Deformations of the drop surface were obtained from the condition that the jump of the normal stress across the drop surface is equal to the pressure increment associated with interfacial tension. It was shown that a drop has the shape of an oblate (in the flow direction) ellipsoid with the ratio of the major to the minor semiaxis equal to

$$\chi = 1 + \varepsilon\, \mathsf{We}. \qquad (2.8.1)$$

Here the dimensionless parameter $\varepsilon$ is given by the expression

$$\varepsilon = \frac{3}{8(\beta+1)^3}\left[\left(\frac{81}{80}\beta^3 + \frac{57}{20}\beta^2 + \frac{103}{40}\beta + \frac{3}{4}\right) - \frac{\gamma-1}{12}(\beta+1)\right],$$

where $\beta$ is the drop–fluid dynamic viscosity ratio and $\gamma$ is the drop–fluid density ratio. The values $\beta \approx 0$ and $\gamma \approx 0$ correspond to a gas bubble.

**Shear flows.** A drop in simple shear flow is considered in [474, 475]. The boundary conditions remote from the drop are

$$V_X \to GY, \quad V_Y \to 0, \quad V_Z \to 0 \quad \text{as} \quad R \to \infty,$$

where $R = (X^2 + Y^2 + Z^2)^{1/2}$. The problem was solved in the Stokes approximation for small values of the dimensionless parameter $Ga_e\mu/\sigma$. It was shown that the drop shape is described by the equation

$$R = a_e\left(1 + \frac{Ga_e\mu}{\sigma} \frac{19\beta + 16}{16\beta + 16}XY\right), \tag{2.8.2}$$

and the drop is a prolate ellipsoid of revolution. Experimental data [474] confirm the validity of Eq. (2.8.2).

Axisymmetric shear flow past a gas bubble was studied numerically in [472].

## 2.8-2. Rise of an Ellipsoidal Bubble at High Reynolds Numbers

Let us consider the motion of a gas bubble at high Reynolds numbers. For small We, the bubble shape is nearly spherical. The Weber numbers of the order of 1 constitute an intermediate range of We, very important in practice, when the bubble, though essentially deformed, conserves its symmetry with respect to the midsection. For such We, the bubble shape is well approximated by an ellipsoid with semiaxes $a$ and $b = \chi a$ oblate in the flow direction; the semiaxis $b$ is directed across the flow, and $\chi \geq 1$.

The requirement that the boundary condition for the normal stress be satisfied at the front and rear critical points, as well as along the boundary of the midsection of the bubble, leads to the following relationship between the Weber number We and the ratio $\chi$ of the major semiaxis to the minor semiaxis of the ellipsoid [291]:

$$\text{We} = 2\chi^{-4/3}(\chi^3 + \chi - 2)\left[\chi^2 \operatorname{arcsec}\chi - (\chi^2 - 1)^{1/2}\right]^2(\chi - 1)^{-3}. \tag{2.8.3}$$

Numerical estimates in [291] show that the maximal deviation of the true curvature from the corresponding value for the approximating ellipsoid does not exceed 5% for We $\leq 1$ ($\chi \leq 1.5$) and 10% for We $\leq 1.4$ ($\chi \leq 2$).

Note that for

$$\text{Re} \geq 0.55\,\text{Mo}^{-1/5} \quad (\text{Mo} = g\rho^3\nu^4\sigma^{-3})$$

the bubble shape differs from the spherical shape by more than 5% ($\nu$ is the kinematic viscosity of the fluid, $g$ is the free fall acceleration, and Mo is the dimensionless Morton number, which depends only on the fluid properties).

For usual liquids like water, we have Mo $\sim 10^{-10}$, and one must take account of the bubble deformation starting from Re $\sim 10^2$. (For oil, Mo $\sim 10^{-2}$, and the bubble deformation is essential even for low Reynolds numbers.)

The rise velocity $U_i$ of an ellipsoidal bubble and the ratio $\chi$ of its axes were obtained in [337, 495] as functions of the equivalent radius $a_e = (ab^2)^{1/3}$. In the general case, the expression for the rise velocity of a bubble has the form

$$U_i = U_0 f(\text{Mo}, a_e/a_0), \qquad (2.8.4)$$

where the dimensionless Morton number $\text{Mo}$ and the dimensional quantities $U_0$ and $a_0$ depend only on the fluid properties,

$$U_0 = \left(\frac{\sigma^2 g}{\rho^2 \nu}\right)^{1/5}, \qquad a_0 = \left(\frac{\sigma \nu^2}{\rho g^2}\right)^{1/5}. \qquad (2.8.5)$$

Under the condition $\text{Mo}^{1/5} \ll 1$, which is usually satisfied, the ratio $U_i/U_0$ and the ellipticity $\chi$ depend only on $a_e/a_0$. These functions are universal and can be represented as [495, 497]

$$a_e/a_0 = \text{We}^{1/5} E^{2/5}, \qquad U_i/U_0 = \text{We}^{2/5} E^{-1/5}. \qquad (2.8.6)$$

Here $\text{We}(\chi)$ is given by formula (2.8.3) or by the more precise formula [495, 497]

$$\text{We}(\chi) = 2\rho a_e \frac{dS}{d\chi} \left(\frac{dm}{d\chi}\right)^{-1}, \qquad (2.8.7)$$

where $S$ is the surface area and $m$ is the virtual mass of the ellipsoid,

$$S = 2\pi a_e^2 \frac{(1+\alpha^2)^{1/3}}{\alpha^{2/3}} \left(1 + \frac{\alpha^2}{\sqrt{1+\alpha^2}} \ln \frac{1+\sqrt{1+\alpha^2}}{\alpha}\right),$$

$$m = \frac{4\pi}{3} a_e^3 \frac{(1+\alpha^2)(1-\alpha \operatorname{arccot} \alpha)}{1-(1+\alpha^2)(1-\alpha \operatorname{arccot} \alpha)}, \qquad \chi = \sqrt{1 + \frac{1}{\alpha^2}}.$$

According to [497], the function $E(\alpha)$ is determined by

$$E(\alpha) = \frac{3(1+\alpha^2)^{2/3}[\alpha + (1-\alpha^2)\operatorname{arccot} \alpha]}{\alpha^{7/3}[(1+\alpha^2)\operatorname{arccot} \alpha - \alpha]^2}. \qquad (2.8.8)$$

For $a_e \le 3a_0$, the following asymptotic estimates hold for the ellipticity and the rise velocity [337]

$$\chi = 1 + \tfrac{1}{288}(a_e/a_0)^5, \qquad U_i = \tfrac{1}{9} U_0 (a_e/a_0)^2. \qquad (2.8.9)$$

It follows from (2.8.6) that the dimensionless rise velocity $U_i/U_0$ attains its maximum 0.6 at $a_e = 3.7\,a_0$, $\chi = 1.9$, which is in a good agreement with experimental data.

If the bubble size continues to grow, $a_e > 3.7\,a_0$, then, by virtue of the increase in $\chi$, the viscous drag grows faster than the buoyancy force, and the bubble velocity decreases. For $a_e/a_0 \ge 8$, the ellipsoidal bubble model cannot be applied any longer.

Many empirical relations for steady-state velocity of deformed drops and bubbles of various shapes, including shapes more complicated than the ellipsoidal shape, are presented in [94]. Laminar flow past nonspherical drops was studied numerically in [98, 517].

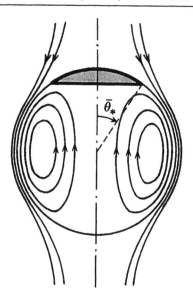

**Figure 2.12.** Qualitative picture of flow past a bubble of spherical segment shape

## 2.8-3. Rise of a Large Bubble of Spherical Segment Shape

As the size of rising bubbles and drops grows, their shape tends to the equilibrium shape, which more and more differs from spherical. If for low and moderate Re and low We, the bubble shape is close to spherical, for moderate $Re = 10^2$ to $10^3$ and We of the order of several units, the bubble shape may approximately be modeled by an oblate ellipsoid, and its trajectory by a helix. If We continues to grow, the "bottom" of the bubble becomes more and more flat. Finally, for We > 10 and high Re the bubble acquires the shape of a "turned-over cup" or a spherical segment and rises along the vertical. A detailed analysis of various regimes is presented in [94].

The steady-state rise velocity for a large bubble can be determined from the following model, confirmed by visual observations. The bubble is a spherical segment (Figure 2.12) with half-opening angle $0 \le \bar{\theta} \le \bar{\theta}_*$, where the angular coordinate $\bar{\theta}$ is measured from the front stagnation point. The remaining part of the sphere is occupied by the toroidal stern vortex thus constituting a complete sphere in the external flow. It is assumed that the flow near the spherical boundary of a gas bubble is potential [270].

In [26, 94] the rise velocity of such a bubble was obtained on the basis of this model:

$$U_i = \tfrac{2}{3}\sqrt{ag}, \qquad (2.8.10)$$

where $a$ is the radius of curvature of the spherical segment. Formula (2.8.10) describes experimental data sufficiently well [94].

We also write out an empirical formula [495], which gives the rise velocity of a spherical-segment bubble in terms of the radius $a_e$ of the volume-equivalent sphere:

$$U_i = 1.01\sqrt{a_e g}.$$

The half-opening angle $\bar{\theta}_*$ can be estimated by the semiempirical formula

$$\bar{\theta}_* = 50 + 190\exp(-0.62\,\text{Re}^{0.4}),$$

where $\bar{\theta}_*$ is measured in degrees. Note that one can set $\bar{\theta}_* \approx 50°$ for $\text{Re} > 10^2$.

Rising (settling) of large low-viscous drops can also be accompanied by strong deformation of their boundary, which takes the form of a spherical segment. The rise velocity of such drops can be estimated by the formula [94]

$$U_i = \frac{2}{3}\sqrt{ga\frac{|\Delta\rho|}{\rho}}, \qquad (2.8.11)$$

where $a$ is the radius of curvature of the spherical segment, $g$ is the free fall acceleration, $\rho$ is the fluid density, and $\Delta\rho$ is the difference between the densities of the phases.

Steady axisymmetric motion of deformable drops falling or rising through a homoviscous fluid in a tube at intermediate Reynolds number was studied numerically in [55].

## 2.8-4. Drops Moving in Gas at High Reynolds Numbers

The dependence of drop deformation on the Weber number and the vorticity inside the drop was studied in [336]. It was shown that the drop is close in shape to an oblate ellipsoid of revolution with semiaxis ratio $\chi > 1$. If there is no vortex inside the drop, then this dependence complies with the function $\text{We}(\chi)$ given in (2.8.3). The ratio $\chi$ decreases as the intensity of the internal vortex increases. Therefore, the deformation of drops moving in gas is significantly smaller than that of bubbles at the same Weber number $\text{We}$. The vorticity inside an ellipsoidal drop, just as that of the Hill vortex, is proportional to the distance $\mathcal{R}$ from the symmetry axis,

$$\omega = |\operatorname{rot}\mathbf{V}_2| = A\mathcal{R}\sin\theta.$$

The intensity parameter of a vortex can be expressed via $\chi$ as follows [336]:

$$A = 3\frac{U_i}{a_e^2}\frac{v_{max}(4+\chi^2)^2}{\chi^2(16-2\chi^2+\chi^4)}, \qquad (2.8.12)$$

where the dependence of $v_{max}$ on $\text{Re}_1$ and $\text{Re}_2$ is described by (2.4.8) and (2.4.9); $a_e$ is the radius of the volume-equivalent sphere.

The steady-state velocity of fall in gas (for example, of a rain drop in air) can be calculated by

$$U_i = \sqrt{\frac{8a_e g\gamma}{3c_f}}, \qquad (2.8.13)$$

where $\gamma$ is the ratio of the density of the drop to the density of gas and the drag coefficient $c_f$ is empirically related to the parameter $\chi$ as

$$c_f = 0.365\,\chi^{1.8}. \tag{2.8.14}$$

Formulas (2.8.12)–(2.8.14) together with the dependence $\chi(\text{We}, A)$ completely determine the motion of a drop in gas. A condition related to the exponential growth of the oscillation amplitude under which the failure of a drop starts, was obtained in [336]. For a rain drop, this condition approximately corresponds to the values $\chi = \frac{5}{3}$, $\text{We} = 5$, and $a_e = 3.8$ mm.

Under strong deformations, drops split into smaller ones, that is, are destroyed. The destruction process for drops is very complicated and is determined by surface tension, viscosity, inertia forces and some other factors. For various characteristic velocities of the relative phase motion, the character of destruction may be essentially different. A comparative analysis of many experimental and theoretical studies of drop destruction was given in [154, 312]. It was pointed out that there are six basic mechanisms of drop destruction, which correspond to different ranges of the Weber number.

# 2.9. Constrained Motion of Particles

The motion of a particle in infinite fluid creates some velocity and pressure fields. Neighboring particles move in already perturbed hydrodynamic fields. Simultaneously, the first particle itself experiences hydrodynamic interaction with the neighboring particles and neighboring moving or fixed surfaces. Since in the majority of actual disperse systems, the existence of an ensemble of particles and the apparatus walls is inevitable, the consideration of the hydrodynamic interaction between these objects is very important. One of the methods for obtaining the required information about the interaction is based on the construction of exact closed-form solutions. However, even within the framework of Stokes hydrodynamics, to describe the motion of an ensemble of particles is a very complicated problem, which admits an exact closed-form solution only in exceptional cases.

## 2.9-1. Motion of Two Spheres

*Motion of two spheres along a line passing through their centers.* In the Stokes approximation, an exact closed-form solution of the axisymmetric problem about the motion of two spheres with the same velocity was obtained in [463]. This solution is practically important and can be used for estimating the accuracy of approximate methods, which are applied for solving more complicated problems on the hydrodynamic interaction of particles.

The force acting on each of the spheres is described by the formula [179]

$$F = 6\pi\mu a U \lambda, \tag{2.9.1}$$

where $a$ is the sphere radius, $U$ is the velocity of the spheres, and $\lambda$ is the correction coefficient depending on the radii and the distance $l$ between the centers of the spheres. In the case of spheres of equal radii, the expression for $\lambda$ has the form

$$\lambda = \frac{4}{3} \sinh \alpha \sum_{n=1}^{\infty} \frac{n(n+1)}{(2n-1)(2n+3)} \left\{ 1 - \frac{4 \sinh^2[(n+\frac{1}{2})\alpha] - (2n+1)^2 \sinh^2 \alpha}{2 \sinh[(2n+1)\alpha] + (2n+1) \sinh 2\alpha} \right\},$$

(2.9.2)

where $\alpha = \ln\left[ \frac{1}{2}(l/a) + \sqrt{\frac{1}{4}(l/a)^2 - 1} \right]$.

For numerical calculations it is convenient to use the approximate formula

$$\lambda = \frac{0.88\,a + l}{2.5\,a + l},$$

(2.9.3)

whose maximal error is less than 1.3% for any $a$ and $l$.

Since $\lambda \leq 1$, it follows from (2.9.1) that the velocity of the steady-state motion of each of the spheres in the ensemble is greater than the velocity of a single sphere.

In the gravitational field, the steady-state velocities of particles and drops of various shapes (or mass) are different [179, 417]. Therefore, the distance between the centers of particles is not constant, and hence, the entire problem about the hydrodynamic interaction is, strictly speaking, nonsteady. It was shown in [417] that for $\text{Re} \ll \frac{1}{2}l/a$ this problem can be treated as quasisteady.

*Motion of two spheres arbitrarily located with respect to each other.* Let us consider two spherical particles of equal radius remote from each other and moving with the same velocity $\mathbf{U}$. The Stokes force acting on each particle is determined by the formula [179]

$$\frac{\mathbf{F}}{6\pi a \mu} = -\mathbf{i}_X \frac{U_X}{1 + \frac{3}{4}(a/l)} + \mathbf{i}_Z \frac{U_Z}{1 + \frac{3}{2}(a/l)},$$

(2.9.4)

where $Z$ is the axis passing through the centers of the spheres and $X$ is the transverse axis.

It follows from (2.9.4) that if spheres fall under the action of the gravity force directed neither along the $X$-axis nor along the $Z$-axis and forming an angle $\beta$ with the $Z$-axis, the particles fall with vertical velocity

$$U_{\parallel} = -\frac{F}{6\pi \mu a} \left[ 1 + \frac{3a}{4l}(1 + \cos^2 \beta) \right],$$

(2.9.5)

and drift horizontally with velocity

$$U_{\perp} = -\frac{F}{6\pi \mu a} \frac{3a}{4l} \sin \beta \cos \beta.$$

(2.9.6)

The books [179, 183, 517] present a detailed review of investigations of the hydrodynamic interaction of two particles of different shape in translational and shear Stokes flows. The cited books also give numerous formulas, tables, and graphs, which show the drag force of particles against the distance between them. The leading terms of asymptotic expansions of the particle drag with respect a small dimensionless distance between their boundaries are given in [183].

The results of numerous papers dealing with the hydrodynamic interaction of two drops were analyzed in [154, 183, 517]. Some results of calculations for the drag force are given there (for various drop radii, values of drop viscosity, and various distances between drops).

The axial and transverse motion of two drops close to each other was considered in [525]. Some leading terms were obtained in the asymptotic expansion of the drag force with respect to the small dimensionless distance between the drop boundaries. The case of interaction between a solid particle and a drop was also investigated.

The deformation of the shapes of drops and bubbles moving close to each other near a plane free surface was analyzed in [89–91, 275, 517].

## 2.9-2. Gravitational Sedimentation of Several Spheres

By the reflection method, some relations between the drag force $F$ and the sedimentation velocity $U$ averaged with respect to various orientations of particles of equal radius in space were obtained in [179]. It was assumed that the maximum distance $l$ between the sphere centers is much larger than their radius $a$. In all considered cases, the drag force is described by formula (2.9.1), where $\lambda$ is the correction coefficient depending on the configuration of the system of particles. In what follows, we write out the correction coefficient for some typical configurations of particles.

For a system of two spheres, one has

$$\lambda = \frac{2}{1 + (a/l)}. \tag{2.9.7}$$

For a system of three spheres arranged in a line, one has

$$\lambda = \frac{3}{1 + \frac{10}{3}(a/l) - \frac{1}{4}(a/l)^2}. \tag{2.9.8}$$

For a system of four spheres arranged in a line, one has

$$\lambda = \frac{4}{1 + \frac{13}{2}(a/l) - \frac{9}{8}(a/l)^2}. \tag{2.9.9}$$

For a system of four spheres arranged in vertices of a square, one has

$$\lambda = \frac{4}{1 + 2.7\,(a/l) - 0.04\,(a/l)^2}. \tag{2.9.10}$$

For a system of eight spheres arranged in vertices of a cube, one has

$$\lambda = \frac{8}{1 + 5.7\,(a/l) - 0.34\,(a/l)^2}. \tag{2.9.11}$$

Flow past periodic arrays of spheres was studied in [87, 517].

TABLE 2.2
Values of the coefficient $k$ in (2.9.12)

| External boundary shape | Particle center position | Motion direction | $k$ |
|---|---|---|---|
| One plane wall | At distance $l$ from the wall | Parallel to the wall | $\frac{9}{16}$ |
| One plane wall | At distance $l$ from the wall | Perpendicular to the wall | $\frac{9}{8}$ |
| Parallel walls at distance $2l$ | At distance $l$ from the wall | Parallel to the walls | 1.004 |
| Circular cylinder of radius $l$ | At distance $b$ from the wall | Along the axis | $2.1044 - 6577\,(b/l)^2$ |
| Sphere of radius $l$ | At the center of the sphere | Radial | $\frac{9}{4}$ |

### 2.9-3. Wall Influence on the Sedimentation of Particles

In actual systems, the particle sedimentation goes, as a rule, in volumes bounded by apparatus walls. If particles move in a fluid of infinite extent, the streamlines of the induced flow close at infinity. Therefore, in the consistent motion of particles in the ensemble, each particle moves in the codirected wake induced by the neighboring particles. As a result, the drag of each particle is smaller than in the case of motion of a single particle, and the sedimentation rate is accordingly larger. In the space bounded by the apparatus walls, the particle motion must induce the counterflow of fluid because of the volume replacement. Therefore, the drag force must be larger than that of a single particle in the space of infinite extent, and the sedimentation rate must be accordingly smaller.

Brenner [61], by using the reflection method, derived the following relation, which allows one to correct the Stokes drag law by taking into account the influence of the walls:

$$F = \frac{F_a}{1 - k(F_a/F_l)}, \qquad (2.9.12)$$

where $l$ is a parameter characterizing the distance between the particle and the walls and $F_a = 6\pi\mu U_i a$ and $F_l = 6\pi\mu U_i l$ are the drag forces of spheres of radii $a$ and $l$ moving with velocity $U_i$ in a stagnant fluid.

The value of the coefficient $k$, calculated for various cases, is given in Table 2.2. Note that formula (2.9.12) can be used for $b/l \ll 1$, where $b$ is the maximal dimension of the particle.

A spherical particle moving parallel to the wall is considered in [323]. It is assumed that the gap $h$ between the particle boundary and the wall is small compared with the radius $a$ of the particle. Several terms of the asymptotic expansion with respect to the small parameter $\varepsilon = h/a$ were obtained for the drag and momentum:

$$F = 6\pi\mu a U_i \left[-0.231 \ln\varepsilon + 0.746 + O(\varepsilon \ln\varepsilon)\right],$$
$$M = -8\pi\mu a^2 U_i \left[0.0434 \ln\varepsilon + 0.232 + O(\varepsilon \ln\varepsilon)\right]. \tag{2.9.13}$$

A similar asymptotic expansion for the force that acts on a spherical bubble moving parallel to a solid plane was obtained in [526]:

$$F = 4\pi\mu a U_i[-0.3 \ln\varepsilon + 0.93 + O(\varepsilon \ln\varepsilon)]. \tag{2.9.14}$$

The problem on finite deformations arising in the motion of a solid sphere toward a free interface and in the motion of a deformed drop to the solid plane wall, which is vital for chemical industry, was studied numerically in [17, 517].

### 2.9-4. Particle on the Interface Between Two Fluids

The motion of a particle across the interface between two fluid phases is an important part of the processes of separation and purification of one of the phases from suspended particles. In this case, in addition to the interfacial excess energy, pure hydrodynamic effects of drag against the phase transition play an important role.

The motion of a disk whose plane coincides with the interface between two fluids with viscosities $\mu_1$ and $\mu_2$ was studied in [387]. The drag forces in the Stokes motion of this disk with velocity $U_i$ along the tangent and the normal to the interface are given by

$$F_\parallel = \tfrac{16}{3} a U_i(\mu_1 + \mu_2), \qquad F_\perp = 8 a U_i(\mu_1 + \mu_2). \tag{2.9.15}$$

In the special case $\mu_1 = \mu_2$, formula (2.9.15) implies (2.6.16) (up to different notation of indices) for a disk moving in a homogeneous medium.

Formulas (2.9.15) can be unified as follows:

$$F = \tfrac{1}{2}(F_1 + F_2),$$

where $F_1$ and $F_2$ are the drag forces acting on the moving disk in homogeneous fluids with viscosities $\mu_1$ and $\mu_2$ (the disk moves in the directions along and transverse to its axis). The last formula can be used for determining the drag force of a plane body of an arbitrary shape situated on the interface between two fluids when it moves along (across) the interface.

### 2.9-5. Rate of Suspension Precipitation. The Cellular Model

For ensembles with large number of particles, it is practically impossible to realize the reflection method, and moreover, to construct pointwise exact solutions in a multiconnected domain. In this case, one of the widely used approximate models for two-phase media is the cell model. To each particle of the disperse phase, this model assigns the corresponding volume of the free liquid. Thus, the entire suspension (or emulsion) is divided into a collection of spherical cells of radius $b$ whose centers coincide with the centers of particles of radius $a$. The geometric parameters of these cells are related to the volume concentration $\phi$ of the disperse phase as follows:

$$b = a\phi^{-1/3}. \tag{2.9.16}$$

The definition of a particle velocity $\mathbf{U}$ determines the axial symmetry of the problem, which it is convenient to consider in spherical coordinates. The general solution of such a system was given in Section 2.1, where arbitrary constants must be determined from the conditions that the solution is bounded at infinity and the velocity is known on the particle surface and from some other conditions on the cell boundary (for $R = b$). The condition that the normal velocity on this boundary is zero, that is, there is no flow across the cell, is undoubtedly obvious. The second condition, which is necessary for the complete identification of the solution, leads to different opinions. Cunningham postulated that the tangential velocity is zero, thus considering a cell as a container with rigid boundary. Happel proposed to use the condition that the tangential stress is zero, thus postulating that the cell is an isolated force unit (is in equilibrium under the action of forces). Finally, Kuwabara proposed to use the condition that the vortex intensity is zero on the cell boundary.

The choice of the boundary condition essentially determines the model of force interaction between the particle at the center of the cell and the other particles. A detailed comparative analysis of various boundary conditions was carried out in [450], where the solutions were obtained for the three above-mentioned versions under the assumption that the particle at the cell center was a drop of a liquid with different viscosity. A new variant of closed model was presented also in [450], where it was proposed to believe that the vortex intensity flux is equal to zero on the cell boundary.

The steady-state velocities of the gravitational sedimentation in suspensions obtained with the help of cell models were compared with numerous experimental data in [450]. It was shown that the most precise results can be obtained by using the Slobodov–Chepura model [450], in which the drag forces lead to formula (2.9.1), where the correction coefficient can be calculated as

$$\lambda = \frac{\beta + \frac{2}{3}}{1 - \frac{3}{5}\phi^{1/3} - \frac{2}{5}\phi^2 + \beta(1 - \frac{9}{10}\phi^{1/3} + \frac{1}{2}\phi + \frac{2}{5}\phi^2)}, \quad \beta = \frac{\mu_2}{\mu_1}. \tag{2.9.17}$$

As $\phi \to 0$ and $\beta \to \infty$, we have $\lambda \to 1$, which corresponds to the Stokes drag law.

## 2.9-6. Effective Viscosity of Suspensions

Suspensions of particles in fluid are widely used in various processes of chemical technology. If the dimensions of particles in the suspension are essentially smaller than those of the apparatus, one can consider the suspension as some continuous medium with properties other than properties of the disperse phase.

Very often, this medium remains Newtonian from the viewpoint of its rheological properties, but the viscosity of this medium is somewhat larger than that of the continuous phase. This viscosity $\mu_{ef}$ is called the effective viscosity. In practice, it is convenient to relate the effective viscosity to the viscosity $\mu$ of the continuous phase and consider the dimensionless effective viscosity $\bar{\mu} = \mu_{ef}/\mu$.

The value $\bar{\mu}$ depends primarily on the volume concentration of the disperse phase $\phi$. The well-known Einstein formula [179]

$$\bar{\mu} = 1 + 2.5\,\phi, \tag{2.9.18}$$

holds for strongly rarefied suspensions of solid spherical particles.

For more concentrated suspensions, one can estimate $\bar{\mu}$ by using the cell model.

The dimensionless effective viscosity of a rarefied emulsion of spherical drops and bubbles moving in a fluid can be determined by the formula

$$\bar{\mu} = 1 + \frac{5\beta + 2}{2\beta + 2}\,\phi, \tag{2.9.19}$$

where $\beta$ is the ratio of the drop viscosity to the fluid viscosity. By passing to the limit as $\beta \to \infty$ in (2.9.19), one obtains the Einstein formula (2.9.18). The value $\beta = 0$ corresponds to gas bubbles.

It was shown in [179] that the effective viscosity is related to the ratio of the velocity of free sedimentation of a single particle according to the Stokes law to the velocity of particles in the suspension, that is, the effective viscosity is related to the correction factor $\lambda$ in the drag force. The expression

$$\bar{\mu} = (1 - \phi)^m \lambda \tag{2.9.20}$$

is obtained for the effective viscosity.

Usually, the following two values are used: $m = 1$ (Kynch's formula) and $m = 2$ (Hawksley's formula). It was shown in [78] that the value $m = 1$ corresponds to the one-velocity suspension model, and $m = 2$ to the two-velocity model, which is considered as two interpenetrating continuous phases with different velocity fields. Since the second model is more precise, the estimates obtained by formula (2.9.20) with $m = 2$ are preferable.

The expressions (2.9.18)–(2.9.20) allow one to estimate the effective viscosity of suspensions and emulsions.

A more appropriate approach to the construction of the mechanics of concentrated disperse systems based on the cell model is developed in [77]. This approach, based on the averaging methods over the ensemble of randomly situated particles, allows one, by using a unique methodology, to obtain both the equations of continuum mechanics of disperse systems and the closing rheological relations theoretically rather than phenomenologically. In particular, the effective viscosity of suspensions is given by a simple formula $\bar{\mu} = (1 - 2.5 \, \phi)^{-1}$, which becomes the Einstein formula (2.9.18) for small $\phi$ and can be used up to the concentrations $\phi = 0.25$. The second approximation for the effective viscosity was also obtained.

The results of experimental and numerical calculations [77] are well approximated by the formula

$$\bar{\mu} = 1 + 2.5 \, \phi + 12.5 \, \phi^2, \tag{2.9.21}$$

which becomes the Einstein formula (2.9.18) as $\phi \to 0$ and can be used for $\phi \le 0.4$.

A jet model of flow around balls in a porous layer was proposed in [153]. Such flow is characterized by a decrease of the drag due to the suppression of wakes. For a sufficiently close packing ($\phi > 0.35$), the layer becomes steady-state. For the drag coefficient of a ball in such a system, the following empirical formula was proposed:

$$c_f = 2\psi \left( 1 + 211 \frac{\psi}{\text{Re}} \right), \qquad \text{Re} = \frac{aU}{\nu}, \tag{2.9.22}$$

where $U$ is the percolation flux and $\psi$ is the relative minimum flow section of the layer, which depending on the volume concentration of particles as follows:

$$\psi = \begin{cases} 1 - 1.16 \, \phi^{2/3} & \text{for } \phi \le 0.6, \\ 0.508 - 0.56 \, \phi & \text{for } \phi > 0.6. \end{cases}$$

Formula (2.9.22) is in a good agreement with experimental data.

Estimates of the rise velocity of a bubble ensemble in sparging devices are given in [256].

Some other theoretical methods for investigating rarefied and concentrated disperse systems, based on equations of mechanics of multiphase systems, are described in the books [86, 183, 205, 312, 313].

# Chapter 3
# Mass and Heat Transfer in Liquid Films, Tubes, and Boundary Layers

So far, we have considered the motion of fluids of homogeneous physical and chemical composition. In practice, one often meets more complicated situations in which the fluid contains dissolved substances (contaminants or reactants) and is a solution or a mixture.

Water solutions of common salt or sugar and water-alcohol mixtures are the simplest examples of such systems.

The composition of solutions and mixtures is usually characterized by the mass density of the substance (the mass of dissolved substance per unit volume) or the dimensionless mass concentration $C$ (the ratio of the mass density of a substance to the total density of the mixture*). The latter is normally used in this book; for brevity, we refer to it simply as the concentration. If there are several solutes $m = 1, \ldots, M$, then for each of them we introduce their own mass density and, accordingly, the mass concentration $C_m$.

The concentration of individual components at each point of the medium depends on convective mass transfer, molecular (or turbulent) diffusion, and the intensity of heterogeneous and homogeneous physical and chemical transformation.

In what follows, heterogeneous transformations are understood as chemical or physical-chemical transformations that take place on some surfaces, for example, on interfaces or surfaces possessing catalytic properties. This wide understanding of the term "heterogeneous transformation" includes surface catalytic reactions, adsorption and desorption on solid and fluid surfaces, dissolving of crystals in fluid, electrochemical reactions on the surface of an electrode in electrolyte, sublimation and condensation, sedimentation of aerosols and colloids, etc. Chemical transformations taking place in the bulk of fluid will be called homogeneous transformations or volume chemical reactions.

---

* Sometimes, the mass density of a substance is called the partial density, and the mass concentration is called the mass fraction of the total mass. Moreover, in the special chemical literature, the molar density is used, which is determined by the number of moles of the solute per unit volume of the solution, as well as its dimensionless analog, the mole concentration or the mole fraction of the total concentration, which is the ratio of the molar density to the total number of moles of all ingredients per unit volume.

# 3.1. Convective Mass and Heat Transfer. Equations and Boundary Conditions

### 3.1-1. Mass Transfer Equation. Laminar Flows

Let us write out the main equations and boundary conditions used in the mathematical statement of physical and chemical hydrodynamic problems. More detailed derivations of these equations and boundary conditions, analysis of their scope, various physical models of numerous related problems, solution methods, as well as applications of the results, can be found in the books [35, 121, 159, 185, 199, 406].

We assume that the medium density and viscosity are independent of concentration and temperature, and hence, the concentration and temperature distributions do not affect the flow field.* This allows one to analyze the hydrodynamic problem about the fluid motion and the diffusion-heat problem of finding the concentration and temperature fields independently. (More complicated problems in which the flow field substantially depends on diffusion-heat factors will be considered later in Chapter 5.) It is assumed that the information about the fluid velocity field necessary for the solution of the diffusion-heat problem is known. We also assume that the diffusion and thermal conductivity coefficients are independent of concentration and temperature. For simplicity, we restrict our consideration to the case of two-component solutions.

In Cartesian coordinates $X, Y, Z$, solute transfer in absence of homogeneous transformations is described by the equation

$$\frac{\partial C}{\partial t} + V_X \frac{\partial C}{\partial X} + V_Y \frac{\partial C}{\partial Y} + V_Z \frac{\partial C}{\partial Z} = D \left( \frac{\partial^2 C}{\partial X^2} + \frac{\partial^2 C}{\partial Y^2} + \frac{\partial^2 C}{\partial Z^2} \right), \quad (3.1.1)$$

where $C$ is the concentration, $D$ is the diffusion coefficient, and $V_X$, $V_Y$, and $V_Z$ are the fluid velocity components, which are assumed to be given.

Equation (3.1.1) reflects the fact that the transfer of a substance in a moving medium is due to two distinct physical mechanisms. First, there is molecular diffusion due to concentration difference in a liquid or gas, which tends to equalize the concentrations. Second, the solute is carried along by the moving medium. The combination of these two processes is usually called convective diffusion [133, 270].

### 3.1-2. Initial Condition and the Simplest Boundary Conditions

To complete the statement of the problem, it is necessary to supplement equation (3.1.1) with initial and boundary conditions. As the initial condition, one usually takes the concentration profile in the flow at time $t = 0$. The boundary conditions are, as a rule, given on some surface and remote from it, in the bulk of the solution.

---

* It is also assumed that mass transfer across the interface does not affect the velocity profiles (this is typical of most problems in chemical engineering sciences).

The latter condition corresponds to prescribing the nonperturbed concentration $C_i$ at infinity:

$$\xi_* \to \infty, \quad C \to C_i, \tag{3.1.2}$$

where $\xi_*$ is the distance measured along the normal to the surface.

In problems of solid dissolution, it is usually assumed in the boundary conditions that the concentration is zero in the bulk of the fluid, $C_i = 0$, and constant on the crystal surface [10],

$$\xi_* = 0, \quad C = C_s, \tag{3.1.3}$$

where $C_s$ is given. The boundary conditions (3.1.2) (with $C_i = 0$) and (3.1.3) are also used in problems of liquid drop evaporation.

The boundary conditions on a surface where a chemical reaction occurs depend on the specific physical statement of the problem. In the special case of an "infinitely rapid" heterogeneous chemical reaction, the corresponding boundary condition has the form

$$\xi_* = 0, \quad C = 0, \tag{3.1.4}$$

and means that the reagent is completely taken up in the reaction on the interface. Such a situation is often called the diffusion regime of reaction. Condition (3.1.4) has the following meaning: the chemical reaction on the interface proceeds vigorously, so that the all substance that approaches the interface takes part in the reaction. Note that condition (3.1.4) is the special case of (3.1.3) for $C_s = 0$.

Condition (3.1.4) also corresponds to diffusion sedimentation of aerosol and colloid particles, and the entrapment phenomenon [139, 499] can be taken into account by assuming that the distance between the surface $\xi_* = 0$ and the sedimentation surface is equal to the mean radius of sediment particles.

### 3.1-3. Mass Transfer Complicated by a Surface Chemical Reaction

*Boundary condition on a surface.* If a surface (heterogeneous) chemical reaction with finite rate occurs on the interface, the (3.1.4) must be replaced by the more complicated boundary condition [133, 270]

$$\xi_* = 0, \quad D\,\frac{\partial C}{\partial \xi_*} = K_s F_s(C), \tag{3.1.5}$$

where $K_s$ is the rate constant of the surface reaction and $K_s F_s(C)$ is the rate of the surface reaction.

The specific form of the function $F_s = F_s(C)$ is determined by the kinetics of the surface chemical reaction. The function $F_s$ must satisfy the condition $F_s(0) = 0$, whose physical meaning is obvious: if the reagent is absent, there is no reaction. For reactions of rate order $n$, in (3.1.5) one must set [270]

$$F_s = C^n \quad \text{(where } n > 0\text{).} \tag{3.1.6}$$

We point out that in a majority of cases the form of the function $F_s(C)$ does not describe the actual kinetics of catalytic chemical transformations but only determines the effective rate of chemical reaction.

If the surface $\xi_* = 0$ is impermeable to the solute, then the boundary condition

$$\xi_* = 0, \quad \frac{\partial C}{\partial \xi_*} = 0$$

is satisfied, which is a special case of (3.1.5) for $K_s = 0$.

**Dimensionless equation and boundary conditions.** Let us introduce a characteristic length $a$ (for example, the radius of particles or of the tube) and a characteristic velocity $U$ (for example, the nonperturbed flow velocity remote from a particle or the fluid velocity on the tube axis). First, we consider the boundary conditions (3.1.2) and (3.1.3). Then it is convenient to rewrite equation (3.1.1) for the convective mass transfer in the following dimensionless form. We introduce dimensionless variables by setting

$$\tau = \frac{Dt}{a^2}, \quad x = \frac{X}{a}, \quad y = \frac{Y}{a}, \quad z = \frac{Z}{a}, \quad \xi = \frac{\xi_*}{a},$$

$$v_x = \frac{V_X}{U}, \quad v_y = \frac{V_Y}{U}, \quad v_z = \frac{V_Z}{U}, \quad c = \frac{C_i - C}{C_i - C_s} \qquad (3.1.7)$$

and substitute them into (3.1.1). As a result, we obtain

$$\frac{\partial c}{\partial \tau} + \mathrm{Pe}\left( v_x \frac{\partial c}{\partial x} + v_y \frac{\partial c}{\partial y} + v_z \frac{\partial c}{\partial z} \right) = \frac{\partial^2 c}{\partial x^2} + \frac{\partial^2 c}{\partial y^2} + \frac{\partial^2 c}{\partial z^2}. \qquad (3.1.8)$$

Here the Peclet number $\mathrm{Pe} = aU/D$ is a dimensionless parameter characterizing the ratio of convective transfer to diffusion transfer and $a$ is the characteristic length (tube radius, film thickness, etc.).

In the new variables (3.1.7), the boundary condition remote from the surface (3.1.2) has the form

$$\xi \to \infty, \quad c \to 0. \qquad (3.1.9)$$

In a similar way, taking into account (3.1.3) and (3.1.7), we obtain the boundary condition

$$\xi = 0, \quad c = 1. \qquad (3.1.10)$$

If the rate of a heterogeneous chemical reaction is finite, it is convenient to pass to the new variables (3.1.7) with $C_s = 0$. As a result, Eq. (3.1.1) and the boundary condition at infinity (3.1.2) turn into (3.1.8) and (3.1.9), and the boundary condition (3.1.5) on the reaction surface acquires the form

$$\xi = 0, \quad -\frac{\partial c}{\partial \xi} = k_s f_s(c). \qquad (3.1.11)$$

In condition (3.1.11), we have used the notation

$$k_s = \frac{aK_s}{DC_i} F_s(C_i), \quad f_s(c) = \frac{F_s(C)}{F_s(C_i)}, \quad c = \frac{C_i - C}{C_i}. \qquad (3.1.12)$$

In the special case (3.1.6) of a surface chemical reaction of order $n$, taking into account (3.1.12), one can state the boundary condition (3.1.11) on the surface as follows:

$$\xi = 0, \quad -\frac{\partial c}{\partial \xi} = k_s (1 - c)^n, \qquad (3.1.13)$$

where $k_s = aK_s C_i^{n-1}/D$ is the dimensionless rate constant of surface reaction.

We divide both sides of (3.1.13) by $k_s$ and let the parameter $k_s$ tend to infinity. As a result, we arrive at the limit boundary condition (3.1.10), corresponding to the diffusion regime of reaction. This passage to the limit adequately illustrates the notion of an "infinitely rapid reaction," which was used previously.

For brevity, we rewrite the convective diffusion equation (3.1.8) in the following frequently used form:

$$\frac{\partial c}{\partial \tau} + \text{Pe} \left( \mathbf{v} \cdot \nabla \right) c = \Delta c, \qquad (3.1.14)$$

where $\nabla$ is the Hamilton operator and $\Delta$ is the Laplace operator; the explicit form of these operators in the Cartesian coordinates $x$, $y$, $z$ follows by comparing (3.1.8) and (3.1.14).

In many specific problems, instead of the Cartesian coordinates $x$, $y$, $z$, it is often convenient to use the spherical coordinates $r$, $\varphi$, $\theta$ or the cylindrical coordinates $\varrho$, $\varphi$, $z$. In these coordinate systems, the differential operators in Eq. (3.1.14) have the following form:

$$\left( \mathbf{v} \cdot \nabla \right) c = v_\varrho \frac{\partial c}{\partial \varrho} + v_z \frac{\partial c}{\partial z} + \frac{v_\varphi}{\varrho} \frac{\partial c}{\partial \varphi},$$

$$\Delta c = \frac{1}{\varrho} \frac{\partial}{\partial \varrho} \left( \varrho \frac{\partial c}{\partial \varrho} \right) + \frac{\partial^2 c}{\partial z^2} + \frac{1}{\varrho^2} \frac{\partial^2 c}{\partial \varphi^2}, \qquad (3.1.15)$$

$$\varrho = \sqrt{x^2 + y^2},$$

in the cylindrical coordinates;

$$\left( \mathbf{v} \cdot \nabla \right) c = v_r \frac{\partial c}{\partial r} + \frac{v_\theta}{r} \frac{\partial c}{\partial \theta} + \frac{v_\varphi}{r \sin \theta} \frac{\partial c}{\partial \varphi},$$

$$\Delta c = \frac{1}{r^2} \frac{\partial}{\partial r} \left( r^2 \frac{\partial c}{\partial r} \right) + \frac{1}{r^2 \sin \theta} \frac{\partial}{\partial \theta} \left( \sin \theta \frac{\partial c}{\partial \theta} \right) + \frac{1}{r^2 \sin^2 \theta} \frac{\partial^2 c}{\partial \varphi^2}, \qquad (3.1.16)$$

$$r = \sqrt{x^2 + y^2 + z^2}$$

in the spherical coordinates.

### 3.1-4. Mass Transfer Complicated by a Volume Chemical Reaction

*Convective diffusion equation with a volume reaction.* If a volume (homogeneous) chemical reaction proceeds in the bulk of flow, the convective diffusion equation can be rewritten in the form

$$\frac{\partial C}{\partial t} + V_X \frac{\partial C}{\partial X} + V_Y \frac{\partial C}{\partial Y} + V_Z \frac{\partial C}{\partial Z} = D \left( \frac{\partial^2 C}{\partial X^2} + \frac{\partial^2 C}{\partial Y^2} + \frac{\partial^2 C}{\partial Z^2} \right) - K_v F_v(C),$$

(3.1.17)

where $K_v$ is the rate constant of a volume chemical reaction and $K_v F_v(C)$ is the rate of a volume chemical reaction.

The form of the function $F_v = F_v(C)$ depends on the reaction kinetics; moreover, $F_v(0) = 0$. The reaction most frequently found in scientific literature is the $n$th-order reaction, for which [18, 103, 270]

$$F_v = C^n.$$

(3.1.18)

For Eq. (3.1.17), one prescribes the following boundary condition in incoming flow:

$$\xi_* \to \infty, \quad C \to 0.$$

(3.1.19)

This condition has the following meaning: the substance that diffuses away from the surface must entirely take part in the reaction while moving away into the bulk of a chemically active medium.

In many practically important cases, condition (3.1.3), saying that the concentration is constant on a given surface, is used.

**Dimensionless equation and boundary conditions.** As previously, it is expedient to rewrite Eq. (3.1.17) and the boundary conditions (3.1.3) and (3.1.19) in a dimensionless form. To this end, we introduce the new variables

$$\tau = \frac{Dt}{a^2}, \quad x = \frac{X}{a}, \quad y = \frac{Y}{a}, \quad z = \frac{Z}{a}, \quad \xi = \frac{\xi_*}{a},$$
$$v_x = \frac{V_X}{U}, \quad v_y = \frac{V_Y}{U}, \quad v_z = \frac{V_Z}{U}, \quad c = \frac{C}{C_s},$$

(3.1.20)

which differ from (3.1.7) only by the definition of dimensionless concentration.

By substituting (3.1.20) into (3.1.17), we obtain

$$\frac{\partial c}{\partial \tau} + \mathrm{Pe} \left( \mathbf{v} \cdot \nabla \right) c = \Delta c - k_v f_v(c),$$

(3.1.21)

where we have used the same notation as in (3.1.14). The variables on the right in the dimensional equation (3.1.17) and the dimensionless equation (3.1.21) are related as follows:

$$k_v = \frac{a^2 K_v F_v(C_s)}{D C_s}, \quad f_v(c) = \frac{F_v(C)}{F_v(C_s)}.$$

(3.1.22)

In the special case of a volume chemical reaction of order $n$ (3.1.18), in (3.1.21) one must set

$$k_v = a^2 K_v C_s^{n-1}/D, \quad f_v = c^n.$$

(3.1.23)

TABLE 3.1

Methods for introducing dimensionless concentration $c$

in problems of convective mass transfer

| No | Physical-chemical process | Concentration on the surface (where the heterogeneous transformation takes place) | Nonperturbed concentration in incoming flow (at the tube or film inlet) | Dimensionless concentration in the fluid phase, $c$ |
|----|---------------------------|------------------------------|------------------------------|------------------------------|
| 1 | Dissolution of solids in pure liquids | $C_s$ | 0 | $\dfrac{C}{C_s}$ |
| 2 | Adsorption of weakly soluble gases on the free boundary of a liquid | $C_s$ | 0 | $\dfrac{C}{C_s}$ |
| 3 | Diffusion in contaminated fluids | $C_s$ | $C_i$ | $\dfrac{C_i - C}{C_i - C_s}$ |
| 4 | Diffusion regime on the reaction surface | 0 | $C_i$ | $\dfrac{C_i - C}{C_i}$ |
| 5 | Finite rate of surface chemical reaction | Determined by the solution of the problem | $C_i$ | $\dfrac{C_i - C}{C_i}$ |
| 6 | Volume chemical reaction | $C_s$ | 0 | $\dfrac{C}{C_s}$ |

*Remark.* The normal derivative of concentration is zero on the interfaces on which no heterogeneous transformations occur.

The boundary conditions (3.1.3) and (3.1.19) in the dimensionless variables (3.1.20) have the form

$$\xi = 0, \quad c = 1; \qquad \xi \to \infty, \quad c \to 0, \qquad (3.1.24)$$

where $\xi = \xi_* / a$ is the dimensionless distance from the interface.

For convenience, in Table 3.1 we give various methods for introducing dimensionless concentration, which are used in this book for the mathematical statement of various problems of convective mass transfer.

It should be noted that in all cases we use the same notation $c$ for the dimensionless concentration. The reason is that the dimensionless concentrations corresponding to rows 1, 2, 4, and 6 in Table 3.1 are special cases of the general formula in row 3 and can be obtained by substituting the appropriate values of $C_i$ and $C_s$. The remaining expression for $c$ (row 5) can also be obtained from the general formula in row 3 by formally setting $C_s = 0$ (in this case, however, one should keep in mind that the concentration $C_s$ on the surface is not known in advance).

## 3.1-5. Diffusion Fluxes and the Sherwood Number

A local (or differential) diffusion flux of a solute to the surface is determined by

$$j_* = D\rho \left( \frac{\partial C}{\partial \xi_*} \right)_{\xi_*=0} \qquad (3.1.25)$$

and in general varies along the surface.

The total (or integral) diffusion flux can be obtained by integrating (3.1.25) over the entire surface $S$:

$$I_* = \iint_S j_* \, dS. \qquad (3.1.26)$$

The total diffusion flux $I_*$ is the amount of substance that reacts on the entire surface per unit time.

In mass transfer problems with the boundary conditions (3.1.2) and (3.1.3), one often replaces (3.1.25) and (3.1.26) by the dimensionless diffusion fluxes

$$j = \frac{aj_*}{D\rho\,(C_i - C_s)}, \qquad I = \frac{I_*}{aD\rho\,(C_i - C_s)}. \qquad (3.1.27)$$

For the diffusion regime of reaction, which corresponds to the boundary conditions (3.1.2) and (3.1.4), as well as for the finite rate of surface chemical reaction in the case of the boundary conditions (3.1.2) and (3.1.5), one must set $C_s = 0$ in (3.1.27).

The main quantity of practical interest is the mean Sherwood number

$$\text{Sh} = \frac{I}{S}, \qquad (3.1.28)$$

where $S = S_*/a^2$ is the dimensionless area of the surface and $S_*$ is the corresponding dimensional area.

The calculation of diffusion fluxes and the mean Sherwood number is usually carried out in three steps. First, the problem of convective mass transfer is solved and the concentration field is determined. Second, the normal derivative $\left( \partial C / \partial \xi_* \right)_{\xi_*=0}$ on the surface is evaluated. Finally, one applies formulas (3.1.25)–(3.1.28).

For brevity, throughout the book (where it cannot lead to a misunderstanding) we use the terms "concentration" and "diffusion flux" instead of "dimensionless concentration" and "dimensionless diffusion flux."

## 3.1-6. Heat Transfer. The Equation and Boundary Conditions

The equation of heat transfer in a moving medium is similar to Eq. (3.1.1) of convective diffusion and has the form

$$\frac{\partial T_*}{\partial t} + V_X \frac{\partial T_*}{\partial X} + V_Y \frac{\partial T_*}{\partial Y} + V_Z \frac{\partial T_*}{\partial Z} = \chi \left( \frac{\partial^2 T_*}{\partial X^2} + \frac{\partial^2 T_*}{\partial Y^2} + \frac{\partial^2 T_*}{\partial Z^2} \right), \qquad (3.1.29)$$

where $T_*$ is temperature and $\chi$ is the thermal diffusivity. The dissipative heating of fluid is neglected in Eq. (3.1.29).

For nonstationary problems, the temperature distribution in the flow at the initial instant of time must be given.

Remote from the surface, one usually has the condition that temperature is constant in the flow region:

$$\xi_* \to \infty, \quad T_* \to T_i. \tag{3.1.30}$$

For problems of body–medium heat exchange, if constant temperature is maintained on the body surface, the second boundary condition is

$$\xi_* = 0, \quad T_* = T_s. \tag{3.1.31}$$

By using the new dimensionless variables

$$\bar{\tau} = \frac{\chi t}{a^2}, \quad x = \frac{X}{a}, \quad y = \frac{Y}{a}, \quad z = \frac{Z}{a}, \quad \mathsf{Pe}_T = \frac{aU}{\chi},$$
$$v_x = \frac{V_X}{U}, \quad v_y = \frac{V_Y}{U}, \quad v_z = \frac{V_Z}{U}, \quad T = \frac{T_i - T_*}{T_i - T_s}, \tag{3.1.32}$$

one can rewrite Eq. (3.1.29) and the boundary conditions (3.1.30) and (3.1.31) in the form

$$\frac{\partial T}{\partial \bar{\tau}} + \mathsf{Pe}_T \left( \mathbf{v} \cdot \nabla \right) T = \Delta T; \tag{3.1.33}$$
$$\xi \to \infty, \quad T \to 0; \quad \xi = 0, \quad T = 1. \tag{3.1.34}$$

Obviously, from the mathematical viewpoint, problem (3.1.33), (3.1.34) describing the body–medium heat exchange is identical to problem (3.1.8)–(3.1.10) describing the flow-particle mass exchange in the case of a diffusion regime of reaction on the particle surface.

## 3.1-7. Some Methods of Theory of Mass and Heat Transfer

**Basic dimensionless parameters.** The diffusion and thermal Peclet numbers in the convective mass and heat transfer equations (3.1.8) and (3.1.33) are related to the Reynolds number $\mathsf{Re} = aU/\nu$ (where $\nu$ is the kinematic viscosity of the fluid) on the right-hand side in the Navier–Stokes equations (1.1.12) by the formulas

$$\mathsf{Pe} = \mathsf{Re}\,\mathsf{Sc}, \quad \mathsf{Pe}_T = \mathsf{Re}\,\mathsf{Pr}. \tag{3.1.35}$$

Here the Schmidt number $\mathsf{Sc} = \nu/D$ and the Prandtl number $\mathsf{Pr} = \nu/\chi$ are dimensionless values depending only on the physical properties of the continuous phase.

For ordinary gases, the diffusion coefficient and the kinematic viscosity are of the same order of magnitude, which corresponds to Schmidt numbers of the order of one ($Sc \sim 1$).

For ordinary liquids like water, the kinematic viscosity is several orders of magnitude larger than the diffusion coefficient ($Sc \sim 10^3$). In extremely viscous liquids like glycerin, the Schmidt number is of the order of $10^6$.

The range of the Prandtl number is narrower than that of the Schmidt number. In gases such as air, $Pr \sim 1$, and in liquids like water, $Pr \sim 10$. In extremely viscous liquids like glycerin, the Prandtl number is of the order of $10^3$. Liquid metals (sodium, lithium, mercury, etc.) are characterized by low Prandtl numbers: $5 \times 10^{-3} \leq Pr \leq 5 \times 10^{-2}$.

The Reynolds number $Re = aU/\nu$ is not a physical constant of a medium and depends on geometric and kinematic factors. Therefore, the range of variation of this number may be arbitrary.

It follows from the considered examples and relation (3.1.35) that the Peclet numbers in problems of physicochemical hydrodynamics can vary in a wide range.

Since diffusion processes in fluids are characterized by very large values of the Schmidt number, we must point out that in problems of convective mass transfer in fluid media, the Peclet number is also large even for low Reynolds numbers at which the Stokes flow law applies (for a creeping flow).

*Methods for solving mass and heat transfer problems.* The convective diffusion equation (3.1.1) is a second-order linear partial differential equation with variable coefficients (in the general case, the fluid velocity depends on the coordinates and time). Exact closed-form solutions of the corresponding problems can be found only in exceptional cases with simple geometry [79, 197, 270, 370, 516]. This is especially true of the nonlinear equation (3.1.17). Exact solutions are important for adequate understanding of the physical background of various phenomena and processes. They can serve as "test" solutions to verify whether the problem is well-posed or to estimate the accuracy of the corresponding numerical, asymptotic, and approximate methods.

To obtain the necessary information about a phenomenon or a process, one usually needs to use various simplifications in the mathematical statement of the problem, various approximations, numerical methods, or combinations of these.

Just as in mechanics of viscous fluids, the approximate solution of convective problems of mass and heat transfer is based on the methods of perturbation theory [96, 224, 258, 485]. In these methods, the dimensionless Peclet number $Pe$ occurring in Eq. (3.1.8) is assumed to be a small (or large) parameter, with respect to which one seeks the solutions in the form of asymptotic series.

We point out that the existence of a small or large parameter is a fundamental characteristic feature of many problems of physicochemical hydrodynamics. Indeed, as was already noted, convective diffusion in fluids is characterized by large Schmidt numbers, which is related to the characteristic values of physical constants. In the corresponding singularly perturbed problems, there exist narrow space-time regions (for example, a diffusion boundary layer or a diffusion wake) in which the solution varies rapidly. Usually, the structure, length, and number of these regions are not known and must be determined in the course of the solution. The vast experience in the application of perturbation methods gives grounds to consider them rather fruitful and most general of all existing analytical methods. These methods allow one to reveal important basic laws and qualitative specific features of complicated linear and nonlinear problems, to obtain asymptotics and construct "test" solutions, and in some cases to develop numerical methods.

It should be noted that for problems in which perturbation methods are highly efficient, numerical methods are, as a rule, of little use.

The asymptotic series provided by perturbation methods are of restricted scope. Moreover, one can usually obtain no more than two or three initial terms of the corresponding expansions. Thus, one cannot estimate the solution behavior for intermediate (finite) parameter values and there are severe limitations on the applicability of asymptotic formulas in the engineering practice. This is the most essential drawback of perturbation methods.

So far, various engineering methods mostly based on intuitive considerations are of practical use; for example, we can cite integral methods [103, 276, 427], the method of equiavailable surface [133], and various modifications of linearization methods for equations and boundary conditions [346]. In many cases, these simple methods are useful for practical needs. Approximate methods are very convenient for obtaining rough estimates at the preliminary stage of any investigation or if it is required to obtain a result in a short time. Low accuracy is typical of approximate engineering methods. This disadvantage can be remedied substantially by combining asymptotic and approximate methods [357, 359].

Many problems of physicochemical hydrodynamics and mass and heat transfer can be solved successfully by numerical methods on computers [99, 220, 227, 305, 327, 328, 344]. These methods are rather universal and effectively solve problems for intermediate values of the characteristic parameter, that is, in the regions where one cannot use asymptotic methods. At present, numerical methods are the main technique for examining applied problems related to the investigation, optimization, and control of various devices and technological industrial processes.

It should be noted that all theoretical methods (exact, asymptotic, approximate, and numerical) complement each other.

### 3.1-8. Mass and Heat Transfer in Turbulent Flows

*Equation of mass and heat transfer.* In mathematically describing transfer of a passive admixture in a turbulent stream, the admixture concentration and the fluid velocity components are represented as

$$C = \overline{C} + C', \quad V_i = \overline{V}_i + V_i', \tag{3.1.36}$$

where the bar and prime denote the time-average and fluctuating components, respectively. The averages of the fluctuations are zero, $\overline{C'} = \overline{V_i'} = 0$.

The introduction of the average and fluctuating components by relations (3.1.36) followed by the application of the averaging operation made it possible, under some assumptions, to obtain the following equation for the average concentration of the admixture [276] from Eq. (3.1.1):

$$\frac{\partial \overline{C}}{\partial t} + (\overline{\mathbf{V}} \cdot \nabla)\overline{C} = D\Delta\overline{C} + \nabla \cdot (-\overline{C'\mathbf{V}'}). \tag{3.1.37}$$

Here $-\overline{C'\mathbf{V}'}$ is the turbulent (eddy) flux of the admixture. This vector can be determined using some closure hypothesis, based, as a rule, on empirical information.

By analogy with Boussinesq's ideas about transfer of momentum, it is often assumed for transfer of a substance that

$$-\overline{C'\mathbf{V}'} = \mathbf{D}_t \cdot \nabla\overline{C}, \tag{3.1.38}$$

where $\mathbf{D}_t$ is the turbulent diffusion tensor [359].

**Turbulent Prandtl number.** In a number of problems, it is important to know, as a rule, only one component of turbulent transfer, namely, the component normal to the wall. In this case, the strength of turbulent diffusion is characterized by a single scalar quantity $D_t$, just as the intensity of turbulent transfer of momentum is characterized by a single scalar quantity $\nu_t$. The ratio of the turbulent viscosity coefficient to the turbulent diffusion coefficient,

$$\mathrm{Pr}_t = \frac{\nu_t}{D_t}, \tag{3.1.39}$$

is referred to as the turbulent Prandtl (Schmidt) number.

The simplest way to close equations (3.1.37) is to use the hypothesis that the turbulent Prandtl number for the examined process is a constant quantity. Then it readily follows from Eq. (3.1.39) that the turbulent diffusion coefficient is proportional to the turbulent viscosity: $D_t = \nu_t/\mathrm{Pr}_t$. By using the expression for $\nu_t$ borrowed from the corresponding hydrodynamic model, one can obtain the desired value of $D_t$. In particular, following Prandtl's or von Karman's model, one can use formula (1.1.21) or (1.1.22) for $\nu_t$.

For a long time, is was assumed that $\mathrm{Pr}_t = 1$ without sufficient justification. With this value of $\mathrm{Pr}_t$, the mechanism of turbulent transfer of momentum and that of any passive scalar substance turns out to be identical (the Reynolds analogy). According to up-to-date knowledge, the Reynolds analogy can be used in some cases for rough estimation of parameters of real flows, but the scope of the predicted results is highly restricted. The turbulent Prandtl number, as well as $\nu_t$ and $D_t$, depends on physical, geometric, and kinematic properties of the turbulent flow [210]. For wall flows, numerous experimental measurements available give the value [212, 289]

$$\mathrm{Pr}_t \approx 0.85. \tag{3.1.40}$$

For jet flows and mixing layers, there are various estimates for $\mathrm{Pr}_t$ ranging from 0.5 to 0.75 [397]. These estimates are helpful for approximate calculations of turbulent heat and mass transfer and can be used for closing equations (3.1.37).

# 3.2. Diffusion to a Rotating Disk

### 3.2-1. Infinite Plane Disk

**Statement of the problem.** Following [270], we consider steady-state mass transfer to the surface of a disk rotating around its axis at a constant angular velocity $\omega$. We assume that remote from the disk the concentration is equal to the constant $C_i$ and that the solute is totally adsorbed by the disk surface. The $z$-axis is normal to the surface of the disk. The solution of the problem about the motion of a fluid carried over by the disk was given in Section 1.2.

It follows from the previous discussion and the results of Section 3.1 that the diffusion equation and the boundary conditions have the following form in the cylindrical coordinates:

$$v_\varrho \frac{\partial c}{\partial \varrho} + v_z \frac{\partial c}{\partial z} + \frac{v_\varphi}{\varrho} \frac{\partial c}{\partial \varphi} = \frac{1}{\mathsf{Sc}} \left[ \frac{1}{\varrho} \frac{\partial}{\partial \varrho} \left( \varrho \frac{\partial c}{\partial \varrho} \right) + \frac{\partial^2 c}{\partial z^2} + \frac{1}{\varrho^2} \frac{\partial^2 c}{\partial \varphi^2} \right]; \quad (3.2.1)$$

$$z = 0, \quad c = 1; \qquad z \to \infty, \quad c \to 0. \tag{3.2.2}$$

Here the dimensionless variables and parameters are related to the initial dimensional variables by (3.1.7) with $C_s = 0$, where the characteristic length and velocity are chosen so that

$$a = \left( \nu/\omega \right)^{1/2}, \quad U = (\nu\omega)^{1/2}, \quad \mathsf{Pe} = aU/D = \mathsf{Sc}. \tag{3.2.3}$$

It follows from the results of Section 1.2 that the dimensionless fluid velocity has the form

$$v_z = v(z), \quad v_\varrho = \varrho u_1(z), \quad v_\varphi = \varrho u_2(z), \tag{3.2.4}$$

where $v$, $u_1$, and $u_2$ are known functions of $z$.

Let us point out some properties of the function $v$. The expansion of $v$ near the disk surface (as $z \to 0$) starts from the second-order term,

$$v = -\alpha z^2 + \cdots, \qquad \text{where} \quad \alpha \approx 0.51. \tag{3.2.5}$$

In the other limit case, we have $v \to -0.89$ as $z \to \infty$.

**Exact solution.** We seek the solution of problem (3.2.1), (3.2.2), (3.2.4) in the form

$$c = c(z). \tag{3.2.6}$$

As a result, we arrive at the second-order ordinary differential equation

$$\mathsf{Sc}\, v(z) \frac{dc}{dz} = \frac{d^2 c}{dz^2} \tag{3.2.7}$$

with the boundary conditions (3.2.2).

One can readily integrate equation (3.2.7), since the substitution $W = dc/dz$ results in a first-order equation with separating variables.

The solution of problem (3.2.7), (3.2.2) is given by the formula

$$
c = \frac{\displaystyle\int_z^\infty \exp\left[\operatorname{Sc}\int_0^z v\left(\bar{z}\right)d\bar{z}\right]dz}{\displaystyle\int_0^\infty \exp\left[\operatorname{Sc}\int_0^z v\left(\bar{z}\right)d\bar{z}\right]dz}. \tag{3.2.8}
$$

By differentiating this expression with respect to $z$ followed by setting $z = 0$, we obtain the dimensionless diffusion flux to the disk surface:

$$
j = -\left(\frac{dc}{dz}\right)_{z=0} = \left\{\int_0^\infty \exp\left[\operatorname{Sc}\int_0^z v\left(\bar{z}\right)d\bar{z}\right]dz\right\}^{-1}. \tag{3.2.9}
$$

***Diffusion boundary layer approximation.*** Now let us take into account the fact that common fluids are characterized by large Schmidt numbers Sc. Obviously, by substituting the leading term of the expansion of $v$ as $z \to 0$ into (3.2.8) and (3.2.9), one can readily obtain the asymptotics of these formulas as $\operatorname{Sc} \to \infty$. By using (3.2.5) and (3.2.8) and carrying out some transformation, we obtain the dimensionless concentration

$$
c = \frac{1}{\Gamma\left(\frac{1}{3}\right)}\, \Gamma\left(\tfrac{1}{3}, \tfrac{1}{3}\alpha\operatorname{Sc} z^3\right). \tag{3.2.10}
$$

Here and in the sequel, we use the following notation:

$$
\Gamma(m, \zeta) = \int_\zeta^\infty e^{-x} x^{m-1}\, dx \quad \text{is the incomplete gamma function,}
$$

$\Gamma(m) = \Gamma(m, 0)$ is the complete gamma function, and $\Gamma\left(\frac{1}{3}\right) \approx 2.679$.

In a similar way, by substituting the first term of the expansion (3.2.5) into (3.2.9), we obtain the dimensionless local diffusion flux

$$
j = \frac{(9\alpha\operatorname{Sc})^{1/3}}{\Gamma(1/3)} \approx 0.62\operatorname{Sc}^{1/3}. \tag{3.2.11}
$$

To obtain a physical interpretation of these results, it is convenient to introduce the dimensionless thickness of the diffusion boundary layer as follows:

$$
\delta = 1/j. \tag{3.2.12}
$$

By using (3.2.11) and (3.2.12), we obtain $\delta \approx 1.6\operatorname{Sc}^{-1/3}$.

We have the following picture of mass transfer near the surface of a rotating disk. The dimensionless concentration exponentially decreases away from the disk. At a distance $z \approx \delta$, this variable is close to its nonperturbed value and practically does not vary any more. At large Schmidt numbers, the concentration mainly varies is a thin layer (of thickness $\operatorname{Sc}^{-1/3}$) adjacent to the disk surface. This region is called a diffusion boundary layer.

### 3.2-2. Disk of Finite Radius

The mean Sherwood numbers for a disk of finite radius $a$ at high Schmidt numbers can be estimated using the formula

$$\mathsf{Sh} = 0.62\,\mathsf{Re}^{1/2}\,\mathsf{Sc}^{1/3}, \tag{3.2.13}$$

where the Reynolds number is determined from the relation $\mathsf{Re} = a^2\omega/\nu$. The corrections that take account of the boundary effects result in quite small variations in the numeric factor in (3.2.13) (within 5% [270]).

Relation (3.2.13) is valid in the region of laminar flow past the disk; the laminar regime occurs until $\mathsf{Re} \approx 10^4$ to $10^5$, depending on the roughness of the surface. For low Reynolds numbers ($\mathsf{Re} \le 10$), this relation is invalid, because the thickness of the hydrodynamic boundary layer becomes comparable with the disk radius and the boundary effects on the hydrodynamic flow and mass transfer become stronger.

In the turbulent regime of flow, the mean Sherwood number for a disk of finite radius $a$ can be approximated by the formula [439]

$$\mathsf{Sh} = 5.6\,\mathsf{Re}^{1.1}\,\mathsf{Sc}^{1/3}, \tag{3.2.14}$$

which is suitable for practical calculations for $6 \times 10^5 < \mathsf{Re} < 2 \times 10^6$ and $120 < \mathsf{Sc} < 1200$.

## 3.3. Heat Transfer to a Flat Plate

### 3.3-1. Heat Transfer in Laminar Flow

*Statement of the problem. Thermal boundary layer.* Let us consider heat transfer to a flat plate in a longitudinal translational flow of a viscous incompressible fluid with velocity $U_i$ at high Reynolds numbers. We assume that the temperature on the plate surface and remote from it is equal to the constants $T_s$ and $T_i$, respectively. The origin of the rectangular coordinates $X, Y$ is at the front edge of the plate, the $X$-axis is tangent, and the $Y$-axis is normal to the plate.

Numerous experiments and numerical calculations show that the laminar hydrodynamic boundary layer occurs for $5 \times 10^2 \le \mathsf{Re}_X \le 5 \times 10^5$ to $10^6$ [427]. In this region the thermal Peclet number $\mathsf{Pe}_X = \mathsf{Pr}\,\mathsf{Re}_X$ is large for gases and common liquids. For liquid metals, there is a range of Reynolds numbers, $10^4 \le \mathsf{Re}_X \le 10^6$, where the Peclet numbers are also large.

Taking into account the previous discussion, we restrict ourselves to the case of high Peclet numbers, for which the longitudinal molecular heat transfer may be neglected. The corresponding equations of the thermal boundary layer and the boundary conditions have the form

$$v_x \frac{\partial T}{\partial x} + v_y \frac{\partial T}{\partial y} = \frac{1}{\mathsf{Pr}} \frac{\partial^2 T}{\partial y^2}; \tag{3.3.1}$$

$$x = 0, \quad T = 0; \qquad y = 0, \quad T = 1; \qquad y \to \infty, \quad T \to 0.$$

Here the dimensionless variables are introduced by formulas (3.1.32) with the characteristic length $a = \nu/U_i$ and $\nu$ is the kinematic viscosity of the fluid.

The fluid velocity in (3.3.1) is given by the Blasius solution

$$v_x = f'(\eta), \quad v_y = \frac{\eta f' - f}{2\sqrt{x}}, \quad \text{where} \quad \eta = \frac{y}{\sqrt{x}}. \tag{3.3.2}$$

The function $f = f(\eta)$ was previously described in Section 1.7 and the prime stands for the derivative with respect to $\eta$.

**Temperature field.** The solution of problem (3.3.1) together with relations (3.3.2) will be sought in the self-similar form $T = T(\eta)$. As a result, we obtain the ordinary differential equation

$$\frac{d^2 T}{d\eta^2} + \frac{1}{2} \Pr f(\eta) \frac{dT}{d\eta} = 0;$$
$$\eta = 0, \quad T = 1; \qquad \eta \to \infty, \quad T \to 0. \tag{3.3.3}$$

By taking into account the relation $f = -f'''/f''$, which follows from the equation for $f$, we can write the solution of problem (3.3.3) as follows [345]:

$$T = \frac{\displaystyle\int_\eta^\infty \left[f''(\eta)\right]^{\Pr} d\eta}{\displaystyle\int_0^\infty \left[f''(\eta)\right]^{\Pr} d\eta}. \tag{3.3.4}$$

For $\Pr = 1$, this formula implies the simple relation $T(\eta) = 1 - f'(\eta) = 1 - v_x$ between temperature and the longitudinal velocity.

**Heat flux and the Nusselt number.** By differentiating (3.3.4) and using the numerical value $f''(0) = 0.332$, we obtain the dimensionless local heat flux to the plate surface:

$$j_T = -\left(\frac{\partial T}{\partial y}\right)_{y=0} = \frac{\varphi(\Pr)}{\sqrt{x}}, \quad \text{where} \quad \varphi(\Pr) = \frac{(0.332)^{\Pr}}{\displaystyle\int_0^\infty \left[f''(\eta)\right]^{\Pr} d\eta}. \tag{3.3.5}$$

It is convenient to seek the asymptotics of the function $\varphi(\Pr)$ at low and high Prandtl numbers starting from Eq. (3.3.3) with the extended variable $\eta = \zeta/\Pr$. As a result, we obtain the equation $T''_{\zeta\zeta} + f(\zeta/\Pr)T'_\zeta = 0$. As $\Pr \to 0$, the argument of the function $f(\zeta/\Pr)$ tends to infinity, which corresponds to a constant velocity inside the thermal boundary layer and $f(\eta) \approx \eta$. In the other limit case as $\Pr \to \infty$, the argument of the function $f(\zeta/\Pr)$ tends to zero, which corresponds to the linear approximation of the velocity inside the boundary layer and $f(\eta) \approx 0.166\,\eta^2$. Substituting the above-mentioned leading terms of

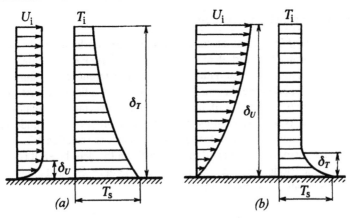

**Figure 3.1.** Velocity and temperature distributions in boundary layers at (a) extremely low Prandtl numbers and (b) very high Prandtl numbers

the asymptotic expansion of $f$ into Eq. (3.3.3) and solving the corresponding problems, we obtain the following expressions for the heat flux (3.3.5):

$$\begin{aligned} \varphi(\text{Pr}) &\to (\text{Pr}/\pi)^{1/2} \quad &(\text{Pr} \to 0), \\ \varphi(\text{Pr}) &\to 0.339\,\text{Pr}^{1/3} \quad &(\text{Pr} \to \infty). \end{aligned} \tag{3.3.6}$$

The both considered limit situations can be encountered in numerous problems of convective heat transfer; they are schematically shown in Figure 3.1. One can see that in the case $\text{Pr} \to 0$, which approximately takes place for liquid metals (e.g., mercury), one can neglect the dynamic boundary layer in the calculation of the temperature boundary layer and replace the velocity profile $v(x, y)$ by the velocity $v_\infty(x)$ of the inviscid outer flow. As $\text{Pr} \to \infty$, which corresponds to the case of strongly viscous fluids (e.g., glycerin), the temperature boundary layer is very thin and lies inside the dynamic boundary layer, where the velocity increases linearly with the distance from the plate surface.

In the entire range of the Prandtl number, the function $\varphi(\text{Pr})$ in formula (3.3.5) is well approximated by the expression

$$\varphi(\text{Pr}) = 0.0817 \left[ (1 + 72\,\text{Pr})^{2/3} - 1 \right]^{1/2}, \tag{3.3.7}$$

whose maximum deviation from the numerical data in [427] is about 0.5%.

Let us write out the local Nusselt number:

$$\text{Nu}_X = -\frac{X}{T_s - T_i} \left( \frac{\partial T_*}{\partial Y} \right)_{Y=0} = \sqrt{\text{Re}_X}\,\varphi(\text{Pr}), \tag{3.3.8}$$

where $\text{Re}_X = XU_i/\nu$ is the local Reynolds number.

## 3.3-2. Heat Transfer in Turbulent Flow

*Temperature profile.* Let us discuss qualitative specific features of convective heat and mass transfer in turbulent flow past a flat plate. Experimental evidence indicates that several characteristic regions with different temperature profiles can be distinguished in the thermal boundary layer on a flat plate. At moderate Prandtl numbers ($0.5 \leq \mathrm{Pr} \leq 2.0$), it can be assumed for rough estimates that the characteristic sizes of these regions are of the same order of magnitude as those of the wall layer and the core of the turbulent stream, see Section 1.7.

For the description of heat and mass transfer in the wall layer, it is common to introduce the friction velocity $U_*$, friction temperature $\Theta_*$, and dimensionless internal coordinate $y^+$ by the relations

$$ U_* = \sqrt{\frac{\tau_s}{\rho}}, \quad \Theta_* = \frac{q_s}{\rho c_p U_*}, \quad y^+ = \frac{Y U_*}{\nu}, \tag{3.3.9} $$

where $\tau_s$ is the shear stress at the wall, $\rho$ the fluid density, $q_s$ the heat flux at the wall, $c_p$ the specific heat of the fluid, $\nu$ the kinematic viscosity, and $Y$ the distance to the plate surface.

In the molecular thermal conduction layer, adjacent to the plate surface, the deviation of the average temperature $\overline{T}$ from the wall temperature $T_s$ depends linearly on the transverse coordinate:

$$ \frac{T_s - \overline{T}(Y)}{\Theta_*} = \mathrm{Pr}\, y^+. \tag{3.3.10} $$

According to [212], at $\mathrm{Pr} > 1$ the linear law (3.3.10) is satisfied for $0 \leq y^+ \leq 9\,\mathrm{Pr}^{-1/3}$.

In the logarithmic layer, it is recommended to calculate the average temperature [212, 289] using the relation*

$$ \frac{T_s - \overline{T}(Y)}{\Theta_*} = 2.12 \ln y^+ + \beta(\mathrm{Pr}), $$
$$ \beta(\mathrm{Pr}) = (3.85\,\mathrm{Pr}^{1/3} - 1.3)^2 + 2.12 \ln \mathrm{Pr}. \tag{3.3.11} $$

It was shown in [209] that formulas (3.3.11) are valid within a wide range of Prandtl numbers, $6 \times 10^{-3} \leq \mathrm{Pr} \leq 10^4$, for $y^+ \geq y^+_{\log}$. The lower bound of the logarithmic layer, $y^+_{\log}$, depends on the Prandtl number as follows:

$$ y^+_{\log} \simeq \begin{cases} 12\,\mathrm{Pr}^{-1/3} & \text{if } \mathrm{Pr} \gg 1 & \text{(for liquids)}, \\ 30 & \text{if } \mathrm{Pr} \sim 1 & \text{(for gases)}, \\ 2/\mathrm{Pr} & \text{if } \mathrm{Pr} \ll 1 & \text{(for liquid metals)}. \end{cases} $$

---

* The universal constant 2.12 before the logarithm is defined as $\mathrm{Pr}_t/\kappa$, where $\mathrm{Pr}_t = 0.85$ is the turbulent Prandtl number for wall flows and $\kappa = 0.4$ the von Karman constant.

In the stream core and a major part of the logarithmic layer, the average temperature profile can be described by the single formula

$$\frac{\overline{T}(Y) - \overline{T}_m}{\Theta_*} = -2.12 \ln \eta + 1.5 (1 - \eta), \qquad (3.3.12)$$

where $\overline{T}_m$ is the average temperature at the stream axis, $\eta = Y/\delta_T$ the dimensionless distance to the plate surface, and $\delta_T$ the thermal boundary layer thickness. The results predicted by Eq. (3.3.12) fairly well agree with experimental data provided by various authors for turbulent flows of air, water, and transformer oil for $10^{-2} < Y/\delta_T < 1$. The experimental data for the verification of formula (3.3.12) are presented in the reference [212] in a systematic form. Also, a similar formula is suggested in [212] with the right-hand side $-2.12 \ln \eta + 1.5$; however, this formula predicts the temperature profile near the edge of the thermal boundary layer worse than (3.3.12).

The thermal boundary layer thickness can be estimated from the relation [212]

$$\delta_T(X) = 0.45 X \sqrt{\tfrac{1}{2} c_f}, \qquad (3.3.13)$$

where $c_f$ is the local friction coefficient determined by formula (1.7.14) or (1.7.15). The empirical coefficient 0.45 was obtained by processing experimental data for $0.7 \le \mathrm{Pr} \le 64$. It is greater than the similar coefficient in relation (1.7.13) for the hydrodynamic turbulent boundary layer thickness. (Qualitatively, this is accounted for by the fact that the turbulent Prandtl number is less than unity, that is, the hydrodynamic boundary layer lies within the thermal boundary layer.)

***Nusselt number.*** For determining the local Nusselt number, it is recommended to use the Kader–Yaglom formula [212, 289]

$$\mathrm{Nu}_X = \frac{\mathrm{Pr}\,\mathrm{Re}_X (\tfrac{1}{2} c_f)^{1/2}}{2.12 \ln(\tfrac{1}{2} c_f\,\mathrm{Pr}\,\mathrm{Re}_X) + (3.85\,\mathrm{Pr}^{1/3} - 1.3)^2 + 1.5}, \qquad (3.3.14)$$

which quite well agrees with numerous experimental data within a wide range of Prandtl numbers ($0.5 \le \mathrm{Pr} \le 100$).

For rough estimates, the simpler formula [253, 254]

$$\mathrm{Nu}_X = 0.0288\,\mathrm{Pr}^{0.4}\,\mathrm{Re}_X^{0.8}$$

can be used, which is valid for $0.5 \le \mathrm{Pr} \le 50$.

More detailed information about heat and mass transfer in turbulent flows past a flat plate, as well as various relations for determining the temperature profile and Nusselt (Sherwood) numbers, and a lot other useful information can be found in the references [184, 185, 212, 289, 406], which contain extensive literature surveys.

# 3.4. Mass Transfer in Liquid Films

## 3.4-1. Mass Exchange Between Gases and Liquid Films

*Statement of the problem.* Dissolution of gases in flowing liquid films is one of the most important methods for dissolving gases, which is widely used in technology. Film absorbers with irrigated walls are used for obtaining water solutions of gases (e.g., absorption of the HCl vapor by water), for separating gas mixtures (e.g., absorption of benzene in the cake and by-product process), for purifying gases from detrimental effluents (e.g., purification of coke-oven from $H_2S$), etc.

Let us consider the absorption of weakly soluble gases on the free surface of a liquid film in a laminar flow on an inclined plane. It follows from the results of Section 1.3 that for moderate velocities of motion, the steady-state distribution of the velocity inside the film has the form of a semiparabola with maximum velocity $U_{max}$ on the free surface, which is one and a half of the mean flow rate velocity $\langle V \rangle$:

$$U_{max} = \frac{3}{2}\langle V \rangle = \frac{gh^2}{2\nu}\sin\alpha.$$

Here $g$ is the gravitational acceleration, $\alpha$ the angle between the plane and the horizon, and $h$ the film thickness, given by the expression

$$h = \left(\frac{3\nu^2}{g}\operatorname{Re}\right)^{1/3},$$

where $\operatorname{Re} = Q/\nu$ is the Reynolds number and $Q$ is the irrigation density (that is, the volume rate flow of liquid per unit width of the film).

The liquid velocity inside the film has a parabolic profile and is given by the formula

$$V = U_{max}(1 - y^2), \qquad y = Y/h,$$

where the $Y$-axis is normal to the film surface (Figure 1.3).

Assume that in the cross-section $X = 0$, fluid flow contacts with a gas, so that constant concentration $C = C_s$ of the absorbed component is attained on the free boundary ($Y = 0$), while no solute is contained in the irrigating liquid. Furthermore, we assume that there is no flow across the wall. We restrict ourselves to the case of high Peclet numbers, in which one can ignore the molecular diffusion along the film.

It follows from these assumptions that the distribution of the concentration inside the film is described by the equation and the boundary conditions [54]

$$(1 - y^2)\frac{\partial c}{\partial x} = \frac{1}{\operatorname{Pe}}\frac{\partial^2 c}{\partial y^2}; \tag{3.4.1}$$

$$x = 0, \quad c = 0 \qquad (0 \le y \le 1); \tag{3.4.2}$$

$$y = 0, \quad c = 1 \qquad (x > 0); \tag{3.4.3}$$

$$y = 1, \quad \partial c/\partial y = 0 \quad (x > 0), \tag{3.4.4}$$

where the following dimensionless variables have been used:

$$x = \frac{X}{h}, \quad y = \frac{Y}{h}, \quad c = \frac{C}{C_s}, \quad \mathsf{Pe} = \frac{hU_{max}}{D}. \tag{3.4.5}$$

Note that for small $x \leq O(\mathsf{Pe}^{-1/2})$, near the input section one should consider the complete mass transfer equation with the term $\partial^2 c / \partial y^2$ replaced by $\Delta c$.

**Diffusion boundary layer approximation.** For $x = O(1)$, the concentration mostly varies on the initial interval in a thin diffusion boundary layer near the free boundary of the film. In this region we expand the transverse coordinate according to the rule

$$y = w/\sqrt{\mathsf{Pe}}. \tag{3.4.6}$$

By substituting (3.4.6) into (3.4.1) and passing to the limit as $\mathsf{Pe} \to \infty$ (it is assumed that the variables $x$ and $w$ and the corresponding derivatives are of the order of 1), we obtain the equation

$$\frac{\partial c}{\partial x} = \frac{\partial^2 c}{\partial w^2}. \tag{3.4.7}$$

In view of (3.4.6), the distance to the wall determined by the coordinate $y = 1$ corresponds to $w = \sqrt{\mathsf{Pe}}$. Therefore, as $\mathsf{Pe} \to \infty$, the value $y = 1$ from the boundary condition (3.4.4) corresponds to $w \to \infty$. This implies that we can rewrite the boundary conditions (3.4.2)–(3.4.4) as follows:

$$x = 0, \quad c = 0; \quad w = 0, \quad c = 1; \quad w \to \infty, \quad \partial c / \partial w \to 0. \tag{3.4.8}$$

The solution of problem (3.4.7), (3.4.8) is given by the formula

$$c = \mathrm{erfc}\left(\frac{w}{2\sqrt{x}}\right), \tag{3.4.9}$$

where $\mathrm{erfc}\, z = \dfrac{2}{\sqrt{\pi}} \displaystyle\int_z^\infty \exp(-t^2)\, dt$ is the complementary error function.

By differentiating (3.4.9), we obtain the local diffusion flux to the film surface [270]:

$$j = -\left(\frac{\partial c}{\partial y}\right)_{y=0} = \left(\frac{\mathsf{Pe}}{\pi x}\right)^{1/2}. \tag{3.4.10}$$

The dimensionless total diffusion flux on the part of the film lying in the interval from 0 to $x$ is equal to

$$I = \int_0^x j\, dx = 2\left(\frac{\mathsf{Pe}}{\pi} x\right)^{1/2}. \tag{3.4.11}$$

Formulas (3.4.10) and (3.4.11) cannot be used for sufficiently large $x$, when the diffusion boundary layer "grows" through the entire film. To estimate the scope of these formulas, let us consider the original problem (3.4.1)–(3.4.4).

*Exact solution.* For $0 \leq x < \infty$, we seek the solution of problem (3.4.1)–(3.4.4) in the form of the series [54, 108]

$$c = 1 - \sum_{m=0}^{\infty} A_m \exp\left(-\frac{\lambda_m^2}{\text{Pe}} x\right) H_m(y), \qquad (3.4.12)$$

where the unknown functions $H_m$ and the coefficients $A_m$ and $\lambda_m$ are independent of the Peclet number.

Substituting the expansion (3.4.12) into (3.4.1) and then separating the variables, we obtain the following ordinary differential equation for the functions $H_m$:

$$\frac{d^2 H_m}{dy^2} + \lambda_m^2 (1 - y^2) H_m = 0. \qquad (3.4.13)$$

Taking into account (3.4.12), we obtain the boundary conditions for $H_m$ from (3.4.3) and (3.4.4):

$$y = 0, \quad H_m = 0; \qquad y = 1, \quad \frac{dH_m}{dy} = 0. \qquad (3.4.14)$$

Problem (3.4.13), (3.4.14) determines the eigenfunctions $H_m$ and the eigenvalues $\lambda_m$.

The general solution of (3.4.13) has the form [108]

$$H_m(y) = \exp(-\tfrac{1}{2}\lambda_m y^2)[B_1 \Phi(a_m, \tfrac{1}{2}; \lambda_m y^2)$$
$$+ B_2 y \Phi(a_m + \tfrac{1}{2}, \tfrac{3}{2}; \lambda_m y^2)], \quad a_m = \tfrac{1}{4}(1 - \lambda_m), \qquad (3.4.15)$$

where $\Phi(a, b, \xi) = 1 + \sum_{m=1}^{\infty} \dfrac{a(a+1)\ldots(a+m-1)}{b(b+1)\ldots(b+m-1)} \dfrac{\xi^m}{m!}$ is the degenerate hypergeometric function.

By satisfying the first boundary condition in (3.4.14), we obtain $B_1 = 0$. Substituting this value into (3.4.15) and setting $B_2 = 1$, we obtain

$$H_m(y) = y \exp\left(-\tfrac{1}{2}\lambda_m y^2\right) \Phi\left(a_m + \tfrac{1}{2}, \tfrac{3}{2}; \lambda_m y^2\right) \qquad (3.4.16)$$

(the functions $H_m$ are determined up to a constant factor). Substituting the function (3.4.16) into the second boundary condition in (3.4.14), we arrive at the following transcendental equation for the eigenvalues $\lambda_m$:

$$\lambda_m \Phi\left(a_m + \tfrac{1}{2}, \tfrac{3}{2}; \lambda_m\right) - \Phi\left(a_m + \tfrac{1}{2}, \tfrac{1}{2}; \lambda_m\right) = 0, \qquad (3.4.17)$$

derived with the help of the relation [28]

$$\frac{d}{d\xi} \Phi(a, b; \xi) = \frac{1-b}{\xi}\left[\Phi(a, b; \xi) - \Phi(a, b-1; \xi)\right].$$

TABLE 3.2

The eigenvalues $\lambda_m$ and the coefficients $A_m$ in the expansion
of the solution (3.4.12) for the concentration distribution
inside the film with gas absorption on the film surface

| $m$ | $\lambda_m$ | $A_m$ | $m$ | $\lambda_m$ | $A_m$ |
|---|---|---|---|---|---|
| 1 | 2.2631 | 1.3382 | 6 | 22.3181 | –0.1873 |
| 2 | 6.2977 | –0.5455 | 7 | 26.3197 | 0.1631 |
| 3 | 10.3077 | 0.3589 | 8 | 30.3209 | –0.1449 |
| 4 | 14.3128 | –0.2721 | 9 | 34.3219 | 0.1306 |
| 5 | 18.3159 | 0.2211 | 10 | 38.3227 | –0.1191 |

Table 3.2 contains the first ten eigenvalues $\lambda_m$ calculated in [408].

Now let us calculate the coefficients $A_m$. By substituting the series (3.4.12) into the boundary condition (3.4.2), we obtain

$$\sum_{m=1}^{\infty} A_m H_m(y) = 1. \tag{3.4.18}$$

We multiply Eq. (3.4.13) by the eigenfunction $H_k$ $(k \neq m)$ and integrate the obtained expression with respect to $y$ from 0 to 1. Taking into account the boundary conditions (3.4.14) and performing some transformations, we obtain the orthogonality conditions with weight function $(1 - y^2)$ for $H_k$ and $H_m$:

$$\int_0^1 (1 - y^2) H_m H_k \, dy = 0 \qquad (k \neq m). \tag{3.4.19}$$

Multiplying both sides of (3.4.18) by the function $(1 - y^2)H_k$ and integrating the obtained series over the entire thickness of the film with regard to (3.4.19), we find the coefficients

$$A_m = \frac{\displaystyle\int_0^1 (1 - y^2) H_m(y) \, dy}{\displaystyle\int_0^1 (1 - y^2)[H_m(y)]^2 \, dy}, \qquad \text{where} \quad m = 1, 2, \ldots \tag{3.4.20}$$

Table 3.2 contains the first ten coefficients $A_m$ calculated in [408].

Taking into account (3.4.12), we obtain the dimensional total diffusion flux to the part of the film boundary from 0 to $x$ in the form

$$I = -\int_0^x \left(\frac{\partial c}{\partial y}\right)_{y=0} dx$$

$$= \text{Pe} \sum_{m=1}^{\infty} \frac{A_m}{\lambda_m^2} \left(\frac{dH_m}{dy}\right)_{y=0} \left[1 - \exp\left(-\frac{\lambda_m^2}{\text{Pe}} x\right)\right]. \tag{3.4.21}$$

The substitution of the function (3.4.16) into (3.4.21) implies

$$I = \text{Pe} \sum_{m=1}^{\infty} \frac{A_m}{\lambda_m^2} \left[ 1 - \exp\left( -\frac{\lambda_m^2}{\text{Pe}} x \right) \right].$$  (3.4.22)

The comparison of this formula with (3.4.11) shows that the diffusion boundary layer approximation can be used in the region $x \leq 0.1\,\text{Pe}$.

## 3.4-2. Dissolution of a Plate by a Laminar Liquid Film

*Statement of the problem.* Now let us consider mass transfer from a solid wall to a liquid film at high Peclet numbers. Such a problem is of serious interest in dissolution, crystallization, corrosion, anodic dissolution of metals in some electrochemical processes, etc. In many practical cases, dissolution processes are rather rapid compared with diffusion. Therefore, we assume that the concentration on the plate surface is equal to the constant $C_s$ and the incoming liquid is pure. As previously, we introduce dimensionless variables according to formulas (3.4.5). In this case, the convective mass transfer in the liquid film is described by Eq. (3.4.1), the boundary condition (3.4.2) imposed on the longitudinal variable $x$, and the following boundary conditions with respect to the transverse coordinate:

$$y = 0, \quad \frac{\partial c}{\partial y} = 0 \quad (x > 0);$$  (3.4.23)

$$y = 1, \quad c = 1 \quad (x > 0).$$  (3.4.24)

Although this problem differs from the previously studied problem (3.4.1)–(3.4.4) only by a rearrangement of the boundary conditions (3.4.3) and (3.4.4), there is a substantial difference in the solutions of there problems.

*Diffusion boundary layer approximation.* The concentration mostly varies on the initial interval $x = O(1)$, that is, in the diffusion boundary layer near the film surface. In this region, the asymptotic solution can be obtained by substituting the expanded coordinate

$$\xi = (1 - y)\,\text{Pe}^{1/3}$$  (3.4.25)

into Eq. (3.4.1) and taking the leading term of the concentration expansion as $\text{Pe} \to \infty$. As a result, we arrive at the diffusion boundary layer equation

$$2\xi \frac{\partial c}{\partial x} = \frac{\partial^2 c}{\partial \xi^2}.$$  (3.4.26)

By analogy with the already considered problem about gas absorption on the film surface, using (3.4.2), (3.4.23), and (3.4.24), we obtain the boundary conditions for Eq. (3.4.26), which coincide with (3.4.8) up to the replacement $\xi \to w$. The corresponding problem has the solution

$$c = \frac{1}{\Gamma(1/3)} \Gamma\left(\frac{1}{3}, \frac{2\xi^3}{9x}\right),$$    (3.4.27)

where $\Gamma(1/3, z)$ is the incomplete gamma function.

By differentiating (3.4.27), we calculate the local diffusion flux [270]

$$j = \frac{6^{1/3}}{\Gamma(1/3)} \frac{Pe^{1/3}}{x^{1/3}} \approx 0.678 \frac{Pe^{1/3}}{x^{1/3}}.$$    (3.4.28)

The corresponding total diffusion flux to the film boundary is

$$I = \int_0^x j \, dx = 1.02 \, Pe^{1/3} x^{2/3}.$$    (3.4.29)

By comparing formulas (3.4.9)–(3.4.11) with (3.4.27)–(3.4.29), we see that for $x \sim 1$ the thickness of the diffusion boundary layer near the free surface of the film, $\delta_0 \sim Pe^{-1/2}$, is considerably less than that of the boundary layer near the solid surface, $\delta_{sol} \sim Pe^{-1/3}$. Accordingly, the diffusion flux to the free surface is larger than that to the solid surface. Moreover, the diffusion flux decreases more rapidly on the free surface than on the solid boundary with the increase of the distance from the input cross-section. These effects are due to the fact that the fluid moves much more rapidly near the free surface than near the solid boundary, where the no-slip condition is satisfied.

All facts established for a fluid film remain valid for a majority of problems on the diffusion boundary layer. Namely, near a gas–fluid or fluid-fluid interface, the dimensionless thickness of the layer is proportional to $Pe^{-1/2}$ (for the diffusion flux we have $j \sim Pe^{1/2}$), and near the fluid–solid interface the thickness of the boundary layer is proportional to $Pe^{-1/3}$ (the diffusion flux is $j \sim Pe^{1/3}$).

***Exact solution.*** As previously, we seek the solution of problem (3.4.1), (3.4.2), (3.4.23), (3.4.24) in the entire region $0 \le x < \infty$ in the form of the series (3.4.12), where the eigenfunctions $H_m$ satisfy Eq. (3.4.13). (The unknown coefficients $\lambda_m$ and $A_m$ and the functions $H_m$ must be determined and differ from those obtained in the problem of gas absorption on the free surface of the film.) We obtain the boundary conditions for $H_m$ by substituting the expansion (3.4.12) into (3.4.23) and (3.4.24). As a result, we obtain

$$y = 0, \quad \frac{dH_m}{dy} = 0; \quad y = 1, \quad H_m = 0.$$    (3.4.30)

The general solution of Eq. (3.4.13) is given by formula (3.4.15). By satisfying the boundary condition (3.4.30), we obtain $B_2 = 0$. On substituting this value into (3.4.15) and setting $B_1 = 1$, we arrive at the expression

$$H_m(y) = \exp\left(-\tfrac{1}{2}\lambda_m y^2\right) \Phi\left(\tfrac{1}{4} - \tfrac{1}{4}\lambda_m, \tfrac{1}{2}; \lambda_m y^2\right).$$    (3.4.31)

The transcendental equation for the eigenvalues $\lambda_m$ is derived by using the second boundary condition (3.4.30) and formula (3.4.31):

$$\Phi\left(a_m, \tfrac{1}{2}; \lambda_m\right) = 0, \qquad \text{where} \quad a_m = \tfrac{1}{4} - \tfrac{1}{4}\lambda_m. \tag{3.4.32}$$

The roots of Eq. (3.4.32) are positive and increase monotonically, so that $\lambda_m \to \infty$ as $m \to \infty$. With regard to this, let us find the asymptotics of the eigenvalues $\lambda_m$ for large $m$.

If the two conditions

$$x \to \infty \quad \text{and} \quad x - 2b + 4a = \text{const} \tag{3.4.33}$$

are satisfied simultaneously, then the asymptotics of the degenerate hypergeometric function has the form [4]

$$\Phi(a, b; x) = \frac{2\Gamma(b)}{3^{2/3}\Gamma(2/3)}(b-2a)^{2/3-b}e^{x/2}\left[\sin\left(a\pi + \frac{\pi}{6}\right) + O(x^{-2/3})\right]. \tag{3.4.34}$$

In our case, $x = \lambda_m$ and $x - 2b + 4a = 0$, that is, conditions (3.4.33) are satisfied. Therefore, according to (3.4.34), as $m \to \infty$ the distribution of the eigenvalues $\lambda_m$ satisfying Eq. (3.4.32) is determined by the trigonometric equation $\sin(a_m\pi + \pi/6) = 0$. This equation has the solution

$$a_m = -m - \tfrac{1}{6}, \qquad \text{where} \quad m = 0, \pm 1, \pm 2, \ldots \tag{3.4.35}$$

By substituting $a_m$ from (3.4.32) into (3.4.35), we obtain the asymptotics of the eigenvalues $\lambda_m$ as $m \to \infty$ [436]:

$$\lambda_m = 4m + \tfrac{5}{3}. \tag{3.4.36}$$

It is remarkable that although formula (3.4.36) was derived for large $m$, it can successfully be used for all $m = 0, 1, 2, \ldots$ The comparison with numerical results [70] (see also [341]) shows that formula (3.4.36) has the maximum error of 0.9% for $m = 0$.

Instead of (3.4.36), one can use a more precise relation for the eigenvalues,

$$\lambda_m = 4m + 1.68 \qquad (m = 0, 1, 2, \ldots), \tag{3.4.37}$$

whose error is less than 0.2%.

As previously, the coefficients $A_m$ in the series (3.4.12) are determined by (3.4.20), where the eigenfunctions $H_m$ are given in (3.4.31).

The coefficients $A_m$, calculated numerically in [70] and asymptotically in [436], can be approximated as follows:

$$A_0 = 1.2; \quad A_m = 2.27\,(-1)^m\lambda_m^{-7/6} \qquad \text{for} \quad m = 1, 2, 3, \ldots, \tag{3.4.38}$$

where the $\lambda_m$ are given in (3.4.37). The maximum error of (3.4.38) is less than 0.1%.

The total diffusion flux on the film surface is given by (3.4.21) with $(dH_m/dy)_{y=0}$ replaced by $(dH_m/dy)_{y=1}$. One can use the expressions (3.4.31), (3.4.37), and (3.4.38) for calculating the eigenfunctions $H_m$ and the coefficients $\lambda_m$ and $A_m$.

# 3.5. Heat and Mass Transfer in a Laminar Flow in a Circular Tube

Many process of convective mass and heat transfer in chemical industry, petroleum chemistry, gas, nuclear, and other branches of industry proceed in pipes (water, gas, and oil pipelines, heat exchangers, etc.).

Starting from the classical work by Graetz and Nusselt [157, 319], many authors considered the problem about the temperature distribution in a fluid moving in a tube under various assumptions on the type of flow, the tube shape, the form of boundary conditions, the value of the Peclet number, and some other simplifications (e.g., see [31, 70, 80, 108, 185, 406]). In this section, we outline the most important results obtained in this field.

---
**3.5-1. Tube With Constant Temperature of the Wall**
---

***Statement of the problem.*** Let us consider laminar steady-state fluid flow in a circular tube of radius $a$ with Poiseuille velocity profile (see Section 1.5). We introduce cylindrical coordinates $\mathcal{R}$, $Z$ with the $Z$-axis in the direction of flow. We assume that for $Z > 0$ the temperature on the wall is equal to the constant $T_2$. In the entry area $Z < 0$, the temperature on the wall is also constant but takes another value $T_1$.

The convective heat transfer in a tube is described by the equation

$$\mathrm{Pe}_T(1-\varrho^2)\frac{\partial T}{\partial z} = \frac{\partial^2 T}{\partial \varrho^2} + \frac{1}{\varrho}\frac{\partial T}{\partial \varrho} + \frac{\partial^2 T}{\partial z^2} \qquad (3.5.1)$$

and the boundary conditions

$$\varrho = 0, \quad \frac{\partial T}{\partial \varrho} = 0; \qquad \varrho = 1, \quad T = \begin{cases} 0 & \text{for } z < 0; \\ 1 & \text{for } z > 0; \end{cases} \qquad (3.5.2)$$

$$z \to -\infty, \quad T \to 0; \qquad z \to \infty, \quad T \to 1. \qquad (3.5.3)$$

The following dimensionless variables have been used:

$$\varrho = \frac{\mathcal{R}}{a}, \quad z = \frac{Z}{a}, \quad T = \frac{T_* - T_1}{T_2 - T_1}, \quad \mathrm{Pe}_T = \frac{aU_{\max}}{\chi},$$

where $T_*$ is the fluid temperature, $\chi$ the thermal diffusivity, $U_{\max} = a^2 \Delta P/(4\mu L)$ the maximal velocity at the center of the tube, $\Delta P$ the pressure increment along the distance $L$, and $\mu$ the dynamic viscosity of the fluid.

***High Peclet numbers (initial region).*** As $\mathrm{Pe}_T \to \infty$, the fluid temperature in the region $z < 0$ is constant and equal to the temperature $T \approx 0$ on the wall. In the region $z > 0$ at $z = O(1)$, a thin boundary layer is being formed near the wall of the tube. In this region, on the left-hand side of Eq. (3.5.1) one may retain only the leading term of the velocity expansion as $\varrho \to 1$ and write $v = 1 - \varrho^2 \approx 2\xi$, where $\xi = 1 - \varrho$. Moreover, in contrast with the first term on the right-hand side in (3.5.1), one can neglect the last two terms, that is, $\Delta T \approx \partial^2 T/\partial \xi^2$. Thus, we arrive at an equation which coincides with (3.4.26) up to notation. Taking into account the boundary conditions (3.5.2), we obtain the temperature distribution in the boundary layer

$$T = \frac{1}{\Gamma(1/3)}\Gamma\left(\frac{1}{3}, \frac{2\,\mathrm{Pe}_T(1-\varrho)^3}{9z}\right). \tag{3.5.4}$$

The corresponding dimensionless local $j_T$ and total $I_T$ heat fluxes have the form [269]

$$j_T = -\left(\frac{\partial T}{\partial \varrho}\right)_{\varrho=1} = \frac{1}{\Gamma(1/3)}\left(\frac{6\mathrm{Pe}_T}{z}\right)^{1/3}, \tag{3.5.5}$$

$$I_T = \int_0^z j_T\,dz = \frac{3(6\,\mathrm{Pe}_T)^{1/3}}{2\Gamma(1/3)}z^{2/3}. \tag{3.5.6}$$

The scope of formulas (3.5.4)–(3.5.6) is bounded by $z \ll \mathrm{Pe}_T$. According to the estimate obtained in [270], this restriction practically always holds in the similar problem about the diffusion boundary layer.

***Arbitrary Peclet numbers.*** In the general case $0 \le \mathrm{Pe}_T < \infty$, we seek the fluid temperature distribution separately on both sides of the inlet cross-section of the tube in the form of the series:

$$T = \sum_{k=0}^{\infty} B_k \exp\left(\frac{\eta_k^2}{\mathrm{Pe}_T}z\right)g_k(\varrho) \qquad \text{for} \quad z < 0, \tag{3.5.7}$$

$$T = 1 - \sum_{m=0}^{\infty} A_m \exp\left(-\frac{\lambda_m^2}{\mathrm{Pe}_T}z\right)f_m(\varrho) \qquad \text{for} \quad z > 0. \tag{3.5.8}$$

By substituting the expansions (3.5.7) and (3.5.8) into Eq. (3.5.1) and the boundary conditions (3.5.2) and (3.5.3) and by separating the variables for the eigenfunctions $g_k$ and $f_m$ and the eigenvalues $\eta_k$ and $\lambda_m$, we obtain the spectral problems

$$\frac{d^2 g_k}{d\varrho^2} + \frac{1}{\varrho}\frac{dg_k}{d\varrho} + \eta_k^2\left(\varrho^2 - 1 + \frac{\eta_k^2}{\mathrm{Pe}_T^2}\right)g_k = 0;$$

$$\varrho = 0, \quad \frac{dg_k}{d\varrho} = 0; \qquad \varrho = 1, \quad g_k = 0; \tag{3.5.9}$$

$$\frac{d^2 f_m}{d\varrho^2} + \frac{1}{\varrho}\frac{df_m}{d\varrho} + \lambda_m^2\left(1 - \varrho^2 + \frac{\lambda_m^2}{\mathrm{Pe}_T^2}\right)f_m = 0;$$

$$\varrho = 0, \quad \frac{df_m}{d\varrho} = 0; \qquad \varrho = 1, \quad f_m = 0. \tag{3.5.10}$$

The eigenfunctions are determined only up to a constant factor. To obtain uniquely chosen solutions of the spectral problems (3.5.9), (3.5.10), we state the following normalization condition on the flow axis:

$$g_k = 1, \quad f_m = 1 \quad \text{at} \quad \varrho = 0. \tag{3.5.11}$$

The temperature $T = T(\varrho, z)$ and its derivative must be continuous across the section $z = 0$:

$$T(\varrho, -0) = T(\varrho, +0); \quad \frac{\partial T}{\partial z}(\varrho, -0) = \frac{\partial T}{\partial z}(\varrho, +0). \tag{3.5.12}$$

The consistency conditions (3.5.12) allow one to find the coefficients $B_k$ and $A_m$ in the series (3.5.7) and (3.5.8). In [20, 204], the following formulas are given:

$$B_k = -\frac{2}{\eta_k \left(\dfrac{\partial g}{\partial \eta}\right)_{\varrho=1, \, \eta=\eta_k}}, \quad A_m = -\frac{2}{\lambda_m \left(\dfrac{\partial f}{\partial \lambda}\right)_{\varrho=1, \, \lambda=\lambda_m}}. \tag{3.5.13}$$

Here $g = g(\varrho, \eta)$ and $f = f(\varrho, \lambda)$ are auxiliary functions determined by problems (3.5.9)–(3.5.11) with omitted indices $k$ and $m$ and omitted boundary conditions on the tube walls at $\varrho = 1$.

In what follows we consider only the region $z > 0$.

Straightforward verification shows that the change of variables $u = \lambda_m \varrho^2$, $F = \exp(u/2)f_m$ transforms (3.5.10) into a degenerate hypergeometric equation for the function $F = F(u)$ [28]. Therefore, the solution of the spectral problem (3.5.10) satisfying the normalization condition (3.5.11) can be expressed in terms of the degenerate hypergeometric function $\Phi(a, b; \xi)$ as

$$f_m = \exp\left(-\tfrac{1}{2}\lambda_m \varrho^2\right) \Phi(a_m, 1; \lambda_m \varrho^2), \tag{3.5.14}$$

$$a_m = \frac{1}{2} - \frac{1}{4}\lambda_m - \frac{\lambda_m^3}{4\,\mathsf{Pe}_T^2},$$

where the eigenvalues $\lambda_m$ satisfy the transcendental equation

$$\Phi(a_m, 1; \lambda_m) = 0. \tag{3.5.15}$$

The auxiliary function $f = f(\varrho, \lambda)$ used to calculate the coefficients $A_m$ (3.5.13) can be obtained from (3.5.14) by omitting the indices $m$.

Let us study the asymptotic behavior of the eigenvalues $\lambda_m$ and coefficients $A_m$ in some limit cases.

On can neglect the left-hand side of Eq. (3.5.1) for small Peclet numbers. The corresponding asymptotic solution must be independent of the Peclet number. Therefore, it follows from (3.5.8) that $\lambda_m = O(\sqrt{\mathsf{Pe}_T})$. Taking into account this fact and the limit relation $\lim \Phi(a, 1; -\xi/a) = J_0(2\sqrt{\xi})$ [28] and letting $\mathsf{Pe}_T \to 0$ in (3.5.13)–(3.5.15), we obtain

$$\lambda_m = (\gamma_m \, \mathsf{Pe}_T)^{1/2}, \quad A_m = -[\gamma_m J_1(\gamma_m)]^{-1}, \quad f_m = J_0(\gamma_m \varrho), \qquad (3.5.16)$$

where $J_0 = J_0(\xi)$ and $J_1 = J_1(\xi)$ are the Bessel functions and the $\gamma_m$ are the roots of the Bessel function, $J_0(\gamma_m) = 0$.

It is convenient to calculate approximate values of $\gamma_m$ by using the approximate expression

$$\gamma_m = 2.4 + 3.13\,m \qquad (m = 0,\ 1,\ 2,\ \ldots), \qquad (3.5.17)$$

whose maximum error is less than 0.2% (compared with the data in [4]).

Now we set $\mathsf{Pe} = \infty$ in formulas (3.5.14) and (3.5.15), which corresponds to the parameter $a_m = \frac{1}{4}(2 - \lambda_m)$. Let us consider large values of the number $m$. In our case, both relations in (3.4.33) hold for $x = \lambda_m$, $b = 1$, and $4a = 2 - \lambda_m$. Therefore, the asymptotics (3.4.34) is satisfied for the degenerate hypergeometric function, and the roots of the transcendental equation (3.5.15) are given by (3.4.35). By substituting $a_m = \frac{1}{4}(2 - \lambda_m)$ into (3.4.35), we obtain the eigenvalues $\lambda_m$ [436] in the form

$$\lambda_m = 4m + \tfrac{8}{3}. \qquad (3.5.18)$$

The asymptotic solution (3.5.18) constructed under the assumption that $m \gg 1$ can be used for all values of $m$. The comparison with numerical data in [341] shows that the maximum error in (3.5.18) is attained at $m = 0$ and is equal to 1.4%.

Instead of (3.5.18), it is convenient to use the more precise formula

$$\lambda_m = 4m + 2.7 \qquad (m = 0,\ 1,\ 2,\ \ldots), \qquad (3.5.19)$$

whose maximum error is 0.3%.

The coefficients $A_m$ of the series (3.5.8) can be calculated by the formula

$$A_m = 2.85\,(-1)^m \lambda_m^{-2/3}, \qquad (3.5.20)$$

which up to 0.5% complies with the asymptotic [436] and numerical [341] results.

For large but finite Peclet numbers, the expressions (3.5.19) and (3.5.20) can be used only for a bounded set of eigenvalues such that $\lambda_m \ll \mathsf{Pe}_T$.

Numerical values of $\lambda_0$, $\lambda_1$, and $\lambda_2$ obtained on a computer for various $\mathsf{Pe}_T$ are presented in [265]. The dependence of the ground eigenvalue $\lambda_0$ on the Peclet number is nicely approximated by the formula

$$\lambda_0 = 2.7 \sqrt{\frac{\exp(0.27\,\mathsf{Pe}_T) - 1}{\exp(0.27\,\mathsf{Pe}_T) - 0.18}}, \qquad (3.5.21)$$

whose maximum error is about 1%.

***Mean mass temperature. Heat flux.*** Taking into account the fact that the dimensionless distribution of fluid velocity in a tube is $u(\varrho) = 1 - \varrho^2$, we have the mean mass temperature in an arbitrary cross-section

$$\langle T \rangle_m = \frac{\int_0^1 Tu(\varrho)2\pi\varrho\,d\varrho}{\int_0^1 u(\varrho)2\pi\varrho\,d\varrho} = 4\int_0^1 T(1-\varrho^2)\varrho\,d\varrho.$$

Substituting the series (3.5.8) into this formula, we obtain

$$\langle T \rangle_m = 1 - \sum_{m=0}^{\infty} E_m \exp\left(-\frac{\lambda_m^2}{\mathrm{Pe}_T}z\right), \quad \text{where} \quad E_m = 4A_m\int_0^1 f_m(1-\varrho^2)\varrho\,d\varrho.$$

$$(3.5.22)$$

By differentiating (3.5.8), we arrive at the dimensionless local heat flux to the tube wall:

$$j_T = \left(\frac{\partial T}{\partial \varrho}\right)_{\varrho=1} = -\sum_{m=0}^{\infty} A_m f_m'(1)\exp\left(-\frac{\lambda_m^2}{\mathrm{Pe}_T}z\right), \qquad (3.5.23)$$

where $f_m'(1) = 2a_m\lambda_m \exp(-\frac{1}{2}\lambda_m)\Phi(a_m + 1, 2; \lambda_m)$.

***Nusselt number.*** The most practically important variable is the Nusselt number

$$\mathrm{Nu} = \frac{2j_T}{1 - \langle T \rangle_m}. \qquad (3.5.24)$$

Here $1 - \langle T \rangle$ is the temperature head equal to the difference between the wall temperature and the mean flow rate temperature.

It follows from (3.5.22)–(3.5.24) that at large distances from the inlet cross-section (as $z \to +\infty$), the Nusselt number tends to the constant value

$$\mathrm{Nu}_\infty = \frac{-f_0'(1)}{2\int_0^1 f_0(\varrho)(1-\varrho^2)\varrho\,d\varrho}. \qquad (3.5.25)$$

Let us consider the case of low Peclet numbers. To this end, we substitute (3.5.16) with $m = 0$ into (3.5.25). The numerator of the fraction (3.5.25) can be calculated as $dJ_0/dx = -J_1(x)$ [29], and the denominator satisfies the recurrent formula

$$\int x^k J_m(x)\,dx = x^k J_{m+1}(x) - (k-m-1)\int x^{k-1}J_{m+1}(x)\,dx,$$

which is derived from the properties of the Bessel function:

$$x^{m+1}J_m(x) = \frac{d}{dx}\left[x^{m+1}J_{m+1}(x)\right].$$

As a result, we obtain the limit Nusselt value at $\text{Pe}_T = 0$ in the form

$$\text{Nu}_\infty = \frac{\gamma_0^3}{4} \frac{J_1(\gamma_0)}{J_2(\gamma_0)} \approx 4.16 \qquad (\text{as } \text{Pe}_T \to 0). \qquad (3.5.26)$$

For high Peclet numbers, the calculations according to (3.5.25) on a computer lead to the value [341]

$$\text{Nu}_\infty = \tfrac{1}{2}\lambda_0^2 \approx 3.66 \qquad (\text{as } \text{Pe}_T \to \infty). \qquad (3.5.27)$$

In the entire range of the Peclet number, the limit Nusselt number is nicely approximated by

$$\text{Nu}_\infty = \frac{4.16 + 1.15\,\text{Pe}_T}{1 + 0.315\,\text{Pe}_T}, \qquad (3.5.28)$$

whose maximum error is about 0.6% (the precision of this expression was estimated by using the data in [341]).

Calculations according to (3.5.24) show that for high Peclet numbers, one can conventionally divide the entire length of a heated (cooled) tube into two parts. On the first part, the temperature profile is being formed with the radial temperature distribution varying from the initial value (at $z = 0$) to some limit value $f_0(\varrho)$. In this region, the number Nu decreases near the inlet cross-section as a power-law function $\text{Nu} \approx 2j_T$, where $j_T$ is described by (3.5.5). On the second part of the tube, the radial distribution of the excess temperature $\delta T = 1 - T$ does not vary along the tube (though the absolute values of temperature do vary), and the number Nu preserves the constant value 3.66. The first part of the tube is called the thermal initial region, and the second one, the region of steady-state heat exchange.

It is conventional to define the length of the thermal initial region as the distance from the inlet cross-section to the point at which the Nusselt number differs from its limit value (3.5.27) by 1%. Calculations show that the dimensional length of the thermal initial region is $l = 0.11a\,\text{Pe}_T$.

## 3.5-2. Tube With Constant Heat Flux at the Wall

*Statement of the problem.* Now let us study the case in which a constant heat flux $q = \varkappa(\partial T/\partial R)_{R=a} = \text{const}$, where $\varkappa$ is the thermal conductivity coefficient of the fluid, is given on the surface of a circular tube for $Z > 0$. The entry part is modeled by the region $Z < 0$ where there is no heat flux across the tube surface and the temperature tends to the constant value $T_1$ as $Z \to -\infty$.

In this case, it is convenient to introduce the dimensionless temperature as

$$\tilde{T} = \frac{\varkappa(T - T_1)}{aq}, \qquad (3.5.29)$$

and the other dimensionless variables are determined just as in problem (3.5.1)–(3.5.3).

The considered process of heat transfer is described by Eq. (3.5.1) with $T$ replaced by $\tilde{T}$ and the boundary conditions

$$\varrho = 0, \quad \frac{\partial \tilde{T}}{\partial \varrho} = 0; \qquad \varrho = 1, \quad \frac{\partial \tilde{T}}{\partial \varrho} = \begin{cases} 0 & \text{for } z < 0; \\ 1 & \text{for } z > 0; \end{cases} \qquad (3.5.30)$$

$$z \to -\infty, \quad \tilde{T} \to 0. \qquad (3.5.31)$$

Note that the temperature distribution as $z \to +\infty$ is not known in advance.

Let us write out the heat balance equation, which we shall need later. To this end, we multiply (3.5.1) (as $T \to \tilde{T}$) by $\varrho$ and integrate the obtained expression first along the radial coordinate from 0 to 1, then along the longitudinal coordinate from $-\infty$ to $z$, so that $z > 0$. Applying the boundary conditions (3.5.30) and (3.5.31) and changing the order of integration when necessary, we arrive at the equation

$$\text{Pe}_T \int_0^1 \varrho(1 - \varrho^2)\tilde{T} \, d\varrho = z + \int_0^1 \varrho \frac{\partial \tilde{T}}{\partial z} \, d\varrho. \qquad (3.5.32)$$

***Temperature field far from the inlet cross-section.*** Let us study the temperature field far from the cross-section at $z \gg 1$. In this region we seek the solution as the sum

$$\tilde{T} = \alpha z + \Psi(\varrho), \qquad (3.5.33)$$

where $\alpha$ is an unknown constant and $\Psi$ is an unknown function.

The substitution of (3.5.33) into (3.5.1) (as $T \to \tilde{T}$) and into the boundary conditions (3.5.30) yields the problem

$$\alpha \, \text{Pe}_T(1 - \varrho^2) = \frac{1}{\varrho} \frac{d}{d\varrho} \left( \varrho \frac{d\Psi}{d\varrho} \right); \qquad (3.5.34)$$

$$\varrho = 0, \quad \frac{d\Psi}{d\varrho} = 0; \qquad \varrho = 1, \quad \frac{d\Psi}{d\varrho} = 1. \qquad (3.5.35)$$

Once integrating Eq. (3.5.34), we obtain

$$\alpha \, \text{Pe}_T \left( \frac{\varrho}{2} - \frac{\varrho^3}{4} + \frac{C_1}{\varrho} \right) = \frac{d\Psi}{d\varrho}, \qquad (3.5.36)$$

where $C_1$ is an arbitrary constant.

The constant $C_1 = 0$ and the parameter $\alpha = 4/\text{Pe}_T$ can be found from the boundary conditions (3.5.35). Thus, integrating (3.5.36), we arrive at the function $\Psi$:

$$\Psi = \varrho^2 - \frac{\varrho^4}{4} + C_2, \qquad \alpha = \frac{4}{\text{Pe}_T}. \qquad (3.5.37)$$

To determine the unknown constant $C_2$, we substitute (3.5.33) into the heat balance equation (3.5.32) and use (3.5.37). The calculations show that

$C_2 = \mathsf{Pe}_T^{-2} - \frac{7}{24}$. Thus, the temperature distribution far from the inlet cross-section has the form

$$\widetilde{T} = 4\frac{z}{\mathsf{Pe}_T} + \varrho^2 - \frac{\varrho^4}{4} + \frac{8}{\mathsf{Pe}_T^2} - \frac{7}{24}. \qquad (3.5.38)$$

The mean flow rate temperature of the fluid and the temperature head in the region of heat stabilization are, respectively, equal to

$$\langle\widetilde{T}\rangle_m = \frac{4}{\mathsf{Pe}_T}z + \frac{8}{\mathsf{Pe}_T^2}, \qquad \widetilde{T}_s - \langle\widetilde{T}\rangle_m = \frac{11}{24},$$

where $\widetilde{T}_s$ is the dimensionless temperature on the tube wall.

Let us calculate the limit Nusselt number:

$$\mathsf{Nu}_\infty = \frac{2(d\widetilde{T}/d\varrho)_{\varrho=1}}{\widetilde{T}_s - \langle\widetilde{T}\rangle} = \frac{48}{11} \approx 4.36. \qquad (3.5.39)$$

Obviously, $\mathsf{Nu}_\infty$ is independent of the Peclet number.

**Temperature field in the initial region of the tube.** For $z < 0$, we seek the solution of the complete problem (3.5.1), (3.5.30), (3.5.31) in the form of the series (3.5.7) (with $T$ replaced by $\widetilde{T}$). For $z > 0$, the temperature field is constructed on the basis of the asymptotic distribution (3.5.38) as follows:

$$\widetilde{T} = 4\frac{z}{\mathsf{Pe}_T} + \varrho^2 - \frac{\varrho^4}{4} + \frac{8}{\mathsf{Pe}_T^2} - \frac{7}{24} - \sum_{m=0}^{\infty} A_m \exp\left(-\lambda_m^2 \frac{z}{\mathsf{Pe}_T}\right) f_m(\varrho).$$

Substituting these series into (3.5.1), (3.5.30), and (3.5.31) and separating the variables, we obtain the same equations (3.5.9) and (3.5.10) for the eigenvalues $\eta_k$ and $\lambda_m$ and the eigenfunctions $g_k$ and $f_m$ with the boundary conditions

$$\frac{dg_k}{d\varrho} = \frac{df_m}{d\varrho} = 0 \qquad \text{for} \quad \varrho = 0 \quad \text{and} \quad \varrho = 1.$$

The solution of the problem for $f_m$ is given by (3.5.14), where the numbers $\lambda_m$ satisfy the transcendental equation

$$\Phi(a_m, 1; \lambda_m) = 2a_m\Phi(a_m + 1, 2; \lambda_m).$$

The coefficients of the expansions $A_m$ and $B_m$ are determined by the condition that the temperature and its derivative are continuous across the section $z = 0$ (3.5.12).

Let us calculate the length $l$ of the thermal initial region on the basis of the formula $\mathsf{Nu} = 1.01\,\mathsf{Nu}_\infty$. As a result, we obtain $l = 0.14\,\mathsf{Pe}\,a$.

# 3.6. Heat and Mass Transfer in a Laminar Flow in a Plane Channel

## 3.6-1. Channel With Constant Temperature of the Wall

*Temperature field.* We shall study the heat exchange in laminar flow of a fluid with parabolic velocity profile in a plane channel of width $2h$. Let us introduce rectangular coordinates $X, Y$ with the $X$-axis codirected with the flow and lying at equal distances from the channel walls. We assume that on the walls (at $Y = \pm h$) the temperature is constant and is equal to $T_1$ for $X < 0$ and to $T_2$ for $X > 0$. Since the problem is symmetric with respect to the $X$-axis, it suffices to consider a half of the flow region, $0 \le Y \le h$.

The temperature distribution $T_*$ is described by the following equation and boundary conditions:

$$\text{Pe}_T(1 - y^2)\frac{\partial T}{\partial x} = \frac{\partial^2 T}{\partial x^2} + \frac{\partial^2 T}{\partial y^2}; \tag{3.6.1}$$

$$y = 0, \quad \frac{\partial T}{\partial y} = 0; \quad y = 1, \quad T = \begin{cases} 0 & \text{for } x < 0; \\ 1 & \text{for } x > 0; \end{cases} \tag{3.6.2}$$

$$x \to -\infty, \quad T \to 0; \quad x \to +\infty, \quad T \to 1, \tag{3.6.3}$$

with the dimensionless variables

$$x = \frac{X}{h}, \quad y = \frac{Y}{h}, \quad T = \frac{T_* - T_1}{T_2 - T_1}, \quad \text{Pe}_T = \frac{hU_{\max}}{\chi}, \quad U_{\max} = \frac{3}{2}\langle V \rangle,$$

where $U_{\max}$ is the maximal fluid velocity on the flow axis and $\langle V \rangle$ is the mean flow rate velocity over the cross-section.

By analogy with the case of a circular tube, we seek the solution of problem (3.6.1)–(3.6.3) in the form of a series for separated variables:

$$T = \sum_{k=0}^{\infty} B_k \exp\left(\frac{\eta_k^2}{\text{Pe}_T}x\right) g_k(y) \qquad \text{for} \quad x < 0, \tag{3.6.4}$$

$$T = 1 - \sum_{m=0}^{\infty} A_m \exp\left(-\frac{\lambda_m^2}{\text{Pe}_T}x\right) f_m(y) \qquad \text{for} \quad x > 0. \tag{3.6.5}$$

The eigenvalues $\eta_k$, $\lambda_m$ and the eigenfunctions $g_k$, $f_m$ can be obtained by solving Eqs. (3.5.9) and (3.5.10) in which we omit the second terms (proportional to the first-order derivatives) and replace $\varrho$ by $y$. The boundary conditions remain the same. The coefficients $A_m$ and $B_k$ can be obtained from the condition that the temperature and its derivatives are continuous at $x = 0$ [110].

In what follows, we present only the basic solutions of the problem for $x > 0$. The eigenfunctions $f_m$ can be written as

$$f_m(y) = \exp\left(-\frac{1}{2}\lambda_m y^2\right) \Phi\left(\frac{1}{4} - \frac{1}{4}\lambda_m - \frac{\lambda_m^3}{4\,\text{Pe}_T^2}, \frac{1}{2}; \lambda_m y^2\right). \tag{3.6.6}$$

Here $\Phi(a, b; \xi)$ is the degenerate hypergeometric function and the $\lambda_m$ satisfy the transcendental equation

$$\Phi\left(a_m, \frac{1}{2}; \lambda_m\right) = 0, \quad \text{where} \quad a_m = \frac{1}{4} - \frac{\lambda_m}{4} - \frac{\lambda_m^3}{4\,\text{Pe}_T^2}. \tag{3.6.7}$$

The coefficients $A_m$ are calculated by the formula (3.5.13) with $\varrho$ replaced by $y$, where the auxiliary function $f$ can be obtained from (3.6.6) after the indices $m$ are omitted.

In the limit case as $\text{Pe}_T \to 0$, we have

$$\lambda_m = \sqrt{\text{Pe}_T\left(\frac{\pi}{2} + \pi m\right)}, \quad A_m = \frac{4(-1)^m}{(\pi + 2\pi m)^2}, \quad f_m = \cos\left[\left(\frac{\pi}{2} + \pi m\right) y\right], \tag{3.6.8}$$

where $m = 0, 1, 2, \ldots$

As $\text{Pe}_T \to \infty$, the eigenvalues $\lambda_m$ and the coefficients $A_m$ can be obtained from formulas (3.4.37) and (3.4.38).

Numerical calculations of the first three eigenvalues $\lambda_0$, $\lambda_1$, $\lambda_2$ were performed in [265] for various Peclet numbers.

***Mean flow rate temperature. Nusselt number.*** The flow rate temperature for a plane channel is given by the formula

$$\langle T \rangle_m = \frac{3}{2} \int_0^1 T(1 - y^2)\, dy. \tag{3.6.9}$$

The local heat flux can be found by using (3.5.23) with $z$ replaced by $x$ and the derivative $f_m'(1)$ calculated from (3.6.6).

Let us substitute formulas (3.6.5), (3.6.9), and (3.5.23) into (3.5.24), and let $x$ tend to infinity. As a result, we obtain the limit Nusselt number

$$\text{Nu}_\infty = \frac{-4 f_0'(1)}{3 \int_0^1 f_0(y)(1 - y^2)\, dy}. \tag{3.6.10}$$

The eigenfunction $f_0(y) = \cos(\pi y/2)$ follows from (3.6.8) for low Peclet numbers. By applying (3.6.10), we obtain

$$\text{Nu}_\infty = \frac{\pi^4}{24} \approx 4.06 \quad (\text{as } \text{Pe}_T \to 0). \tag{3.6.11}$$

For high Peclet numbers, the limit Nusselt number is [341]

$$\text{Nu}_\infty = \frac{4}{3}\lambda_0^2 \approx 3.77 \quad (\text{as } \text{Pe}_T \to \infty). \tag{3.6.12}$$

Over the entire range of Peclet numbers, $\text{Nu}_\infty$ is nicely approximated by the formula

$$\text{Nu}_\infty = \frac{4.06 + 3.66\,\text{Pe}_T}{1 + 0.97\,\text{Pe}_T}, \tag{3.6.13}$$

whose maximal error is about 0.5% (this estimate is based on the comparison with the data from [341]).

### 3.6-2. Channel With Constant Heat Flux at the Wall

Now let us consider the case in which a constant heat flux $q$ is given on the walls of a plane channel for $X > 0$. We assume that for $X < 0$ the walls are thermally isolated and the temperature tends to a constant $T_1$ as $X \to -\infty$.

For $a \equiv h$, we introduce the dimensionless temperature $\widetilde{T}$ according to (3.5.29). The heat exchange in a plane channel is described by Eq. (3.5.1) (as $T \to \widetilde{T}$) and the boundary conditions (3.5.30), (3.5.31), where $z$ and $\varrho$ are, respectively, replaced by $x$ and $y$.

We seek the asymptotic solution in the heat stabilization region (for $x \gg 1$) in the form of the sum $\widetilde{T} = \alpha x + \Psi(y)$. The unknown variable $\alpha$ and the function $\Psi$ are determined by analogy with the case of a circular tube. As a result, we have

$$\widetilde{T} = \frac{3}{2}\frac{x}{\mathsf{Pe}_T} + \frac{3}{4}y^2 - \frac{1}{8}y^4 + \frac{9}{4\,\mathsf{Pe}_T^2} - \frac{39}{280}. \tag{3.6.14}$$

From the temperature distribution (3.6.14) far from the input cross-section, one can find the limit Nusselt number

$$\mathsf{Nu}_\infty = \frac{70}{17} \approx 4.12. \tag{3.6.15}$$

Formulas for calculating the Nusselt number along the channel in the case of high Peclet numbers are given in [341].

## 3.7. Turbulent Heat Transfer in Circular Tube and Plane Channel

### 3.7-1. Temperature Profile

Let us discuss qualitative specific features of convective heat and mass transfer in a turbulent flow through a circular tube and plane channel in the region of stabilized flow. Experimental evidence indicates that several characteristic regions with different temperature profiles can be distinguished. At moderate Prandtl numbers ($0.5 \leq \mathsf{Pr} \leq 2.0$), the structure and sizes of these regions are similar to those of the wall layer and the core of the turbulent stream considered in Section 1.6.

In the molecular thermal conduction layer, adjacent to the tube wall, the deviation of the average temperature $\overline{T}$ from the wall temperature $T_s$ satisfies the linear dependence (3.3.10). In the logarithmic layer, the average temperature can be estimated using relations (3.3.11), which are valid for liquids, gases, and liquid metals within a wide range of Prandtl numbers, $6 \times 10^{-3} \leq \mathsf{Pr} \leq 10^4$ [209, 212, 289].

In the stream core and major part of the logarithmic layer, the average temperature profile can be described by the single formula

$$\frac{\overline{T}(Y) - \overline{T}_{\mathrm{m}}}{\Theta_*} = -2.12 \ln \eta + 0.3\,(1 - \eta^2), \tag{3.7.1}$$

where $\overline{T}_m$ is the average temperature at the stream axis, $\eta = Y/a$ the dimensionless distance to the tube wall, and $a$ the tube radius. The results predicted by Eq. (3.7.1) fairly well agree with experimental data provided by various authors for turbulent flows of water, air, and liquid metals through circular tubes and plane channels for $3 \times 10^{-2} < Y/a < 1$. The experimental data for the verification of formula (3.7.1) are presented in the reference [212, 289] in a systematic form. Also, a similar formula is suggested in [212, 289] with the right-hand side $-2.12 \ln \eta + 0.3$; this formula predicts the temperature profile near the stream axis a bit worse than (3.7.1).

### 3.7-2. Nusselt Number for the Thermal Stabilized Region

In the region of thermal stabilization of turbulent flow through a smooth tube, it is recommended to calculate the limiting Nusselt number by the formula [212, 289]

$$\mathsf{Nu}_\infty = \frac{\mathsf{Pr}\,\mathsf{Re}_d\,\xi}{2.12\ln(\mathsf{Pr}\,\mathsf{Re}_d\,\xi) + (3.85\,\mathsf{Pr}^{1/3} - 1.3)^2 - 4.3 + 6.7\,\xi}. \tag{3.7.2}$$

Here the following notation is used:

$$\mathsf{Nu}_\infty = \frac{\alpha d}{c_p \rho \chi}, \quad \alpha = \frac{q_s}{T_s - \overline{T}_m}, \quad \mathsf{Re}_d = \frac{d\langle V \rangle}{\nu}, \quad \mathsf{Pr} = \frac{\nu}{\chi}, \quad \xi = \sqrt{\tfrac{1}{8}\lambda},$$

where $\alpha$ is the heat transfer coefficient, $d = 2a$ the tube diameter, $c_p$ the specific heat at constant pressure, $\rho$ the fluid density, $\chi$ the thermal diffusivity, $q_s$ the heat flux at the wall, $\nu$ the kinematic viscosity, and $\langle V \rangle$ the mean flow rate velocity. The drag coefficient $\lambda$ can be found from Eq. (1.6.12) or (1.6.13). Formula (3.7.2) unifies the results of more than fifty experimental studies and applies to liquids and gases within wide ranges of Reynolds and Prandtl numbers, $5 \times 10^3 \le \mathsf{Re}_d \le 2 \times 10^6$ and $0.6 \le \mathsf{Pr} \le 4 \times 10^4$. In the case of mass exchange, the Nusselt number must be replaced by the Sherwood number, and the thermal Prandtl number by the diffusion one.

In engineering calculations, the Nusselt number is often determined from simple two-term (or one-term) formulas like

$$\mathsf{Nu}_\infty = A + B\,\mathsf{Pr}^n\,\mathsf{Re}_d^m \tag{3.7.3}$$

having limited scope in Prandtl and Reynolds numbers [80, 184, 185, 267, 406]. For liquid metals, in the case of $10^4 < \mathsf{Re}_d < 10^6$ and constant temperature or heat flux at the wall, it is recommended in Eq. (3.7.3) to set [41, 254]

$$\begin{aligned} A = 5, \quad &B = 0.025, \quad n = m = 0.8 \quad (\text{if } T_s = \text{const}), \\ A = 6.8, \quad &B = 0.044, \quad n = m = 0.75 \quad (\text{if } q_s = \text{const}). \end{aligned} \tag{3.7.4}$$

### 3.7-3. Intermediate Domain and the Entry Region of the Tube

In the intermediate domain ($2200 \leq \mathrm{Re}_d \leq 4000$) at moderate Prandtl numbers, the Nusselt number can be estimated using the formula [254]

$$\mathrm{Nu} = 3 \times 10^{-4} \, \mathrm{Pr}^{0.35} \, \mathrm{Re}_d^{1.5} \qquad \text{(for } 0.5 \leq \mathrm{Pr} \leq 5\text{)}.$$

For the entry region of the tube in the case of simultaneous development of the hydrodynamic and thermal boundary layers, the local Nusselt number $\mathrm{Nu} = \mathrm{Nu}(X)$ at constant temperature or heat flux at the wall can be calculated from the formulas [267]

$$\mathrm{Nu} = \begin{cases} 0.02 \, \mathrm{Re}_d^{0.75} \, \mathrm{Pr}^{0.4} \, \zeta(\zeta - 1)^{-0.2} & \text{if } T_s = \text{const}, \\ 0.021 \, \mathrm{Re}_d^{0.75} \, \mathrm{Pr}^{0.4} \, \zeta^{0.8} (\ln \zeta)^{-0.2} & \text{if } q_s = \text{const}. \end{cases} \qquad (3.7.5)$$

Here the following notation is used:

$$\mathrm{Nu} = \frac{\alpha d}{c_p \rho \chi}, \qquad \mathrm{Re}_d = \frac{dU_i}{\nu}, \qquad \zeta = (1.18)^b, \qquad b = \mathrm{Re}_d^{-0.25} \frac{X}{a},$$

where $U_i$ is the inlet velocity and $\alpha = \alpha(X)$ the local heat transfer coefficient (based on the inlet temperature). Formulas (3.7.5) are valid for $0.5 \leq \mathrm{Pr} \leq 200$ in the entire entry region, whose length $x_L$ is estimated as $x_L = 1.3 \, \mathrm{Re}^{0.25}$.

More detailed information about heat and mass transfer in turbulent flows through a circular tube or plane channel, as well as various relations for determining the temperature profile and Nusselt (Sherwood) numbers, and other useful information can be found in the references [185, 289, 406, 427, 457], which contain extensive literature surveys.

## 3.8. Limit Nusselt Numbers for Tubes of Various Cross-Section

### 3.8-1. Laminar Flows

Heat exchange in fully developed laminar flow of fluids in tubes of various cross-sections was studied in many papers (e.g., see [80, 253, 341]). In what follows, we present some definitive results for the limit Nusselt numbers corresponding to the region of heat stabilization in the flow in the case of high Peclet numbers (when the molecular heat transfer can be neglected).

Let us introduce the equivalent (or "hydraulic") diameter $d_e$ by the formula

$$d_e = \frac{4S_*}{\mathcal{P}}, \qquad (3.8.1)$$

where $S_*$ is the area of the tube cross-section and $\mathcal{P}$ is the perimeter of the cross-section. For tubes of circular cross-section, $d_e$ coincides with the diameter, and for a plane channel, $d_e$ is twice the height of the channel.

Let us consider a tube of arbitrary shape and denote the contour of the cross-section by $\Gamma$. Generally speaking, the Nusselt number varies along the contour $\Gamma$. We define the perimeter-average Nusselt number $\overline{\text{Nu}}$ as follows:

$$\overline{\text{Nu}} = \frac{q_s}{T_s - \langle T_* \rangle_m} \frac{d_e}{\varkappa}. \tag{3.8.2}$$

Here $T_s$ is the temperature on the wall of the tube, $\langle T_* \rangle_m$ is the mean flow rate temperature of the fluid, $\varkappa$ is the thermal conductivity coefficient, and $q_s$ is the perimeter-average heat flux given by the formula

$$q_s = -\frac{1}{\mathcal{P}} \varkappa \int_\Gamma \left( \frac{\partial T_*}{\partial \xi} \right)_\Gamma d\Gamma, \tag{3.8.3}$$

where $\partial T_* / \partial \xi$ is the derivative of the temperature $T_*$ along the normal to the contour of the tube cross-section.

For a tube of elliptic cross-section with constant temperature on the wall, the perimeter-average Nusselt number in the heat stabilization region (far from the input cross-section) is determined by [341]

$$\overline{\text{Nu}}_\infty = \left[ \frac{3\pi}{E(\vartheta)} \right]^2 \frac{(1 + \omega^2)(1 + 6\omega^2 + \omega^4)}{17 + 98\omega^2 + 17\omega^4}, \tag{3.8.4}$$

where $E(\vartheta)$ is the complete elliptic integral of the second kind (the function $E$ is tabulated in [202]), $\vartheta = \sqrt{1 - \omega^2}$, and $\omega = a/b$ is the ratio of semiaxes of the ellipse. In the special case of a circular tube, we have $\omega = 1$, $E(0) = \pi/2$, and $\overline{\text{Nu}}_\infty = 48/11$.

Table 3.3 presents the perimeter-average Nusselt numbers for tubes with various shapes of the cross-section (according to [80]).

At constant temperature on the wall of a tube of rectangular cross-section with sides $a$ and $b$, the value $\overline{\text{Nu}}_\infty$ for $a \geq b$ is nicely approximated by the formula

$$\overline{\text{Nu}}_\infty = 7.5 - 17.5\,\epsilon + 23\,\epsilon^2 - 10\,\epsilon^3, \qquad \epsilon = b/a, \tag{3.8.5}$$

whose maximal error is 3%.

At constant temperature on the wall of a tube whose cross-section is a regular $N$-gon, the limit Nusselt number is given by the approximate formula

$$\overline{\text{Nu}}_\infty = 3.65 - 0.18\,N^{-1} - 10\,N^{-2}. \tag{3.8.6}$$

The comparison with Table 3.3 shows that the maximal error in (3.8.6) is less than 0.5% at $N = 3, 4, 6, \infty$.

Problems of convective heat transfer in tubes of more complicated profiles were considered in [418].

TABLE 3.3

Values of the limit numbers $\overline{\text{Nu}}_\infty$ for fully developed flow in tubes of various shape for high Peclet numbers (the subscript $T$ corresponds to the constant temperature of the wall and the subscript $q$, to the constant heat flux)

| Tube profile | | $\overline{\text{Nu}}_{\infty T}$ | $\overline{\text{Nu}}_{\infty q}$ | Equivalent diameter $d_e$ |
|---|---|---|---|---|
| Circular tube of diameter $d$ | | 3.658 | 4.364 | $d$ |
| Flat tube of width $2h$ | | 7.541 | 8.235 | $4h$ |
| Elliptic tube with semiaxes $a$ and $b$ | $b/a =$ 1.00 | 3.658 | 4.364 | $\dfrac{\pi b}{E\left(\sqrt{1-b^2/a^2}\right)}$, where $E(\vartheta)$ is the complete elliptic integral of the second kind |
| | 0.80 | 3.669 | 4.387 | |
| | 0.50 | 3.742 | 4.558 | |
| | 0.25 | 3.792 | 4.880 | |
| | 0.125 | 3.725 | 5.085 | |
| | 0.0625 | 3.647 | 5.176 | |
| | 0 | 3.488 | 5.225 | |
| Tube of rectangular cross-section with sides $a$ and $b$ | $b/a =$ 1.00 | 2.976 | 3.608 | $\dfrac{2ab}{a+b}$ |
| | 0.714 | 3.077 | 3.734 | |
| | 0.50 | 3.391 | 4.123 | |
| | 0.25 | 4.439 | 5.331 | |
| | 0.125 | 5.597 | 6.490 | |
| | 0.05 | | 7.451 | |
| | 0 | 7.541 | 8.235 | |
| Equilateral triangle with side $a$ | | 2.47 | 3.111 | $\dfrac{a\sqrt{3}}{3}$ |
| Regular hexagon with side $a$ | | 3.34 | 4.002 | $a\sqrt{3}$ |
| Semi-circle of diameter $d$ | | | 4.089 | $\dfrac{\pi d}{\pi+2}$ |

### 3.8-2. Turbulent Flows

According to the results of [267], the strength of heat transfer in turbulent flows of fluids through tubes of noncircular cross-section can be estimated using relations (3.7.2)–(3.7.4) suggested for circular tubes. In this case, the characteristic length on which the Nusselt numbers $\text{Nu}_\infty$ are based is taken to be the equivalent diameter, $d_e$, defined by Eq. (3.8.1). The quantity $\mathcal{P}$ in this formula is the wettable perimeter, regardless of what part of the perimeter exchanges heat with the fluid.

Detailed information about heat transfer in turbulent flows in tubes and channels, as well as various relations for determining the mean flow rate temperature and Nusselt number, can be found in the references [185, 196, 406], which contain extensive literature surveys.

# Chapter 4
# Mass and Heat Exchange Between Flow and Particles, Drops, or Bubbles

The analysis of many technological processes involving dissolution, extraction, vaporization, combustion, chemical transformations in dispersions, sedimentation of colloids, etc. are based on the solution of the problem of mass exchange between particles, drops, or bubbles and the ambient medium. For example, in industry one often deals with processes of extraction from drops or bubbles or with heterogeneous transformations on the surface of catalyst particles suspended in a fluid. The rate of extraction and the intensity of a catalytic process to a large extent are determined by the value of the total diffusion flux of a reactant to the surface of particles of the disperse phase, which, in turn, depends on the character of flow and the particle shape, the influence of neighboring particles, the kinetics of the surface chemical reaction, and some other factors.

The description of a number of meteorological phenomena is also based on the study of Brownian diffusion of aerosols to single solid and liquid particles. The increasing atmospheric pollution is a problem that requires understanding and description of the processes of atmospheric self-purification of chemical and mechanical pollutants and radioactive contaminants. The problem of settling aerosol particles on various collectors also arises in the analysis of filter efficiency.

The results of investigation of similar convective heat transfer processes can be used in the design and analysis of heat exchangers.

A number of physical and mathematical statements of problems on mass and heat exchange between particles, drops, or bubbles and flow, as well as various formulas and experimental data, can be found, for example, in the books [94, 166, 255, 270].

## 4.1. The Method of Asymptotic Analogies in Theory of Mass and Heat Transfer

### 4.1-1. Preliminary Remarks

The most important stage in the study of specific mass and heat transfer problems is to find general quantitative laws valid for a class of qualitatively similar problems. In many cases, general results of this type can be obtained by the method of asymptotic analogies [357, 359]. The method is based on the passage from the usual dimensionless variables to special asymptotic coordinates and can be used for constructing wide-scope approximate formulas. (One and the same formula can be used for describing a variety of qualitatively similar problems that differ in surface shape and flow structure.)

Suppose that there is a class of problems that differ in geometric characteristics and depend on a dimensionless parameter $\tau$ ($0 \leq \tau < \infty$). We assume that the dependence of the basic desired variable $w$ on the parameter $\tau$ is known for some specific (say, the simplest) geometry:

$$w = F(\tau), \tag{4.1.1}$$

where $F$ is a monotone function.

In problems of mass and heat transfer, $w$ is usually the Sherwood (Nusselt) number or the volume-average concentration; the parameter $\tau$ is dimensionless time, the Peclet number, or the dimensionless rate constant of a reaction.

### 4.1-2. Transition to Asymptotic Coordinates

Let us transform (4.1.1) as follows. Let the leading terms of the asymptotic expansions of $w$ for small and large $\tau$ have the form

$$w_0 = A\tau^k \qquad \left(\lim_{\tau \to 0} w/w_0 = 1\right), \tag{4.1.2}$$

$$w_\infty = B\tau^m \qquad \left(\lim_{\tau \to \infty} w/w_\infty = 1\right), \tag{4.1.3}$$

respectively, where $A$, $B$, $k$, and $m$ are some constants and $k \neq m$.

Note that the original dependence (4.1.1), as well as the asymptotics (4.1.2) and (4.1.3) can be determined either theoretically or experimentally.

In the sequel, we assume that the asymptotics as $\tau \to 0$ and $\tau \to \infty$ for the entire class of considered problems is given by formulas (4.1.2) and (4.1.3), respectively, where the constants $k$ and $m$ remain the same but the parameters $A$ and $B$ can vary.

Using (4.1.1)–(4.1.3), we obtain

$$\frac{w}{w_0} = \frac{F(\tau)}{A\tau^k}, \qquad \frac{w_\infty}{w_0} = \frac{B\tau^m}{A\tau^k}. \tag{4.1.4}$$

The variables $w/w_0$ and $w_\infty/w_0$ are called asymptotic coordinates.

By expressing the parameter $\tau$ from the second equation in (4.1.4) and by substituting it into the first equation, we obtain

$$\frac{w}{w_0} = \frac{1}{A}\left(\frac{A}{B}\frac{w_\infty}{w_0}\right)^{\frac{k}{k-m}} F\left(\left(\frac{A}{B}\frac{w_\infty}{w_0}\right)^{\frac{1}{m-k}}\right), \tag{4.1.5}$$

or, equivalently,

$$\frac{w}{w_\infty} = \frac{1}{B}\left(\frac{A}{B}\frac{w_\infty}{w_0}\right)^{\frac{m}{k-m}} F\left(\left(\frac{A}{B}\frac{w_\infty}{w_0}\right)^{\frac{1}{m-k}}\right). \tag{4.1.6}$$

## 4.1-3. Description of the Method

The basic idea of the method of asymptotic analogies is to use the expression (4.1.5) (or (4.1.6)) to approximate similar characteristics for a wider class of problems describing qualitatively similar phenomena or processes. Specifically, after the relation (4.1.5) has been constructed with the help of (4.1.1) for some specific (say, the simplest) case, we can evaluate $w$ for other problems of this class by finding the asymptotics $w_0$ (as $\tau \to 0$) and $w_\infty$ (as $\tau \to \infty$) and then by substituting these asymptotics into (4.1.5). The approximate formulas thus obtained are asymptotically sharp in both limit cases $\tau \to 0$ and $\tau \to \infty$.

In [357, 359, 367, 368], formulas obtained by the method of asymptotic analogies were compared with already known formulas obtained by exact, numerical, and approximate methods for a large number of specific cases. These investigations confirmed the high accuracy and wide capabilities of the method of asymptotic analogies. In other words, the final functional relation (4.1.5) between $w$ and its asymptotics remains the same (or varies slightly) in a wide class of problems of the same type, and the specific features of geometric distinctions between these problems (like the surface shape and the flow structure) are sufficiently well taken into account by the corresponding asymptotic parameters $w_0$ and $w_\infty$.

As a result, the scope of the final formula (4.1.5) is substantially wider than that of the original formula (4.1.1). In this sense, one can say that formulas of the type of (4.1.6) are more informative than the original formula (4.1.1).

# 4.2. Interior Heat Exchange Problems for Bodies of Various Shape

## 4.2-1. Statement of the Problem

Let us consider a class of problems concerning transient heat exchange between convex bodies of various shape and the environment. At the initial time $t = 0$ the temperature is the same throughout the body and is equal to $T_i$, and for $t > 0$ the temperature on the surface $\Gamma$ of the body is maintained constant and is equal to $T_s$. The temperature distribution inside the body is described by the heat equation

$$\frac{\partial T}{\partial \bar{\tau}} = \Delta T, \tag{4.2.1}$$

the initial condition

$$T = 0 \quad \text{at} \quad \bar{\tau} = 0, \tag{4.2.2}$$

and the boundary condition

$$T = 1 \quad \text{on } \Gamma. \tag{4.2.3}$$

The dimensionless variables are introduced according to formulas (3.1.32), where the characteristic length can be taken arbitrarily.

The problem of transient diffusion into a cavity filled with a stagnant medium can be stated in a similar manner.

In this section, attention is chiefly paid to the study of the bulk body temperature

$$\langle T \rangle = \frac{1}{V} \int_v T \, dv, \tag{4.2.4}$$

where $V = \int_v dv$ is the dimensionless volume of the body.

## 4.2-2. General Formulas for the Bulk Temperature of the Body

To approximate the dependence of the bulk temperature on time, we use the method of asymptotic analogies. The simplest original problem is taken to be the one-dimensional (with respect to spatial coordinates) heat exchange problem for a sphere of radius $a$. The solution of this problem is well known [277] and results in the following expression for bulk temperature:

$$\langle T \rangle = 1 - \frac{6}{\pi^2} \sum_{k=1}^{\infty} \frac{1}{k^2} \exp(-\pi^2 k^2 \bar{\tau}). \tag{4.2.5}$$

The asymptotic expressions (for small and large $\bar{\tau}$) of (4.2.5) have the form

$$\langle T \rangle_0 = 6\pi^{-1/2}\sqrt{\bar{\tau}} \quad (\bar{\tau} \to 0); \qquad \langle T \rangle_\infty = 1 \quad (\bar{\tau} \to \infty) \tag{4.2.6}$$

and are a special case of (4.1.2) and (4.1.3) for $w_0 = \langle T \rangle_0$ and $w_\infty = \langle T \rangle_\infty$, where $A = 6\pi^{-1/2}$, $B = 1$, $k = \frac{1}{2}$, and $m = 0$. By substituting these values into (4.1.6) with $F = \langle T \rangle$, we can rewrite (4.2.5) as follows:

$$\frac{\langle T \rangle}{\langle T \rangle_\infty} = 1 - \frac{6}{\pi^2} \sum_{k=1}^{\infty} \frac{1}{k^2} \exp\left[ -\frac{\pi^3}{36} k^2 \left( \frac{\langle T \rangle_0}{\langle T \rangle_\infty} \right)^2 \right]. \tag{4.2.7}$$

Following the method of asymptotic analogies, we shall use formula (4.2.7) for the calculation of bulk temperature for nonspherical bodies. To this end, for a body of a given shape, we must first calculate the asymptotics of bulk temperature for small and large $\bar{\tau}$ and then substitute these asymptotics into (4.2.7).

For a bounded body of arbitrary shape, the solution of problem (4.2.1)–(4.2.3) tends as $\bar{\tau} \to \infty$ to the limit value (equal to 1) determined by the boundary condition on the surface of the body. By setting $T = 1$ in (4.2.4), we find the asymptotics of bulk temperature for large $\bar{\tau}$:

$$\langle T \rangle_\infty = 1. \tag{4.2.8}$$

Now let us consider the initial stage of the process, corresponding to small values of dimensionless time. Let us integrate Eq. (4.2.1) over the volume $v$ occupied by the body. Taking into account the identity $\Delta T = \text{div}\,(\text{grad}\,T)$ and applying the Gauss divergence theorem, we replace the volume integral on the right-hand side by a surface integral. As a result, we obtain

$$\frac{\partial}{\partial \bar{\tau}} \int_{v} T \, dv = - \int_{\Gamma} \frac{\partial T}{\partial \xi} \, d\Gamma, \tag{4.2.9}$$

where $\xi$ is the coordinate along the inward normal on $\Gamma$.

For small $\bar{\tau}$, the temperature mainly varies in a thin region adjacent to the surface. In this region, the derivatives along the surface can be neglected compared with the normal derivatives. Therefore, the temperature distribution as $\bar{\tau} \to 0$ is described by the equation

$$\frac{\partial T}{\partial \bar{\tau}} = \frac{\partial^2 T}{\partial \xi^2} \tag{4.2.10}$$

with the initial and boundary conditions,

$$\bar{\tau} = 0, \quad T = 0; \qquad \xi = 0, \quad T = 1,$$

where the value $\xi = 0$ corresponds to the surface of the body.

The solution of problem (4.2.10) is given by the complementary error function

$$T = \operatorname{erfc}\left(\frac{\xi}{2\sqrt{\bar{\tau}}}\right). \tag{4.2.11}$$

By differentiating this formula with respect to $\xi$ and by setting $\xi = 0$, we obtain the asymptotics as $\bar{\tau} \to 0$ of local heat flux to the surface of the body:

$$\left(\frac{\partial T}{\partial \xi}\right)_{\Gamma} = -\frac{1}{\sqrt{\pi \bar{\tau}}}. \tag{4.2.12}$$

Let us substitute (4.2.12) into (4.2.9). After the integration, we obtain

$$\frac{\partial}{\partial \bar{\tau}} \int_{v} T \, dv = \frac{1}{\sqrt{\pi \bar{\tau}}} S, \tag{4.2.13}$$

where $S$ is the dimensionless surface area of the body.

Let us integrate both sides of (4.2.13) with respect to $\bar{\tau}$ from 0 to $\bar{\tau}$. With regard to the initial condition (4.2.2) and relation (4.2.4), we obtain the desired asymptotic expression for bulk temperature as $\bar{\tau} \to 0$:

$$\langle T \rangle_0 = 2\frac{S}{V}\sqrt{\frac{\bar{\tau}}{\pi}}. \tag{4.2.14}$$

The substitution of (4.2.8) and (4.2.14) into (4.2.7) gives an approximate dependence of bulk temperature of an arbitrarily shaped body on time:

$$\langle T \rangle = 1 - \frac{6}{\pi^2} \sum_{k=1}^{\infty} \frac{1}{k^2} \exp\left(-\frac{\pi^2 k^2 S^2}{9V^2}\bar{\tau}\right).$$

This expression can be rewritten as follows [362, 518]:

$$\langle T \rangle = 1 - \frac{6}{\pi^2} \sum_{k=1}^{\infty} \frac{1}{k^2} \exp\left(-\frac{\pi^2}{9} k^2 \frac{S_*^2 \chi t}{V_*^2}\right), \qquad (4.2.15)$$

where $S_*$ and $V_*$ are, respectively, the dimensional surface area and volume of the body.

For practical calculations, it is expedient to replace the infinite series by the simpler formula

$$\langle T \rangle = \sqrt{1 - e^{-1.27\,\omega}} + 0.6 \left(e^{-1.5\,\omega} - e^{-1.1\,\omega}\right), \qquad \omega = \frac{S_*^2 \chi t}{V_*^2}, \qquad (4.2.16)$$

whose maximum deviation from (4.2.15) is about 1.7% (see Table 4.1).

---

### 4.2-3. Bulk Temperature for Bodies of Various Shape

Let us compare the approximate dependence (4.2.15) with some well-known exact results on heat exchange for nonspherical bodies.

First, we consider a parallelepiped with sides $L_1$, $L_2$, and $L_3$. The solution of the corresponding three-dimensional problem (4.2.1)–(4.2.3) can be constructed by separation of variables and results in the following formula for bulk temperature [277]:

$$\langle T \rangle = 1 - \left(\frac{8}{\pi^2}\right)^3 \sum_{k=1}^{\infty} \sum_{m=1}^{\infty} \sum_{l=1}^{\infty} \frac{1}{(2k-1)^2(2m-1)^2(2l-1)^2}$$
$$\times \exp\left\{-\pi^2 \left[\frac{(2k-1)^2}{L_1^2} + \frac{(2m-1)^2}{L_2^2} + \frac{(2l-1)^2}{L_3^2}\right] \chi t\right\}. \qquad (4.2.17)$$

Since the surface area and the volume of the parallelepiped are given by the formulas $S_* = 2(L_1 L_2 + L_1 L_3 + L_2 L_3)$ and $V_* = L_1 L_2 L_3$, we can rewrite (4.2.17) as

$$\langle T \rangle = 1 - \left(\frac{8}{\pi^2}\right)^3 \sum_{k=1}^{\infty} \sum_{m=1}^{\infty} \sum_{l=1}^{\infty} \frac{1}{(2k-1)^2(2m-1)^2(2l-1)^2}$$
$$\times \exp\left[-\frac{\pi^2}{4} \frac{\left(\frac{2k-1}{L_1}\right)^2 + \left(\frac{2m-1}{L_2}\right)^2 + \left(\frac{2l-1}{L_3}\right)^2}{\left(\frac{1}{L_1} + \frac{1}{L_2} + \frac{1}{L_3}\right)^2} \frac{S_*^2 \chi t}{V_*^2}\right]. \qquad (4.2.18)$$

In Table 4.1, the approximation (4.2.15) is compared with the exact bulk temperature (4.2.18) for six distinct values of $L_1$, $L_2$, and $L_3$. The maximum error of formulas (4.2.15) and (4.2.16) is about 5% for $0.25 \leq L_3/L_1 \leq 4.0$ and $L_2/L_1 = 1$.

TABLE 4.1

Comparison of exact and approximate values of
bulk temperature $\langle T \rangle$ for bodies of various shape

| Bodies of various shape | | Dimensionless time $S_*^2 \chi t / V_*^2$ | | | | | | | |
|---|---|---|---|---|---|---|---|---|---|
| | | 0.05 | 0.1 | 0.2 | 0.3 | 0.5 | 1.0 | 1.5 | 2.0 |
| Sphere, formula (4.2.15) | | 0.236 | 0.323 | 0.438 | 0.518 | 0.631 | 0.795 | 0.882 | 0.932 |
| Approximate formula (4.2.16) | | 0.237 | 0.324 | 0.437 | 0.514 | 0.623 | 0.782 | 0.870 | 0.923 |
| Paralle-lepiped, formula (4.2.18); $E_i = L_i/L_1$ | $E_2 = 1, E_3 = 0.25$ | 0.237 | 0.326 | 0.443 | 0.527 | 0.647 | 0.821 | 0.907 | 0.951 |
| | $E_2 = 1, E_3 = 0.5$ | 0.233 | 0.318 | 0.429 | 0.506 | 0.615 | 0.774 | 0.862 | 0.915 |
| | $E_2 = 1, E_3 = 1$ | 0.232 | 0.316 | 0.425 | 0.499 | 0.604 | 0.757 | 0.843 | 0.897 |
| | $E_2 = 1, E_3 = 2$ | 0.232 | 0.318 | 0.427 | 0.503 | 0.610 | 0.767 | 0.854 | 0.920 |
| | $E_2 = 1, E_3 = 4$ | 0.234 | 0.320 | 0.432 | 0.510 | 0.620 | 0.782 | 0.871 | 0.952 |
| | $E_2 = 2, E_3 = 4$ | 0.234 | 0.321 | 0.435 | 0.514 | 0.628 | 0.794 | 0.882 | 0.932 |
| Cylinder, formula (4.2.20); $E = 2a/L$ | $E = 0.25$ | 0.236 | 0.325 | 0.440 | 0.522 | 0.638 | 0.807 | 0.894 | 0.942 |
| | $E = 0.5$ | 0.234 | 0.321 | 0.434 | 0.513 | 0.624 | 0.787 | 0.875 | 0.926 |
| | $E = 1$ | 0.233 | 0.319 | 0.429 | 0.506 | 0.613 | 0.770 | 0.857 | 0.910 |
| | $E = 2$ | 0.234 | 0.320 | 0.431 | 0.509 | 0.619 | 0.780 | 0.868 | 0.920 |
| | $E = 4$ | 0.237 | 0.326 | 0.444 | 0.528 | 0.649 | 0.823 | 0.909 | 0.952 |

Now let us consider heat exchange for a cylinder of finite length. Let $a$ be the radius and $L$ the length of the cylinder. By solving problem (4.2.1)–(4.2.3), we obtain the following expression for bulk temperature [277]:

$$\langle T \rangle = 1 - \frac{32}{\pi^2} \sum_{k=1}^{\infty} \sum_{m=1}^{\infty} \frac{1}{\vartheta_k^2 (2m-1)^2} \exp\left\{ -\left[ \frac{\vartheta_k^2}{a^2} + \frac{\pi^2 (2m-1)^2}{L^2} \right] \chi t \right\}, \quad (4.2.19)$$

where the $\vartheta_k$ are the roots of the Bessel function of the first kind of index zero: $J_0(\vartheta_k) = 0$ (the first sixty roots $\vartheta_k$ are tabulated in the book [202]).

Formula (4.2.19) can be rewritten in the form

$$\langle T \rangle = 1 - \frac{32}{\pi^2} \sum_{k=1}^{\infty} \sum_{m=1}^{\infty} \frac{1}{\vartheta_k^2 (2m-1)^2} \exp\left[ -\frac{L^2 \vartheta_k^2 + \pi^2 a^2 (2m-1)^2}{4(a+L)^2} \frac{S_*^2 \chi t}{V_*^2} \right],$$

$$(4.2.20)$$

where $S_* = 2\pi a(a + L)$ is the surface area and $V_* = \pi a^2 L$ the volume of the cylinder.

In Table 4.1, the exact value (4.2.20) is compared with the approximation (4.2.15) for various ratios of the cylinder dimensions. The maximum error in formula (4.2.15) is about 3.5% for $0.25 \leq 2a/L \leq 4.0$.

# 4.3. Mass and Heat Exchange Between Particles of Various Shape and a Stagnant Medium

### 4.3-1. Stationary Mass and Heat Exchange

*Statement of the problem.* Following [367, 368], let us consider stationary diffusion to a particle of finite size in a stagnant medium, which corresponds to the case $Pe = 0$. We assume that the concentration on the surface of the particle and remote from it is constant and equal to $C_s$ and $C_i$, respectively. The concentration field outside the particle is described by the Laplace equation

$$\Delta c = 0 \qquad (4.3.1)$$

and the boundary conditions

$$c = 1 \quad \text{on the surface } \Gamma \text{ of the particle,} \qquad (4.3.2)$$
$$c = 0 \quad \text{remote from the particle,} \qquad (4.3.3)$$

where the dimensionless concentration $c$ was introduced in (3.1.7).

The unknown quantity which is of most practical interest in these problems is the mean Sherwood number, which is determined by (3.1.28) and is related to the mass transfer coefficient $\alpha_c$ by

$$Sh = \frac{a\alpha_c}{D}, \qquad (4.3.4)$$

where $a$ is the characteristic length.

From the mathematical viewpoint, the diffusion problem (4.3.1)–(4.3.3) is equivalent to the problem on the electric field of a charged conductive body in a homogeneous charge-free dielectric medium. Therefore, the mean Sherwood number in a stagnant fluid coincides with the dimensionless electrostatic capacitance of the body and can be calculated or measured by methods of electrostatics.

*Shape factor.* In what follows, it is convenient to introduce a shape factor $\Pi$, which has the dimension of length, as follows:

$$\Pi = \frac{\alpha_c S_*}{D} = Sh \frac{S_*}{a}, \qquad (4.3.5)$$

where $S_*$ is the dimensional surface area of the particle. Note that sometimes $\Pi$ is referred to as "conductance" [94].

Tables 4.2 shows the values of $\Pi$ for particles of various shape (according to [94, 166]). It follows from (4.3.5) that the mean Sherwood number can be obtained from the data in this table by dividing the shape factor by the surface area of the particle and then multiplying by the characteristic length.

One can interpret the table data as follows. Let us project a body of revolution on the plane perpendicular to the symmetry axis. The projection is a disk of

TABLE 4.2

Shape factor for particles in stagnant medium

| No | Shape of particle | Shape factor $\Pi = \mathrm{Sh}\,\dfrac{S_*}{a}$ |
|----|-------------------|--------------------------------------------------|
| 1 | Sphere of radius $a$ | $4\pi a$ |
| 2 | Oblate ellipsoid of revolution with semiaxes $a$ and $b$, $\chi = b/a < 1$ | $\dfrac{4\pi a\sqrt{1-\chi^2}}{\arccos\chi}$ |
| 3 | Prolate ellipsoid of revolution with semiaxes $a$ and $b$, $\chi = b/a > 1$ | $\dfrac{4\pi a\sqrt{\chi^2-1}}{\ln\left(\chi+\sqrt{\chi^2-1}\right)}$ |
| 4 | Circular cylinder of radius $a$ and length $L$ $(0 \leq L/a \leq 16)$ | $\left[8 + 4.1\,(L/a)^{0.76}\right]a$ |
| 5 | Tangent spheres of equal radius $a$ | $2\ln 2\,(4\pi a)$ |
| 6 | Tangent spheres of radii $a_1$ and $a_2$ | $-\dfrac{4\pi a_1 a_2}{a_1 + a_2}\left[\psi\left(\dfrac{a_1}{a_1+a_2}\right) + \psi\left(\dfrac{a_2}{a_1+a_2}\right) + 2\ln\gamma\right],$ where $\psi(x) = \frac{d}{dx}\Gamma(x)$ is the logarithmic derivative of the gamma function and $\ln\gamma = -\psi(1) = 0.5772\ldots$ is the Euler constant |
| 7 | Orthogonally intersecting spheres with radii $a_1$ and $a_2$ | $4\pi\left(a_1 + a_2 - \dfrac{a_1 a_2}{\sqrt{a_1^2 + a_2^2}}\right)$ |
| 8 | Cube with edge $a$ | $0.654\,(4\pi a)$ |
| 9 | Thin rectangular plate with sides $L_1$ and $L_2$ $(L_1 \geq L_2)$ | $\dfrac{2\pi L_1}{\ln(4L_1/L_2)}$ |

radius $a_\mathrm{p}$. The sphere of radius $a_\mathrm{p}$ is called a perimeter-equivalent sphere. We introduce the perimeter-equivalent factor [94]

$$\Sigma = \frac{S_*}{4\pi a_\mathrm{p}^2} = \frac{\text{surface area of particle}}{\text{surface area of perimeter-equivalent sphere}} \qquad (4.3.6)$$

and consider the corresponding shape factor ratio

$$\widetilde{\Pi} = \frac{\Pi}{4\pi a_\mathrm{p}} = \frac{\text{shape factor of particle}}{\text{shape factor of perimeter-equivalent sphere}}. \qquad (4.3.7)$$

The dimensionless variables (4.3.6) and (4.3.7) are invariant with respect to the choice of the characteristic length. The dependence of $\widetilde{\Pi}$ on $\Sigma$ is shown in Figure 4.1. One can see that particles of various geometric shape fairly well fit in a universal curve which can be approximated by

$$\widetilde{\Pi} = 0.637 + 0.327\,(2\Sigma - 1)^{0.76} \qquad (0.5 \leq \Sigma \leq 8.5). \qquad (4.3.8)$$

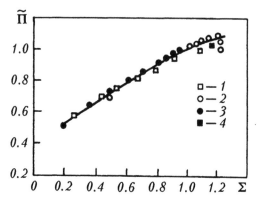

**Figure 4.1.** Shape factor ratio against perimeter-equivalent factor for particles of various shape in a stagnant medium: 1, circular cylinder; 2, oblate ellipsoid of revolution; 3, prolate ellipsoid of revolution; 4, cube

It is expedient to use this approximate formula for the analysis of mass exchange between particles of complicated shape and a stagnant medium if the solution of problem (4.3.1)–(4.3.3) is unknown.

To calculate the shape factor of a particle, it is also convenient to use the simple approximate formula

$$\Pi = 5.25\, S_*^{1/4} V_*^{1/6}, \qquad (4.3.9)$$

where $S_*$ is the surface area and $V_*$ the volume of the particle.

For spheres and cubes, the error in formula (4.3.9) is at most 0.13%. For circular cylinders of radius $a$ and length $L = a$, $L = 2a$, or $L = 3a$, the error in formula (4.3.9) is respectively 1.1%, 0.6%, and 2.1%.

It is useful to rewrite formula (4.3.9) as follows:

$$\Pi = \sqrt{S_*}\, f(\xi), \qquad f(\xi) = A_s \sqrt{\xi}, \qquad A_s = 2\sqrt{\pi} \approx 3.545,$$

where $\xi$ is a dimensionless geometric parameter,

$$\xi = \lambda \frac{V_*^{1/3}}{S_*^{1/3}}, \qquad \lambda = (36\pi)^{1/6} \approx 2.199.$$

The parameter $\xi$ characterizes the shape of a particle and ranges in the interval $0 \leq \xi \leq 1$. The value $\xi = 1$ corresponds to a spherical particle. Formula (4.3.9) gives good results for weakly and moderately deformed convex particles with $0.88 \leq \xi \leq 1.0$ (for example, this condition is satisfied for cubic particles, regular polyhedra, and circular cylinders of radius $a$ and length $L$ with $a \leq L \leq 3a$).

Let us write out some lower and upper bounds for the shape factor [94]. The lower bound for an arbitrary particle is determined by the shape factor of the sphere of the same volume $V_*$:

$$\Pi \geq (48\pi^2 V_*)^{1/3}. \qquad (4.3.10)$$

Another lower bound has the form

$$\Pi \geq 8(S_{max}/\pi)^{1/2},$$

where $S_{max}$ is the maximum projected area of the particle. The last formula becomes an equality for a disk. An upper bound is given by the shape factor of any surface (say, a sphere or an ellipsoid of revolution) surrounding the particle.

### 4.3-2. Transient Mass and Heat Exchange

Suppose that at the initial time $t = 0$ the concentration in the continuous medium is constant and is equal to $C_i$ and that a constant surface concentration $C_s$ is maintained for $t > 0$. The transient mass exchange between a particle and a stagnant medium is described by the equation

$$\frac{\partial c}{\partial \tau} = \Delta c \tag{4.3.11}$$

with the boundary conditions (4.3.2) and (4.3.3) and the initial condition

$$\tau = 0, \quad c = 0, \tag{4.3.12}$$

where $\tau = tD/a^2$ is dimensionless time.

For a spherical particle, the solution of problem (4.3.11), (4.3.12), (4.3.2), (4.3.3) can be expressed via the complementary error function and has the form

$$c = \frac{1}{r} \, \text{erfc} \left( \frac{r-1}{2\sqrt{\tau}} \right), \tag{4.3.13}$$

where $r$ is the dimensionless radial coordinate normalized to the radius of the particle.

The mean Sherwood number for a sphere is

$$\text{Sh} = 1 + \frac{1}{\sqrt{\pi \tau}}. \tag{4.3.14}$$

For nonspherical particles, the mean Sherwood number can be approximated by the expression

$$\text{Sh} = \text{Sh}_{st} + \frac{1}{\sqrt{\pi \tau}}, \tag{4.3.15}$$

where $\text{Sh}_{st}$ is the Sherwood number corresponding to the solution of the stationary problem (4.3.1)–(4.3.3). In (4.3.15), Sh, $\text{Sh}_{st}$, and $\tau$ are assumed to be reduced to a dimensionless form by using the same characteristic length.

For nonspherical particles, one can obtain $\text{Sh}_{st}$ by using Table 4.2 and the expression (4.3.8).

# 4.4. Mass Transfer in Translational Flow at Low Peclet Numbers

### 4.4-1. Statement of the Problem

Following [367, 368], let us consider steady-state diffusion to a particle in a laminar flow. We assume that on the surface of the particle and remote from it, the concentration is constant and equal to $C_s$ and $C_i$, respectively. In the dimensionless variables (3.1.7), the mass transfer process in the continuous medium is described by the equation

$$\mathsf{Pe}(\mathbf{v} \cdot \nabla)c = \Delta c \qquad (4.4.1)$$

with the boundary conditions (4.3.2) and (4.3.3). The velocity distribution $\mathbf{v}$ depends on the shape of the particle and the structure of the nonperturbed flow at infinity, and in the sequel we shall present specific distributions where necessary.

In this section, we study the one-dimensional translational flow with velocity $U_i$ remote from the particle.

### 4.4-2. Spherical Particle

*Statement of the Problem.* Let us consider mass exchange between a spherical particle of radius $a$ and a translational flow. In the Stokes flow ($\mathsf{Re} \to 0$), one can represent the dimensionless (divided by $U_i$) fluid velocity components as

$$v_r = \frac{1}{r^2 \sin \theta} \frac{\partial \psi}{\partial \theta}, \qquad v_\theta = -\frac{1}{r \sin \theta} \frac{\partial \psi}{\partial r},$$
$$\psi = \frac{1}{4}(r-1)^2 \left(2 + \frac{1}{r}\right) \sin^2 \theta. \qquad (4.4.2)$$

Here $r$ is the dimensionless (divided by $a$) radial coordinate, $\theta$ is the angular coordinate (measured from the incoming flow direction), and $\psi$ is the dimensionless (divided by $a^2 U_i$) stream function.

By using the stream function (4.4.2), one can rewrite the convective diffusion equation (4.4.1) as follows:

$$\Delta c = \frac{\mathsf{Pe}}{r^2 \sin \theta} \left(\frac{\partial \psi}{\partial \theta} \frac{\partial c}{\partial r} - \frac{\partial \psi}{\partial r} \frac{\partial c}{\partial \theta}\right), \qquad (4.4.3)$$

where $\mathsf{Pe} = aU_i/D$.

The boundary conditions read

$$r = 1, \qquad c = 1 \qquad \text{(on the surface of the sphere)}, \qquad (4.4.4)$$
$$r \to \infty, \qquad c \to 0 \qquad \text{(at infinity)}. \qquad (4.4.5)$$

*Asymptotic solution.* We seek an approximate closed-form solution of problem (4.4.3)–(4.4.5) at small Peclet numbers by the method of matched asymptotic expansions [96, 258, 485]. To this end, we divide the flow region into two parts, the inner region $\Omega = \{1 \leq r \leq O(\text{Pe}^{-1})\}$ and the outer region $\overline{\Omega} = \{O(\text{Pe}^{-1}) \leq r\}$. In the inner region we preserve the variables $r$, $\theta$, and in the outer region, instead of $r$ we introduce the compressed radial coordinate $\bar{r} = \text{Pe}\, r$.

We seek the solution separately in each region in the form of the inner and outer expansions

$$c = \sum_{k=0}^{\infty} \varepsilon_k(\text{Pe}) c_k(r, \theta) \qquad \text{in} \quad \Omega, \tag{4.4.6}$$

$$\bar{c} = \sum_{k=0}^{\infty} \bar{\varepsilon}_k(\text{Pe}) \bar{c}_k(\bar{r}, \theta) \qquad \text{in} \quad \overline{\Omega}. \tag{4.4.7}$$

The dependence of the coefficients $\varepsilon_k$ and $\bar{\varepsilon}_k$ on the Peclet number is unknown and must be found in the course of solution. It is only assumed a priori that

$$\frac{\varepsilon_{k+1}}{\varepsilon_k} \to 0, \qquad \frac{\bar{\varepsilon}_{k+1}}{\bar{\varepsilon}_k} \to 0 \qquad \text{as} \quad \text{Pe} \to 0.$$

The terms of the inner expansion (4.4.6) are determined by solving Eq. (4.4.3) with the boundary conditions (4.4.4) on the surface of the particle. The terms of the outer expansion (4.4.7) tend to zero as $\bar{r} \to \infty$ and satisfy Eq. (4.4.3) after the change of variables $r = \bar{r}/\text{Pe}$ with regard to (4.4.2). The arbitrary constants arising in this problem are determined from the matching conditions

$$c(r \to \infty) = \bar{c}(\bar{r} \to 0). \tag{4.4.8}$$

The leading term of the inner expansion (4.4.6) corresponds to the mass exchange between a sphere and a stagnant medium and is determined by the solution of problem (4.4.3)–(4.4.5) at Pe = 0. Therefore, we have

$$c_0 = \frac{1}{r}, \qquad \varepsilon_0(\text{Pe}) = 1. \tag{4.4.9}$$

Let us find the explicit form of the coefficient $\bar{\varepsilon}_0(\text{Pe})$ in the outer expansion. To this end, we pass to the outer variable $c_0 = \text{Pe}/\bar{r}$ in (4.4.9). It follows from the matching condition (4.4.8) that $\bar{\varepsilon}_0 = \text{Pe}$. We substitute $r = \bar{r}/\text{Pe}$ and $c = \text{Pe}\,\bar{c}_0 + \cdots$ into (4.4.2), (4.4.3), and (4.4.5) and omit the terms of the order $o(\text{Pe})$. As a result, we obtain the following problem for the leading term of the outer expansion:

$$\bar{\Delta} \bar{c}_0 = \cos\theta \frac{\partial \bar{c}_0}{\partial \bar{r}} + \frac{\sin\theta}{\bar{r}} \frac{\partial \bar{c}_0}{\partial \theta}; \qquad \bar{r} \to \infty, \quad \bar{c} \to 0. \tag{4.4.10}$$

Here $\bar{\Delta}$ is the axisymmetric Laplace operator with $\bar{r}$ as the radial coordinate.

The general solution of problem (4.4.10) has the form

$$\bar{c}_0 = \left(\frac{\pi}{\bar{r}}\right)^{1/2} \exp\left(\frac{\bar{r}\cos\theta}{2}\right) \sum_{m=0}^{\infty} A_m K_{m+1/2}\left(\frac{\bar{r}}{2}\right) P_m(\cos\theta),$$

$$K_{m+1/2}\left(\frac{\bar{r}}{2}\right) = \left(\frac{\pi}{\bar{r}}\right)^{1/2} \exp\left(-\frac{\bar{r}}{2}\right) \sum_{k=0}^{m} \frac{(k+m)!}{(m-k)!\,k!\,\bar{r}^k},$$

$$P_m(x) = \frac{1}{2^m m!} \frac{d^m}{dx^m}(x^2-1)^m,$$

where the $K_{m+1/2}(x)$ are the Macdonald functions and the $P_m(x)$ are Legendre polynomials. The constants $A_m$ must be determined by the matching procedure, that is, by comparing the behavior of the function $\bar{c} = \text{Pe}\,\bar{c}_0 + \cdots$ as $\bar{r} \to 0$ with that of the function (4.4.9) as $r \to \infty$. One can readily see that $A_0 = 1/\pi$ and $A_m = 0$ ($m = 1, 2, \ldots$). Therefore,

$$\bar{c}_0 = \frac{1}{\bar{r}} \exp\left[\frac{1}{2}\bar{r}(\cos\theta - 1)\right], \quad \bar{\varepsilon}_0 = \text{Pe}. \tag{4.4.11}$$

Let us find the first approximation for the inner expansion. To this end, we substitute formulas (4.4.11) into (4.4.7) and pass to the inner variable $r$. By expanding the obtained expression into a series in Pe, we obtain $\varepsilon_1(\text{Pe}) = \text{Pe}$ from the matching condition (4.4.8). Hence, the first approximation for the inner expansion must be sought with regard to (4.4.9) in the form

$$c = \frac{1}{r} + \text{Pe}\,c_1(r,\theta) + o(\text{Pe}). \tag{4.4.12}$$

We substitute (4.4.12) into (4.4.3) and (4.4.4) and use formulas (4.4.2) for the stream function. By collecting the terms of the order of Pe, we obtain the following equation and boundary condition for $c_1$:

$$\Delta c_1 = -\frac{1}{r^2}\left(1 - \frac{3}{2r} + \frac{1}{2r^3}\right)\cos\theta, \tag{4.4.13}$$

$$r = 1, \quad c_1 = 0. \tag{4.4.14}$$

The general solution of Eq. (4.4.13) is given by

$$c_1 = \left(\frac{1}{2} - \frac{3}{4r} - \frac{1}{8r^3}\right)\cos\theta + \sum_{m=0}^{\infty}(a_m r^m + b_m r^{-m-1})P_m(\cos\theta). \tag{4.4.15}$$

The boundary condition (4.4.14) allows us to obtain linear relations between the constants $a_m$ and $b_m$:

$$a_1 = \tfrac{3}{8} - b_1; \quad a_m = -b_m \quad \text{for} \quad m = 0, 2, 3, 4, \ldots \tag{4.4.16}$$

To determine the coefficients in (4.4.15), we match the expressions (4.4.12) and (4.4.15) as $r \to \infty$ and the expressions (4.4.7) and (4.4.11) as $\bar{r} \to 0$. As a result, we obtain

$$a_0 = -\tfrac{1}{2}, \quad b_0 = \tfrac{1}{2}, \quad a_1 = 0, \quad b_1 = \tfrac{3}{8};$$
$$a_m = b_m = 0 \quad \text{for} \quad m = 2, 3, 4, \dots$$

Hence,

$$c_1 = -\frac{1}{2} + \frac{1}{2r} + \left( \frac{1}{2} - \frac{3}{4r} + \frac{3}{8r^2} - \frac{1}{8r^3} \right) \cos \theta. \qquad (4.4.17)$$

**Sherwood number.** For solid particles, drops, and bubbles of spherical shape, the mean Sherwood number can be calculated according to the formula

$$\text{Sh} = \frac{1}{2} \int_0^\pi \sin \theta \left( \frac{\partial c}{\partial r} \right)_{r=1} d\theta. \qquad (4.4.18)$$

On substituting the two-term expansion (4.4.12) into (4.4.18) and taking account of (4.4.17), we obtain

$$\text{Sh} = 1 + \tfrac{1}{2} \text{Pe} + o(\text{Pe}). \qquad (4.4.19)$$

The next three terms of the expansion of the Sherwood number corresponding to the asymptotic solution of problem (4.4.2)–(4.4.5) as $\text{Pe} \to 0$ were obtained in [7]. These results were generalized in [401], where the solution found in [382] was used to describe the velocity field of the fluid around the sphere. Here we write out the final expression for the Sherwood number [401]:

$$\text{Sh} = 1 + \frac{1}{2}\text{Pe} + \frac{1}{2}\text{Pe}^2 \ln \text{Pe} + \frac{1}{2}Q(\text{Sc})\text{Pe}^2 + \frac{1}{4}\text{Pe}^3 \ln \text{Pe} + O(\text{Pe}^3),$$
$$Q(\text{Sc}) = -\frac{173}{160} + \ln \gamma + \frac{\text{Sc}^2}{2} - \frac{\text{Sc}}{4} - (\text{Sc} + 1)^2 \left( \frac{\text{Sc}}{2} - 1 \right) \ln \left( 1 + \frac{1}{\text{Sc}} \right),$$
$$(4.4.20)$$

where $\ln \gamma = 0.5772 \dots$ is the Euler constant.

Formula (4.4.20) can be used for $0.4 \leq \text{Sc} < \infty$. By passing to the limit as $\text{Sc} \to \infty$ in (4.4.20), we arrive at the result presented in [7].

### 4.4-3. Particle of an Arbitrary Shape

*Asymptotic formulas for the Sherwood number and diffusion flux.* For low Peclet numbers, the problem of mass exchange between a particle of arbitrary shape and a uniform translational flow were studied by the method of matched asymptotic expansions in [62]. The following expression was obtained for the mean Sherwood number up to first-order infinitesimals with respect to $\text{Pe}$:

$$\frac{\text{Sh}}{\text{Sh}_0} = 1 + \frac{1}{8\pi} \text{Pe}_M, \qquad \text{Pe}_M = \frac{\Pi U_i}{D}, \qquad (4.4.21)$$

where $Sh_0$ is the Sherwood number corresponding to a stagnant medium. The influence of the fluid motion is characterized by the modified Peclet number in which the shape factor $\Pi$ of the particle plays the role of the characteristic length.

Formula (4.4.21) is quite general and holds for solid particles, drops, and bubbles of arbitrary shape in a uniform translational flow at any $Re$ as $Pe \to 0$.

It gives a good approximation of the Sherwood number ratio for $Pe_M < 5$. In the special case of a spherical particle, (4.4.21) coincides with (4.4.19). For nonspherical particles, in (4.4.21) one must use the values of $\Pi$ from Table 4.2.

For a particle of arbitrary shape in a translational flow, the first three terms of the asymptotic expansion of the dimensionless total diffusion flux as $Pe \to 0$ have the form [62]

$$I = I_0 + \frac{1}{8\pi} Pe\, I_0^2 + \frac{1}{8\pi} Pe^2 \ln Pe\, I_0^2 (\mathbf{f} \cdot \mathbf{e}) + O(Pe^2). \qquad (4.4.22)$$

Here $I_0 = \Pi/a$ is the total flux to the particle in a stagnant fluid, $\mathbf{f}$ is the dimensionless vector equal to the ratio of the drag of the particle to the Stokes drag of a solid sphere of radius $a$ ($a$ is the characteristic length using which the dimensionless variables $Pe$, $I$, and $I_0$ were introduced), and $\mathbf{e}$ is the unit vector codirected with the fluid velocity at infinity. The Sherwood number is given by $Sh = I/S$, where $S$ is the dimensionless surface area of the particle.

To calculate $I_0 = \Pi/a$, one can use the results of Section 4.3.

*Some special cases.* In the case of a spherical drop of radius $a$, in (4.4.22) one must set

$$(\mathbf{f} \cdot \mathbf{e}) = \frac{2 + 3\beta}{3 + 3\beta}, \qquad I_0 = 4\pi, \qquad (4.4.23)$$

where $\beta$ is the ratio of the dynamic viscosity of the drop to that of the ambient fluid (the value $\beta = 0$ corresponds to a gas bubble, and $\beta = \infty$ to a solid sphere).

For a spherical particle covered by a viscous film, we have [166]

$$(\mathbf{f} \cdot \mathbf{e}) = \frac{2}{3} + \frac{1}{3}\left[1 + \frac{1}{\beta}\frac{1-\delta}{1+\delta}\left(1 + \frac{5}{2}\frac{\delta}{2+\delta+2\delta^2}\right)\right]^{-1}, \qquad I_0 = 4\pi,$$

where $\delta$ is the particle-film radius ratio ($\delta = 1$ corresponds to a solid particle; $\delta = 0$, to a drop).

For an ellipsoid of revolution with semiaxes $a$ and $b$ ($a$ is the equatorial radius) whose symmetry axis is directed along flow, the drag is given by the formula [179]

$$(\mathbf{f} \cdot \mathbf{e}) = \begin{cases} \dfrac{4}{3}(\chi^2 + 1)^{-1/2}\left[\chi - (\chi^2 - 1)\operatorname{arccot}\chi\right]^{-1} & \text{for } a \geq b, \\[4mm] \dfrac{8}{3}(\chi^2 - 1)^{-1/2}\left[(\chi^2 + 1)\ln\dfrac{\chi+1}{\chi-1} - 2\chi\right]^{-1} & \text{for } a \leq b, \end{cases}$$

where $\chi = \left|(a/b)^2 - 1\right|^{-1/2}$.

For a body of revolution whose axis is inclined at an angle $\omega$ to the incoming flow direction, the following formula [166] holds in the Stokes approximation (as $\mathrm{Re} \to 0$):

$$(\mathbf{f} \cdot \mathbf{e}) = f_\parallel \cos^2 \omega + f_\perp \sin^2 \omega, \qquad (4.4.24)$$

where $f_\parallel$ and $f_\perp$ are the values of the dimensionless drag of the body for the cases in which its axis is parallel ($\omega = 0$) and perpendicular ($\omega = \pi/2$), respectively, to the flow direction.

In particular, for a thin circular disk, one must set $f_\parallel = 8/(3\pi)$ and $f_\perp = 16/(9\pi)$ in (4.4.24); for a dumbbell-like particle consisting of two adjacent spheres of equal radius, $f_\parallel \approx 0.645$ and $f_\perp \approx 0.716$ [179].

The logarithmic term sharply restricts the practical value of the expansion (4.4.22). The two-term expression (4.4.21) has a wider range of applicability with respect to the Peclet number (although this expression is less accurate for very small $\mathrm{Pe}$).

## 4.4-4. Cylindrical Bodies

*Circular cylinder.* The mass exchange between a circular cylinder of radius $a$ and a uniform translational flow whose direction is perpendicular to the generatrix of the cylinder was considered in [186, 218] for low Peclet and Reynolds numbers $\mathrm{Pe} = \mathrm{Sc}\,\mathrm{Re}$ and $\mathrm{Re} = aU_i/\nu$. For the mean Sherwood number (per unit length of the cylinder) determined with respect to the radius, the following two-term expansions were obtained:

as $\mathrm{Re} \to 0$ and $\mathrm{Sc}$ is fixed, one has

$$\mathrm{Sh} = \epsilon - \epsilon^3 q(\mathrm{Sc}), \quad \epsilon = \frac{1}{2\ln 2 - \ln(\gamma \mathrm{Sc}\,\mathrm{Re})},$$
$$q(0.72) = 1.38, \quad q(1) = 1.63, \quad q(6.82) = 3.42; \qquad (4.4.25a)$$

as $\mathrm{Re} \to 0$ and $\mathrm{Sc} = \mathrm{Re}^{-\alpha}$ ($0 < \alpha < 1$), one has

$$\mathrm{Sh} = \delta - \delta^3 p(\alpha);$$
$$\delta = \frac{1}{2\ln 2 - \ln\left[\gamma(1-\alpha)\mathrm{Sc}\,\mathrm{Re}\right]}, \qquad (4.4.25b)$$
$$p(\alpha) = \frac{3-\alpha}{2} + \ln(1-\alpha) + \alpha \ln \frac{\gamma}{4},$$

where $\ln \gamma$ is the Euler constant. The error of both expressions (4.4.25) is of the order of $(\ln \mathrm{Re})^{-4}$ as $\mathrm{Re} \to 0$.

The difference between experimental data on the cylinder-gas flow heat exchange ($\mathrm{Sc} = 0.72$) and the results obtained by the first formula in (4.4.25a) is less than 3% for $\mathrm{Re} < 0.2$ [186].

*Cylinder of arbitrary shape.* Let us consider mass exchange for cylindrical bodies of arbitrary shape in a uniform translational flow of viscous fluid at small

Peclet numbers. To obtain the leading term of the expansion as $\mathsf{Pe} \to 0$, we proceed as follows. Consider the auxiliary equation

$$\mathsf{Pe}(\mathbf{w} \cdot \nabla)c = \Delta c \qquad (4.4.26)$$

with the boundary conditions (4.3.2), (4.3.3). The vector field $\mathbf{w}$ in (4.4.26) is independent of the Peclet number and is related to the actual distribution of the fluid velocity $\mathbf{v}$ only by the limit relation

$$\mathbf{e} = \lim_{\varrho \to \infty} \mathbf{v} = \lim_{\varrho \to \infty} \mathbf{w}. \qquad (4.4.27)$$

We rewrite Eq. (4.4.26) in Oseen's approximation

$$\mathsf{Pe}(\mathbf{e} \cdot \nabla)c = \Delta c. \qquad (4.4.28)$$

For each $\mathbf{w}$ satisfying condition (4.4.27), the leading terms of the asymptotic expansions for Eqs. (4.4.26) and (4.4.28) with the same boundary conditions coincide in the inner and outer regions. Therefore, as $\mathsf{Pe} \to 0$, in the diffusion equation one can replace the actual fluid velocity field $\mathbf{v}$ by $\mathbf{w}$. This fact allows one to use the results presented later on in Section 4.11. Namely, as $\mathbf{w}$ we take the velocity field for the potential flow of ideal fluid past the cylinder. This approximation yields an error of the order of $\mathsf{Pe}$ in the inner expansion. By retaining only the leading terms in (4.11.15), we obtain the dimensionless diffusion flux at small Peclet numbers in the form

$$I = -4\pi \left[ \ln \left( \frac{\gamma \overline{\mathsf{Pe}}}{8} \right) \right]^{-1}, \qquad \overline{\mathsf{Pe}} = \frac{\varphi_{max} - \varphi_{min}}{2D}, \qquad (4.4.29)$$

where $\varphi_{max}$ and $\varphi_{min}$ are, respectively, the maximum and minimum values of the potential on the cylinder surface (for some cylindrical bodies of various shape, these values can be found, e.g., in [26, 259, 431]); here $\ln \gamma$ is the Euler constant.

For an elliptic cylinder with semiaxes $a$ and $b$ ($a \geq b$), regardless of its orientation in the flow, we have (see Section 4.11)

$$I \approx 4\pi \left[ -\ln \mathsf{Pe} + \ln \frac{8a}{\gamma(a+b)} \right]^{-1}, \qquad \mathsf{Pe} = \frac{aU_i}{D}. \qquad (4.4.30)$$

For a circular cylinder, one must set $a = b$ in (4.4.30).

# 4.5. Mass Transfer in Linear Shear Flows at Low Peclet Numbers

Mass transfer to a particle in a translational flow, considered in Section 4.4, is a good model for many actual processes in disperse systems in which the velocity of the translational motion of particles relative to fluid plays the main role in convective transfer and the gradient of the nonperturbed velocity field can be neglected.

In Section 1.1 we briefly described the velocity field for some cases of gradient flows with nonuniform structure. For particles whose size is much less than the characteristic scale of spatial nonuniformity of the flow, the velocity distribution (1.1.15) can be used as the velocity distribution remote from the particle in problems about mass transfer to a particle in a linear shear flow.

## 4.5-1. Spherical Particle in a Linear Shear Flow

In practice, one often deals with the case in which particles are completely entrained by the flow and convective transfer due to the shear flow plays the main role. In studying the corresponding diffusion processes, it is convenient to attach the frame of reference to the particle center of gravity; then this frame moves translationally at the particle velocity, and the particle itself can freely rotate around the origin. For linear shear flow, the fluid velocity components have the form

$$R \to \infty, \quad V_k \to G_{km} X_m, \tag{4.5.1}$$

where the $G_{km}$ are the shear matrix entries.

For a spherical particle in an arbitrary linear shear flow (4.5.1), the first four terms of the asymptotic expansion as $Pe \to 0$ of the mean Sherwood number have the form [5]

$$Sh = 1 + \alpha\, Pe^{1/2} + \alpha^2\, Pe + \alpha^3\, Pe^{3/2} + O(Pe^2), \quad Pe = \frac{a^2 G}{D}. \tag{4.5.2}$$

Here the parameter $\alpha = \alpha(G_{km})$ depends on the type of the shear flow and can be calculated via an improper integral [27, 166].

Note that the first two terms of the expansion (4.5.2) were first obtained in [27].

The value of the parameter $\alpha$ remains unchanged if all entries of the shear matrix change their signs simultaneously; that is, $\alpha(G_{km}) = \alpha(-G_{km})$.

Table 4.3 shows numerical values of the coefficient $\alpha$ and the variable $G$ in formula (4.5.2) for some types of shear flow that are of practical interest. The sum in the third column of the last row can be calculated by the formula $G_{km} G_{km} = E_1^2 + E_2^2 + E_3^2$, where $E_1$, $E_2$, and $E_3$ are the diagonal entries of the symmetric tensor $[G_{km}]$ reduced to the principal axes.

In [71, 166] the values of the coefficient $\alpha$ for a plane shear flow of the form

$$G_{12} = G, \quad G_{21} = \omega G; \quad \text{the other } G_{km} = 0 \tag{4.5.3}$$

are given for $-1 \le \omega \le 1$. The coefficient $\alpha = \alpha(\omega)$ is monotone increasing from $\alpha = 0$ at $\omega = -1$ (purely rotational motion) to the maximum value $\alpha = 0.428$ at $\omega = 1$ (purely straining flow). For weakly straining flow, which corresponds to $\omega \to -1$, we have $\alpha \approx \frac{1}{15}(1 + \omega)^2$.

TABLE 4.3

The values of $\alpha$ and $G$ for some shear flows (according to [27, 132])

| No | Flow | Coefficients $G_{km}$ | $\alpha$ | $G$ |
|---|---|---|---|---|
| 1 | Simple shear flow | $G_{12} \neq 0$, the other $G_{km} = 0$ | 0.257 | $\|G_{12}\|$ |
| 2 | Axisymmetric shear flow | $G_{11} = G_{22} = -\frac{1}{2}G_{33}$, $G_{km} = 0$ for $k \neq m$ | 0.399 | $\|G_{33}\|$ |
| 3 | Plane shear flow | $G_{11} = -G_{22}$, the other $G_{km} = 0$ | 0.428 | $\|G_{11}\|$ |
| 4 | Arbitrary linear straining flow | $G_{km} = G_{mk}$ | 0.36 | $(G_{km}G_{km})^{1/2}$ (the sum is over both indices) |

### 4.5-2. Particle of Arbitrary Shape in a Linear Shear Flow

The mass exchange for a particle of arbitrary shape freely suspended in a linear shear flow described by (4.5.1) was considered in [5]. The three-term expansion

$$I = I_0 + \frac{\alpha}{4\pi}I_0^2 \, \text{Pe}^{1/2} + \frac{\alpha^2}{(4\pi)^2}I_0^3 \, \text{Pe} + O(\text{Pe}^{3/2}) \qquad (4.5.4)$$

as $\text{Pe} \to 0$ was obtained for the dimensionless total diffusion flux to the surface of the particle. Here $I_0$ is the total flux to the particle in a stagnant fluid, $\text{Pe} = a^2 G/D$, $a$ is a unified characteristic length (the dimensionless variables $I$ and $I_0$ are also normalized to $a$), and the values of the parameters $G$ and $\alpha = \alpha(G_{km})$ are given in Table 4.3 for some types of shear flow.

For a spherical particle of radius $a$, we have $I = 4\pi \, \text{Sh}$, and the expansion (4.5.4) turns into (4.5.2) with accuracy to terms of the order of $\text{Pe}$ inclusively. In the case of nonspherical particles, one can calculate $I_0 = \Pi/a$ by using the results of Section 4.3 (e.g., see Table 4.2).

### 4.5-3. Circular Cylinder in a Simple Shear Flow

Mass transfer between a circular cylinder of radius $a$ and a simple shear flow ($G_{12} = \pm 1$, the other $G_{km} = 0$) was studied in [132]. For the dimensionless total diffusion flux per unit length of the cylinder, the following expression was obtained as $\text{Pe} \to 0$:

$$I \approx \frac{4\pi}{2.744 - \ln \text{Pe}}, \qquad \text{Pe} = \frac{a^2|G_{12}|}{D}. \qquad (4.5.5)$$

One can proceed to the Sherwood number (radius-based) by using the formula $I = 2\pi \, \text{Sh}$.

## 4.6. Mass Exchange Between Particles or Drops and Flow at High Peclet Numbers

### 4.6-1. Diffusion Boundary Layer Near the Surface of a Particle

Following [270], we first consider steady-state diffusion to the surface of a solid spherical particle in a translational Stokes flow (Re $\to$ 0) at high Peclet numbers. In the dimensionless variables, the mathematical statement of the corresponding problem for the concentration distribution is given by Eq. (4.4.3) with the boundary conditions (4.4.4) and (4.4.5), where the stream function is determined by (4.4.2).

As Pe increases, a diffusion boundary layer is formed near the surface of the sphere. The ratio of the thickness of this layer to the radius of the particle is of the order of $\mathrm{Pe}^{-1/3}$. In this region, the radial component of molecular diffusion to the surface of the particle is essential, and tangential diffusion may be neglected. Convective mass transfer due to the motion of the fluid must also be taken into account.

Assuming that $\varepsilon = \mathrm{Pe}^{-1/3}$ is a small parameter, we introduce a stretched coordinate $y$ in the diffusion boundary layer region according to the formula $r = 1 + \varepsilon y$. By substituting this coordinate into (4.2.2) and (4.4.3) and by retaining the leading terms of the expansion with respect to $\varepsilon$, we obtain

$$\frac{\partial^2 c}{\partial y^2} = \frac{1}{\sin^2 \theta} \left( \frac{\partial \psi}{\partial \theta} \frac{\partial c}{\partial y} - \frac{\partial \psi}{\partial y} \frac{\partial c}{\partial \theta} \right), \tag{4.6.1}$$

where $\psi = \frac{3}{4} y^2 \sin^2 \theta$.

Further, by passing from $\theta, y$ to the von Mises variables $\theta, \psi$, we obtain

$$-\frac{\partial c}{\partial \theta} = \sqrt{3} \sin^2 \theta \frac{\partial}{\partial \psi} \sqrt{\psi} \frac{\partial c}{\partial \psi}. \tag{4.6.2}$$

The transformation

$$\zeta = \sqrt{\psi} = \frac{\sqrt{3}}{2} \mathrm{Pe}^{1/3} (r - 1) \sin \theta,$$

$$\tau = \frac{\sqrt{3}}{4} \int_\theta^\pi \sin^2 \theta \, d\theta = \frac{\sqrt{3}}{8} \left( \pi - \theta + \frac{1}{2} \sin 2\theta \right) \tag{4.6.3}$$

reduces Eq. (4.6.2) to the form

$$\frac{\partial c}{\partial \tau} = \zeta^{-1} \frac{\partial^2 c}{\partial \zeta^2}. \tag{4.6.4}$$

The boundary conditions (4.4.4) and (4.4.5) in the variables (4.6.3) can be represented in the form

$$\tau = 0, \quad c = 0; \qquad \zeta = 0, \quad c = 1; \qquad \zeta \to \infty, \quad c \to 0. \tag{4.6.5}$$

The solution of problem (4.6.4), (4.6.5) can be expressed as follows via the incomplete gamma function:

$$c = \frac{1}{\Gamma(1/3)}\Gamma\left(\frac{1}{3}, \frac{\zeta^3}{9\tau}\right) = \frac{1}{\Gamma(1/3)}\Gamma\left(\frac{1}{3}, \frac{\text{Pe}}{3}\frac{(r-1)^3\sin^3\theta}{\pi - \theta + \frac{1}{2}\sin 2\theta}\right). \qquad (4.6.6)$$

By differentiating this equation, we obtain the dimensionless local diffusion flux to the surface of the sphere in the form

$$j = -\left(\frac{\partial c}{\partial r}\right)_{r=1} = 0.766 \sin\theta \left(\pi - \theta + \frac{1}{2}\sin 2\theta\right)^{-1/3}\text{Pe}^{1/3}. \qquad (4.6.7)$$

One can see that the local diffusion flux attains its maximum at the front stagnation point on the surface of the sphere (at $\theta = \pi$) and monotonically decreases with the angular coordinate to the minimum value, which is equal to zero and is attained at $\theta = 0$.

The corresponding mean Sherwood number is [270]

$$\text{Sh} = 0.625\,\text{Pe}^{1/3}. \qquad (4.6.8)$$

This formula was refined in [6], where the following two-term expansion was obtained for the Sherwood number:

$$\text{Sh} = 0.625\,\text{Pe}^{1/3} + 0.461. \qquad (4.6.9)$$

Formula (4.6.9) can be used for practical calculations if $\text{Pe} \geq 10$.

---

### 4.6-2. Diffusion Boundary Layer Near the Surface of a Drop (Bubble)

Now let us consider the exterior problem about mass exchange between a spherical drop (bubble) of radius $a$ and a translational Stokes flow with limiting diffusion resistance of the continuous phase.

The process of convective diffusion to the liquid–liquid (liquid–gas) interface substantially differs from the diffusion to the fluid–solid interface. This is due to the difference between the hydrodynamic conditions on the interfaces. The fluid velocity on the surface of a solid is always zero by virtue of the no-slip condition. On the contrary, the interface between two fluids can move, and the tangential velocity on the interface differs from zero.

Convective transfer of a substance by a moving fluid to the fluid–solid interface takes place under the condition that the flow is somewhat retarded, so that the transfer rate close to the surface is considerably lower than that in the bulk of the solution. On the contrary, diffusion to a liquid–liquid (liquid–gas) interface takes place under the more favorable conditions of nonretarded flow. That is why convective diffusion of a substance to a liquid–liquid interface is much more intensive than that to a fluid–solid interface.

The mathematical statement of the problem on the concentration distribution outside a drop is described by Eq. (4.4.3) and the boundary conditions (4.4.4) and (4.4.5), in which the dimensionless stream function satisfies the Hadamard–Rybczynski solution (see Section 2.2)

$$\psi = \frac{1}{2}(r-1)\left[r - \frac{1}{2}\frac{\beta}{\beta+1}\left(1+\frac{1}{r}\right)\right]\sin^2\theta, \tag{4.6.10}$$

where $\beta$ is the drop-medium ratio of dynamic viscosities. The corresponding dimensional stream function can be obtained by multiplying (4.6.10) by $a^2 U_i$. In the problem on diffusion to a drop falling in a stagnant medium, the characteristic velocity is chosen to be

$$U_i = \frac{2(\rho - \rho_i)ga^2}{3\mu_i}\frac{\beta+1}{3\beta+1},$$

where $\rho_i$ and $\rho$ are the fluid densities, respectively, outside and inside the drop, $g$ is the free fall acceleration, and $\mu_i$ is the dynamic viscosity of the fluid outside the drop.

Assuming $\varepsilon = \mathrm{Pe}^{-1/2}$ to be a small parameter, we pass in Eq. (4.4.3) and formula (4.6.10) from the radial coordinate $r$ to the stretched variable $\xi = \varepsilon^{-1}(r-1)$. By retaining the leading terms of the expansion with respect to the parameter $\varepsilon$, we obtain Eq. (4.6.1) with

$$\psi = \frac{1}{2(\beta+1)}\xi\sin^2\theta. \tag{4.6.11}$$

Next, we use (4.6.11) to express $\xi$ via $\psi$ in (4.6.1), thus obtaining

$$-\frac{\partial c}{\partial\theta} = \frac{\sin^3\theta}{2(\beta+1)}\frac{\partial^2 c}{\partial\psi^2}. \tag{4.6.12}$$

By the change of variable

$$\tau = \frac{1}{2(\beta+1)}\int_\theta^\pi \sin^3\theta\, d\theta = \frac{1}{2(\beta+1)}\left(\frac{2}{3} + \cos\theta - \frac{\cos^3\theta}{3}\right), \tag{4.6.13}$$

we reduce (4.6.12) to the standard heat equation

$$\frac{\partial c}{\partial\tau} = \frac{\partial^2 c}{\partial\psi^2}. \tag{4.6.14}$$

The boundary conditions (4.4.4) and (4.4.5) in the variables (4.6.11) and (4.6.13) can be represented in the form (4.6.5) with $\zeta$ replaced by $\psi$. The corresponding solution of Eq. (4.6.14) can be expressed via the complementary error function as follows:

$$c = \mathrm{erfc}\left(\frac{\psi}{2\sqrt{\tau}}\right) = \mathrm{erfc}\left(\frac{1}{4}\sqrt{\frac{6\,\mathrm{Pe}}{\beta+1}}\,(r-1)\frac{1-\cos\theta}{\sqrt{2-\cos\theta}}\right). \tag{4.6.15}$$

Let us calculate the dimensionless local diffusion flux to the surface of the drop:

$$j = -\left(\frac{\partial c}{\partial r}\right)_{r=1} = \sqrt{\frac{3\,\text{Pe}}{\pi(\beta+1)}}\,\frac{1-\cos\theta}{\sqrt{2-\cos\theta}}. \tag{4.6.16}$$

The mean Sherwood number is given by the formula [270]

$$\text{Sh} = \sqrt{\frac{2\,\text{Pe}}{3\pi(\beta+1)}} = 0.461\left(\frac{\text{Pe}}{\beta+1}\right)^{1/2}. \tag{4.6.17}$$

The following two-term expansion of the Sherwood number with respect to the small parameter $\varepsilon = \text{Pe}^{-1/2}$ was obtained in [166]:

$$\text{Sh} = 0.461\left(\frac{\text{Pe}}{\beta+1}\right)^{1/2} + 0.41\left(\frac{3}{4}\beta+1\right); \tag{4.6.18}$$

this is a refinement of (4.6.17).

Formula (4.6.18) can be used for practical calculations at $\text{Pe} \geq 100$ for $0 \leq \beta \leq 0.82\,\text{Pe}^{1/3} - 1$ (this follows from the comparison with the numerical solution [359]). The value $\beta = 0$ in (4.6.17) and (4.6.18) corresponds to a gas bubble.

### 4.6-3. General Formulas for Diffusion Fluxes

*Structure of the flow near the surface of particles, drops, and bubbles.* By analogy with the case of spherical drops and solid particles in a translational flow, one can consider the more general problem of steady-state mass exchange between drops (bubbles) or nonspherical particles and an arbitrary given laminar flow of an incompressible fluid. Without going into detail, we present some final formulas for calculating the dimensionless total diffusion fluxes corresponding to the asymptotic solutions of plane and axisymmetric problems of convective mass transfer (4.4.1), (4.3.2), and (4.3.3) at high Peclet numbers.

We use a local orthogonal curvilinear system of dimensionless coordinates $\xi$, $\eta$, $\varphi$, where $\eta$ varies along and $\xi$ is normal to the surface of the particle. In the axisymmetric case, the azimuth coordinate $\varphi$ varies from 0 to $2\pi$; in the plane case, it is supposed that $0 \leq \varphi \leq 1$. We assume that there are no closed streamlines in the flow and the surface of the particle is given by a constant value $\xi = \xi_s$. The dimensionless fluid velocity components can be expressed via the dimensionless stream function $\psi$ as follows:

$$v_\xi = -\left(\frac{g_{\xi\xi}}{g}\right)^{1/2}\frac{\partial\psi}{\partial\eta}, \qquad v_\eta = \left(\frac{g_{\eta\eta}}{g}\right)^{1/2}\frac{\partial\psi}{\partial\xi}, \tag{4.6.19}$$

where $g_{\xi\xi}$, $g_{\eta\eta}$, and $g_{\varphi\varphi}$ are the metric tensor components and $g = g_{\xi\xi}g_{\eta\eta}g_{\varphi\varphi}$; in the plane case, $g_{\varphi\varphi} = 1$.

In the case of a viscous flow, on the surface of the particle we have either the no-slip condition (for a solid particle) or the no-flow condition (for a fluid particle); therefore, one can represent the stream function in the form

$$\psi \to (\xi - \xi_s)^m f(\eta) \qquad \text{as} \quad \xi \to \xi_s. \qquad (4.6.20)$$

For a viscous flow past drops (bubbles) and for an ideal fluid flow past solid particles, we have $m = 1$. For a laminar viscous flow past smooth solid particles, one usually has $m = 2$; there also exist some examples in which $m = 3$ [166]. It follows from the preceding that in the leading approximation, the tangential velocity $v_\eta$ (4.6.19) in the diffusion boundary layer near the drop surface is constant and is equal to the fluid velocity on the interface, whereas in the diffusion boundary layer close to the surface of a solid particle, the tangential velocity in the leading approximation depends on the distance to the surface linearly (sometimes, even quadratically) and is zero on the surface.

Let us consider the flow geometry near a drop or a solid particle in more detail. The coordinates $\eta_k$ of stagnation points and lines on the interface are determined by the equation

$$f(\eta_k) = 0. \qquad (4.6.21)$$

If there are stagnation lines, then the coordinate surfaces $\eta = \eta_k$ separate regions where the leading term of the expansion of the stream function (4.6.20) preserves its sign. Stagnation points and lines play an important role in the theory of diffusion boundary layer. They may be of two types: in a small neighborhood of these points and lines, the normal component of the fluid velocity is directed either toward the interface ("onflow" points and lines) or away from the interface ("run-off" points and lines). In Figure 4.2, the onflow lines are determined by the value of $\eta_k$, and the run-off lines, by $\eta_{k+1}$. By virtue of the mass conservation law, in a neighborhood of the onflow (respectively, run-off) point or line, the tangential fluid velocity component near the interface is directed from (respectively, to) this point or line, and the onflow and run-off points or lines themselves must alternate. In a neighborhood of a onflow point (line), a diffusion boundary layer arises, whose thickness is minimal here. In a neighborhood of a run-off point (line), the thickness of the diffusion boundary layer increases sharply.

Note that in axisymmetric problems, on the interface there are always two isolated stagnation points on the symmetry axis.

*Formulas for calculating diffusion fluxes.* The dimensionless total diffusion flux to a part of the surface of a particle (drop or bubble) between neighboring stagnation lines (points) $\eta_k$ and $\eta_{k+1}$ can be calculated by the formula [166]

$$I(k, k+1) = \frac{\Lambda(m+1)^{\frac{2m}{m+1}}}{\Gamma\left(\frac{1}{m+1}\right)} \left[F(k, k+1)\right]^{\frac{m}{m+1}} \mathrm{Pe}^{\frac{1}{m+1}}, \qquad (4.6.22)$$

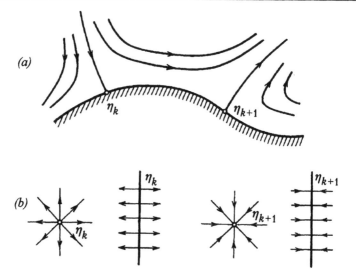

**Figure 4.2.** (a) The flow pattern near the surface of a particle in a neighborhood of stagnation points or lines: onflow (with coordinate $\eta_k$) and run-off (with coordinates $\eta_{k+1}$); arrows show the direction of the fluid velocity vector. (b) Distribution of the tangential component of the fluid velocity near stagnation points or lines on the surface of the body

where

$$F(k,\ k+1) = \frac{1}{m}\left| \int_{\eta_k}^{\eta_{k+1}} \frac{\sqrt{g^s}}{g_{\xi\xi}^s} |f(\eta)|^{\frac{1}{m}}\, d\eta \right|,$$

$$\Lambda = \begin{cases} 2\pi & \text{in the axisymmetric case,} \\ 1 & \text{in the plane case.} \end{cases} \tag{4.6.23}$$

In the expression (4.6.23), the superscript "s" refers to values on the interface at $\xi = \xi_s$. The value of the constant $\Lambda$ corresponds to the conventional practice to determine the plane total diffusion flux per unit length of the cylinder ($0 \le \varphi \le 1$). The value $m = 1$ corresponds to drops and bubbles, and $m = 2$, to solid particles in a viscous fluid.

In specific axisymmetric and plane problems, it is useful to have some expressions for the function $F(k,\ k+1)$ in the spherical and cylindrical coordinates.

Suppose that in the spherical (or cylindrical) coordinates, the surface of a particle (drop, bubble) is described by the equation $r = R(\theta)$, where $r$ is the dimensionless (referred to the characteristic length) radial coordinate and $\theta$ is the angular coordinate. Then the velocity field near the interface is determined by the dimensionless stream function $\psi = [r - R(\theta)]^m f(\theta)$, and the value $F(k,\ k+1)$ in (4.6.22) is calculated by the following formulas [166]:

in the axisymmetric case, $0 \le \theta \le \pi$ and

$$F(k,\ k+1) = \frac{1}{m}\left| \int_{\theta_k}^{\theta_{k+1}} \sin\theta \left[ R^2 + \left(\frac{dR}{d\theta}\right)^2 \right] |f(\theta)|^{\frac{1}{m}}\, d\theta \right|; \tag{4.6.24}$$

in the plane case, $0 \leq \theta \leq 2\pi$ and

$$F(k, k+1) = \frac{1}{m} \left| \int_{\theta_k}^{\theta_{k+1}} R \left[ 1 + \frac{1}{R^2} \left( \frac{dR}{d\theta} \right)^2 \right] |f(\theta)|^{\frac{1}{m}} \, d\theta \right|. \qquad (4.6.25)$$

Here $\theta_k$ and $\theta_{k+1}$ are the angular coordinates of stagnation lines (points) on the interface; it is assumed that there are no stagnation lines and points in the interval $\theta_k < \theta < \theta_{k+1}$.

Note that in the case of axisymmetric linear shear flow to the surface of a spherical particle (drop or bubble), there are two isolated stagnation points $\theta = 0$ and $\theta = \pi$, as well as the stagnation line $\theta = \pi/2$.

To determine the total dimensionless diffusion flux $I_\Sigma$, one must first find the coordinates $\eta_1 < \eta_2 < \cdots < \eta_k < \eta_{k+1} < \cdots < \eta_M$ of all stagnation lines and points on the surface of the particle (drop or bubble) and then, by using any of the formulas (4.6.23)–(4.6.25), calculate the fluxes (4.6.22) to the parts of the surface between adjacent stagnation lines (points). It remains to calculate the sum

$$I_\Sigma = \sum_{k=1}^{M-1} I(k, k+1). \qquad (4.6.26)$$

The mean Sherwood number is obtained by dividing (4.6.26) by the dimensionless surface area of the particle (drop or bubble).

A method for solving three-dimensional problems on the diffusion boundary layer based on a three-dimensional analog of the stream function, was proposed in [348, 350]. In [27, 166, 353], this method was used for studying mass exchange between spherical particles, drops, and bubbles and three-dimensional shear flow.

# 4.7. Particles, Drops, and Bubbles in Translational Flow. Various Peclet and Reynolds Numbers

In this section, some interpolation formulas are presented (see [367, 368]) for the calculation of the mean Sherwood number for spherical particles, drops, and bubbles of radius $a$ in a translational flow with velocity $U_i$ at various Peclet numbers $Pe = aU_i/D$ and Reynolds numbers $Re = aU_i/\nu$. We denote the mean Sherwood number by $Sh_b$ for a gas bubble and by $Sh_p$ for a solid sphere.

### 4.7-1. Mass Transfer at Low Reynolds Numbers

***Spherical particle as*** $Re \to 0$, $0 \leq Pe < \infty$. The problem of mass transfer to a solid spherical particle in a translational Stokes flow ($Re \to 0$) was studied in the entire range of Peclet numbers by finite-difference methods in [1, 60, 281]. To find the mean Sherwood number for a spherical particle, it is convenient to use the following approximate formula [94]:

$$Sh_p = 0.5 + (0.125 + 0.243 \, Pe)^{1/3}. \qquad (4.7.1)$$

The interpolation formula (4.7.1) gives exact asymptotic results both as $Pe \to 0$ and as $Pe \to \infty$. The maximum deviation of (4.7.1) from the data given in [1, 60, 281] in the entire range of Peclet numbers is about 2%.

*Spherical bubble as* $Re \to 0$, $0 \le Pe < \infty$. The problem of mass transfer to a spherical bubble in a translational flow as $Re \to 0$ was studied numerically in [321]. The results for the mean Sherwood number can be approximated well by the expression

$$Sh_b = 0.6 + (0.16 + 0.213\,Pe)^{1/2}, \tag{4.7.2}$$

whose maximum error is about 3%.

*Spherical drop as* $Re \to 0$, $0 \le Pe < \infty$. In the range $0 \le Pe \le 200$, the results of numerical calculations of mean Sherwood numbers for a spherical drop in a translational flow under a limiting resistance of the continuous phase is well described by the approximate formula [68]

$$Sh = \frac{1}{\beta+1}\,Sh_b + \frac{\beta}{\beta+1}\,Sh_p, \tag{4.7.3}$$

where $\beta$ is the ratio of dynamic viscosities of the drop and the ambient fluid ($\beta = 0$ corresponds to a gas bubble, and $\beta \to \infty$, to a solid sphere), and $Sh_b$ and $Sh_p$ are the Sherwood numbers for a bubble and for a solid particle, which can be calculated by formulas (4.7.2) and (4.7.1), respectively.

It is important to note that the expression (4.7.3) gives three valid terms of the asymptotic expansion of $Sh$ as $Pe \to 0$ for any $\beta$ [359].

In the interval $200 \le Pe < \infty$, for any values of the phase viscosities, the mean Sherwood number for a drop can be calculated by solving the cubic equation

$$Sh^3 - 0.212\frac{Pe}{\beta+1}Sh - (0.624)^3\,Pe = 0. \tag{4.7.4}$$

## 4.7-2. Mass Transfer at Moderate and High Reynolds Numbers

*Spherical particle at various Reynolds numbers.* In the case of a spherical particle in a translational flow at $0.5 \le Re \le 200$ and $0.125 \le Sc \le 50$, numerical results concerning the mean Sherwood number (e.g., see [114, 281]) can be described [94] by the approximate formula

$$Sh_p = 0.5 + 0.527\,Re^{0.077}(1 + 2\,Re\,Sc)^{1/3}, \tag{4.7.5}$$

whose error is about 3%.

The analysis of available experimental data on heat and mass transfer to a solid sphere in a translational flow results in the following correlations [94].

Heat exchange with air at $Pr = 0.7$:

$$Nu_p = 0.5 + 0.47\,Re^{0.47} \quad \text{for} \quad 50 \le Re \le 2 \times 10^3,$$
$$Nu_p = 0.5 + 0.2\,Re^{0.58} \quad \text{for} \quad 2 \times 10^3 \le Re \le 5 \times 10^4.$$

Mass transfer with a liquid for large Schmidt numbers ($Sc > 100$):

$$Sh_p = 0.5 + 0.5\,Re^{0.48}\,Sc^{1/3} \qquad \text{for} \quad 50 \le Re \le 10^3,$$

$$Sh_p = 0.5 + 0.31\,Re^{0.55}\,Sc^{1/3} \qquad \text{for} \quad 10^3 \le Re \le 5 \times 10^4.$$

The experimental data for $0.5 < Re < 50$ are well described by the expression (4.7.5).

*Spherical bubble at any Peclet numbers for* $Re \ge 35$. For a spherical bubble in a translational flow at moderate and high Reynolds numbers and high Peclet numbers, the mean Sherwood number can be calculated by the formula [94]

$$Sh_b = \left(\frac{2}{\pi}Pe\right)^{1/2}\left(1 - \frac{2}{\sqrt{Re}}\right)^{1/2}, \qquad (4.7.6)$$

whose error is less than 7% for $Re \ge 35$.

For $0 \le Pe < \infty$ and $Re \ge 35$, to calculate the mean Sherwood number for a spherical bubble, one can use the approximate formula

$$Sh_b = 0.6 + \left[0.16 + 0.637\left(1 - \frac{2}{\sqrt{Re}}\right)Pe\right]^{1/2}, \qquad (4.7.7)$$

which is exact for $Pe = 0$ and passes into (4.7.6) as $Pe \to \infty$. For $Re = \infty$, the error of formula (4.7.7) is about 3% [360].

*Spherical drop at high Peclet numbers for* $Re \ge 35$. For high $Re$, the fluid velocity distribution in the boundary layer near the drop surface was obtained in [180]. These results were used in [504], where mass transfer to a spherical drop in a translational flow was investigated. The results for the mean Sherwood number are well approximated by the formula [94]

$$Sh = \left(\frac{2}{\pi}Pe\right)^{1/2}\left(1 - \frac{2 + 1.49\,\beta^{0.64}}{\sqrt{Re}}\right)^{1/2}, \qquad (4.7.8)$$

which passes into (4.7.6) for $\beta = 0$. Formula (4.7.8) can be applied for $0 \le \beta \le 2$ and $Re \ge 35$.

### 4.7-3. General Correlations for the Sherwood Number

*Particles and bubbles.* Using the method of asymptotic analogies, we shall derive formulas for the calculation of the Sherwood number in a laminar flow past spherical particles, drops, and bubbles for an arbitrary structure of the nonperturbed flow at infinity. We assume that closed streamlines are lacking in the flow.

We start from the approximate formula (4.7.1). Let us transform (4.7.1) by using the procedure described in Section 4.1. We take into account the fact that the asymptotic expansions of $Sh_p$ at small and large $Pe$ have the form

$$Sh_{p0} = 1 \quad (Pe \to 0); \qquad Sh_{p\infty} = 0.624\,Pe^{1/3} \quad (Pe \to \infty).$$

These formulas coincide with (4.1.2) and (4.1.3) up to notation ($w \implies \mathsf{Sh}_p$, $\tau \implies \mathsf{Pe}$) for $A = 1$, $B = 0.624$, $k = 0$, and $m = \frac{1}{3}$. Let us substitute these values into (4.1.5), where the function $F$ is determined from (4.7.1). We take into account the fact that the equation $\mathsf{Sh}_{p0} = 1$ holds for spherical particles, and obtain the following result [359]:

$$\mathsf{Sh}_p = 0.5 + (0.125 + \mathsf{Sh}_{p\infty}^3)^{1/3} \quad \text{(solid particle).} \quad (4.7.9)$$

Formula (4.7.9) can be used for the calculation of the mean Sherwood number in laminar flows of various types without closed streamlines past a solid spherical particle. The auxiliary value $\mathsf{Sh}_{p\infty}$ must be chosen equal to the leading term of the asymptotic expansion of the Sherwood number as $\mathsf{Pe} \to \infty$.

Similarly, for an arbitrary flow past a spherical bubble, one can derive the approximate formula

$$\mathsf{Sh}_b = 0.6 + (0.16 + \mathsf{Sh}_{b\infty}^2)^{1/2} \quad \text{(bubble),} \quad (4.7.10)$$

where $\mathsf{Sh}_{b\infty}$ is the asymptotic value of the mean Sherwood number as $\mathsf{Pe} \to \infty$, which can be calculated by solving the diffusion boundary layer equation in a given flow field.

Formulas (4.7.9) and (4.7.10) after the substitution of the values of $\mathsf{Sh}_{p\infty}$ and $\mathsf{Sh}_{b\infty}$ can be used for the calculation of the Sherwood number in the entire range of Peclet numbers. We note that formula (4.7.7) was derived from (4.7.10), where the right-hand side of (4.7.6) was substituted for the asymptotic value $\mathsf{Sh}_{b\infty}$. Other specific examples of application of formulas (4.7.9) and (4.7.10) will be presented in Section 4.8.

*Drops in the entire range of phase viscosities.* For low and moderate Peclet numbers in an arbitrary laminar flow past a spherical drop under limiting resistance of the continuous phase, it is expedient to calculate the mean Sherwood number by using formula (4.7.3), where $\mathsf{Sh}_p$ and $\mathsf{Sh}_b$ are the Sherwood numbers for the limit cases of a solid particle and a bubble. These quantities can be calculated by formulas (4.7.9) and (4.7.10). For high Peclet numbers, in the entire range of phase viscosities, the mean Sherwood number can be found by solving the cubic equation [359]

$$\mathsf{Sh}^3 - \mathsf{Sh}_\beta^2 \, \mathsf{Sh} - \mathsf{Sh}_{p\infty}^3 = 0 \quad \text{(drop),} \quad (4.7.11)$$

where $\mathsf{Sh}_\beta$ is the asymptotic value of the mean Sherwood number obtained in the diffusion boundary layer approximation for a drop of moderate viscosity $\beta = O(1)$ as $\mathsf{Pe} \to \infty$, and $\mathsf{Sh}_{p\infty}$ is the corresponding asymptotic value for a solid particle ($\beta = \infty$) as $\mathsf{Pe} \to \infty$.

In a Stokes flow ($\mathsf{Re} \to 0$), one can use the approximate formula

$$\mathsf{Sh}_\beta = \frac{\mathsf{Sh}_{b\infty}}{\sqrt{\beta + 1}}, \quad (4.7.12)$$

TABLE 4.4
The mean Sherwood number for spherical solid particles,
drops, and bubbles in a linear straining shear flow ($G_{km} = 0$
for $k \neq m$) at low Reynolds numbers and high Peclet numbers

| Type of particle | Type of flow | Coefficients $G_{kk}$ | Sherwood number Sh | Peclet number Pe | References |
|---|---|---|---|---|---|
| Solid particle | Axisymmetric shear flow | $G_{11} = G_{22}$, $G_{33} = -2G_{11}$ | $0.968 \, Pe^{1/3}$ | $\dfrac{a^2 \lvert G_{33} \rvert}{D}$ | [164] |
| Drop, bubble | Axisymmetric shear flow | $G_{11} = G_{22}$, $G_{33} = -2G_{11}$ | $\left( \dfrac{3}{2\pi} \dfrac{Pe}{\beta+1} \right)^{1/2}$ | $\dfrac{a^2 \lvert G_{33} \rvert}{D}$ | [164] |
| Solid particle | Plane shear flow | $G_{11} = -G_{22}$, $G_{33} = 0$ | $1.01 \, Pe^{1/3}$ | $\dfrac{a^2 \lvert G_{11} \rvert}{D}$ | [27] |
| Drop, bubble | Plane shear flow | $G_{11} = -G_{22}$, $G_{33} = 0$ | $0.731 \left( \dfrac{Pe}{\beta+1} \right)^{1/2}$ | $\dfrac{a^2 \lvert G_{11} \rvert}{D}$ | [353] |

where $Sh_{b\infty}$ is the asymptotic value of the Sherwood number for a gas bubble ($\beta = 0$) as $Pe \to \infty$. For an translational or arbitrary straining Stokes flow past a spherical drop, formula (4.7.12) is exact.

For the special case of a translational Stokes flow past a spherical drop, Eq. (4.7.11) passes into (4.7.4).

# 4.8. Particles, Drops, and Bubbles in Linear Shear Flows. Arbitrary Peclet Numbers

We assume that the fluid velocity distribution remote from the interface is given by Eq. (4.5.1). The mean Sherwood number for a spherical particle, drop, or bubble does not change if we change all signs of the shear coefficients, that is, $Sh(G_{km}) = Sh(-G_{km})$.

| **4.8-1. Linear Straining Shear Flow. High Peclet Numbers** |
|---|

The solution of hydrodynamic problems for an arbitrary straining linear shear flow ($G_{km} = G_{mk}$) past a solid particle, drop, or bubble in the Stokes approximation (as $Re \to 0$) is given in Section 2.5. In the diffusion boundary layer approximation, the corresponding problems of convective mass transfer at high Peclet numbers were considered in [27, 164, 353]. In Table 4.4, the mean Sherwood numbers obtained in these papers are shown.

For a solid spherical particle in an arbitrary linear straining shear flow, the following interpolation formula was suggested in [27] for the mean Sherwood number:

$$\text{Sh} = 0.9\,\text{Pe}_M^{1/3}, \tag{4.8.1}$$

where the modified Peclet number $\text{Pe}_M$ is defined via the second invariant $J_2$ of the shear tensor by

$$\text{Pe}_M = \frac{a^2 J_2}{D}, \qquad \text{where} \quad J_2 = (G_{km}G_{km})^{1/2}. \tag{4.8.2}$$

Here the summation is carried out over both indices $k$ and $m$. If $G_{km} = 0$ for $k \neq m$, then $J_2 = \sqrt{(G_{11})^2 + (G_{22})^2 + (G_{33})^2}$.

For the axisymmetric and plane shear flows (see Table 4.4), the error of formula (4.8.1) does not exceed 1%.

For a spherical drop in an arbitrary straining linear shear flow under limiting resistance of the continuous phase, one can use the interpolation formula [353]:

$$\text{Sh} = 0.62 \left( \frac{\text{Pe}_M}{\beta + 1} \right)^{1/2}, \tag{4.8.3}$$

where $\text{Pe}_M$ is defined in (4.8.2); the value $\beta = 0$ corresponds to a gas bubble.

For axisymmetric and plane shear flows (see Table 4.4), the error of formula (4.8.3) does not exceed 1%.

---

### 4.8-2. Linear Straining Shear Flow. Arbitrary Peclet Numbers

*Spherical particle.* First, we consider an axisymmetric shear flow, where the dimensional fluid velocity components remote from the particle have the following form in the Cartesian coordinates $X_1, X_2, X_3$:

$$\mathbf{V} = (V_1, V_2, V_3) = \left( -\tfrac{1}{2}GX_1, \ -\tfrac{1}{2}GX_2, \ GX_3 \right).$$

Here, by definition, $G = G_{33}$.

For a spherical particle in an axisymmetric shear Stokes flow ($\text{Re} \to 0$), numerical results for the mean Sherwood number can be well approximated in the entire range of Peclet numbers by the expression (4.7.9), where the asymptotic value $\text{Sh}_{p\infty}$ must be taken from the first row in Table 4.4. As a result, we obtain the formula

$$\text{Sh}_p = 0.5 + (0.125 + 0.745\,\text{Pe})^{1/3}, \tag{4.8.4}$$

whose error does not exceed 3%.

For an arbitrary straining shear flow ($G_{km} = G_{mk}$), the mean Sherwood number for a solid sphere can be expressed by the similar formula

$$\text{Sh}_p = 0.5 + (0.125 + 0.729\,\text{Pe}_M)^{1/3}, \tag{4.8.5}$$

where the modified Peclet number $\text{Pe}_M$ is introduced in (4.8.2). For an axisymmetric shear, (4.8.5) passes into (4.8.4).

***Spherical bubble.*** The problem of mass transfer to a spherical bubble in an axisymmetric shear flow at low Reynolds numbers was solved numerically in the entire range of Peclet numbers in [251]. The results for the mean Sherwood number can be approximated by formula (4.7.10), where the corresponding value from the second row in Table 4.4 at $\beta = 0$ must be substituted into the right-hand side. Thus, we obtain the formula

$$Sh_b = 0.6 + (0.16 + 0.48\,Pe)^{1/2}, \qquad (4.8.6)$$

whose error does not exceed 3%.

By generalizing this expression to an arbitrary straining flow ($G_{km} = G_{mk}$), for a spherical bubble we obtain

$$Sh_b = 0.6 + (0.16 + 0.384\,Pe_M)^{1/2}, \qquad (4.8.7)$$

where $Pe_M$ is defined in (4.8.2).

***Spherical drop.*** For moderate Peclet numbers, the mean Sherwood number for a drop in an axisymmetric shear flow can be calculated as $Re \to 0$ in the entire range of phase viscosities, in accordance with [251], by using formula (4.7.3), where $Sh_b$ and $Sh_p$ are the Sherwood numbers for a bubble and a solid particle, given by (4.8.6) and (4.8.4), respectively. For $0 \le Pe \le 100$ ($0 \le \beta < \infty$), the maximum error of this formula is observed at $Pe = 100$ and is about 1%. In the wider range $0 \le Pe \le 500$, the maximum error is about 5%.

In the case of an arbitrary straining shear flow past a spherical drop for $0 \le Pe_M \le 200$, the expressions (4.8.5) and (4.8.7) must be substituted into the expression (4.7.3) of the mean Sherwood number.

For high Peclet numbers ($Pe \ge 100$), the mean Sherwood number for a drop in an axisymmetric shear flow is well approximated by the positive root of the cubic equation

$$Sh^3 - 0.478\frac{Pe}{\beta + 1}Sh - 0.745\,Pe = 0, \qquad (4.8.8)$$

whose maximum error for $Pe \ge 100$ in the entire range of the phase viscosities ($0 \le \beta < \infty$) is 7% [251].

The generalization of this equation to an arbitrary straining shear flow has the form

$$Sh^3 - 0.384\frac{Pe_M}{\beta + 1}Sh - 0.729\,Pe_M = 0. \qquad (4.8.9)$$

Both equations (4.8.8) and (4.8.9) were derived from (4.7.11) with regard to relation (4.7.12).

## 4.8-3. Simple Shear and Arbitrary Plane Shear Flows

Let us investigate convective mass transfer to the surface of a solid sphere freely suspended in an arbitrary plane shear Stokes flow. In this case, the fluid velocity distribution remote from the particle is given by formulas (4.5.1) with

$G_{k3} = G_{3k} = 0$ ($k = 1, 2, 3$). Since the fluid is incompressible, we can represent the shear tensor as the sum of a symmetric and antisymmetric tensor, which correspond to the purely straining and purely rotational components of the fluid motion at infinity:

$$\begin{bmatrix} G_{11} & G_{12} & 0 \\ G_{21} & G_{22} & 0 \\ 0 & 0 & 0 \end{bmatrix} = \begin{bmatrix} E_1 & E_2 & 0 \\ E_2 & -E_1 & 0 \\ 0 & 0 & 0 \end{bmatrix} + \begin{bmatrix} 0 & -\Omega & 0 \\ \Omega & 0 & 0 \\ 0 & 0 & 0 \end{bmatrix}, \qquad (4.8.10)$$

where

$$E_1 = G_{11} = -G_{22}, \quad E_2 = \tfrac{1}{2}(G_{12} + G_{21}), \quad \Omega = (G_{21} - G_{12}).$$

In the case of general plane shear, the tensor $\mathbf{G} = [G_{km}]$ is determined by three independent quantities $E_1$, $E_2$, and $\Omega$. A simple shear flow (a Couette flow) is characterized by the values $E_1 = 0$, $E_2 = -\Omega = \tfrac{1}{2}G_{12}$.

By virtue of the no-slip condition on the surface, a sphere freely suspended in a plane shear flow will rotate at a constant angular velocity $\Omega$ equal to the flow rotation velocity at infinity. The solution of the corresponding three-dimensional hydrodynamic problem on a particle in a Stokes flow is given in [343].

To describe the solution of the mass transfer problem for an arbitrary plane shear flow at high Peclet numbers, we introduce dimensionless quantities by the formulas

$$\Omega_E = \frac{\Omega}{E}, \quad \mathrm{Pe} = \frac{a^2 \overline{E}}{D}, \quad \overline{E} = (E_1^2 + E_2^2)^{1/2}. \qquad (4.8.11)$$

For $0 < |\Omega_E| \leq 1$, there are both closed and open streamlines in the flow; moreover, there is a region adjacent to the particle filled with closed streamlines, whereas the streamlines far from the sphere are open. It is important that the diffusion boundary layer is not formed at high Peclet numbers in the region of closed streamlines near the sphere (the diffusion boundary layer is always produced by critical streamlines that come from infinity to the surface of the body).

For $0 < |\Omega_E| \leq 1$ and as $\mathrm{Pe} \to \infty$, the concentration is uniform in the region with open streamlines and equal to its value at infinity; the concentration distribution in the region filled with closed streamlines can be represented as a regular expansion in powers of the inverse Peclet number:

$$c = c_0 + \mathrm{Pe}^{-1} c_1 + \cdots \quad (\mathrm{Pe} \to \infty). \qquad (4.8.12)$$

By substituting this series into the convective diffusion equation (4.4.1) and by matching the coefficients of like powers of the small parameter $\mathrm{Pe}^{-1}$, one can show that the leading term of this expansion satisfies the equation $(\mathbf{v} \cdot \nabla)c_0 = 0$. Therefore, the concentration $c_0$ is constant along the streamlines. But this information is insufficient for determining $c_0$. By writing the equation for the next term of the expansion, $c_1$, and by integrating it over the closed streamlines [343], one can obtain an elliptic equation for $c_0$. Taking account of the structure of the expansion of $c$, one can draw a very important qualitative conclusion about the properties of the function $c_0$: if a particle (drop) is surrounded by a flow region filled with closed streamlines, then the mean Sherwood number tends as $\text{Pe} \to \infty$ to a finite constant,

$$\lim_{\text{Pe} \to \infty} \text{Sh} = \text{const} \neq \infty. \qquad (4.8.13)$$

This limit property of the mean Sherwood number differs essentially from the corresponding behavior of $\text{Sh}$ in the presence of singular hydrodynamic points, where the mean Sherwood number increases infinitely as $\text{Pe} \to \infty$ (e.g., see formulas (4.8.4) and (4.8.6)).

The analysis of the mass transfer problem for a sphere freely suspended in an arbitrary shear flow at high Peclet numbers leads to the following two-term asymptotics for the mean Sherwood number at low values of the angular velocity [343]

$$\text{Sh} = 10.35\,|\Omega_E|^{-1} - 3.5 + O(\Omega_E) \qquad \text{for} \quad |\Omega_E| \to 0. \qquad (4.8.14)$$

The numerical results [343] in the range $0 < |\Omega_E| \le 1$ as $\text{Pe} \to \infty$ are well approximated by the formula

$$\text{Sh} = 10.35\,|\Omega_E|^{-1} - 3.5 + |\Omega_E| - 3.4\,\Omega_E^2, \qquad (4.8.15)$$

whose maximum error does not exceed 3%.

For a simple shear flow ($|\Omega_E| = 1$), formula (4.8.15) yields $\text{Sh} = 4.45$, which coincides with the results in [343].

## 4.9. Mass Transfer in a Translational-Shear Flow and in a Flow With Parabolic Profile

### 4.9-1. Diffusion to a Sphere in a Translational-Shear Flow

Let us consider mass transfer for a translational flow past a solid spherical particle, where the flow field remote from the particle is the superposition of a translational flow with velocity $U_i$ and an axisymmetric straining shear flow, the translational flow being directed along the axis of the straining flow. The dimensional fluid velocity components in the Cartesian coordinates relative to the center of the particle have the form

$$\mathbf{V} = (V_1, V_2, V_3) = \left(-\tfrac{1}{2}GX_1, \ -\tfrac{1}{2}GX_2, \ U_i + GX_3\right). \qquad (4.9.1)$$

In the Stokes approximation, the stream function for the flow (4.9.1) is equal to the sum of the stream functions of the constituent flows.

Mass exchange between a spherical particle and the translational-shear flow (4.9.1) at high Peclet numbers was studied in [175]. For the mean Sherwood number depending on the parameters

$$\mathsf{Pe} = aU_i/D, \qquad \omega = 5a|G|/U_i, \tag{4.9.2}$$

the following expressions were obtained:

$$\mathsf{Sh} = 0.206\,(\omega + 1)^{1/3} f\left(\sqrt{\frac{2\omega}{\omega + 1}}\right) \mathsf{Pe}^{1/3} \tag{4.9.3}$$

for $0 \leq \omega \leq 1$;

$$\mathsf{Sh} = \frac{0.103}{\omega}\left[(\omega - 1)^{4/3} f\left(\sqrt{\frac{\omega - 1}{2\omega}}\right) + (\omega + 1)^{4/3} f\left(\sqrt{\frac{\omega + 1}{2\omega}}\right)\right] \mathsf{Pe}^{1/3} \tag{4.9.4}$$

for $1 \leq \omega$. Here

$$f(k) = \left[\frac{8}{15}\frac{(1 - k^2)(2 - k^2)}{k^4} K(k) - \frac{16}{15}\frac{k^4 - k^2 + 1}{k^4} E(k)\right]^{2/3},$$

where $K(k)$ and $E(k)$ are the complete elliptic integrals of the first and the second kind, respectively.

The mean Sherwood number for the translational-shear flow (4.9.1) past a spherical drop under limiting resistance of the continuous phase at high Peclet numbers can be calculated by the formulas [164]

$$\mathsf{Sh} = \left[\frac{2\,\mathsf{Pe}}{3\pi(\beta + 1)}\right]^{1/2}, \tag{4.9.5}$$

for $0 \leq \omega \leq 5/3$;

$$\mathsf{Sh} = \left[\frac{\mathsf{Pe}}{8\pi(\beta + 1)}\right]^{1/2}\left[\left(1 + \frac{5}{3\omega}\right)^{3/2}\left(\frac{3\omega}{5} - \frac{1}{3}\right)^{1/2}\right.$$
$$\left. + \left(1 - \frac{5}{3\omega}\right)^{3/2}\left(\frac{3\omega}{5} + \frac{1}{3}\right)^{1/2}\right] \tag{4.9.6}$$

for $5/3 \leq \omega$, where the parameters $\mathsf{Pe}$ and $\omega$ are defined in (4.9.2).

Thus the Sherwood number remains constant (is equal to the Sherwood number for a uniform translational flow) as $\omega$ varies in the range $0 \leq \omega \leq 5/3$, and grows with $\omega$ for $\omega > 5/3$.

To construct approximate formulas for the Sherwood number in the case of a translational shear flow past particles and bubbles in the entire range of Peclet numbers, one can use formulas (4.7.9) and (4.7.10) where $\mathsf{Sh}_{p\infty}$ and $\mathsf{Sh}_{b\infty}$ must be replaced by the right-hand sides of Eqs. (4.9.3), (4.9.4) and (4.9.5), (4.9.6), respectively, with $\beta = 0$.

### 4.9-2. Diffusion to a Sphere in a Flow With Parabolic Profile

Let us consider diffusion to the surface of a solid spherical particle of radius $a$ entrained by a Poiseuille flow along the axis of a round tube of radius $L$. We assume that the particle velocity coincides with the fluid velocity on the stream axis and that the inequality $a \ll L$ holds. Then the fluid velocity remote from the sphere has the parabolic profile

$$\mathbf{V} \to \mathbf{e}_3 H (X_1^2 + X_2^2), \tag{4.9.7}$$

where $X_1$, $X_2$, $X_3$ are the Cartesian coordinates relative to the center of the particle; the $X_3$-axis is directed along the tube axis; $\mathbf{e}_3$ is the unit vector along the $X_3$-axis; the parameter $H$ characterizes the curvature of the velocity profile on the symmetry axis and depends on the flow rate.

At high Peclet numbers, the mean Sherwood number corresponding to the parabolic Stokes ($\mathsf{Re} \to 0$) flow (4.9.7) past a sphere is given [166] by the formula

$$\mathsf{Sh} = 0.957 \, \mathsf{Pe}^{1/3}, \tag{4.9.8}$$

where $\mathsf{Pe} = a^3 H / D$.

## 4.10. Mass Transfer Between Nonspherical Particles or Bubbles and Translational Flow

### 4.10-1. Ellipsoidal Particle

Let us consider diffusion to the surface of a solid ellipsoidal particle in a homogeneous translational Stokes flow ($\mathsf{Re} \to 0$). The particle is an ellipsoid of revolution with semiaxes $a$ and $b$ oriented along and across the flow, respectively ($b$ is the equatorial radius). We introduce the following notation:

$$\chi = b/a, \quad a_e = a\chi^{2/3}, \quad \mathsf{Pe}_e = a_e U_i / D, \tag{4.10.1}$$

where $a_e$ is the radius of the volume-equivalent sphere serving as the characteristic length.

Dimensionless total diffusion flux to the surface of an ellipsoidal particle at high Peclet numbers is determined by the formula [433]

$$I = 7.85 \, K(\chi) \, \mathsf{Pe}_e^{1/3}, \tag{4.10.2}$$

where the shape coefficient $K$ is given by

$$K(\chi) = \left(\frac{4}{3}\right)^{1/3} \chi^{-2/9} (\chi^2 - 1)^{1/3} \left(1 + \frac{\chi^2 - 2}{\sqrt{\chi^2 - 1}} \arctan \sqrt{\chi^2 - 1}\right)^{-1/3}$$
$$\text{for} \quad \chi \geq 1,$$

$$K(\chi) = \left(\frac{4}{3}\right)^{1/3} \chi^{-2/9} (1 - \chi^2)^{1/3} \left(\frac{2 - \chi^2}{2\sqrt{1 - \chi^2}} \ln \frac{1 + \sqrt{1 - \chi^2}}{1 - \sqrt{1 - \chi^2}} - 1\right)^{-1/3}$$
$$\text{for} \quad \chi \leq 1. \tag{4.10.3}$$

For $\chi = 1$, we have $K = 1$, and formula (4.10.2) after the division by $4\pi$ gives the Sherwood number (4.6.8) for the solid sphere.

For $0.5 \leq \chi \leq 3.0$, the shape coefficient is well approximated by the expression [166]

$$K(\chi) = 1 + \tfrac{2}{45}(\chi - 1), \tag{4.10.4}$$

whose maximum error is 0.8%.

The mean Sherwood number is given by the formula $\mathsf{Sh} = I/S$, where $S$ is the dimensionless surface area of the ellipsoid of revolution:

$$S = \frac{2\pi}{\chi^{1/3}} \left( \chi + \frac{1}{2\sqrt{\chi^2 - 1}} \ln \frac{\chi + \sqrt{\chi^2 - 1}}{\chi - \sqrt{\chi^2 - 1}} \right) \quad \text{for} \quad \chi \geq 1,$$

$$S = \frac{2\pi}{\chi^{1/3}} \left( \chi + \frac{1}{\sqrt{1 - \chi^2}} \arcsin \sqrt{1 - \chi^2} \right) \quad \text{for} \quad \chi \leq 1. \tag{4.10.5}$$

The dimensionless quantity $S$ is related to the dimensional surface area $S_*$ of the ellipsoid by $S = S_*/a_e^2$.

For arbitrary Peclet numbers, the mean Sherwood number (corresponding to the characteristic length $a_e$) for a translational Stokes flow past an ellipsoidal particle can be approximated by the formula [94]

$$\mathsf{Sh} = 0.5 \frac{1}{S} \left( \frac{\Pi}{a_e} \right) + \frac{1}{S} \left\{ 0.125 \left( \frac{\Pi}{a_e} \right)^3 + [7.85 \, K(\chi)]^3 \, \mathsf{Pe}_e \right\}^{1/3}, \tag{4.10.6}$$

where the shape factor $\Pi$ is given in the second and third rows of Table 4.2, and the quantities $\mathsf{Pe}_e$, $K$, and $S$ are defined in Eqs. (4.10.1), (4.10.3), and (4.10.5), respectively.

The axisymmetric problem on mass exchange between an ellipsoidal particle and a translational Stokes flow was numerically studied in [281] by the finite-difference method. Two cases were considered, in which the length of the particle semiaxis oriented along the flow was, respectively, five times greater and five times smaller than the length of the semiaxis perpendicular to the flow. According to [94], it follows from the results of the numerical solution in [281] that the maximum error of formula (4.10.6) for an ellipsoidal particle does not exceed 10% in the cases under consideration.

Formulas (4.10.2) and (4.10.4) cannot be used in the case of a strongly oblate ($\chi \gg 1$) or a strongly prolate ($\chi \ll 1$) ellipsoid of revolution.

### 4.10-2. Circular Thin Disk

The case $\chi \to \infty$ (that is, $a \to 0$ and $b = \text{const}$) corresponds to diffusion to the surface of a thin circular disk of radius $b$ perpendicular to a uniform translational Stokes flow.

Let us indicate two essential differences between the diffusion flux distribution on the surface of a disk and that on the surface of a sphere at high Peclet numbers. First, the local diffusion flux in the latter case monotonically increases (rather than decreases, as is the case for the sphere) with distance from the front stagnation point (onflow point) along the disk surface. Second, the diffusion flux to the disk is proportional to the Peclet number raised to the power of $1/4$ (rather than $1/3$, as was previously shown for the solid sphere). The reduced diffusion flux is due to the much more intensive flow retardation near the disk.

As $\text{Pe} \to \infty$, the dimensionless total diffusion flux to the front part of the disk is described by the asymptotic formula [166]

$$I = 3.66 \, \text{Pe}_b^{1/4}, \qquad \text{Pe}_b = bU_i/D. \qquad (4.10.7)$$

The results of numerical calculations of the mean Sherwood number at various Peclet and Reynolds numbers are given in [281].

### 4.10-3. Particles of Arbitrary Shape

The following general statement was proved in [63] for the case of a uniform translational Stokes flow ($\text{Re} \to 0$) or a potential flow past a particle of an arbitrary shape: the mean Sherwood number remains the same if the flow direction is changed to the opposite.

Suppose that the axis of revolution makes an angle $\omega$ with the translation flow velocity at infinity. In [358] the following approximate formula for the mean Sherwood number was obtained:

$$\text{Sh} = \text{Sh}_\parallel \cos^2 \omega + \text{Sh}_\perp \sin^2 \omega, \qquad (4.10.8)$$

where $\text{Sh}_\parallel$ and $\text{Sh}_\perp$ are the mean Sherwood numbers corresponding to the case of a parallel ($\omega = 0$) and perpendicular ($\omega = \pi/2$) position of the body in flow, respectively.

At low Peclet numbers, for the translational Stokes flow past an arbitrarily shaped body of revolution, formula (4.10.8) coincides with the exact asymptotic expression in the first three terms of the expansion [358]. Since (4.10.8) holds identically for a spherical particle at all Peclet numbers, one can expect that for particles whose shape is nearly spherical, the approximate formula (4.10.8) will give good results for low as well as moderate or high Peclet numbers.

For a steady-state viscous flow (without closed streamlines) past arbitrarily shaped smooth particles, one can calculate the mean Sherwood number by the approximate formula [359]

$$\text{Sh} = 0.5 \, \text{Sh}_0 + (0.125 \, \text{Sh}_0^3 + \text{Sh}_\infty^3)^{1/3}, \qquad (4.10.9)$$

which can be derived from (4.7.1) by the method of asymptotic analogies.

The auxiliary variables $\text{Sh}_0$ and $\text{Sh}_\infty$ in (4.10.9) are the leading terms of the asymptotic expansions of the mean Sherwood number at small and large Peclet numbers, respectively. (In (4.10.9), $\text{Sh}$, $\text{Sh}_0$, and $\text{Sh}_\infty$ are defined on the basis of the same characteristic length.)

For spherical particles, we have $Sh_0 = 1$ (the radius is taken as the characteristic length), and (4.10.9) turns into (4.7.9). The substitution of the values of $Sh_0$ and $Sh_\infty$ corresponding to ellipsoidal particles in a translational Stokes flow into (4.10.9) results in (4.10.6).

For a translational Stokes flow past a convex body of revolution of sufficiently smooth shape with symmetry axis parallel to the flow, the error $\mathcal{E}$ (in percent) in formula (4.10.9) for the mean Sherwood number can be approximately estimated as follows:

$$\mathcal{E} < 2\left(\frac{a}{b} + \frac{b}{a}\right),$$

where $a$ and $b$ are the maximum longitudinal and transverse dimensions of the particle. This estimate agrees well with the results described previously for an ellipsoidal particle.

For a particle of a given shape, the auxiliary quantities $Sh_0$ and $Sh_\infty$ occurring in (4.10.9) can be determined either theoretically (see Section 4.3) or experimentally. In the last case, the parameter $Sh_0$ must be found from experiments on diffusion to the particle in a stagnant fluid. (Recall that the value $Sh_0$ corresponds to the dimensionless capacity of the body; the electrostatic method for measuring this capacity is widely used in electrical engineering.) For a solid particle, the asymptotics of the mean Sherwood number as $Pe \to \infty$ has the form $Sh_\infty = B\,Pe^{1/3}$, where $B$ is a constant [166]. Therefore, to find the parameter $B$ and hence $Sh_\infty$, it suffices to carry out a single experiment at high Peclet numbers (high Peclet numbers at low Reynolds numbers, $Re < 0.5$, can easily be achieved in water solutions of glycerin). Thus, to find $Sh_0$ and $Sh_\infty$, it suffices to carry out two fairly simple experiments.

For smooth particles of an arbitrary shape in ideal fluid (this model is used, say, to describe heat exchange between particles and liquid metals at $Pr \ll 1$ and $Re \gg 1$) and in the absence of regions with closed streamlines, the mean Nusselt number can be calculated by the formula

$$Nu = 0.6\,Nu_0 + (0.16\,Nu_0^2 + Nu_\infty^2)^{1/2}, \qquad (4.10.10)$$

where $Nu_0$ and $Nu_\infty$ are the asymptotics of the Nusselt number as $Pe_T \to 0$ and $Pe_T \to \infty$.

For a noncirculatory translational flow of an ideal fluid past a sphere, the maximum error in (4.10.10) is about 3%.

Formula (4.10.10) can be used for calculating the mean Sherwood number (one replaces $Nu$ by $Sh$) for nonspherical bubbles moving in a viscous fluid.

### 4.10-4. Deformed Gas Bubble

Let us consider diffusion to a bubble rising in a fluid at high Reynolds numbers. The shape of the bubble substantially depends on the Weber number

$$We = a_e \rho U_i^2/\sigma, \qquad (4.10.11)$$

where $a_e$ is the radius of the volume-equivalent sphere, $U_i$ is the steady-state velocity of the bubble, and $\sigma$ is the coefficient of surface tension.

At small We, the shape of the bubble is nearly spherical; at large We, the bubble becomes a spherical segment; this is partly due to flow separation in the rear area.

The Weber numbers We of the order of 1 constitute a practically important intermediate region in which the bubble is strongly deformed but still preserves its symmetry with respect to the midsection. For such We, the shape of the bubble is well approximated by an oblate ellipsoid of rotation with semiaxes $a$ and $b = \chi a$, where the semiaxis $b$ is perpendicular to the flow and $\chi \geq 1$.

Since the boundary condition for the normal stresses must be satisfied at the front and rear stagnation points as well as along the boundary of the midsection of the bubble, we have the following relationship between the Weber number We and the ratio $\chi$ of the major to the minor semiaxis of the ellipsoid [291]:

$$\text{We} = 2\chi^{-4/3}(\chi^3 + \chi - 2)\left[\chi^2 \operatorname{arcsec} \chi - (\chi^2 - 1)^{1/2}\right]^2 (\chi - 1)^{-3}.$$

Numerical estimates [291] show that the maximum difference between the actual curvature and the corresponding value for the approximating ellipsoid does not exceed 5% for We $\leq 1$ ($\chi \leq 1.5$) and 10% for We $\leq 1.4$ ($\chi \leq 2$).

For common fluids such as water, one must take into account the bubble deformation starting from Re $\sim 10^2$, where Re $= a_e U_i / \nu$ is the Reynolds number and $\nu$ is the kinematic viscosity.

The dimensionless total diffusion flux corresponding to a potential flow past an ellipsoidal bubble (Re $= \infty$) is calculated at high Peclet numbers by the formula [174]

$$I = 4(2\pi)^{1/2}\Omega(\chi)\chi^{-1/3}\,\text{Pe}^{1/2}, \qquad \text{Pe} = a_e U_i / D, \tag{4.10.12}$$

where

$$\Omega(\chi) = \left(\frac{2}{3}\right)^{1/2} \frac{(\chi^2 - 1)^{3/4}}{\chi^{2/3}} \left[\arctan(\chi^2 - 1)^{1/2} - \frac{(\chi^2 - 1)^{1/2}}{\chi^2}\right]^{-1/2}.$$

For a spherical bubble, one must set $\chi = 1$ and $\Omega(1) = 1$ in (4.10.12). For $1 \leq \chi \leq 2$, the function $\Omega(\chi)$ can be approximated by the simple expression $\Omega = 0.5\,(\chi + 1)$ with an error of at most 3%.

In the diffusion boundary layer approximation with allowance for the corrections (with respect to the Reynolds number) to the potential flow past the bubble, one can obtain the following two-term expansion of the dimensionless total flux $I$:

$$I = 4(2\pi)^{1/2}\Omega(\chi)\chi^{-1/3}\left[1 - \text{Re}^{-1/2}\,\Omega_1(\chi)\chi^{1/3}\right]^{1/2}\text{Pe}^{1/2}. \tag{4.10.13}$$

Here $\Omega_1$ is a function of the ratio of the semiaxes of the bubble, which was calculated numerically in [174]. Formula (4.10.13) passes into (4.10.12) as

$\mathsf{Re} \to \infty$. For a spherical bubble ($\chi = 1$), one must set $\Omega_1(1) = 2.05$ [510] in (4.10.13). In the region $1 \le \chi \le 2$ ($\mathsf{We} \le 1.4$), one can use the approximate expression

$$\Omega_1(\chi) = 0.2\,(\chi^2 + 3\chi + 6), \qquad (4.10.14)$$

whose maximal deviation from the exact value does not exceed 3%.

The mean Sherwood number for an ellipsoidal bubble is calculated with the help of (4.10.13) by the formula $\mathsf{Sh} = I/S$, where the dimensionless surface area $S$ is obtained from the first expression in (4.10.5).

Convective heat and mass transfer to spherical-cap bubbles was studied in [222].

# 4.11. Mass and Heat Transfer Between Cylinders and Translational or Shear Flows

## 4.11-1. Diffusion to a Circular Cylinder in a Translational Flow

Let us consider diffusion to the surface of a circular cylinder of radius $a$ in a flow with velocity $U_i$ directed along the normal to the cylinder axis. This is a model problem used in chemical engineering for calculating mass transfer to prolate particles; it is used even more widely in mechanics of aerosols for analyzing diffusion sedimentation of aerosols on fibrous filters [139, 461].

At low Reynolds numbers $\mathsf{Re} = aU_i/\nu$, the analytical solution of this problem results in the following two-term expansion of the mean Sherwood number (based on the radius of the cylinder) with respect to a high Peclet number $\mathsf{Pe} = aU_i/D$ [461]:

$$\mathsf{Sh} = \frac{0.580}{(2.00 - \ln 2\,\mathsf{Re})^{1/3}}\,\mathsf{Pe}^{1/3} + 0.0993. \qquad (4.11.1)$$

The leading term in (4.11.1) was calculated in [134, 307].

Diffusion to an elliptic cylinder in a translational flow at high Peclet numbers was considered in [166].

If $\mathsf{Sc} > 0.5$, then the mean Sherwood number for cylinders of various cross-section perpendicular to the flow direction in a wide range of Reynolds numbers can be determined by using the following formula, derived from experimental data [253]:

$$\mathsf{Sh} = A\,\mathsf{Sc}^{0.37}\,\mathsf{Re}^m, \qquad (4.11.2)$$

where the coefficients $A$ and $m$ are given in Table 4.5.

## 4.11-2. Diffusion to a Circular Cylinder in Shear Flows

*Fixed cylinder.* Let us consider diffusion to the surface of a fixed circular cylinder in a steady-state linear shear Stokes ($\mathsf{Re} \to 0$) flow in the plane normal to the cylinder axis. The velocity distribution of such a flow remote from the

TABLE 4.5

Coefficients $A$ and $m$ in formula (4.11.2) for rods of
various shapes perpendicular to the flow direction

| Cross-section of the rod (flow direction is from left to right) | The range of Re | $A$ | $m$ |
|---|---|---|---|
| ◯ | 0.05 to 2 | 0.640 | 0.305 |
|  | 2 to 4 | 0.556 | 0.41 |
|  | 4 to 500 | 0.381 | 0.47 |
|  | 500 to $2.5 \times 10^3$ | 0.430 | 0.47 |
|  | $2.5 \times 10^3$ to $2.5 \times 10^4$ | 0.142 | 0.60 |
|  | $2.5 \times 10^4$ to $10^5$ | 0.0168 | 0.80 |
| ◇ | $2.5 \times 10^3$ to $5 \times 10^4$ | 0.162 | 0.588 |
| ▢ | $1.25 \times 10^3$ to $2.5 \times 10^3$ | 0.116 | 0.699 |
|  | $2.5 \times 10^3$ to $5 \times 10^4$ | 0.0672 | 0.675 |
| ⬡ | $2.5 \times 10^3$ to $5 \times 10^4$ | 0.101 | 0.638 |
| ⬢ | $2.5 \times 10^3$ to $9.8 \times 10^3$ | 0.105 | 0.638 |
|  | $9.8 \times 10^3$ to $5 \times 10^4$ | 0.0255 | 0.782 |

cylinder can in general be represented by formula (2.7.8) and is determined by the three variables $E_1$, $E_2$, and $\Omega$. The parameter $\overline{E} = \sqrt{E_1^2 + E_2^2}$ characterizes the intensity of the purely straining component of motion, whereas $\Omega$ characterizes the fluid rotation. The qualitative pattern of the shear flow past the cylinder depends on the parameter ratio $\Omega_E = \Omega/\overline{E}$.

An arbitrary shear Stokes flow past a fixed cylinder is described by the stream function (2.7.9). We restrict our discussion to the case $0 \leq |\Omega_E| \leq 1$, in which there are four stagnation points on the surface of the cylinder. Qualitative streamline patterns for a purely straining flow (at $\Omega_E = 0$) and a purely shear flow (at $\Omega_E = 1$) are shown in Figure 2.10.

The solution of the corresponding mass exchange problem for a circular cylinder and an arbitrary shear flow was obtained in [353] in the diffusion boundary layer approximation. It was shown that an increase in the absolute value of the angular velocity $\Omega$ of the shear flow results in a small decrease in the intensity of mass and heat transfer between the cylinder and the ambient medium. The corresponding mean Sherwood number is nicely approximated by the formula

$$Sh = (0.92 - 0.012\,|\Omega_E|)\,Pe^{1/3}, \qquad Pe = a^2\overline{E}/D, \qquad (4.11.3)$$

whose maximum error is about 0.5% for $0 \le |\Omega_E| \le 1$.

It follows from (4.11.3) that in the region $-1 \le \Omega_E \le +1$, the mean Sherwood number varies only slightly (the relative increment in the mean Sherwood number as $|\Omega_E|$ varies from 0 to 1 is at most 1.3%). In the special cases of purely straining ($\Omega_E = 0$) and purely shear ($|\Omega_E| = 1$) linear Stokes flow past a circular cylinder, formula (4.11.3) turns into those given in [342, 343].

*Freely rotating cylinder.* Now let us consider convective mass transfer to the surface of a circular cylinder freely suspended in an arbitrary linear shear Stokes flow (Re → 0). In view of the no-slip condition, the cylinder rotates at a constant angular velocity equal to the angular velocity of the flow at infinity. The fluid velocity distribution is described by formulas (2.7.11). The streamline pattern qualitatively differs from that for the case of a fixed cylinder. For $\Omega \ne 0$, there are no stagnation points on the surface of the cylinder and there exist two qualitatively different types of flow. For $0 < |\Omega_E| < 1$, there are both closed and open streamlines in the flow, the region filled with closed streamlines is adjacent to the surface of the cylinder, and streamlines far from the cylinder are open (Figure 2.11). For $|\Omega_E| > 1$, all streamlines are open.

In the mass exchange problem for a circular cylinder freely suspended in linear shear flow, no diffusion boundary layer is formed as Pe → ∞ near the surface of the cylinder. The concentration distribution is sought in the form of a regular asymptotic expansion (4.8.12) in negative powers of the Peclet number. The mean Sherwood number remains finite as Pe → ∞. This is due to the fact that mass and heat transfer to the cylinder is blocked by the region of closed circulation. As a result, mass and heat transfer to the surface is mainly determined by molecular diffusion in the direction orthogonal to the streamlines. In this case, the concentration is constant on each streamline (but is different on different streamlines).

For simple shear flow, the asymptotic value of the mean Sherwood number was calculated at high Peclet numbers in [132]:

$$Sh = 2.87 \qquad (|\Omega_E| = 1). \qquad (4.11.4)$$

For small angular velocities of the flow, the following two-term expansion accurate up to the terms of the order of $\Omega_E$ was obtained in [343]:

$$Sh = 7.79\,|\Omega_E|^{-1} - 2.97 \qquad (|\Omega_E| \to 0). \qquad (4.11.5)$$

The asymptotic expressions (4.11.4) and (4.11.5) correspond to the case of infinite Peclet numbers, and the expansion (4.11.5) has a singularity at $\Omega_E = 0$. Further, the case $\Omega_E = 0$ corresponds to a purely straining flow, in which a cylinder remains immovable regardless of whether it is fixed or not. Therefore, at $\Omega_E = 0$ one can use formula (4.11.3), which in this case gives an asymptotic expression for the mean Sherwood number at $Pe \gg 1$. The comparison of this formula with the expression (4.11.5) shows that (4.11.5) can be applied for angular velocities of flow such that $O(Pe^{-1/3}) < |\Omega_E| \leq 1$.

The following dependence of the mean Sherwood number on the Peclet number and the parameter $\Omega_E$ corresponds to the above-mentioned asymptotic solutions (4.11.3) (at $\Omega_E = 0$), (4.11.4), and (4.11.5):

$$\text{Sh} = \frac{7.8}{8.46\,Pe^{-1/3} + |\Omega_E|} - 2.97 - 1.94\,|\Omega_E|^3. \qquad (4.11.6)$$

This dependence can be treated as an approximate formula for Sh at high Peclet numbers and for all $|\Omega_E| \leq 1$.

The comparison with numerical results [343] shows that the maximum error of (4.11.6) does not exceed 5% at $Pe = \infty$.

Note that, just as in the case of a fixed cylinder, an increase in the angular flow velocity results in a decrease in the intensity of mass exchange.

An experimental verification [405] of the fact that the leading term of the asymptotic expansion of the mean Sherwood number for $Pe \gg 1$ is independent of the Peclet number for a freely rotating circular cylinder in a simple shear flow ($|\Omega_E| = 1$) showed good qualitative and quantitative agreement with the theoretical results [132]. The measured mean Sherwood number was 2.65, which is close to the corresponding asymptotic value (4.11.4).

It follows from the expressions for the stream function that for $|\Omega_E| > 1$ all streamlines are closed and surround the cylinder. This case is characterized by small values of the Sherwood number and was considered in [343]. This means that if there are only closed streamlines in the flow, then practically no mass is transferred to the surface of the cylinder.

### 4.11-3. Heat Exchange Between Cylindrical Bodies and Liquid Metals

*Statement of the problem.* In the theory of heat exchange between liquid metals ($Pr \ll 1$), the fluid field is usually considered on the basis of the ideal fluid model [48], since the hydrodynamic boundary layer lies deep inside the thermal boundary layer. In this case, generally speaking, the Peclet number need not be sufficiently large for the thermal boundary layer approximation to be applicable.

Let us consider a plane problem on heat exchange between a cylinder with arbitrary boundary $\Gamma$ of the cross-section in a plane-parallel flow of an ideal incompressible fluid with velocity $U_i$ normal to the cylinder axis. The temperature of the cylinder is supposed to be constant and equal to $T_s$, and the temperature of the fluid at infinity is equal to $T_i$. We shall use rectangular coordinates $X, Y$ with the $X$-axis directed along the flow (Figure 4.3).

We analyze this problem following the lines of [49].

Let $\Phi$ and $\Psi$ be the potential and the stream function of the potential flow of a fluid. Since $\Phi$ and $\Psi$ are determined up to additive constants, we can assume

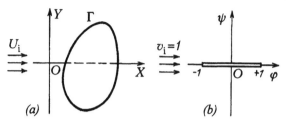

**Figure 4.3.** Heat exchange between a cylinder of an arbitrary shape and a translational flow: (a) the original system of rectangular coordinates; (b) the plane of the new variables $\varphi$, $\psi$

that $\Psi = 0$ and $-\varphi_0 < \Phi < \varphi_0$ on $\Gamma$. We write $\varphi = \Phi/\varphi_0$ and $\psi = \Psi/\varphi_0$ and introduce dimensionless variables by formulas (3.1.32), where $a = \varphi_0/U_i$ is the characteristic length.

The dimensionless fluid velocity components can be expressed via $\varphi$ and $\psi$ by

$$v_x = \frac{\partial\varphi}{\partial x} = -\frac{\partial\psi}{\partial y}, \qquad v_y = \frac{\partial\varphi}{\partial y} = \frac{\partial\psi}{\partial x}. \tag{4.11.7}$$

The function $\varphi$ satisfies the Laplace equation $\Delta\varphi = 0$ with the boundary conditions

$$\frac{\partial\varphi}{\partial n} = 0 \quad \text{on} \quad \Gamma; \qquad \frac{\partial\varphi}{\partial x} \to 1 \quad \text{as} \quad x^2 + y^2 \to \infty, \tag{4.11.8}$$

where $\partial/\partial n$ is the derivative along the normal to the surface of the cylinder.

For a given contour $\Gamma$, one can find $\varphi$ and $\psi$ as some functions of $x$ and $y$ by using methods of the theory of functions of a complex variable [26, 261, 431]; in the sequel, these functions are supposed to be known.

**Exact solution.** Taking into account (4.11.7), we obtain the equation and the boundary conditions for temperature in the dimensionless variables,

$$\frac{\partial\varphi}{\partial x}\frac{\partial T}{\partial x} + \frac{\partial\varphi}{\partial y}\frac{\partial T}{\partial y} = \frac{1}{\text{Pe}_T}\Delta T, \qquad \text{Pe}_T = \frac{\varphi_0}{\chi},$$
$$T = 1 \quad \text{on} \quad \Gamma; \qquad T \to 0 \quad \text{as} \quad x^2 + y^2 \to \infty. \tag{4.11.9}$$

In problem (4.11.9), we pass from the coordinates $x$, $y$ to the new coordinates $\varphi$, $\psi$ (the Boussinesq variables). By using the differential relations (4.11.7) between $\varphi$ and $\psi$ and by applying some transformations, we obtain [53]

$$\text{Pe}_T\frac{\partial T}{\partial\varphi} = \frac{\partial^2 T}{\partial\varphi^2} + \frac{\partial^2 T}{\partial\psi^2};$$
$$\psi = 0, \quad T = 1 \; (|\varphi| < 1); \qquad \varphi^2 + \psi^2 \to \infty, \quad T \to 0. \tag{4.11.10}$$

Thus, in the plane of the variables $\varphi$, $\psi$, we arrive at the problem on convective heat transfer from a heated plane of length 2 to a longitudinal flow of heat-conducting ideal fluid with velocity $v_i = 1$ (Figure 4.3).

We make the substitution

$$u = T \exp\left(-\tfrac{1}{2} \operatorname{Pe}_T \varphi\right). \tag{4.11.11}$$

As a result, problem (4.11.10) acquires the form

$$\frac{\partial^2 u}{\partial \varphi^2} + \frac{\partial^2 u}{\partial \psi^2} = \frac{1}{4} \operatorname{Pe}_T^2 u;$$

$$\psi = 0, \quad u = \exp\left(-\tfrac{1}{2} \operatorname{Pe}_T \varphi\right) \quad (|\varphi| < 1); \tag{4.11.12}$$

$$\varphi^2 + \psi^2 \to \infty, \quad u \to 0.$$

In the elliptic coordinates $\varsigma$, $\eta$ introduced by the formulas (see Supplement 3.6)

$$\varphi = \cosh \varsigma \cos \eta, \qquad \psi = \sinh \varsigma \sin \eta, \tag{4.11.13}$$

the general solution of problem (4.11.12) satisfying the decay condition at infinity can be represented by the sum

$$u = \sum_{m=0}^{\infty} \alpha_m \operatorname{ce}_m(\eta, -q) \operatorname{Fek}_m(\varsigma, -q), \qquad q = -\frac{1}{16} \frac{1}{\operatorname{Pe}_T^2}, \tag{4.11.14}$$

where $\operatorname{ce}_m(\eta, -q)$ are the Mathieu functions [30, 284] with the corresponding eigenvalues $h = h_m(q)$ and $\operatorname{Fek}_m(\varsigma, -q)$ are the modified Mathieu functions. One has $\operatorname{ce}_m(\eta, 0) = \cos(2m\eta)$. The Mathieu functions are tabulated in [284, 308]. The modified Mathieu functions are described in detail in [30, 284].

By passing to the variables $\varsigma$, $\eta$ (4.11.13) in the boundary condition at $\psi = 0$ in (4.11.12), by taking into account the representation (4.11.14), and by expanding this condition in a series with respect to the functions $\operatorname{ce}_m(\eta, -q)$, one can find the coefficients $\alpha_m$.

The corresponding expressions for temperature and the dimensionless total heat flux to the surface of the cylinder were obtained in [49]; we do not write them out here because of their awkwardness. Let us cite the most important final results [49] that can be used in practice.

**Heat flux.** At small Peclet numbers, the dimensionless total heat flux, up to terms of the order of $\operatorname{Pe}_T^2$ inclusively, is given by

$$I_T = -4\pi \left(1 + \frac{9}{64} \operatorname{Pe}_T^2\right) \left[\ln\left(\frac{\gamma \operatorname{Pe}_T}{8}\right)\right]^{-1} - \frac{\pi}{2} \operatorname{Pe}_T^2, \tag{4.11.15}$$

where $\ln \gamma = 0.5772\ldots$ is the Euler constant and $\operatorname{Pe}_T = \varphi_0 / \chi$.

At large Peclet numbers, we have the asymptotics

$$I_T = 4(2 \operatorname{Pe}_T / \pi)^{1/2}, \tag{4.11.16}$$

which corresponds to the thermal boundary layer approximation.

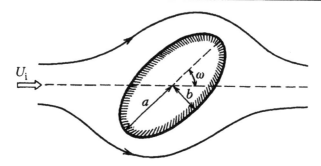

**Figure 4.4.** A noncirculatory ideal fluid flow past an elliptic cylinder

The heat flux in the entire range of Peclet numbers can be approximated by the expression

$$I_T = -\oint \frac{\partial T}{\partial n}\, d\Gamma = 4\pi \frac{|F|^{-1.3} + 20\,G^{1.02}}{|F|^{-2.3} + 20\,G^{0.02}}, \qquad (4.11.17)$$

where the auxiliary functions $F$ and $G$ are given by the formulas

$$F = \left(1 + \tfrac{9}{64}\,\mathrm{Pe}_T^2\right)\left[\ln\left(\tfrac{1}{8}\gamma\,\mathrm{Pe}_T\right)\right]^{-1} + \tfrac{1}{8}\,\mathrm{Pe}_T^2, \qquad G = (2\,\mathrm{Pe}_T/\pi^3)^{1/2}.$$

The function (4.11.17) for $\mathrm{Pe}_T \ll 1$ and $\mathrm{Pe}_T \gg 1$ behaves as the corresponding exact asymptotic solutions (4.11.15) and (4.11.16). The maximum error of (4.11.17) does not exceed 2% for all Peclet numbers.

*Elliptic cylinder.* By way of example, let us consider heat exchange for an elliptic cylinder with semiaxes $a$ and $b$ whose surface is described by the equation $(X/a)^2 + (Y/b)^2 = 1$, $a \ge b$. We assume that the velocity at infinity forms an angle $\omega$ with the major semiaxis (Figure 4.4).

Let us introduce new coordinates $\sigma$, $\nu$ as follows:

$$X = \sigma\left(1 + \frac{a^2 - b^2}{4\sigma^2}\right)\cos\nu, \qquad Y = \sigma\left(1 - \frac{a^2 - b^2}{4\sigma^2}\right)\sin\nu. \qquad (4.11.18)$$

The velocity potential and the stream function for a noncirculatory flow of ideal fluid past an elliptic cylinder has the form [26]

$$\Phi = -U_i\left[\sigma + \frac{(a+b)^2}{4\sigma}\right]\cos(\nu + \omega),$$

$$\Psi = -U_i\left[\sigma - \frac{(a+b)^2}{4\sigma}\right]\sin(\nu + \omega). \qquad (4.11.19)$$

For $\sigma = (a+b)/2$, the potential on the surface $\Gamma$ of the cylinder is given by $\Phi = -U_i(a+b)\cos\nu$. On the contour $\Gamma$ in the plane $\Phi$, $\Psi$, we have the relations $\Psi = 0$ and $-\varphi_0 < \Phi < \varphi_0$, where $\varphi_0 = U_i(a+b)$. Therefore, the dimensionless

TABLE 4.6
The thermal Peclet number for a noncirculatory flow past bodies of various shape

| No | Shape of the body | Peclet number |
|----|-------------------|---------------|
| 1 | Plate of length $2a$ | $\mathbf{Pe}_T = aU_i/\chi$ |
| 2 | Circular cylinder of radius $a$ | $\mathbf{Pe}_T = 2aU_i/\chi$ |
| 3 | Elliptic cylinder with semiaxes $a$ and $b$ | $\mathbf{Pe}_T = (a+b)U_i/\chi$ |

total heat flux to the surface of an elliptic cylinder can be calculated by formulas (4.11.15)–(4.11.17), where $\mathbf{Pe}_T = U_i(a+b)/\chi$, where $\chi$ is the thermal diffusivity. Obviously, the result for $I_T$ is independent of the orientation of the cylinder in the flow.

Table 4.6 gives expressions for the thermal Peclet number occurring in (4.11.15)–(4.11.17) for noncirculatory flow of an ideal fluid past bodies of various shape.

# 4.12. Transient Mass Transfer in Steady-State Translational and Shear Flows

### 4.12-1. Statement of the Problem

We consider a laminar steady-state flow past a solid spherical particle (drop or bubble) of radius $a$ and study transient mass transfer to the particle surface. At the initial time $t = 0$, the concentration in the continuous phase is constant and equal to $C_i$, whereas for $t > 0$ a constant concentration $C_s$ is maintained on the particle surface.

In the particle-centered spherical coordinate system $(R, \theta, \varphi)$, the nonstationary problem for the concentration $C$ in dimensionless variables comprises the convective diffusion equation

$$\frac{\partial c}{\partial \tau} + \mathbf{Pe}(\mathbf{v} \cdot \nabla)c = \Delta c \qquad (4.12.1)$$

and the initial and boundary conditions

$$\tau = 0, \quad c = 0; \qquad r = 1, \quad c = 1; \qquad r \to \infty, \quad c \to 0, \qquad (4.12.2)$$

where $c = (C_i - C)/(C_i - C_s)$, $\tau = Dt/a^2$, $r = R/a$, $\mathbf{Pe} = aU/D$, and $U$ is the characteristic flow velocity. The steady-state flow field $\mathbf{v}$ is given.

---

### 4.12-2. Spherical Particles and Drops at High Peclet Numbers

In what follows we restrict ourselves to the case of high Peclet numbers, in which there are no closed streamlines in the flow.

The diffusion boundary layer in problem (4.12.1), (4.12.2) is first adjacent to the particle surface and then rapidly spreads over the flow region with the subsequent exponential relaxation to a steady state. The characteristic relaxation time $\tau_r$ is of the order of $Pe^{-2/3}$ for a solid particle and of the order of $Pe^{-1}$ for bubbles and drops of moderate viscosity.

The following approximate formula for the time dependence of the mean Sherwood number for an arbitrary steady-state flow past spherical particles, drops, and bubbles was obtained by the method of asymptotic analogies in [357]:

$$\frac{Sh}{Sh_{st}} = \sqrt{\coth(\pi\, Sh_{st}^2\, \tau)}, \qquad (4.12.3)$$

where $Sh_{st} = \lim_{\tau \to \infty} Sh$ is the Sherwood number for the steady-state diffusion mode; $Sh_{st}$ depends on the Peclet number and can be determined by the solution of the corresponding stationary problem (4.4.1), (4.3.2), (4.3.3); $\tau = Dt/a^2$.

Equation (4.12.3) gives a valid asymptotic result for any flow field in both limit cases $\tau \to 0$ and $\tau \to \infty$.

For a linear shear flow past a bubble, Table 4.4 gives $Sh_{st} = \sqrt{\dfrac{3\,Pe}{2\pi}}$. By substituting this value into (4.12.3), we arrive at the exact expression

$$Sh = \sqrt{\frac{3\,Pe}{2\pi} \coth\left(\frac{3}{2}\,Pe\,\tau\right)}, \qquad (4.12.4)$$

which was obtained in the diffusion boundary layer approximation in [170]. Note that Eq. (4.12.4) served as a basis for deriving the general dependence (4.12.3) in [354, 357].

Table 4.7 presents $Sh_{st}$ for various flows past spherical particles, drops, and bubbles of radius $a$. The parameter $\beta$ is the ratio of the dynamic viscosity of the drop to that of the ambient fluid and varies in the range $0 \le \beta \le 2$ (the value $\beta = 0$ corresponds to a gas bubble).

In the case of nonstationary mass transfer in a steady-state translational Stokes flow past a spherical drop with limiting resistance of the continuous phase, the steady-state value $Sh_{st}$ is presented in the first row of Table 4.7. By substituting this value into (4.12.3), we obtain

$$Sh = \left[\frac{2\,Pe}{3\pi(\beta + 1)} \coth\left(\frac{2}{3}\frac{Pe\,\tau}{\beta + 1}\right)\right]^{1/2}. \qquad (4.12.5)$$

The comparison with the results of [84, 271, 410] carried out in [359] shows that the maximum error of Eq. (4.12.5) does not exceed 0.7%. Note that the

TABLE 4.7
The expressions for $\mathsf{Sh}_{st}$ in Eq. (4.12.3) for various types
of flow past spherical particles, drops, and bubbles

| No | Disperse phase | Flow type | $\mathsf{Sh}_{st}$ | Notation, $\mathrm{Pe} = aU/D$ |
|----|----------------|-----------|--------------------|--------------------------------|
| 1 | Drop, bubble | Translational Stokes flow | $\left[\dfrac{2\,\mathrm{Pe}}{3\pi(\beta+1)}\right]^{1/2}$ | $U = U_i$ is the fluid velocity at infinity |
| 2 | Drop, bubble | Arbitrary straining linear shear flow $(G_{ij} = G_{ji})$ | $0.62\left(\dfrac{\mathrm{Pe}}{\beta+1}\right)^{1/2}$ | $U = a\left(\sum\limits_{i,j=1}^{3} G_{ij}G_{ij}\right)^{1/2}$, $G_{ij}$ are the shear matrix coefficients |
| 3 | Bubble | Laminar translational flow at high Reynolds numbers | $\left(\dfrac{2\,\mathrm{Pe}}{\pi}\right)^{1/2}$ | $U = U_i$ is the fluid velocity at infinity |
| 4 | Solid particle | Translational Stokes flow | $0.624\,\mathrm{Pe}^{1/3}$ | $U = U_i$ is the fluid velocity at infinity |
| 5 | Solid particle | Arbitrary straining linear shear flow $(G_{ij} = G_{ji})$ | $0.9\,\mathrm{Pe}^{1/3}$ | $U = a\left(\sum\limits_{i,j=1}^{3} G_{ij}G_{ij}\right)^{1/2}$, $G_{ij}$ are the shear matrix coefficients |

results of [84, 271, 410] are written in the form of a rather complicated integral, which cannot be represented in a simple analytical form similar to (4.12.5).

Approximate formulas for other transient problems can be obtained in a similar manner.

Table 4.8 presents a comparison of the mean Sherwood numbers calculated according to Eq. (4.12.3) with available data for various flows past spherical drops, bubbles, and solid particles at high Peclet numbers (in this table, we use the abbreviation DBLA for "diffusion boundary layer approximation").

### 4.13-3. Spherical Particles and Drops at Arbitrary Peclet Numbers

To calculate the mean Sherwood number for an arbitrary laminar flow past spherical particles, drops, and bubbles in the entire range of Peclet numbers, one can use the interpolation formula

$$\mathsf{Sh} = (\mathsf{Sh}_{st} - 1)\sqrt{\coth\left[\pi(\mathsf{Sh}_{st} - 1)^2\tau\right]} + 1. \qquad (4.12.6)$$

Let us consider the behavior of this function in various limit cases. Since $\mathsf{Sh}_{st} \to 1$ as $\mathrm{Pe} \to 0$, we see that Eq. (4.12.6) yields the exact result (4.3.14)

TABLE 4.8

Maximum error of Eq. (4.12.3) for various types of
flow past spherical drops, bubbles, and solid particles

| No | Dispersed phase | Flow type | Solution method | Error, % | Sources |
|---|---|---|---|---|---|
| 1 | Drop, bubble | Axisymmetric shear Stokes flow | Analytical, DBLA | 0 | [170] |
| 2 | Drop, bubble | Translational Stokes flow | Analytical, DBLA | 0.7 | [84, 271, 410] |
| 3 | Drop, bubble | Two-dimensional shear Stokes flow | Analytical, DBLA | 1.8 | [354] |
| 4 | Bubble | Laminar translational flow at high Reynolds numbers | Analytical, DBLA | 0.7 | [84, 410] |
| 5 | Bubble | Axisymmetric shear flow at high Reynolds numbers | Analytical, DBLA | 0 | [359] |
| 6 | Drop, bubble | Flow caused by an electric field | Analytical, DBLA | 0 | [293] |
| 7 | Solid particle | Translational flow of an ideal (inviscid) liquid | Analytical, DBLA | 0.7 | [84, 410] |
| 8 | Solid particle | Translational Stokes flow | Interpolation of numerical and analytical results | 1.4 | [94] |
| 9 | Solid particle | Translational Stokes flow | Finite-difference numerical method (at $Pe = 500$) | 4 | [68] |

for a stagnant medium. As $Pe \to \infty$, we have $Sh_{st} \to \infty$, and the expression
(4.12.6) passes into (4.12.3). For small $\tau$, Eq. (4.12.6) yields the exact result
$Sh \approx (\pi\tau)^{-1/2}$. As $\tau \to \infty$, we have $Sh \to Sh_{st}$, which follows from (4.12.6).

### 4.12-4. Nonspherical Particles, Drops, and Bubbles

The dependence (4.12.3) can also be used to estimate the intensity of transient
mass transfer for nonspherical particles, drops, and bubbles at $Pe \gg 1$. In this
case, all dimensionless variables $\tau$, $Sh$, $Sh_{st}$, and $Pe$ must be defined on the
basis of the same characteristic length $a$. Under this condition, the expression
(4.12.3) provides valid asymptotic results for small as well as large times.

Equation (4.12.3) can be rewritten as follows:

$$\frac{Sh}{Sh_{st}} = \sqrt{\coth\left(\frac{Sh_{st}^2}{Sh_{in}^2}\right)}, \qquad (4.12.7)$$

where $\text{Sh}_{\text{in}}$ and $\text{Sh}_{\text{st}}$ are the leading terms of asymptotic expansions of the mean Sherwood number as $\tau \to 0$ and $\tau \to \infty$, respectively, that is,

$$\lim_{\tau \to 0}(\text{Sh}/\text{Sh}_{\text{in}}) = 1, \qquad \lim_{\tau \to \infty}(\text{Sh}/\text{Sh}_{\text{st}}) = 1.$$

Later on in Section 5.6 it will be shown that the expression (4.12.7) is also suitable for the description of a broad class of more complicated nonlinear problems of transient diffusion boundary layer.

# 4.13. Qualitative Features of Mass Transfer Inside a Drop at High Peclet Numbers

### 4.13-1. Limiting Diffusion Resistance of the Disperse Phase

*Statement of the problem. Preliminary remarks.* Let us consider the transient convective mass and heat transfer between a spherical drop of radius $a$ and a translational Stokes flow where the resistance to the transfer exists only in the disperse phase. We assume that at the initial time $t = 0$ the concentration inside the drop is constant and equal to $C_0$, whereas for $t > 0$ the concentration on the interface is maintained constant and equal to $C_s$.

Mass transfer inside a drop is described by Eq. (4.12.1) and the first two conditions (4.12.2). The fluid velocity field $\mathbf{v} = (v_r, v_\theta)$ at low Reynolds numbers is given by the Hadamard–Rybczynski stream function and, in the dimensionless variables, has the form

$$\psi = -\frac{1}{4(\beta+1)}r^2(1-r^2)\sin^2\theta;$$

$$v_r = \frac{1}{r^2\sin\theta}\frac{\partial\psi}{\partial\theta}, \quad v_\theta = -\frac{1}{r\sin\theta}\frac{\partial\psi}{\partial r}.$$

To analyze the asymptotic solution, it is convenient to introduce an auxiliary Peclet number by setting

$$\text{Pe}_\beta = \frac{\text{Pe}}{\beta+1}, \quad \text{where} \quad \text{Pe} = \frac{aU_i}{D} \quad \text{and } U_i \text{ is the incoming flow velocity.}$$

Further, we assume that the condition $\text{Pe}_\beta \gg 1$ holds. To be more descriptive in the physical interpretation of the process, we shall use a terminology corresponding to complete absorption of the substance at the drop surface for $C_s = 0$.

The inner problem of convective mass and heat transfer is essentially different from the similar outer problem, primarily, by the streamline pattern. This leads to a corresponding qualitative distinction between the dynamics of processes of transient mass transfer inside and outside a drop. In the outer problem considered in Section 4.12, all streamlines are open. The lines near the flow axis carry the

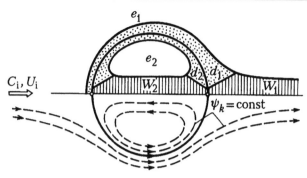

**Figure 4.5.** Flow pattern inside the drop and the concentration field structure; here $d_1$ and $d_2$ are regions of the diffusion boundary layer, $W_1$ and $W_2$ are regions of the diffusion wake, $e_1$ and $e_2$ are regions of the stream core (when the resistance of the disperse phase is limiting; we consider only regions inside the drop)

nondepleted concentration from infinity, pass near the drop surface (here the solution is substantially depleted by the complete absorption of the reactant at the drop surface), and return to infinity. Since the concentrations at infinity and on the drop surface are kept constant, the solution of the outer problem tends to a steady-state profile (4.6.15) of the stationary diffusion boundary layer, which is reached exponentially rapidly.

In the inner problem (see Figure 4.5), all flow streamlines are closed. Therefore, a substance dissolved in the liquid is partially absorbed near the drop surface, and the remaining part goes inward the drop along the streamlines near the flow axis. Here the solution is enriched by mixing with the liquid inside the drop. But no complete renewal of the solution is produced, since the concentration in the drop volume decreases due to the absence of the reactant influx. The streamlines issuing from the near-axis region begin to pass once more near the surface of the drop, where the solution is depleted even more than before (since it was not completely recovered, etc.). Thus, all the substance dissolved at the initial moment gets reacted on the drop surface as $\tau \to \infty$.

A more detailed analysis [352] shows that mass transfer inside the drop at high Peclet numbers develops through three consecutive stages. Each of them has its special features and different duration.

*First (fast) stage of the process.* This stage is characterized by the formation of unstable diffusion boundary layer near the drop surface; its thickness is proportional to $Pe^{-1/2}$. At this stage, the inner diffusion boundary layer is qualitatively analogous to the self-similar transient boundary layer of the corresponding outer problem. Here the mean Sherwood number can be calculated by formula (4.12.5), and the results of [84, 271, 410] are valid for the concentration fields. The transient boundary layer rapidly enters an intermediate stationary regime, where the graph of the mean Sherwood number has a typical flat part beginning at $\tau \approx 2/Pe_\beta$. The initial stage of the process is determined by the time interval on which the diffusion boundary layer approximation (first transient and then steady-state) is valid with the nondepleted input concentration. The concentration in the drop core at this stage is equal to the nonperturbed concentration at the initial time.

The inner diffusion boundary layer generates an inner diffusion wake of thickness proportional to $\mathrm{Pe}^{-1/4}$ near the flow axis. In the diffusion wake, the dissolved substance from the endpoint of the boundary layer is carried by the fluid along the streamlines without change. Since the flow velocity is finite, in the beginning, for small times $\tau < \tau_*$, a nondepleted concentration is carried to the region of the front stagnation point of the boundary layer from the bulk of fluid. This occurs until the depleted solution from the endpoint of the boundary layer that happened to be inside the diffusion wake meets the origin of the diffusion boundary layer. The results of [66] show that the characteristic time $\tau_*$ for transfer of the reactant in the diffusion wake of a drop is of the order of $(\ln \mathrm{Pe}_\beta)/\mathrm{Pe}_\beta$, and this defines the scope of the self-similar solution [84, 271, 410], which does not correctly describe the concentration distribution in the diffusion boundary layer for $\tau > \tau_*$ (in view of the changed "onflow" conditions).

More accurate numerical estimates [151] show that the initial stage of the process occurs for $0 \leq \tau \leq 0.5 \,(\ln \mathrm{Pe}_\beta)/\mathrm{Pe}_\beta$.

*Intermediate stage of the process.* At this stage the diffusion boundary layer still exists near the interface, and the concentration within the drop core is constant and equal to the initial value. However, as was mentioned above, the concentration at the entrance to the boundary layer is nonuniform; one can find it from the conditions of matching with the concentration field in the inner diffusion layer. The problem is complicated by the fact that this concentration, in turn, depends on the concentration distribution in the diffusion boundary layer. In view of these effects, an integral equation was derived in [151, 355] for $\mathrm{Pe}_\beta \geq 10^4$ in order to obtain the "onflow" conditions at the entrance to the boundary layer, which leads to a nonself-similar solution.

Figure 4.6 shows the results of calculation of the mean Sherwood number obtained in [151] by a numerical solution of the corresponding integral equation for various values of dimensionless time and the Peclet number. One can see that, after the inner diffusion wake has been developed, the complete substance flux to the inner surface of the drop decreases rapidly.

At the intermediate stage of the process, the diffusion wake interacts with the boundary layer and strongly "erodes" it, producing an increase in the boundary layer thickness (here the boundary layers for the inner and outer problems differ considerably). Gradually, as a result of absorption of the substance dissolved in the liquid on the interface, the diffusion boundary layer spreads over the entire drop and begins to decay.

We note that for $\mathrm{Pe}_\beta = 10^2$ to $10^3$ the intermediate stage of the process is not very pronounced.

*Final (slow) stage of the process.* At this stage, as a result of a multiple circulation of the fluid along closed streamlines, the concentration has been leveled out and become uniform on the streamlines (each streamline has its own concentration, which depends on $\tau$). The diffusion boundary layer and the

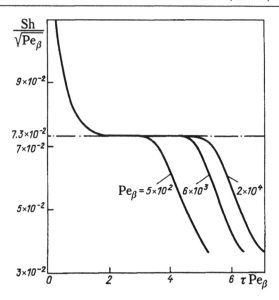

**Figure 4.6.** The mean Sherwood number against dimensionless time for the inner problem

diffusion wake have dispersed completely and ceased to exist.

Here, the solution can be found in the form of a regular asymptotic expansion (4.8.12) in powers of the small parameter $Pe^{-1}$. The leading term of this series satisfies the equation derived in [238]. The numeric solution leads to the following formula for the mean-volume dimensionless concentration inside the drop:

$$\langle c \rangle = 1 - \frac{3}{2} \sum_{k=1}^{\infty} A_k \exp(-\lambda_k \tau); \qquad (4.13.1)$$

$A_1 = 0.4554;$   $A_2 = 0.0654;$   $A_3 = 0.0542;$   $A_4 = 0.0412;$   $A_5 = 0.0038;$
$\lambda_1 = 26.844;$   $\lambda_2 = 137.91;$   $\lambda_3 = 315.66;$   $\lambda_4 = 724.98;$   $\lambda_5 = 1205.2.$

Here the coefficients $A_k$ and $\lambda_k$ are given according to the data in [68]; for $k = 1, 2,$ close values of these coefficients were calculated earlier in [238].

The expression for the mean concentration can be used for $Pe_\beta \geq 10^2$ starting from $\tau \geq 5 \times 10^{-4}$.

It is important to note that although formula (4.13.1) was derived for the Stokes flow regime ($Re \to 0$), it can be used successfully also for high Reynolds numbers ($Re \leq 10^2$) provided that the shape of the drop is nearly spherical. A review of experimental data on mass transfer under limiting resistance of the continuous phase in liquid–liquid systems for $10^2 \leq Re \leq 4 \times 10^2$ is presented in [68]. Comparison shows that the experimental data in the range $4 \times 10^{-4} \leq \tau \leq 10^{-1}$ are in a good agreement with the results of calculation by (4.13.1) (the extraction percentage is between 10% and 70%). The results of numerical solution of this problem [67] are in a good agreement with formula (4.13.1).

## 4.13-2. Comparable Diffusion Phase Resistances

Let us consider a transient solute concentration field in a liquid outside and inside a spherical drop of radius $a$ moving at a constant velocity $U_i$ in an infinite fluid medium. We assume that the fluid velocity fields for the continuous and disperse phases are determined by the Hadamard–Rybczynski solution [177, 420], obtained for low Reynolds numbers. The concentration far from the drop is maintained constant and equal to $C_i$. At the initial time $t = 0$, the concentration outside the drop is everywhere uniform and is equal to $C_i$; inside the drop, it is also uniform, but is equal to $C_0$.

The following boundary conditions hold at the drop surface:

$$C_2 = F(C_1), \qquad D_1 \frac{\partial C_1}{\partial R} = D_2 \frac{\partial C_2}{\partial R} \quad \text{for} \quad R = a, \qquad (4.13.2)$$

where the subscript 1 corresponds to the continuous phase and 2, to the disperse phase. The first condition in (4.13.2) is the condition of phase equilibrium on the drop surface. It is usually assumed [84, 271, 410], that the function $F$ is linear in the concentration (Henry's law), $F(C_1) = \alpha C_1$, where the distribution coefficient $\alpha$ depends on physical properties of the fluids outside and inside of the drop. The extreme cases of limiting resistance for the disperse and continuous phases correspond to $\alpha \to 0$ and $\alpha \to \infty$. If the values of the phase resistances are comparable, then we have $\alpha \sim 1$. It was shown in [68] that sometimes the power dependence $F(C_1) = \alpha C_1^m$ must be used, where the exponent $m$ lies between 0.5 and 2.0. The second condition in (4.13.2) accounts for the continuity of diffusion fluxes at the drop surface.

The concentration distribution is described by Eq. (3.1.1), where the variables $C$, $\mathbf{V}$, $D$ are marked by the subscript 1 for the continuous phase (for $R > a$) and by the subscript 2 for the disperse phase (for $R < a$).

It was shown in [66, 355] that at high Peclet numbers and for comparable values of phase resistances, the diffusion process can be characterized by three stages with different mechanisms of mass transfer. The duration of these stages is the same as in the case of limiting resistance of the disperse phase. At the initial stage of the process, transient diffusion boundary layers (qualitatively similar to each other) are formed on either side of the drop surface. The inner boundary layer generates a diffusion wake near the stream axis (see Figure 4.5). At the intermediate stage of the process, the developing inner diffusion wake begins to interact with the boundary layer and strongly "erodes" it (here the boundary layers inside and outside the drop are substantially different, and as a result, the thickness of the inner boundary layer gradually increases considerably). At the final stage of the process, the concentration field is deformed to a greater extent, so that practically the boundary layers cease to exist. At the same time, the concentration outside the drop becomes constant and equal to the nonperturbed concentration at infinity $C_i$, whereas an essentially transient process sets in inside the drop, where the concentration on each given streamline is practically leveled out (as a result of multiple circulation of the fluid along closed lines) and

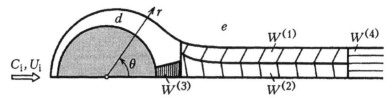

**Figure 4.7.** Partition of the concentration field outside the drop into regions with different structure of asymptotic solutions

mass transfer is produced by molecular diffusion in the direction normal to the streamlines.

In view of the discussion above, we conclude that at the final stage of the process the concentration on the interface is constant by virtue of the first boundary condition in (4.13.2) and is equal to

$$C = C_s, \quad \text{where} \quad C_s = F(C_i). \tag{4.13.3}$$

The mean concentration inside the drop can be calculated by formula (4.13.2), where the dimensionless concentration is given by $c = (C_0 - C)/(C_0 - C_s)$, and the value of $C_s$ is found in (4.13.3).

# 4.14. Diffusion Wake. Mass Exchange of Liquid With Particles or Drops Arranged in Lines

### 4.14-1. Diffusion Wake at High Peclet Numbers

The method of matched asymptotic expansions (with respect to large Peclet number) was used in [166, 167, 446] to study problems of steady-state convective diffusion to a solid sphere [446] and a drop [167] in a translational Stokes flow for diffusion regime of reaction on the interface. Six regions with different structure of asymptotic solutions corresponding to different mass transfer mechanisms were singled out in the flow (Figure 4.7). Let us give a brief qualitative description of these regions by using a dimensionless spherical coordinate system $r$, $\theta$, relative to the center of a solid particle (drop).

In the outer region $e$, the concentration is constant and is equal to its undisturbed values at infinity.

In the diffusion boundary layer $d$, in the mass transfer equation one can neglect molecular tangential diffusion transfer compared with the diffusion in the radial direction; the convective terms are retained (but somewhat simplified by linearization near the interface). The concentration distribution in that region was obtained earlier in Section 4.6.

Four subregions $W^{(i)}$ ($i = 1, 2, 3, 4$) lying behind a drop or a solid particle near the flow axis constitute the diffusion wake region (Figure 4.7).

In the convective-boundary layer region $W^{(1)}$ of the diffusion wake, molecular diffusion can be neglected. The concentration here depends only on the stream function and is constant along the streamlines, its values being equal to those at the exit from the diffusion boundary layer.

In the internal wake region $W^{(2)}$, one can neglect molecular mass transfer in the radial direction.

In the rear stagnation region $W^{(3)}$, the mass transfer equation can be somewhat simplified. In doing so, one must take into account convective terms as well as the radial and tangential molecular diffusion components.

In the mixture region $W^{(4)}$, the convective terms and tangential mass transfer due to molecular diffusion play the primary role (molecular diffusion along the radial coordinate can be neglected).

It is worth mentioning that in all diffusion wake regions $W^{(i)}$ ($i = 1, 2, 3, 4$), one must take into account convective mass transfer due to the motion of the fluid. In the regions $W^{(i)}$ ($i = 2, 3, 4$), the molecular diffusion component normal to the streamlines plays an important role.

In the case of a drop (or a bubble), the concentration distribution was obtained in a closed form in all diffusion wake regions $W^{(i)}$ [166, 167], and in the case of solid sphere, closed-form analytical expressions were obtained in all regions except for the rear stagnation region [166, 446]. The concentration field in $W^{(3)}$ for the cases of a solid sphere and a circular cylinder was analyzed by numerical methods in [309].

The order of magnitude of the characteristic dimensions of diffusion wake regions past a spherical drop and a solid particle in a translational flow is shown in Table 4.9. These estimates remain valid at moderate Reynolds numbers, when the stagnation zones past a drop or a solid particle are absent.

For a solitary drop or solid particle, the wake diffusion region contributes to mean Sherwood number starting from the third term of the asymptotic expansion as Pe → ∞.

It was shown in [166] that in the two-dimensional problem of mass exchange between cylindrical bodies and a viscous flow, the diffusion layer consists of only two subregions $W^{(3)}$ and $W^{(4)}$ of total length $L \sim a\,\text{Pe}^{-1/9}$ (as Pe → ∞); the regions $W^{(1)}$ and $W^{(2)}$ are absent. The diffusion wake in the vicinity of the stagnation lines on the surface of a solid particle has a similar structure.

## 4.14-2. Diffusion Interaction of Two Particles or Drops

*Solid particles.* Let us consider stationary diffusion to two axisymmetric solid particles arranged one after the other on the axis of a translational Stokes flow. We assume that the solid particles are symmetric with respect to some plane $z = \text{const}$ (see Figure 4.8) and each of them has only two stagnation points on the surface, which lie on the flow axis (closed streamlines are absent). The surfaces of solid particles completely absorb the solute.

At high Peclet numbers, the diffusion flux to the surface of the first solid particle can be found by solving the conventional diffusion boundary layer equation; the presence of the second solid particle influences only by changing the velocity

TABLE 4.9

Order of magnitude of dimensionless (related to the radius of drops
or solid particles) characteristic sizes of regions of diffusion
wake in translational Stokes flow at high Peclet number

| Diffusion wake regions | Dimensionless distance from the interface, $y = r - 1$ | Dimensionless distance from the flow axis, $h$ |
|---|---|---|
| *Bubbles or drops of moderate viscosity $0 \leq \beta \leq 1$* | | |
| Convective-boundary layer regions $W^{(1)}$ | $O(\mathrm{Pe}^{-1/2}) \leq y \leq O(\mathrm{Pe}^{1/2})$ | $O(\mathrm{Pe}^{-1/2}) \leq h \leq O(\mathrm{Pe}^{-1/4})$ |
| Inner region $W^{(2)}$ | $O(\mathrm{Pe}^{-1/2}) \leq y \leq O(\mathrm{Pe}^{1/2})$ | $0 \leq h \leq O(\mathrm{Pe}^{-1/2})$ |
| Region of rear stagnation point $W^{(3)}$ | $0 \leq y \leq O(\mathrm{Pe}^{-1/2})$ | $0 \leq h \leq O(\mathrm{Pe}^{-1/2})$ |
| Mixture region $W^{(4)}$ | $y \geq O(\mathrm{Pe}^{1/2})$ | $0 \leq h \leq O(\mathrm{Pe}^{-1/4})$ |
| *Solid particles* | | |
| Convective-boundary layer region $W^{(1)}$ | $O(\mathrm{Pe}^{-1/3}) \leq y \leq O(\mathrm{Pe}^{1/3})$ | $O(\mathrm{Pe}^{-1/2}) \leq h \leq O(\mathrm{Pe}^{-1/3})$ |
| Inner region $W^{(2)}$ | $O(\mathrm{Pe}^{-1/3}) \leq y \leq O(\mathrm{Pe}^{1/3})$ | $0 \leq h \leq O(\mathrm{Pe}^{-1/2})$ |
| Region of rear stagnation point $W^{(3)}$ | $0 \leq y \leq O(\mathrm{Pe}^{-1/3})$ | $0 \leq h \leq O(\mathrm{Pe}^{-1/3})$ |
| Mixture region $W^{(4)}$ | $y \geq O(\mathrm{Pe}^{1/3})$ | $0 \leq h \leq O(\mathrm{Pe}^{-1/3})$ |

profile near the surface of the first particle. (Mass exchange of the second solid particle exerts no influence on the mass exchange of the first particle.)

Mass exchange of the second solid particle is more complicated. The main role is played by the interaction of the diffusion boundary layer of the second particle with the diffusion wake of the first particle.

The following limit relation for the total diffusion fluxes on the surface of two identical solid particles arranged on the axis of a translational Stokes flow (Figure 4.8) was derived in [169]:

$$\lim_{\mathrm{Pe}\to\infty} \frac{I_2}{I_1} = \lim_{\mathrm{Pe}\to\infty} \frac{\mathrm{Sh}_2}{\mathrm{Sh}_1} = 4^{1/3} - 1 \approx 0.587. \tag{4.14.1}$$

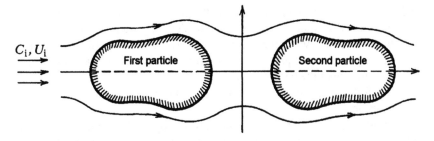

**Figure 4.8.** Translational Stokes flow past two identical solid particles

The passage to the limit in this formula is carried out under a constant distance between the solid particles, and the Peclet number is determined on the basis of the characteristic size of the solid particle.

Relation (4.14.1), in particular, holds for spheres of equal radius arranged on the axis of a translational Stokes flow (the velocity distribution for this case is presented in [179, 463]. It also holds for a three-dimensional Stokes flow past two identical ellipsoids of rotation whose axes are parallel and perpendicular to the undisturbed flow. The direction of the line passing through their centers coincides with the direction of translational flow.

As follows from formula (4.14.1), substantial retardation of mass exchange of the second solid particle takes place compared with that of the first.

***Drops and bubbles.*** For two spherical drops (bubbles) of equal radius arranged one behind the other on the axis of a translational Stokes flow, the following limit equation holds [169]:

$$\lim_{\text{Pe}\to\infty} \frac{I_2}{I_1} = \lim_{\text{Pe}\to\infty} \frac{\text{Sh}_2}{\text{Sh}_1} = 2^{1/2} - 1 \approx 0.414. \qquad (4.14.2)$$

One can see that the interaction of the diffusion wake of the first drop with the boundary layer of the second drop is more intensive than for the case of a solid particle. In addition, the total mass exchange of the second drop with the liquid is less than half that of the first.

Formula (4.14.2) will hold for the case of an noncirculatory flow past two identical solid particles on the axis of a translational flow of an ideal fluid (see Figure 4.8).

### 4.14-3. Chains of Particles or Drops at High Peclet Numbers

***Drops and bubbles.*** Let us consider diffusion to the surfaces of drops (bubbles) arranged one after another on the axis of a translational Stokes flow of a viscous incompressible fluid. In such systems, further referred to as chains, the flow field is arranged so that the singular streamline issuing from an isolated stagnation point on the surface of the first drop enters the surface of the second drop, the singular streamline issuing from the surface of the second drop enters the surface of the third drop, and so on (i.e., the drops are strung on the singular streamline).

Such a situation occurs in practice, say, in the process of extraction of substances from drops and dissolution of gases from bubbles. In particular, it takes place in the extraction process when the drops are introduced into the extraction column at the same points with equal time intervals and in the case of barbotage, for a constant flow rate of the barbotage gas.

In the following, assume that the main resistance to mass exchange is in the continuous phase.

The diffusion boundary layer of any given drop in chain interacts with the diffusion wake of the previous drop (located upstream). The concentration field in it is substantially nonuniform and is depleted because of the absorption of the solute at the surfaces of all preceding drops. By virtue of such interaction, the inner mass exchange will be appreciably retarded (the "shielding" phenomenon) compared with the case of isolated drops.

The concentration distribution in the diffusion boundary layer of each drop in the chain was obtained in [348]. Depending on the distance between the drops, one must distinguish between two situations: (1) the diffusion boundary layer of a given drop interacts with the convective-boundary layer region of the diffusion wake of the previous drop (close interaction); (2) interaction proceeds with the mixture region. In particular, in the case of close interaction of spherical drops (bubbles) of equal radius located at a dimensionless distance $l$: $\{O(1) < l < O(\sqrt{Pe})\}$ (the drop radius is taken to be the characteristic length) one by one on the axis of a uniform Stokes flow, the total diffusion flux to the surface of the $k$th drop in the chain (the numbering starts from the drop going ahead) is determined by the formula [168, 348]

$$I_k = I_1(\sqrt{k} - \sqrt{k-1}).$$ (4.14.3)

It follows from this expression that the total diffusion flux to the second drop is less than half that for the first drop, and we have $I_k/I_1 \to 0$ as $k \to \infty$. The total diffusion flux for all drops of the chain is equal to

$$I_\Sigma = \sum_{i=1}^{k} I_i = I_1\sqrt{k},$$ (4.14.4)

and is substantially lower than the similar total diffusion flux for a system of randomly located drops of equal radius with no diffusion interaction between them (in the latter case, the total fluxes are simply summed, so that $I_\Sigma = I_1 k$).

*Solid particles.* The problem of convective diffusion to a chain of solid reacting particles was studied in [168, 350]. The retardation mechanism (shielding) of mass exchange in a chain of solid particles and the quantitative behavior of such a system are the same as for chains of drops.

In the case of close interaction of reacting solid spherical particles of equal radius located at a dimensionless distance $l$: $\{O(1) < l < O(Pe^{1/3})\}$ one by one on the axis of a translational Stokes flow, the total diffusion flux on the surface of $k$th sphere in the chain is given by the formula [168, 350]

$$I_k = I_1 \left[ k^{2/3} - (k-1)^{2/3} \right]. \tag{4.14.5}$$

The particles located upstream are shielding the subsequent particles. As a result, the total diffusion flux on their surfaces decreases monotonically so that $I_1 > I_2 > \cdots > I_k > I_{k+1} > \cdots$, and the ratio $I_k/I_1$ tends to zero as $k \to \infty$. As follows from (4.14.5), the total diffusion flux on the second sphere is nearly half the total diffusion flux on the first sphere, and that on the seventh sphere is less than one-third of that on the first.

The total diffusion flux on all particles in the chain is calculated with the help of (4.14.5):

$$I_\Sigma = \sum_{i=1}^{k} I_i = I_1 k^{2/3}, \tag{4.14.6}$$

which is substantially less then the total flux calculated without allowance for the interaction between diffusion wakes and boundary layers on solid particles.

## 4.15. Mass and Heat Transfer Under Constrained Flow Past Particles, Drops, or Bubbles

In Section 2.9, various aspects were considered of the hydrodynamics of a constrained flow past a system of particles based on the cell model. Here we briefly describe mass and heat transfer in such systems at high Peclet numbers. We investigate either sufficiently rarefied systems of particles or systems with an irregular structure, where the diffusion interaction of isolated particles can be neglected. (Regular disperse systems, where the interaction between diffusion wakes and boundary layers must be taken into account, were investigated in [172, 365].)

For the sedimentation of rarefied monodisperse systems of spherical particles, drops, or bubbles, the mean Sherwood number can be calculated by using formulas (4.6.8) and (4.6.17), where the Peclet number must be determined on the basis of the constrained flow velocity.

### 4.15-1. Monodisperse Systems of Spherical Particles

To find mass and heat transfer coefficients at high Peclet numbers, just as for isolated particles, it suffices to know the vortex distribution over the surfaces of solid spheres. Therefore, one can use the results of Section 4.6 in the calculations.

Let us consider mass and heat transfer for a monodisperse system of spherical particles of radius $a$ with volume density $\phi$ of the solid phase. We use the fluid velocity field obtained at low Reynolds numbers from the Happel cell model (see Section 2.9) to find the mean Sherwood number [74, 76]

$$\mathrm{Sh} = 0.625 \left[ \frac{2(1 - \phi^{5/3})}{2 - 3\phi^{1/3} + 3\phi^{5/3} - 2\phi^2} \right]^{1/3} \mathrm{Pe}_\phi^{1/3}. \tag{4.15.1}$$

Here $Pe_\phi = aU_\phi/D$ is the Peclet number calculated from the constrained flow velocity, which can be calculated by the formula

$$U_\phi = \frac{3}{2}\left(\frac{2 - 3\phi^{1/3} + 3\phi^{5/3} - 2\phi^2}{3 + 2\phi^{5/3}}\right)U_i, \qquad (4.15.2)$$

where $U_i$ is the velocity of an isolated sphere falling in an infinite liquid.

The evaluation of mass and heat transfer coefficients at moderate and high Reynolds numbers encounters some difficulties related to the description of a constrained flow with allowance for inertia forces at $Pe \gg 1$. However, it is important to note that in concentrated disperse systems, the flow field depends of the Reynolds number weaker than for isolated particles. For example, the formation of regions with closed fluid circulation, affecting mass and heat transfer, is prolonged and is completed at $Re \simeq 10^1$ to $10^2$. The above-mentioned perturbation smoothing in a flow with a sufficiently large concentration of the disperse phase allows one to estimate the process of the mass and heat transfer for higher $Re$ by using the corresponding data for the Stokes regime.

Experimental data on mass and heat transfer in a constrained flow are often treated as the dependence of the Kollborn factor $Ko = Sh/(Sc\,Re_\phi)$ on the Reynolds number. A comparison of experimental data for the Kollborn factor for solid spheres [76] with $0.5 \le \phi \le 0.7$ with the theoretical values for $Re < 1$ showed that the results of calculations for low $Re$ remain valid for $Re \le 50$.

To calculate the mean Sherwood (Nusselt) number in the case of a free-poured layer of particles of various shape, one can use the following empiric formulas [254] in a wide range of Reynolds numbers:

$$
\begin{aligned}
Sh &= 0.46\,Sc^{0.33}\,Re_{ef}^{0.85} && \text{for} \quad 0.1 \le Re_{ef} \le 1, \\
Sh &= 0.50\,Sc^{0.33}\,Re_{ef}^{0.47} && \text{for} \quad 1 \le Re_{ef} \le 15, \qquad (4.15.3) \\
Sh &= 0.30\,Sc^{0.33}\,Re_{ef}^{0.64} && \text{for} \quad 15 \le Re_{ef} \le 4\times 10^4,
\end{aligned}
$$

where the effective Reynolds number is determined by

$$Re_{ef} = \frac{a_e\langle U\rangle}{\nu}, \qquad a_e = \frac{2(1-\phi)}{s\phi}, \qquad \langle U\rangle = \frac{U_i}{1-\phi}. \qquad (4.15.4)$$

Here $a_e$ is the equivalent particle radius, $\langle U\rangle$ is the mean flow velocity, $s = S_*/V_*$ is the specific surface area of a particle, and $U_i$ is the incoming flow velocity (at $\phi = 0$).

For a monodisperse layer of spherical particles of radius $a$, one must put $s = 3/a$ in formulas (4.15.3) and (4.15.4).

---

### 4.15-2. Polydisperse Systems of Spherical Particles

For the analysis of mass and heat transfer processes in polydisperse systems, the distribution function $f(a)$ of particles with respect to their size is introduced satisfying the normalization condition

$$\int_0^\infty f(a)\,da = 1. \qquad (4.15.5)$$

The total number $N$ of particles in the system is calculated according to the formulas

$$N = \frac{3\phi V_*}{32\pi\bar{a}^3}, \qquad \bar{a} = \left[\int_0^\infty a^3 f(a)\,da\right]^{1/3}, \qquad (4.15.6)$$

where $V_*$ is the total volume of the system and $\bar{a}$ is the mean particle radius.

The dimensional total mass flux is defined as follows:

$$I_* = -\int_0^\pi \int_0^\infty ND\left(\frac{\partial C}{\partial R}\right)_{R=a} 2\pi a^2 \sin\theta\, f(a)\,da\,d\theta. \qquad (4.15.7)$$

Using the results of [473] obtained for the flow field in the point force model, one can find the mean Sherwood number for a polydisperse system of particles [74]:

$$\text{Sh} = 0.625\,(A\,\text{Pe}_\phi)^{1/3}, \qquad \text{Pe}_\phi = aU_\phi/D, \qquad (4.15.8)$$

where

$$A = 1 + \frac{a}{2-3\phi}\left\{\left[9\phi(2-3\phi)\frac{b_1}{b_3} + \frac{81}{4}\phi^2\left(\frac{b_2}{b_3}\right)^2\right]^{1/2} + \frac{9}{2}\phi\frac{b_2}{b_3}\right\},$$

$$b_m = \int_0^\infty a^m f(a)\,da \qquad (m = 1,\,2,\,3).$$

If the distribution function is known, then the calculation of the moments $b_m$ is easy. Usually, the function $f(a)$ is assumed to be exponential or Maxwellian. Some methods for the experimental determination of $f(a)$ are described in [454].

| **4.15-3. Monodisperse Systems of Spherical Drops or Bubbles** |

For the Stokes motion of spherical drops and bubbles, the cell Happel model (see Section 2.9) results in the following expression for the mean Sherwood number [503]:

$$\text{Sh} = 0.461\left\{\frac{2(1-\phi^{5/3})\,\text{Pe}_\phi}{(1-\phi^{1/3})[3\beta + 2 + 2(\beta-1)\phi^{5/3}] - \beta(1-\phi^{5/3})}\right\}^{1/2}, \qquad (4.15.9)$$

where $\text{Pe}_\phi = aU_\phi/D$ is the Peclet number determined by the constrained flow velocity $U_\phi$ and $\beta$ is the ratio of dynamic viscosities of the disperse and continuous phases (the value $\beta = 0$ corresponds to a gas bubble).

For high Reynolds numbers $\text{Re}_\phi = aU_\phi/\nu \geq 500$, the Sherwood number in the constrained flow of gas bubbles can be evaluated by the formula [263]

$$\text{Sh} = 0.8\frac{\sqrt{\text{Pe}_\phi}}{\sqrt{1-\phi}}. \qquad (4.15.10)$$

We note that in [421], the cell flow model was used for the investigation of mass and heat transfer in monodisperse systems of spherical drops, bubbles, or solid particles for $\text{Re}_\phi < 250$ and $0 < \phi < 0.5$.

| 4.15-4. Packets of Circular Cylinders |
|---|

Let us consider mass and heat transfer of under a transverse flow of packets of cylinders with unstaggered chess arrangement. At sufficiently high Reynolds numbers, the tubes in the first row of a packet are in conditions close to the conditions of mass transfer for an isolated cylinder (if the gap between tubes is of the order of the cylinder radius), while the mass transfer considerably increases in the subsequent rows. This effect is produced by the fact that the first rows serve as flow turbulizers. The stabilization of mass and heat transfer is about 10% after the fourth row, and is complete after the 14th row. In what follows, we take a tube of radius $a$ as the characteristic length, and the velocity $U = U_i/\psi$ as the characteristic flow velocity, where $U_i$ is the flow velocity remote from the cylinder and $\psi$ is the maximum narrowing coefficient for the packet cross-section downstream.

The mean Sherwood number for the unstaggered packets in "deep" rows (for $k \geq 14$, where $k$ is the row number) is given by the formulas [254]

$$Sh_{max} = 0.59 \, Sc^{0.36} \, Re^{0.4} \quad \text{for} \quad 1 < Re < 50,$$
$$Sh_{max} = 0.37 \, Sc^{0.36} \, Re^{0.5} \quad \text{for} \quad 50 < Re < 200,$$
$$Sh_{max} = 0.21 \, Sc^{0.36} \, Re^{0.63} \quad \text{for} \quad 200 < Re < 10^5.$$

For the chess arrangement of tubes in the packet, the mean Sherwood number is determined by the expressions [254]

$$Sh_{max} = 0.69 \, Sc^{0.36} \, Re^{0.4} \quad \text{for} \quad 1 < Re < 20,$$
$$Sh_{max} = 0.50 \, Sc^{0.36} \, Re^{0.5} \quad \text{for} \quad 20 < Re < 150,$$
$$Sh_{max} = 0.28 \, Sc^{0.36} \, Re^{0.6} \quad \text{for} \quad 150 < Re < 10^5.$$

Mass and heat transfer in the front rows of the packet can be calculated by the approximate formula

$$Sh_k = \frac{k}{k + \alpha} \, Sh_{max} \quad (k \geq 2),$$

in which one must set $\alpha = 0.3$ for the unstaggered tube arrangement and $\alpha = 0.5$ for the chess arrangement.

# Chapter 5
# Mass and Heat Transfer Under Complicating Factors

In the last two chapters, relatively simple linear problems of mass and heat transfer were discussed. However, no processes of mass transfer complicated by surface (heterogeneous) or volume (homogeneous) chemical reactions with finite rates have been considered so far. Moreover, it was assumed that the basic parameters of the fluid are temperature- and concentration-independent. This assumption permitted the hydrodynamic part of the problem to be solved first and then the linear thermal or diffusion problem to be considered for a known velocity field.

In this chapter, some problems of mass and heat transfer with various complicating factors are discussed. The effect of surface and volume chemical reactions of any order on the convective mass exchange between particles or drops and a translational or shear flow is investigated. Linear and nonlinear nonstationary problems of mass transfer with volume chemical reaction are studied. Universal formulas are given which can be used for estimating the intensity of the mass transfer process for arbitrary kinetics of the surface or volume reaction and various types of flow.

A wide class of nonlinear problems of convective mass and heat exchange is considered taking into account the dependence of the transfer coefficients on concentration (temperature). The results are presented in the form of simple unified formulas for the Sherwood number.

Nonisothermal flows through tubes and channels accompanied by dissipative heating of liquid are studied. Qualitative features of heat transfer in liquids with temperature-dependent viscosity are discussed. Some issues of film condensation are considered.

Various thermal hydrodynamic phenomena are analyzed, which are related to the dependence of the surface tension coefficient on temperature. Thermogravitational and thermocapillary convection in a fluid layer is studied. The problem of thermocapillary drift of a drop in an external temperature-gradient field is considered, as well as other, more complicated problems.

# 5.1. Mass Transfer Complicated by a Surface Chemical Reaction

## 5.1-1. Particles, Drops, and Bubbles

*Statement of the problem.* In the preceding chapters we considered processes of mass transfer to surfaces of particles and drops for the case of an infinite rate of chemical reaction (adsorption or dissolution.) Along with the cases considered in the preceding chapters, finite-rate surface chemical reactions (see Section 3.1) are of importance in applications. Here the concentration on the surfaces is a priori unknown and must be determined in the course of the solution. Let us consider a laminar fluid flow with velocity $U$ past a spherical particle (drop or bubble) of radius $a$. Let $R$ be the radial coordinate relative to the center of the particle. We assume that the concentration is uniform remote from the particle and is equal to $C_i$. Next, the rate of chemical reaction on the surface is given by $W_s = K_s F_s(C)$, where $K_s$ is the surface reaction rate constant and the function $F_s$ is defined by the reaction kinetics and satisfies the condition $F_s(0) = 0$.

The corresponding problem for the concentration distribution in the continuous medium is stated as follows:

$$\mathsf{Pe}(\mathbf{v} \cdot \nabla)c = \Delta c; \tag{5.1.1}$$

$$r = 1, \quad \frac{\partial c}{\partial r} = -k_s f_s(c); \tag{5.1.2}$$

$$r \to \infty, \quad c \to 0. \tag{5.1.3}$$

Here the dimensionless functions and parameters are related to the original dimensional variables by the formulas

$$c = \frac{C_i - C}{C_i}, \quad r = \frac{R}{a}, \quad \mathsf{Pe} = \frac{aU}{D}, \quad k_s = \frac{aK_s F_s(C_i)}{DC_i}, \quad f_s(c) = \frac{F_s(C)}{F_s(C_i)}.$$

In particular, for an $n$th-order surface reaction we have $F_s = C^n$ and $f_s = (1-c)^n$. In the general case, the function $f_s$ has the properties

$$f_s(1) = 0, \quad f_s(0) = 1. \tag{5.1.4}$$

*General correlations for the Sherwood number.* In [364], the following approximate equation was suggested for the mean Sherwood number:

$$\mathsf{Sh} = k_s f_s \left( \frac{\mathsf{Sh}}{\mathsf{Sh}_\infty} \right). \tag{5.1.5}$$

Equation (5.1.5) can be successfully used to determine $\mathsf{Sh}$ for arbitrary flows past spherical particles, drops, or bubbles under any dependence of the rate of the surface reaction on the concentration in the entire range $0 \leq \mathsf{Pe} < \infty$ of Peclet numbers.

The quantity $Sh_\infty = Sh_\infty(Pe)$ in (5.1.5) corresponds to the diffusion regime of the reaction (i.e., the limit case $k_s \to \infty$) and must be determined for the solution of the auxiliary problem (5.1.1), (5.1.3) with the simplest boundary condition $r = 1$, $c = 1$ on the surface. The corresponding formulas for $Sh_\infty(Pe)$ for various types of flow were given in Sections 4.7 and 4.8.

For an $n$th-order surface reaction, Eq. (5.1.5) acquires the form

$$Sh = k_s \left(1 - \frac{Sh}{Sh_\infty}\right)^n, \qquad k_s = \frac{aK_s C_i^{n-1}}{D}.$$

By solving it for $Sh$ in the special cases $n = 1/2$, 1, 2, one obtains, respectively,

$$Sh = k_s \left[\left(\frac{k_s^2}{4\,Sh_\infty^2} + 1\right)^{1/2} - \frac{k_s}{2\,Sh_\infty}\right] \qquad \text{for} \quad n = \frac{1}{2},$$

$$Sh = \left(\frac{1}{k_s} + \frac{1}{Sh_\infty}\right)^{-1} \qquad \text{for} \quad n = 1,$$

$$Sh = \frac{Sh_\infty^2}{4k_s}\left[\left(\frac{4k_s}{Sh_\infty} + 1\right)^{1/2} - 1\right]^2 \qquad \text{for} \quad n = 2.$$

It was shown in [166, 351] that Eq. (5.1.5) provides several valid initial terms of the asymptotic expansion of the Sherwood number as $Pe \to 0$ for any kinetics of the surface chemical reaction. (Specifically, one obtains three valid terms for the translational Stokes flow and four valid terms for an arbitrary shear flow.)

The approximate expression (5.1.5) was tested at moderate Peclet numbers $Pe = 10$, 20, 50 (the corresponding values of $Re$ are $Re = 10$, 20, and 0.5) for a translational flow past a solid sphere by comparison with the results of a numerical solution for a first-order surface chemical reaction. For the data taken from [2, 68], the error of Eq. (5.1.5) in these cases does not exceed 1.5%.

At high Peclet numbers, for an $n$th-order surface reaction with $n = 1/2$, 1, 2, Eq. (5.1.5) was tested in the entire range of the parameter $k_s$ by comparing its root with the results of numerical solution of appropriate integral equations for the surface concentration (derived in the diffusion boundary layer approximation) in the case of a translational Stokes flow past a sphere, a circular cylinder, a drop, or a bubble [166, 171, 364]. The comparison results for a second-order surface reaction ($n = 2$) are shown in Figure 5.1 (for $n = 1/2$ and $n = 1$, the accuracy of Eq. (5.1.5) is higher than for $n = 2$). Curve $I$ (solid line) corresponds to a second-order reaction ($n = 2$). One can see that, the maximum inaccuracy is observed for $0.5 \le k_s/Sh_\infty \le 5.0$ and does not exceed 6% for a solid sphere (curve 2), 8% for a circular cylinder (curve 3), and 12% for a spherical bubble (curve 4).

The validity of Eq. (5.1.5) for $n = 1/2$, 1, 2 was studied in [235], in the entire range of the parameters $k_s$ and $Pe$ for a shear Stokes flow past a sphere. In all cases considered, the maximum error did not exceed 5%.

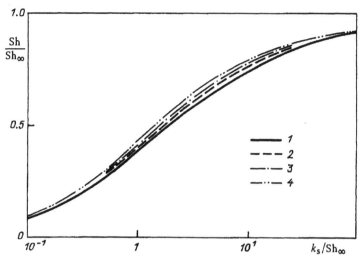

**Figure 5.1.** The Sherwood number against the rate constant of second-order surface chemical reaction: *1*, by formula (5.1.5); *2*, for a solid sphere; *3*, for a circular cylinder; and *4*, for a spherical drop or bubble

Note that to calculate the mean Sherwood number for particles of an irregular form, the following more general equation must be used:

$$\frac{\text{Sh}}{\text{Sh}_0} = f_s\left(\frac{\text{Sh}}{\text{Sh}_\infty}\right),\tag{5.1.6}$$

where $\text{Sh}_0$ is the asymptotics of $\text{Sh}$ as $k_s \to 0$.

---

### 5.1-2. Rotating Disk and a Flat Plate

*Rotating disk.* The mass exchange between a thin rotating disk and the surrounding fluid is governed by Eq. (3.2.1), the boundary conditions at infinity (3.2.2), and a boundary condition at the disk surface of the form (5.1.2) which determines the kinetics of the reaction. It is assumed that the dimensionless concentration is introduced in the same manner as in Subsection 5.1-1. The velocity field of the fluid near a disk is presented in Subsection 1.2-1.

The solution of this problem shows that relation (5.1.5) is exact in this case for arbitrary kinetics of the surface chemical reaction for any Peclet numbers [270].

*Flat plate.* Let us investigate convective diffusion to the surface of a flat plate in a longitudinal translational flow of a viscous incompressible fluid at high Reynolds numbers. The velocity field of the fluid near a flat plate is presented in Subsection 1.7-2. We assume that mass transfer is accompanied by a surface reaction.

Concentration distribution is described by the steady-state equation (3.1.1) and boundary conditions (3.1.2) and (3.1.5), where $\xi_* = Y$ is the distance from plate surface. In the diffusion boundary layer approximation, the exact analytical solution of the problem for the first-order reaction in a complicated integral form was obtained by Levich [270].

In the case of the first-order reaction, the power-series expansion [81]

$$C_s(\zeta) = C_i(1 - 0.731\zeta + 0.453\zeta^2 - 0.252\zeta^3 + \cdots),$$

$$\zeta = 2.81(K_s/U_i)\,\mathrm{Sc}^{2/3}\,\mathrm{Re}_X^{1/2}, \quad \mathrm{Re}_X = XU_i/\nu,$$

can be used for the surface concentration. The coordinate $X$ is directed along the plate (the origin is placed at the front edge). For the calculation of the surface concentration, the following simple approximate relation was suggested

$$C_s(\zeta) = \frac{C_i}{1 + \zeta}$$

The maximum inaccuracy of the this formula does not exceed 5% for any value of the surface reaction rate constant $K_s$.

The local diffusion flux for the first-order reaction is calculated by the formula $j_* = K_s C_s(\zeta)$.

The dimensionless local diffusion flux is determined by

$$j = -(\partial c/\partial y)_{y=0}, \quad y = Y/L,$$

where $L = \nu/U_i$ is a characteristic length.

For arbitrary kinetics of the chemical reaction, the diffusion flux distribution can estimated by the approximate formula [133, 359]

$$j = k_s f_s(j/j_\infty), \tag{5.1.7}$$

where $j_\infty = 0.399\,\mathrm{Sc}^{1/3}\,\mathrm{Re}_X^{-1/2}$ is the dimensionless local diffusion flux on the plate for the diffusion regime reaction (as $k_s \to \infty$).

The surface concentration can be calculated by the formula $C_s = C_i(1 - j/j_\infty)$.

The results of numerical solution of appropriate integral equations for the surface concentration (derived in the diffusion boundary layer approximation) in the case of power-law surface reaction for $n = 1/2$ and $n = 2$ was indicated in [133]. The maximum inaccuracy of formula (5.1.7) in these cases is about 10% for any value of the surface reaction rate constant.

In [81], several more complicated surface chemical reactions were considered.

### 5.1-3. Circular Tube

Let us briefly consider convective mass transfer accompanied by a surface reaction in a circular tube. Laminar steady-state fluid flow in a circular tube of radius $a$ with Poiseuille velocity profile is outlined in Subsection 1.5-3. For arbitrary kinetics of the chemical reaction at high Peclet numbers, the diffusion flux distribution can be estimated by the same equation (5.1.7), where

$$j_\infty = 0.678 \, \text{Pe}^{1/3}(X/a)^{-1/3}, \qquad \text{Pe} = aU_{\max}/D,$$

is the dimensionless local diffusion flux at the wall of a tube for the diffusion regime reaction (as $k_s \to \infty$).

In [81, 133], the results of numerical solutions of the problem of mass transfer in a tube for various surface chemical reactions were represented. The maximum inaccuracy of formula (5.1.7) for power-law reactions with $1/2 \le n \le 2$ is about 12% for any value of the surface reaction rate constant.

# 5.2. Diffusion to a Rotating Disk and a Flat Plate Complicated by a Volume Reaction

## 5.2-1. Mass Transfer to the Surface of a Disk Rotating in a Fluid

*Statement of the problem.* Let us consider diffusion on the surface of a disk rotating in a fluid at a constant angular velocity $\omega$. The velocity field of the fluid near a disk is presented in Subsection 1.2-1. We assume that the process is accompanied by an irreversible volume chemical reaction with rate $W_v = K_v F_v(C)$.

At high Peclet numbers, distribution of the substance concentration in the fluid is described by the equation

$$\frac{d^2c}{dy^2} + \text{Pe} \, y^2 \frac{dc}{dy} = k_v f_v(c) \tag{5.2.1}$$

with the boundary conditions

$$y = 0, \quad c = 1; \qquad y \to \infty, \quad c \to 0. \tag{5.2.2}$$

Here the dimensionless variables and parameters are introduced as follows:

$$c = \frac{C}{C_s}, \quad y = \frac{Y}{a}, \quad \text{Pe} = 0.51\frac{\nu}{D}, \quad a = \left(\frac{\nu}{\omega}\right)^{1/2},$$

$$k_v = \frac{a^2 K_v F_v(C_s)}{DC_s}, \quad f_v(c) = \frac{F_v(C)}{F_v(C_s)},$$

where $Y$ is the distance from the disk surface in the normal direction, $a$ is the characteristic length, $C_s$ is the concentration at the disk surface, and $\nu$ is the kinematic viscosity of the fluid.

We assume that $W_v \ge 0$ and $F_v(0) = 0$. Therefore, the function $f_v(c)$ satisfies the conditions $f_v(0) = 0$ and $f_v(1) = 1$.

***Diffusion flux.*** To approximate the dimensionless local diffusion flux $j = -(dc/dy)_{y=0}$ on the disk surface, it is convenient to use the cubic equation

$$j^3 - 2k_v\langle f_v\rangle j - 6[\Gamma(1/3)]^{-3}\,\text{Pe} = 0, \tag{5.2.3}$$

where the angle brackets denote the mean integral value of the kinetic function $f_v$,

$$\langle f_v\rangle = \int_0^1 f_v(c)\,dc. \tag{5.2.4}$$

If the volume reaction is absent ($k_v = 0$), then Eq. (5.2.3) yields the exact answer (3.2.11). For large $k_v \to \infty$ and fixed $\text{Pe}$, the approximate equation (5.2.3) provides a valid asymptotic result for any kinetic function $f_v = f_v(c)$.

For power-law volume reactions, we have $f_v(c) = c^n$. In this case, we must set

$$\langle f_v\rangle = \frac{1}{n+1} \tag{5.2.5}$$

in Eq. (5.2.3).

One can see from (5.2.5) that the diffusion flux decreases with the increase of the exponent $n$ and increases with the decrease of the dimensionless reaction rate constant $k_v$.

For $n$th-order volume reactions with $n = 1/2$, 1, 2, a numerical solution of problem (5.2.1), (5.2.2) was obtained in [357]. The maximum error of the cubic equation for the diffusion flux (5.2.3) in the entire range of the dimensionless reaction constant $k_v$ does not exceed 3%.

---

### 5.2-2. Mass Transfer to a Flat Plate in a Translational Flow

Let us investigate steady-state convective diffusion on the surface of a flat plate in a longitudinal translational flow of a viscous incompressible fluid at high Reynolds numbers (the Blasius flow). We assume that mass transfer is accompanied by a volume reaction. In the diffusion boundary layer approximation, the concentration distribution is described by the equation

$$\frac{1.33}{4}\frac{y}{x^{1/2}}\frac{\partial c}{\partial x} + \frac{1.33}{16}\frac{y^2}{x^{3/2}}\frac{\partial c}{\partial y} = \frac{\partial^2 c}{\partial y^2} - k_v f_v(c) \tag{5.2.6}$$

with the boundary condition

$$x = 0, \quad c = 0; \qquad y = 0, \quad c = 1; \qquad y \to \infty, \quad c \to 0. \tag{5.2.7}$$

Here the dimensionless variables are defined by the formulas

$$c = \frac{C}{C_s}, \quad x = \frac{X}{a}, \quad y = \frac{Y}{a}, \quad a = \frac{\nu^{1/3}D^{2/3}}{U_i},$$

$$k_v = \frac{a^2 K_v F_v(C_s)}{DC_s}, \quad f_v(c) = \frac{F_v(C)}{F_v(C_s)},$$

where $U_i$ is the nonperturbed fluid velocity remote from the plate, $X$ is the distance along the plate from the front edge, and $Y$ is the distance from the plate surface.

The distribution of the dimensionless local diffusion flux along the plate, $j = -(\partial c/\partial y)_{y=0}$, can be approximately found by solving the cubic equation

$$j^3 - 2k_v\langle f_v\rangle j - (0.399)^3 x^{-3/2} = 0, \qquad (5.2.8)$$

which gives a valid asymptotic result in both limit cases $k_v \to 0$ and $k_v \to \infty$ for any kinetics of the volume chemical reaction. For power-law reactions, one must substitute (5.2.5) into (5.2.8).

## 5.3. Mass Transfer Between Particles, Drops, or Bubbles and Flows With Volume Reaction

### 5.3-1. Statement of the Problem

Let us consider steady-state mass transfer on a spherical particle (drop or bubble) of radius $a$ in a laminar fluid flow. We assume that a volume chemical reaction proceeds in the continuous phase with $W_v = K_v F_v(C)$. The reactant transfer in the continuous phase is described in dimensionless variables by the equation

$$\text{Pe}(\mathbf{v} \cdot \nabla)c = \Delta c - k_v f_v(c) \qquad (5.3.1)$$

with the boundary conditions

$$r = 1, \quad c = 1; \qquad r \to \infty, \quad c \to 0, \qquad (5.3.2)$$

where $r = R/a$, $\text{Pe} = aU/D$, $k_v = a^2 K_v F_v(C_s)/(DC_s)$, $R$ is the radial coordinate relative to the particle center, and $U$ is the characteristic flow velocity; the dimensionless concentration $c$ and the kinetic function $f_v$ are introduced by analogy with (5.2.1).

### 5.3-2. Particles in a Stagnant Medium

*Spherical particle.* At $\text{Pe} = 0$, problem (5.3.1), (5.3.2) admits an exact closed-form solution for a first-order volume reaction, which corresponds to the linear function $f_v = c$. In this case, we have

$$c = \frac{1}{r}\exp\left[k_v^{1/2}(1-r)\right]. \qquad (5.3.3)$$

The mean Sherwood number for the solution (5.3.3) is given by the formula

$$\text{Sh} = 1 + \sqrt{k_v}. \qquad (5.3.4)$$

For an arbitrary dependence of the kinetic function on the concentration, the mean Sherwood number for a spherical particle in a stagnant fluid can be calculated [360] by using the expression

$$ \mathrm{Sh} = 1 + \left[ 2k_v \int_0^1 f_v(c)\, dc \right]^{1/2}. \tag{5.3.5} $$

Formula (5.3.5) guarantees an exact asymptotic result in both limit cases $k_v \to 0$ and $k_v \to \infty$ for any function $f_v(c)$. For a first-order volume reaction ($f_v = c$), the approximate formula (5.3.5) is reduced to the exact result (5.3.4). The maximum error of formula (5.3.5) for a chemical volume reaction of the order $n = 1/2$ ($f_v = \sqrt{c}$) in the entire range of the dimensionless reaction rate constant $k_v$ is 5%; for a second-order volume reaction ($f_v = c^2$), the error of (5.3.5) is 7% [360]. The mean Sherwood number decreases with the increase of the rate order $n$ and increases with $k_v$.

*Nonspherical particles.* For nonspherical particles in a stagnant medium with the first-order volume chemical reaction taken into account, the mean Sherwood number can be calculated by using the approximate expression

$$ \mathrm{Sh} = \mathrm{Sh}_0 + \sqrt{k_v}. \tag{5.3.6} $$

Here $\mathrm{Sh}_0$ is the Sherwood number corresponding to mass transfer of a particle in a stagnant medium without the reaction. Each summand in (5.3.6) must be reduced to a dimensionless form on the basis of the same characteristic length. The value of $\mathrm{Sh}_0$ can be determined by the formula $\mathrm{Sh}_0 = a\Pi/S_*$, where $a$ is the value chosen as the length scale and, $S_*$ is the surface area of the particle; the shape factor $\Pi$ is shown in Table 4.2 for some nonspherical particles.

For nonspherical particles in the case of a more complicated kinetic function $f_v(c)$, to calculate the Sherwood number, one can use formula (5.3.5), where the first summand on the right-hand side (equal to 1) must be replaced by $\mathrm{Sh}_0$.

## 5.3-3. Particles, Drops, and Bubbles. First-Order Reaction

*Moderate Peclet Numbers.* For spherical particles, drops, and bubbles (under limiting resistance of the continuous phase), in the case of a first-order volume reaction, the mean Sherwood number can be calculated [358] by the formula

$$ \mathrm{Sh} = 1 + \left[ (\mathrm{Sh}_0 - 1)^2 + k_v \right]^{1/2}. \tag{5.3.7} $$

Here $\mathrm{Sh}_0 = \mathrm{Sh}_0(\mathrm{Pe})$ is the Sherwood number in the absence of chemical reactions ($k_v = 0$).

The expression (5.3.7) gives exact asymptotic results for all four limit cases $k_v \to 0$, $k_v \to \infty$, $\mathrm{Pe} \to 0$, and $\mathrm{Pe} \to \infty$ (it is assumed that there are stagnant points on the interface). For a spherical drop in a translational Stokes flow, the maximum error of Eq. (5.3.7) is about 7%. For a translational or linear straining shear Stokes flow past a solid sphere, one must set $\mathrm{Sh}_0 = \mathrm{Sh}_p$ in (5.3.7), where the value of $\mathrm{Sh}_p$ is calculated with the aid of (4.7.9) and (4.8.5).

TABLE 5.1

Maximum errors of formula (5.3.8) and of the cubic equation (5.3.9) for different flows past particles, drops, or bubbles at high Peclet numbers in the presence of a first-order volume chemical reaction

| No | Disperse phase | Flow type | Error of formula (5.3.8), % | Error of Eq. (5.3.9), % |
|----|----------------|-----------|------------------------------|--------------------------|
| 1 | Drop, bubble | Axisymmetric shear Stokes flow | 2 | 1 |
| 2 | Drop, bubble | Translational Stokes flow | 2.6 | 1.6 |
| 3 | Drop, bubble | Plane shear Stokes flow | 3.8 | 2.8 |
| 4 | Bubble | Translational flow at high Reynolds numbers | 2.6 | 1.6 |
| 5 | Bubble | Axisymmetric flow at high Reynolds numbers | 2 | 1 |
| 6 | Solid particle | Translational Stokes flow | 3.4 | 2.4 |

**High Peclet Numbers.** At high Peclet numbers, to calculate the mean Sherwood number, one can use the approximate formula [363]

$$Sh = \sqrt{k_v}\, coth\left(\frac{\sqrt{k_v}}{Sh_0}\right), \tag{5.3.8}$$

where $Sh_0 = \lim_{k_v \to 0} Sh$.

The dependence of the auxiliary Sherwood number $Sh_0$ on the Peclet number $Pe$ for a translational Stokes flow past a spherical particle or a drop is determined by the right-hand sides of (4.6.8) and (4.6.17). In the case of a linear shear Stokes flow, the values of $Sh_0$ are shown in the fourth column in Table 4.4.

To calculate the mean Sherwood number, one can replace formula (5.3.8) by the cubic equation [359]

$$Sh^3 - k_v\, Sh - Sh_0^3 = 0, \tag{5.3.9}$$

which leads to more accurate results.

In Table 5.1, the maximum error of formulas (5.3.8) and (5.3.9) are shown in the entire range of the parameter $k_v$ for six different kinds of spherical particles, drops, or bubbles. All these estimates were found by comparison with the closed-form solution of problem (5.3.1), (5.3.2) obtained in the diffusion boundary layer approximation [363].

Formula (5.3.8) and Eq. (5.3.9) can be used for the calculation of the mean Sherwood number for nonspherical particles, drops, and bubbles at high Peclet numbers.

| 5.3-4. Particles, Drops, and Bubbles.  Arbitrary Rate of Reaction |
|---|

For an arbitrary rate of volume chemical reaction, the mean Sherwood number at high Peclet numbers can be calculated according to the approximate formula

$$Sh = (2k_v \langle f_v \rangle)^{1/2} \coth \left[ \frac{(2k_v \langle f_v \rangle)^{1/2}}{Sh_0} \right] \qquad (5.3.10)$$

or by solving the cubic equation [356]

$$Sh^3 - 2k_v \langle f_v \rangle Sh - Sh_0^3 = 0. \qquad (5.3.11)$$

In the general case, $\langle f_v \rangle$ is given by (5.2.4). For an $n$th-order reaction, the expression (5.2.5) must be used.

## 5.4. Mass Transfer Inside a Drop (Cavity) Complicated by a Volume Reaction

Let us now study the inner mass transfer problems involving a volume chemical reaction. We assume that the diffusion process is quasi-stationary and takes place inside a solid spherical inclusion or a drop of radius $a$ filled with a stagnant or moving medium.

The concentration distribution in the region $0 \leq r \leq 1$ is described in the dimensionless variables by Eq. (5.3.1) and the first boundary condition in (5.3.2).

| 5.4-1. Spherical Cavity Filled by a Stagnant Medium |
|---|

For a first-order volume reaction at $Pe = 0$, the exact closed-form solution of our problem reads

$$c = \frac{1}{r} \frac{\sinh(r\sqrt{k_v})}{\sinh(\sqrt{k_v})}, \qquad (5.4.1)$$

and the corresponding mean Sherwood number on the inner surface of the cavity is given by the formula

$$Sh = -1 + k_v^{1/2} \coth k_v^{1/2}. \qquad (5.4.2)$$

For an $n$th-order volume reaction, one can calculate the mean Sherwood number [360] by the approximate formula

$$Sh = -\frac{2}{n+1} + \left( \frac{2k_v}{n+1} \right)^{1/2} \coth \left( \frac{n+1}{2} k_v \right)^{1/2}, \qquad (5.4.3)$$

which guarantees a valid asymptotic result for both small and large values of the parameter $k_v$.

For $n = 1$, formula (5.4.3) gives the exact result (5.4.2). The comparison of the approximate expression (5.4.3) with the results of the numerical solution of the inner problem (5.3.1), (5.3.2) for $n$th-order reactions with $n = 1/2$ and $n = 2$ reveals an error of 5% in the entire range of the parameter $k_v$.

For an arbitrary dependence of the rate of volume chemical reaction on the concentration, it is expedient to use the approximate formula

$$\mathrm{Sh} = -2\langle f_v \rangle + (2k_v\langle f_v \rangle)^{1/2} \coth \left( \frac{k_v}{2\langle f_v \rangle} \right)^{1/2}, \qquad (5.4.4)$$

where the mean value $\langle f_v \rangle$ is defined by the integral (5.2.4).

For a power-law reaction of any order, formula (5.4.4) is reduced to (5.4.3).

### 5.4-2. Nonspherical Cavity Filled by a Stagnant Medium

The method of asymptotic analogies (see Section 4.1) permits one to generalize formulas (5.4.2)—(5.4.4) to cavities of an arbitrary shape. In the special case of a first-order volume reaction, we obtain the formula

$$\mathrm{Sh} = -\frac{S}{3V} + \sqrt{k_v} \coth \left( \frac{3V}{S} \sqrt{k_v} \right), \qquad (5.4.5)$$

where $S$ and $V$ are, respectively, the dimensionless surface area and volume of the cavity. (All dimensionless quantities are defined on the basis of an arbitrary characteristic length the same for each quantity.) For a spherical cavity, by substituting $S = 4\pi$ and $V = 4\pi/3$ into (5.4.5), we obtain the exact expression (5.4.2).

For an arbitrary rate of volume chemical reaction, the mean Sherwood number may be calculated according to the formula

$$\mathrm{Sh} = -\frac{2S}{3V}\langle f_v \rangle + (2k_v\langle f_v \rangle)^{1/2} \coth \left( \frac{9V^2 k_v}{2S^2 \langle f_v \rangle} \right)^{1/2},$$

which generalizes the approximate formula (5.4.4) to nonspherical cavities.

For a first-order volume reaction, the mean concentration over the volume is calculated as follows:

$$\bar{c} = \frac{\mathrm{Sh}}{k_v} \frac{S}{V}, \qquad \text{where} \quad \bar{c} = \frac{1}{V} \int_v c \, dv. \qquad (5.4.6)$$

We note that relation (5.4.6) between $\bar{c}$ and $\mathrm{Sh}$ is exact. Substituting (5.4.5) into (5.4.6), one can obtain a formula for the mean concentration.

### 5.4-3. Convective Mass Transfer Within a Drop (Cavity)

Let us consider mass transfer within a drop (or a cavity) of an arbitrary shape taking into account the circulation of the fluid (the motion of the fluid can be caused by the external flow about the drop or other factors).

By integrating Eq. (5.3.1) over the cavity volume $V$, after a number of transformations, we obtain [166]

$$\mathsf{Sh} = \frac{k_v}{S} \int_V f_v(c)\, dV. \tag{5.4.7}$$

For a monotone kinetic function $f_v = f_v(c)$, by taking into account the inequality $f_v(c) \leq f_v(1) = 1$ for $0 \leq c \leq 1$, from (5.4.7) we obtain a rough estimate

$$\mathsf{Sh} \leq k_v V/S \tag{5.4.8}$$

of the mean Sherwood number. In case of a zero-order volume reaction, the equality sign in (5.4.8) corresponds to the exact result. It is important to note that the estimate (5.4.8) is independent of the Peclet number.

In the inner problems of the convective mass transfer for $k_v = O(1)$ as $\mathsf{Pe} \to \infty$, the concentration is leveled out along each streamline. The mean Sherwood number, by virtue of the estimate (5.4.8), is bounded above uniformly with respect to the Peclet number: $\mathsf{Sh} \leq \mathrm{const}\, k_v$. This means that the inner diffusion boundary layer cannot be formed by increasing the circulation intensity alone (i.e., by increasing the fluid velocity, which corresponds to $\mathsf{Pe} \to \infty$) for moderate values of $k_v$. This property of the mean Sherwood number is typical of all inner problems. For outer problems of mass transfer, the behavior of this quantity is essentially different: here a thin diffusion boundary layer is usually produced near the interface as $\mathsf{Pe} \to \infty$, and $\lim\limits_{\mathsf{Pe} \to \infty} \mathsf{Sh} = \infty$.

For a first-order volume reaction and a translational Stokes flow past a spherical drop, the asymptotic solution of the inner problem (5.3.1), (5.3.2) as $\mathsf{Pe} \to \infty$ results in the following expression for the mean Sherwood number [104]:

$$\mathsf{Sh} = \frac{1}{3}k_v \left( 1 - \frac{3}{2}k_v \sum_{m=1}^{\infty} \frac{A_m}{k_v + \lambda_m} \right), \tag{5.4.9}$$

where the first five coefficients $A_m$ and $\lambda_m$ are defined in (4.13.1).

In Figure 5.2, the dependence of the mean Sherwood number on the dimensionless parameter $k_v$ is shown for a first-order volume chemical reaction in the problem of quasi-steady-state mass transfer within a drop for the extreme values $\mathsf{Pe} = 0$ (formula (5.4.2)) and $\mathsf{Pe} = \infty$ (formula (5.4.9)) of the Peclet number. The dashed line corresponds to the rough upper bound (5.4.8). For moderate Peclet numbers ($0 < \mathsf{Pe} < \infty$), the mean Sherwood number gets into the dashed region bounded by the limit curves corresponding to $\mathsf{Pe} = 0$ and $\mathsf{Pe} = \infty$. One can see that the variation of the parameter $\mathsf{Pe}$ (for $k_v = O(1)$) only weakly affects the mean influx of the reactant to the drop surface, i.e., one cannot achieve a substantial increase in the Sherwood number by any increase in the Peclet number. In the special case $k_v = 10$, the maximum relative increment of the mean Sherwood number caused by the increase in the Peclet number from zero to infinity is only

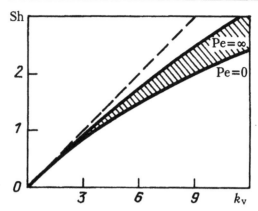

**Figure 5.2.** The mean Sherwood number against the dimensionless rate constant of a first-order volume chemical reaction for the inner problem (continuous lines: the lower line corresponds to $Pe = 0$, the upper line, to $Pe = \infty$). The dashed line corresponds to a zero-order reaction

about 25%. This means that the main factor affecting the behavior of basic characteristics of mass transfer intensity within a drop is the chemical reaction, whereas the influence of the fluid circulation rate and of the flow geometry on the behavior of these characteristics is weak.

The dashed line in Figure 5.2 correspond to a zero-order volume reaction. The mean Sherwood number is a monotone decreasing function of the rate order $n$. Therefore, for $0 < n < 1$ the curves describing the limit Sherwood number as $Pe = \infty$ lie between the dashed line and the upper continuous line. As the rate order $n$ decreases, the curves corresponding to the Sherwood number for $Pe = 0$ and $Pe = \infty$ gradually come closer to each other and rise to the dashed line. In the limit case $n = 0$, all three curves merge into a single curve, i.e., the mean Sherwood number does not depend on the Peclet number at all.

For large rate constants $k_v$ of the volume chemical reaction, a thin diffusion boundary layer is produced near the drop surface; its thickness is of the order of $k_v^{-1/2}$ at low and moderate Peclet numbers, and the solute in this layer has time to react completely. As the Peclet number is increased further, because of the intensive liquid circulation within the drop, there is not enough time to complete the reaction in the boundary layer. The nonreacted solute begins to get out of the boundary layer and penetrate into the depth of the drop along the streamlines near the flow axis. If the circulation within the drop is well developed, a complete diffusion wake is produced with essentially nonuniform concentration distribution that "pierces" the entire drop and joins the endpoint and the origin of the diffusion boundary layer. In case of a first-order volume chemical reaction, an appropriate analysis of convective mass transfer within the drop for $Pe \gg 1$ and $k_v \gg 1$ was carried out in [150, 151]. It should be said that in this case, in view of the estimate (5.4.8), which is uniform with respect to the Peclet number, the mass transfer intensity within the drop is bounded by the rate of volume chemical reaction.

# 5.5. Transient Mass Transfer Complicated by Volume Reactions

## 5.5-1. Statement of the Problem

Let us consider transient mass transfer between a gas and a stagnant medium in which an irreversible volume chemical reaction proceeds with rate $W_v = K_v F_v(C)$. We assume that the concentration of the solute at the initial instant $t = 0$ is zero, and for $t > 0$ the concentration on the surface is constant and is equal to $C_s$.

In dimensionless variables, the process is described by the equation

$$\frac{\partial c}{\partial \tau} = \frac{\partial^2 c}{\partial x^2} - k_v f_v(c) \tag{5.5.1}$$

with initial and boundary conditions

$$\tau = 0, \quad c = 0; \qquad x = 0, \quad c = 1; \qquad x \to \infty, \quad c \to 0, \tag{5.5.2}$$

where $\tau = Dt/a^2$; $x = X/a$, and $X$ is the coordinate measured from the surface towards the bulk of fluid; $a$ is a dimensional characteristic length; the remaining dimensionless functions and parameters in (5.5.1) are introduced in the same way as in (5.2.1).

## 5.5-2. Irreversible First-Order Reaction

For a first-order reaction ($f_v = c$), the exact closed-form solution of problem (5.5.1), (5.5.2) is given by

$$c = \frac{1}{2}\left[\exp\left(x\sqrt{\tau}\right)\operatorname{erfc}\left(\frac{x+2\tau\sqrt{k_v}}{2\sqrt{\tau}}\right) + \exp\left(-x\sqrt{\tau}\right)\operatorname{erfc}\left(\frac{x-2\tau\sqrt{k_v}}{2\sqrt{\tau}}\right)\right],$$
$$\tag{5.5.3}$$

where erfc $z$ is the complementary error function,

$$\operatorname{erfc} z = 1 - \operatorname{erf} z, \quad \operatorname{erf} z = \frac{2}{\sqrt{\pi}}\int_0^z \exp(-z^2)\,dz.$$

By differentiating (5.5.3) with respect to $x$ and by setting $x = 0$, we find the expression for the dimensionless diffusion flux of the substance through the interface:

$$j = (\pi\tau)^{-1/2}\exp(-k_v\tau) + k_v^{1/2}\operatorname{erf}(k_v\tau)^{1/2}. \tag{5.5.4}$$

## 5.5-3. Irreversible Reactions With Nonlinear Kinetics

For arbitrary kinetics of volume chemical reaction, the diffusion flux can be calculated according to the approximate formula [369]

$$j = (\pi\tau)^{-1/2}\exp(-2k_v\langle f_v\rangle\tau) + (2k_v\langle f_v\rangle)^{1/2}\operatorname{erf}(2k_v\langle f_v\rangle\tau), \tag{5.5.5}$$

where the mean value $\langle f_v \rangle$ is defined in (5.2.4).

Formula (5.5.5) guarantees a valid asymptotic result for any function $f_v = f_v(c)$ in the four limit cases $k_v \to 0$, $k_v \to \infty$, $\tau \to 0$, and $\tau \to \infty$ and is reduced to the exact solution (5.5.4) for a first-order reaction.

The specific values of $\langle f_v \rangle$ for some typical reactions [331] are shown below:

| Reaction | $n$th-order | Fermentative | Autocatalytic |
|---|---|---|---|
| Kinetic function $f_v(c)$ | $c^n$ | $\dfrac{c}{1+Mc}$ | $\dfrac{c}{(1+Mc)^2}$ |
| $\langle f_v \rangle$ | $\dfrac{1}{n+1}$ | $\dfrac{2}{M^2}[M-\ln(1+M)]$ | $\dfrac{2}{M^2}\left[\ln(1+M)-\dfrac{M}{1+M}\right]$ |

In the entire range of the dimensionless rate constant of volume chemical reaction, the maximum error of formula (5.5.5) in the above four cases for $n = 0.5$, $n = 2$; $M = 0.5$, and $M = 2$ is about 3%.

## 5.5-4. Reversible First-Order Reaction

We now consider a reaction described by the equation $A \rightleftarrows B$. Let $K_1$ and $K_{-1}$ be the rate constants of the direct and the reverse reactions. In our case, one mole of dissolved gas $A$, as a result of the reaction, yields one mole of product $B$. We denote by $C_A$ the gas concentration and by $C_B$ the product concentration.

Mass transfer in liquid is described by the system of equations

$$D_A \frac{\partial^2 C_A}{\partial X^2} = \frac{\partial C_A}{\partial t} + K_1\left(C_A - \frac{1}{q}C_B\right), \tag{5.5.6}$$

$$D_B \frac{\partial^2 C_B}{\partial X^2} = \frac{\partial C_B}{\partial t} - K_1\left(C_A - \frac{1}{q}C_B\right) \tag{5.5.7}$$

with the initial and boundary conditions

$$C_A = C_A^{(i)}, \quad C_B = qC_A^{(i)} \qquad \text{for} \quad t = 0, \tag{5.5.8}$$

$$C_A = C_A^{(s)}, \quad \partial C_B/\partial X = 0 \qquad \text{for} \quad X = 0, \tag{5.5.9}$$

$$C_A = C_A^{(i)}, \quad C_B = qC_A^{(i)} \qquad \text{as} \quad X \to \infty. \tag{5.5.10}$$

In the statement of problem (5.5.6)—(5.5.10), we assumed that the initial concentration of the dissolved gas is $C_A^{(i)}$ at each point of the liquid and that the corresponding equilibrium concentration of product $B$ is $C_B^{(i)} = qC_A^{(i)}$, where $q = K_1/K_{-1}$ is the equilibrium constant. The second boundary condition in (5.5.9) means that the product does not intersect the fluid surface.

We introduce the enhancement coefficient by setting $\mathcal{E} = j_A(K_1)/j_A(0)$, where $j_A$ is the diffusion flux of the gas through the interface $X = 0$. Below is the solution of problem (5.5.6)–(5.5.10) for $\mathcal{E}$ [103].

In the case $q > 1$,

$$\mathcal{E} = 1 + \frac{q^2}{q^2 - 1} \frac{\sqrt{\pi}}{2\alpha} \exp(\alpha^2) \left[ \text{erf}(\alpha q) - \text{erf}(\alpha) \right]$$

$$- \frac{q}{2\alpha} \left( \frac{\pi}{q^2 - 1} \right)^{1/2} \text{erf}\left( \alpha \sqrt{q^2 - 1} \right), \qquad \alpha = \left[ \frac{K_1 t}{q(q-1)} \right]^{1/2}. \quad (5.5.11)$$

In the case $q < 1$,

$$\mathcal{E} = 1 - \frac{q^2}{\gamma(1 - q^2)} \exp(-\gamma^2) \left[ \int_{\gamma}^{q\gamma} \exp(z^2)\, dz \right]$$

$$+ \frac{q}{2\gamma} \left( \frac{\pi}{1 - q^2} \right)^{1/2} \text{erf}\left( \gamma \sqrt{1 - q^2} \right), \qquad \gamma = \left[ \frac{K_1 t}{q(1 - q)} \right]^{1/2}. \quad (5.5.12)$$

In practice, reversible reactions of true first order in both directions usually do not occur. However, one often deals with first-order reactions with respect to the dissolved gas concentration, where the reactant concentration is uniform in the volume, and hence, the direct reaction is of pseudo-first order. At the same time, the concentration of the products also can be practically invariable in the entire volume of the liquid, and therefore, the reverse reaction rate also proves to be invariable. Then, instead of the ratio $C_B/q$, one can substitute the constant $C_A^{(e)}$ into (5.5.6); this constant is a characteristic of the equilibrium concentration of dissolved gas $A$ in the bulk of fluid. As a result, we obtain the problem for the relative concentration $c = \dfrac{C_A - C_A^{(e)}}{C_A^{(s)} - C_A^{(e)}}$, which coincides with the linear problem (5.5.1), (5.5.2) for $f_v = c$ in the dimensionless variables.

# 5.6. Mass Transfer for an Arbitrary Dependence of the Diffusion Coefficient on Concentration

## 5.6-1. Preliminary Remarks. Statement of the Problem

It is usually assumed that the diffusion coefficient is independent of the concentration. However, experimental data [64, 120, 272, 388, 393, 439, 491] show that the diffusion coefficients in liquids often strongly depend on the concentration. In dilute solutions, an increase in the concentration always produces a decrease in the diffusion coefficient. For example, two grams of sodium chloride dissolved in one liter of water decreases the diffusion coefficient by 10%. Often the diffusion coefficient linearly decreases with the increase of the diffusing substance concentration (sugar, raffinase, etc.) in the water solution [64].

In solutions of a number of monovalent salts (NaCl, KCl, KI, LiCl, etc.), the dependence of the diffusion coefficient on concentration (for $C \le 0.1$ mole/liter) is fairly well described by the expression [332, 393]

$$D/D_0 = 1 - \gamma \sqrt{C},$$

where $D_0$ is the diffusion coefficient for infinite dilution (zero concentration), $C$ is the mole concentration, and $\gamma \simeq 0.5$ to $0.6$ is a numerical factor.

For hemoglobin and gray albumin diffusing in salt solutions, we have [13, 393]

$$D/D_0 = (1 - \bar{c})^{6.5},$$

where $\bar{c}$ is the mole fraction of the solute.

Dissolving 0 to $2 \times 10^{-4}$ mole/liter of $KMnO_4$ in water decreases the diffusion coefficient by 25%. A very large change in the diffusion coefficient is observed in solutions of methylen-blue (the molecular weight $m = 317$): the presence of $6 \times 10^{-4}$ mole/liter of this substance decreases the diffusion coefficient to half the original value at room temperature.

We also note that in some systems (for example, water solutions of acetone, ethanol, or methanol), the diffusion coefficient first decreases and then increases as the concentration increases [64, 388, 439]. For example, the mutual diffusion coefficient for the acetone–water system at $25° C$ in the range 0.45 to 1.0 of acetone mole fraction can be described by the correlation $D/D_0 = \exp(\kappa \bar{c})$ with $\kappa = 3.83$ and $D_0 = 0.109 \times 10^{-5}$ cm$^2$/s [439].

Obviously, these examples show that even for dilute solutions of substances (at concentrations of the order of 0.1%), the variation of the diffusion coefficient must be taken into account. At the same time, as a rule, the dependence of the viscosity and the density of the mixture on the diffusing substance concentration can be neglected. For example, one can see from the data in [64] that for dilute solutions of the monovalent salts, the relative change of the diffusion coefficient is two orders of magnitude higher than that of the solution viscosity.

In view of this, let us consider steady-state convective mass transfer to a solid particle, drop, or bubble for an arbitrary dependence $D = D(C)$ of the diffusion coefficient on concentration. We assume that the concentrations at the particle surface and remote from it are uniform and are equal to $C_s$ and $C_i$, respectively ($C_s \neq C_i$). We also assume that the concentration nonuniformity does not affect the flow parameters. In dimensionless variables, this nonlinear problem is described by the equation

$$\text{Pe}(\mathbf{v} \cdot \nabla)c = \text{div}(\overline{D}\nabla c); \qquad (5.6.1)$$

with the boundary conditions

$$c = 1 \qquad \text{(on the particle surface } \Gamma\text{)},$$
$$c \to 0 \qquad \text{(far from the particle)},$$

where $c = \dfrac{C_i - C}{C_i - C_s}$, $\overline{D}(c) = \dfrac{D(C)}{D(C_i)}$, and $\text{Pe} = \dfrac{aU}{D(C_i)}$; here $a$ and $U$ are the characteristic length and velocity, respectively.

We define the mean Sherwood number corresponding to the solution of problem (5.6.1) as follows:

$$\text{Sh} = \text{Sh}(\overline{D}, \text{Pe}) = -\frac{1}{S} \int_{\Gamma} \overline{D}(c) \frac{\partial c}{\partial \xi} \, d\Gamma, \qquad (5.6.2)$$

where $S$ is the dimensionless surface area of the particle and $\partial/\partial\xi$ is the normal derivative on $\Gamma$.

---

### 5.6-2. Steady-State Problems. Particles, Drops and Bubbles

*Low Peclet numbers.* In [347] it was proved that for an arbitrary dependence of the diffusion coefficient on concentration and for any shape of particles and drops, the following asymptotic formula is valid for the mean Sherwood number at low Peclet numbers:

$$\text{Sh}(\overline{D}, \text{Pe}) = \langle \overline{D} \rangle \, \text{Sh}(1, \text{Pe}), \qquad \text{where} \quad \langle \overline{D} \rangle = \int_0^1 \overline{D}(c)\, dc. \qquad (5.6.3)$$

Here $\text{Sh}(1, \text{Pe})$ is the auxiliary Sherwood number corresponding to the solution of the linear problem (5.6.1) with $\overline{D} = 1$.

For a translational Stokes flow, the expression (5.6.3) gives three terms of the expansion (up to terms of the order of $\text{Pe}^2 \ln \text{Pe}$ inclusive). In this case, the quantity $\text{Sh}(1, \text{Pe})$ is determined by the ratio of the right-hand side of (4.4.22) to the dimensionless surface area of the particle.

For an arbitrary linear shear flow, the expression (5.6.3) gives two terms of the expansion (up to terms of the order of $\sqrt{\text{Pe}}$ inclusive). In this special case, $\text{Sh}(1, \text{Pe})$ is determined by the ratio of the right-hand side of (4.5.4) to the dimensionless surface area of the particle with $O(\text{Pe})$ terms neglected.

Formula (5.6.3) can also be used for other more complicated flows as $\text{Pe} \to 0$ [347].

*Mass exchange for a particle freely suspended in a simple shear flow.* Let a particle be surrounded by a region with closed streamlines. In this case, the leading term of the asymptotic expansion of the mean Sherwood number at high Peclet numbers is described by formula (5.6.3), which was derived in [359].

In the special case of a spherical particle freely suspended in a simple shear flow, one must set $\text{Sh}(1, \text{Pe}) = 4.45$ in (5.6.3). To calculate $\text{Sh}(1, \text{Pe})$ for spherical particles in a simple shear flow, the expression (4.8.15) can be used.

Apparently, formula (5.6.3) can be successfully used in the approximate calculation of the mean Sherwood number for a spherical particle freely suspended in a simple shear flow in the entire range $0 \le \text{Pe} < \infty$ of Peclet numbers. (We recall that formula (5.6.3) gives a valid asymptotic result in both limit cases $\text{Pe} \to 0$ and $\text{Pe} \to \infty$.)

*Diffusion boundary layer approximation.* For high Peclet numbers, problem (5.6.1) was investigated in [236]. The solution was obtained by using the diffusion boundary layer method, and the following formula was derived for the mean Sherwood number:

$$\text{Sh}(\overline{D}, \text{Pe}) = \alpha_m(\overline{D}) \, \text{Sh}(1, \text{Pe}). \qquad (5.6.4)$$

Here $\alpha_m$ is the nonlinearity coefficient given by the formula

$$\alpha_m = (m+1)^{\frac{m-1}{m+1}} \Gamma\left(\frac{1}{m+1}\right) \left[-\overline{D}(c)\frac{dc}{dz}\right]_{z=0}, \qquad (5.6.5)$$

where the function $c = c(z)$ is the solution of the auxiliary problem for the ordinary differential equation

$$\frac{d}{dz}\left[\overline{D}(c)\frac{dc}{dz}\right] + \frac{z^m}{m+1}\frac{dc}{dz} = 0; \qquad (5.6.6)$$
$$z = 0, \quad c = 1; \qquad z \to \infty, \quad c \to 0.$$

The value $m = 2$ corresponds to solid particles, and $m = 1$, to bubbles and drops of moderate viscosity $(0 \le \beta \le 2)$. If the diffusion coefficient is constant, then we have $\alpha_m(1) = 1$.

Formula (5.6.4) is valid for an arbitrary laminar flow without closed streamlines for particles and drops of an arbitrary shape. The quantity $\mathrm{Sh}(1, \mathrm{Pe})$ corresponds to the asymptotic solution of the linear problem (5.6.1) at $\mathrm{Pe} \gg 1$. For spherical particles, drops, and bubbles in a translational or linear straining shear flow, the values of $\mathrm{Sh}(1, \mathrm{Pe})$ are shown in the fourth column in Table 4.7.

In [236], an exact analytical solution of problem (5.6.6) for any value of $m$ was obtained in the case of a hyperbolic dependence $\overline{D}(c) = (\alpha c + \beta)^{-1}$ of the diffusion coefficient on the concentration, where $\alpha$ and $\beta$ are some constants. In [277], a solution is given for $m = 1$ and $\overline{D}(c) = (\alpha c^2 + \beta c + \gamma)^{-1}$.

To evaluate the coefficient $\alpha_m$ in (5.6.4) for an arbitrary dependence of the diffusion coefficient on the concentration, it is expedient to use the approximate formula [361]

$$\alpha_m = \left[\frac{m+1}{m}\int_0^1 c^{\frac{1}{m}}\,\overline{D}(c)\,dc\right]^{\frac{m}{m+1}}. \qquad (5.6.7)$$

The nonlinearity coefficients obtained by the numerical solution of problem (5.6.6) according to formula (5.6.5) with the use of the approximate expression (5.6.7) for seven typical functions $\overline{D} = \overline{D}(c)$ are compared with each other in Table 5.2 (the relative errors are shown in the last three columns). One can see that formula (5.6.7) is highly accurate.

For solid particles $(m = 2)$, to find the nonlinearity coefficient $\alpha_m$ approximately, one can use the formula

$$\alpha_2 = \left[2\int_0^1 c\overline{D}(c)\,dc\right]^{2/3}, \qquad (5.6.8)$$

which is simpler but less accurate than (5.6.7). Let $\overline{D}_{\max} = \max_{0 \le c \le 1} \overline{D}(c)$ and $\overline{D}_{\min} = \min_{0 \le c \le 1} \overline{D}(c)$. For various functions $\overline{D} = \overline{D}(c)$ shown in the first column of Table 5.2, the maximum error of formula (5.6.8) for $1 \le \overline{D}_{\max}/\overline{D}_{\min} \le 2$ (these inequalities determine the range of the parameter $b$) does not exceed 3.5%.

## 5.6-3. Transient Problems. Particles, Drops, and Bubbles

Now let us consider the outer problem of transient mass exchange between a drop (bubble) and a laminar steady-state flow. We assume that the concentration

TABLE 5.2

Maximum error (in per cent) of formula (5.6.7) for various forms of
dependence of the diffusion coefficient on the concentration

| Dependence $\bar{D} = \bar{D}(c)$ | Range of the parameter $b$ | Drops, bubbles $m = 1$ | Solid particles $m = 2$ | $m = 3$ |
|---|---|---|---|---|
| $\bar{D} = 1 - bc$ | $-3 \leq b \leq 0.8$ | 1.9 | 0.8 | 1.6 |
| $\bar{D} = 1 - b\sqrt{c}$ | $-3 \leq b \leq 0.8$ | 2.0 | 0.7 | 1.2 |
| $\bar{D} = (1 + bc)^{-1}$ | $-0.8 \leq b \leq 3$ | 2.4 | 0.7 | 2.0 |
| $\bar{D} = (1 + bc)^{-2}$ | $-0.8 \leq b \leq 3$ | 4.8 | 1.3 | 3.2 |
| $\bar{D} = (1 + b\sqrt{c})^{-1}$ | $-0.8 \leq b \leq 3$ | 1.9 | 0.3 | 1.8 |
| $\bar{D} = \exp(-bc)$ | $-2 \leq b \leq 3$ | 3.4 | 1.4 | 2.3 |
| $\bar{D} = (1 + bc)^{-1/2}$ | $-0.8 \leq b \leq 3$ | 1.2 | 0.3 | 1.1 |

in the liquid is uniform at the initial moment and is equal to $C_i$; on the drop
surface, the concentration is constant and is equal to $C_s$. Equation of transient
mass transfer in the continuous medium with allowance for the dependence of
the diffusion coefficient on the concentration can be written out in the form

$$\frac{\partial c}{\partial \tau} + \text{Pe}(\mathbf{v} \cdot \nabla)c = \text{div}[\bar{D}(c)\nabla c], \tag{5.6.9}$$

where $\tau = tD(C_i)/a^2$, and the remaining dimensionless quantities are defined in
the same way as in Eq. (5.6.1). We seek the solution under the initial conditions
$\tau = 0$, $c = 0$ and the same boundary conditions as in (5.6.1).

It was shown in [349] that at high Peclet numbers (in the diffusion boundary layer approximation), by solving the corresponding nonlinear problem on
transient mass exchange between drops or bubbles and the flow, one obtains the
following expression for the mean Sherwood number:

$$\text{Sh}(\bar{D}, \text{Pe}, \tau) = \alpha_1(\bar{D})\,\text{Sh}(1, \text{Pe}, \tau), \tag{5.6.10}$$

where, just as above, the nonlinearity coefficient $\alpha_1$ is determined from (5.6.5)
by solving Eq. (5.6.6) with $m = 1$. The quantity $\text{Sh}(1, \text{Pe}, \tau)$ can be found from
Eq. (5.6.9) with $\bar{D} = 1$.

By replacing both factors on the right-hand side in (5.6.10) by the approximate expressions (4.12.3) and (5.6.8), we obtain

$$\text{Sh}(\bar{D}, \text{Pe}, \tau) = \text{Sh}_{\text{st}}[\coth(\pi\,\text{Sh}_{\text{st}}^2\,\tau)]^{1/2} \left[ 2\int_0^1 c\bar{D}(c)\,dc \right]^{1/2}, \tag{5.6.11}$$

where $Sh_{st} = Sh(1, Pe, \infty)$ is the Sherwood number corresponding to the solution of the linear steady-state problem with constant diffusion coefficient. For spherical drops and bubbles in a translational or straining flow, the values of $Sh_{st}$ are shown in Table 4.7; the diffusion coefficient used there is $D(C_i)$.

# 5.7. Film Condensation

## 5.7-1. Statement of the Problem

Suppose that on a vertical wall whose temperature is constant and equal to $T_s$, stagnant dry saturated vapor is condensing. Let us consider the steady-state problem under the assumption that we have laminar waveless flow in the condensate film. According to [200], we make the following assumptions: the film motion is determined by gravity and viscosity forces; the heat transfer is only across the film due to heat conduction; there is no dynamic interaction between the liquid and vapor phases; the temperature on the outer surface of the condensate film is constant and equal to the saturation temperature $T_g$; the physical parameters of the condensate are independent of temperature and the vapor density is small compared with the condensate density; the surface tension on the free surface of the film does not affect the flow.

In a rectangular Cartesian coordinate system with the $X$-axis directed along the plane along which the film flows, the only nonzero velocity component is $V_X$. If we ignore the pressure gradient along the $X$-axis and assume that the film liquid velocity $V_X$ and its temperature $T_*$ depend only on the transverse coordinate $Y$, then we obtain the system of equations

$$\mu \frac{d^2 V_X}{dY^2} = -\rho g, \qquad \frac{d^2 T_*}{dY^2} = 0 \qquad (5.7.1)$$

with the boundary conditions

$$V_X = 0 \quad \text{and} \quad T_* = T_s \qquad \text{at} \quad Y = 0,$$
$$dV_X/dY = 0 \quad \text{and} \quad T_* = T_g \qquad \text{at} \quad Y = \delta(X), \qquad (5.7.2)$$

where $\mu$ is the viscosity of the liquid, $\rho$ is the density, $g$ is the free fall acceleration, and $\delta(X)$ is the thickness of the condensate film, which depends on the longitudinal coordinate $X$.

It is necessary to consider Eqs. (5.7.1) simultaneously, since the heat flux across the film determines the quantity of vapor condensing on its surface (we assume that the entire heat transferred to the wall is the heat of the phase transition) and thus affects the velocity $V_X$, thus varying the thickness $\delta(X)$ of the condensate film. To determine $\delta(X)$, we use the equation of liquid balance in the film with allowance for the vapor condensation on the surface,

$$\rho \frac{d}{dX} \int_0^{\delta(X)} V_X \, dY = \frac{\varkappa}{\Delta H} \frac{dT_*}{dY}\bigg|_{Y=\delta(X)}, \qquad (5.7.3)$$

where $\varkappa$ is the thermal conductivity coefficient of the condensate and $\Delta H$ is the specific condensation (evaporation) heat.

## 5.7-2. Equation for the Thickness of the Film. Nusselt Solution

By solving problem (5.7.1)–(5.7.3), we arrive at the following equation for the thickness of the condensate film:

$$\frac{g\rho^2 \Delta H}{\varkappa\mu} \delta^3 \frac{d\delta}{dX} = T_g - T_s \tag{5.7.4}$$

with the boundary condition

$$\delta = 0 \quad \text{at} \quad X = 0. \tag{5.7.5}$$

It follows from (5.7.4) and (5.7.5) that

$$\delta(X) = \left[ \frac{4\varkappa\mu(T_g - T_s)X}{g\rho^2\Delta H} \right]^{1/4}. \tag{5.7.6}$$

On introducing the coefficient of convective heat transfer by the formula $\alpha = \varkappa/\delta$, we finally obtain

$$\alpha = \left[ \frac{g\rho^2 \varkappa^3 \Delta H}{4\mu(T_g - T_s)X} \right]^{1/4}. \tag{5.7.7}$$

Note that this result was first obtained by Nusselt.

Because of the above-mentioned assumptions, one must treat the solution (5.7.6) and (5.7.7) as an approximate solution. Following [200], we generalize this solution to some practically important cases. If we take into account the inertial term in the equation of motion and the convective heat transfer in the film, then the solution of the problem shows that for the Kutateladze number $\mathsf{Ku} = \dfrac{\Delta H}{c_p \Delta T} > 5$ (here $c_p$ is the specific heat of the condensate and $\Delta T$ is the temperature head) and $1 < \mathsf{Pr} < 100$, the difference between the solution and (5.7.7) is at most several percent. However, for large temperature heads (small $\mathsf{Ku}$), formula (5.7.7) gives a very underestimated $\alpha$; on the contrary, for liquid metals formula (5.7.7) gives an overestimated $\alpha$.

## 5.7-3. Some Generalizations

*Film condensation on the wall.* If physical parameters of the condensate (the thermal conductivity coefficient $\varkappa$ and the viscosity $\mu$) depend on temperature and if we have wave motion of the film, then we need to use the following relation [200] for the coefficient of heat transfer $\alpha_1$:

$$\alpha_1 = \alpha \epsilon_T \epsilon_V, \tag{5.7.8}$$

where $\epsilon_T$ and $\epsilon_V$ are the corresponding correction coefficients. Thus, to take into account the dependence of physical parameters on temperature, one can use the relation $\epsilon_T = [(\varkappa_s/\varkappa_g)^3 \mu_g/\mu_s]^{1/8}$, where the indices "s" and "g" mean that this particular coefficient must be taken at the wall temperature or at the saturation temperature, respectively. In this case, the parameters in formula (5.7.7) must correspond to the saturation temperature. It is assumed that the correction to the wave motion depends only on the Reynolds number $\epsilon_V = \text{Re}^{0.04}$. At $\text{Re} = 0.1$ to 10, we have the correction $\epsilon_V \approx 1$, and its value grows with increasing Re. Here the Reynolds number must be calculated with respect to the parameter values corresponding to the lowest downstream cross-section of the condensate film, namely, at $\text{Re} = \dfrac{g\delta^3(L)}{3\nu^2}$, where $L$ is the length of the condensate film.

For $\text{Re} \geq 400$, the flow in the film becomes turbulent. Then to describe the local coefficient $\alpha$ of convective heat transfer for the parameters $1 \leq \text{Pr} \leq 25$ and $1.5 \times 10^3 \leq \text{Re} \leq 6.9 \times 10^4$ it is suggested to use the following approximate relation [200]:

$$\frac{\alpha}{\varkappa}\left(\frac{\nu}{g}\right)^{1/3} = 0.0325\,\text{Re}^{0.25}\,\text{Pr}^{0.5}. \tag{5.7.9}$$

If it is required to take into account the dependence of physical parameters on temperature, then the coefficient $0.0325$ in this formula must be multiplied by the correction coefficient $\epsilon_T = (\text{Pr}_g/\text{Pr}_s)^{0.25}$ and all parameters in formula (5.7.9) must be taken at the saturation temperature.

If the vapor pressure is large and its density is comparable with the condensate density on the right-hand side in Eq. (5.7.1), then the term $g\rho$ must be replaced by $g\Delta\rho$, where $\Delta\rho$ is the difference between the condensate and vapor densities.

If the wall along which the condensate film flows makes an angle $\theta$ with the vertical, then one must replace $g$ by $g\cos\theta$ in the original equation of motion (5.7.1) and in all subsequent relations.

**Film condensation on a horizontal tube.** For a curvilinear surface, in particular, for a horizontal circular cylinder along which a condensate film flows, the angle $\theta$ is a nonconstant variable. By taking into account the fact that $\delta(\theta) \ll d$, where $d$ is the diameter of a circular cylinder, and proceeding by analogy with (5.7.7), one can readily obtain the following formula for the heat transfer coefficient averaged over the external surface of the tube provided that the flow in the condensate film is laminar [200]:

$$\bar{\alpha} = 0.728\left[\frac{g\rho^2\varkappa^3\Delta H}{4\mu(T_g - T_s)d}\right]^{1/4}. \tag{5.7.10}$$

Note that this formula was obtained under the same assumptions as (5.7.7) and can be generalized to all above-mentioned cases. In particular, this formula can be generalized to the case of wave motion by introducing the usual correction $d > 20(\delta/g\rho)^{0.5}$.

In [200] the following relation is proposed for calculating the convective heat transfer in moving vapor that is condensing on a horizontal tube (the vapor moves from top to bottom) provided that the flow in the condensate film is laminar:

$$\frac{\bar{\alpha}}{\bar{\alpha}_0} = \left(1 + 3.62\gamma^4\,\frac{\text{Fr}}{\text{Pr}\,\text{Ku}}\right)^{1/4}, \tag{5.7.11}$$

where $\bar{\alpha}_0$ is given by (5.7.10) and presents the mean coefficient of convective heat transfer in condensing stagnant vapor, $\mathsf{Fr} = U_g^2/(gd)$ is the Froude number, and $U_g$ is the velocity of the incoming vapor. The coefficient $\gamma$ must be determined by

$$\gamma = 0.9 \left(1 + \mathsf{Pr}\,\mathsf{Ku}\,\sqrt{\frac{\rho_g \mu_g}{\rho_f \mu_f}}\right)^{1/3}.$$

The subscript "g" corresponds to the vapor and "f", to the condensate.

Note that at $U_g = 0$ relation (5.7.11) turns into formula (5.7.10).

**Film condensation in a vertical tube.** To present a theoretical description of film condensation in tubes is much more difficult, since there may arise a strong dynamic interaction between the moving vapor and the flowing condensate film. If the direction of the vapor motion coincides with the direction of the condensate motion due to gravity, then, owing to viscous friction on the phase boundary, the velocity of the film flow increases, its thickness decreases, and the coefficient of convective heat transfer also increases. If the direction of the vapor motion is opposite to that of the condensate flow, then we have the opposite situation. If the vapor velocity increases, then the film may partially separate from the wall and convective heat transfer can increase sharply.

For the calculation of the local coefficient of convective heat transfer $\alpha$ when the condensation takes place in tubes provided that flow in the condensate film is laminar, the following relation based on experimental data was proposed in [200]:

$$\alpha = \alpha_0 \sqrt{0.005\psi + \sqrt{(0.005\psi)^2 + 1}},$$

where $\alpha_0$ is determined by formula (5.7.7). The correction parameter $\psi$ can be calculated by the formula

$$\psi = \epsilon_T \frac{\rho_g}{\rho_f} \left(\frac{\nu_g}{\nu_f}\right)^2 \mathsf{Re}_g^2 \, \mathsf{Re}_{fX}^{-0.28} \, \mathsf{Ga}_f^{-0.67},$$

where $\mathsf{Re}_g = \bar{U} d/\nu_g$, $\mathsf{Ga}_f = gd^3/\nu_f^2$, and $\mathsf{Re}_{fX} = QX/(\mu \Delta H)$. Here $Q$ is the heat flux to the wall and $\bar{U}$ is the vapor velocity averaged over the cross-section $X$. The subscript "g" corresponds to the vapor and "f", to the condensate. The values of all physical parameters are taken at the saturation temperature. This dependence was derived by using experimental data in the range $1.8 \times 10^3 \leq \mathsf{Re}_g \leq 1.7 \times 10^4$. It should be noted that the velocity $\bar{U}$ averaged over the cross-section $X$ is, generally speaking, not constant but depends on the coordinate $X$, since, because of the condensation, the vapor discharge decreases along the tube (but the condensate discharge increases).

# 5.8. Nonisothermal Flows in Channels and Tubes

## 5.8-1. Heat Transfer in Channel. Account of Dissipation

Let us consider the problem of dissipative heating of a fluid in a plane channel of width $2h$ with isothermal walls on which the same constant temperature is maintained:

$$T_* = T_s \quad \text{at} \quad Y = 0, \qquad T_* = T_s \quad \text{at} \quad Y = 2h. \tag{5.8.1}$$

If the fluid temperature at the entry cross-section is equal to the wall temperature $T_s$, then, along some initial part of the tube, the fluid is gradually heated by internal friction until a balance is achieved between the heat withdrawal through the walls and the dissipative heat release. In the region where such an equilibrium is established, the fluid temperature does not vary along the channel, that is, the temperature field is stabilized (provided that the velocity profile is also stabilized). In what follows we just study this thermally and hydrodynamically stabilized flow.

Following [208, 276], we assume that the heat release does not affect the physical properties of the fluid (i.e., viscosity, density, and thermal conductivity coefficient are temperature independent). In this case, the velocity profile can be found independently of the heat problem (for laminar flow, see Subsection 1.5-2).

The equation for the temperature distribution in the region of heat stabilization can be obtained from the equation of heat transfer in Supplement 6 (where the temperature depends only on the transverse coordinate $Y$ and the convective terms are equal to zero, since $V_X = V(Y)$ and $V_Y = V_Z = 0$). This equation has the form

$$\varkappa \frac{d^2 T_*}{dY^2} + \mu \left( \frac{dV}{dY} \right)^2 = 0, \tag{5.8.2}$$

where $\varkappa$ is the thermal conductivity coefficient and $\mu$ is the viscosity.

The exact solution of problem (5.8.1)–(5.8.2) leads to the following temperature distribution in the channel [427]:

$$T_* - T_s = \frac{\mu U_{max}^2}{3\varkappa} \left[ 1 - \left( 1 - \frac{Y}{h} \right)^4 \right], \tag{5.8.3}$$

where $U_{max}$ is determined by the last formula in Subsection 1.5-2.

The maximum temperature difference is given by

$$T_{max} - T_s = \frac{\mu U_{max}^2}{3\varkappa}.$$

The heat flux to the tube wall can be found by using the expression

$$q_T = \varkappa \left( \frac{dT_*}{dY} \right)_{Y=0} = \frac{4\mu U_{max}^2}{3h}.$$

---

### 5.8-2. Heat Transfer in Circular Tube. Account of Dissipation

Under the same assumptions (the tube temperature is constant and is equal to $T_s$, and the physical properties of the medium are temperature independent), the temperature distribution in the region of heat stabilization in a circular tube of radius $a$ has the form [158]

$$T_* - T_s = \frac{\mu}{\varkappa}\langle V \rangle^2 \left[ 1 - \left( \frac{\mathcal{R}}{a} \right)^4 \right],$$

where the mean flow rate velocity in the tube, $\langle V \rangle = \frac{1}{2}U_{max}$, is defined by formula (1.5.12).

The maximum temperature difference for a fluid flow in a circular tube can be expressed as follows:

$$T_{max} - T_s = \frac{\mu}{\varkappa}\langle V \rangle^2. \tag{5.8.4}$$

The formulas obtained in this section can be used for a majority of common fluids. Flows of extremely viscous liquids possess specific qualitative features, which are described in the next section.

### 5.8-3. Qualitative Features of Heat Transfer in Highly Viscous Liquids

Dissipative heating in highly viscous liquids leads to a considerable rise in temperature even at moderate velocities of motion. For example, according to Table 5.3 (according to [427, 487]), the viscosity and the thermal conductivity coefficient of motor oil at indoor temperature ($T_s = 20°C$) are, respectively, $\mu = 0.8\,\text{kg/(m} \cdot \text{s)}$ and $\varkappa = 0.15\,\text{N/(s} \cdot \text{K)}$. By substituting these values into (5.8.4), we obtain

$$T_{max} - T_s = \begin{cases} 5.5°C & \text{for } \langle V \rangle = 1 \text{ m/s,} \\ 22°C & \text{for } \langle V \rangle = 2 \text{ m/s,} \\ 49.5°C & \text{for } \langle V \rangle = 3 \text{ m/s.} \end{cases}$$

Thus, the rise in the oil temperature is so large that one must take account of the dependence of viscosity on temperature (it follows from Table 5.3 that the viscosities at 20°C and 60°C differ by one order of magnitude). In this case, the variability of the specific heat and thermal conductivity coefficient of oil is inessential and can be neglected in the first approximation. In the following we study the nonlinear effects related to the dependence of viscosity on temperature.

For highly viscous liquids (such as glycerin), an exponential dependence of viscosity on temperature is usually assumed [52, 133, 253]:

$$\mu = \mu_0 \exp\left[-\omega(T_* - T_0)\right], \tag{5.8.5}$$

where $\mu_0$, $\omega$, and $T_0$ are empirical constants.

The temperature distribution in the tube for a nonisothermal flow of a liquid in the case an exponential dependence of viscosity on temperature (5.8.5) has the form [52]

$$T_* = T_s + \frac{1}{\omega}\ln\frac{8}{\varepsilon} - \frac{2}{\omega}\ln\left[b\left(\frac{\mathcal{R}}{a}\right)^4 + \frac{1}{b}\right]. \tag{5.8.6}$$

TABLE 5.3.
Physical characteristics of some substances

| Substance (under a pressure of 1 atm) | Temperature $T$, °C | Specific heat $c_p$, $\frac{m^2}{s^2 \cdot K}$ | Thermal conductivity $\varkappa \cdot 10^6$, $\frac{N}{s \cdot K}$ | Thermal diffusivity $\chi \cdot 10^6$, $\frac{m^2}{s}$ | Viscosity $\mu \cdot 10^6$, $\frac{kg}{m \cdot s}$ | Kinematic viscosity $\nu \cdot 10^6$, $\frac{m^2}{s}$ | Prandtl number $Pr = \frac{\mu c_p}{\varkappa}$ |
|---|---|---|---|---|---|---|---|
| Water | 20 | 4183 | 0.598 | 0.143 | 1000 | 1.006 | 7.03 |
| | 40 | 4179 | 0.627 | 0.151 | 654 | 0.658 | 4.35 |
| | 60 | 4191 | 0.650 | 0.159 | 470 | 0.478 | 3.03 |
| | 80 | 4199 | 0.670 | 0.164 | 354 | 0.364 | 2.22 |
| Motor oil "Rotling" | 20 | 1840 | 0.145 | 0.088 | 796000 | 892 | 10100 |
| | 40 | 1920 | 0.143 | 0.084 | 204000 | 231 | 2750 |
| | 60 | 2000 | 0.141 | 0.081 | 71300 | 82 | 1020 |
| | 80 | 2100 | 0.140 | 0.078 | 31500 | 37 | 471 |
| Mercury | 20 | 138 | 9.3 | 5 | 1560 | 0.115 | 0.023 |
| Air | 0 | 1006 | 0.0242 | 19.2 | 17.1 | 13.6 | 0.71 |
| | 50 | 1006 | 0.0278 | 26.2 | 19.6 | 18.6 | 0.71 |
| | 100 | 1009 | 0.0310 | 33.6 | 21.8 | 23.8 | 0.71 |
| | 200 | 1028 | 0.0368 | 49.7 | 25.9 | 35.9 | 0.71 |

where the parameters $\varepsilon$ and $b$ are given by

$$\varepsilon = \frac{a^2\omega}{16\mu_0\varkappa}\left(\frac{\Delta P}{L}\right)^2 \exp[\omega(T_s - T_0)], \quad b = \left(\frac{2}{\varepsilon}\right)^{1/2} - \left(\frac{2}{\varepsilon} - 1\right)^{1/2}. \quad (5.8.7)$$

The liquid velocity profile in a tube is determined by the formula

$$V = \frac{16\varkappa bL}{a^2\omega\Delta P}\left[\frac{b}{1+b^2} + \arctan b - \frac{by}{1+b^2y^2} - \arctan(by)\right], \quad y = \left(\frac{R}{a}\right)^2. \quad (5.8.8)$$

By setting $y = 0$ in (5.8.8), we calculate the velocity on the axis of the tube:

$$U_{\max} = \frac{16\varkappa bL}{a^2\omega\Delta P}\left(\frac{b}{1+b^2} + \arctan b\right).$$

By passing to the limit as $\omega \to 0$ in Eq. (5.8.8), we obtain an isothermal velocity profile of liquid, which is considered in Subsection 1.5-3.

The critical condition of liquid flow in a tube corresponds to $\varepsilon = 2$. For $\varepsilon > 2$, there cannot be a steady-state rectilinear flow in the tube. In this case, the heat due to viscous friction cannot be completely released through the tube walls and results in a rapid increase in temperature (that is, a thermal explosion).

## 5.8-4. Nonisothermal Turbulent Flows in Tubes

For dropping liquids whose viscosity is temperature-dependent, the nonisothermicity can be taken into account by the relation [254, 267]

$$\frac{\mathsf{Nu}}{\mathsf{Nu}_\infty} = \left(\frac{\mu}{\mu_s}\right)^\gamma, \quad (5.8.9)$$

where $\mu_s$ is the dynamic viscosity of the liquid at the wall temperature. The number $\mathsf{Nu}_\infty$ is given by Eq. (3.7.2), with the values of the physical-chemical parameters that characterize the liquid properties (first of all, $\mu$) being determined on the basis of the mean temperature of the liquid at the examined cross-section of the tube. The exponent $\gamma$ is taken to be

$$\gamma = \begin{cases} 0.11 & \text{in heating } (\mu_s/\mu < 1), \\ 0.25 & \text{in cooling } (\mu_s/\mu > 1). \end{cases}$$

Relation (5.8.9) is valid for $0.08 \leq \mu_s/\mu \leq 40$, $10^4 \leq \mathsf{Re}_d \leq 1.25 \times 10^5$, and $2 \leq \mathsf{Pr} \leq 140$.

The drag coefficient $\lambda$ for nonisothermal flows of dropping liquids can be estimated as [254]

$$\frac{\lambda}{\lambda_0} = \left(\frac{\mu_s}{\mu}\right)^\sigma, \quad (5.8.10)$$

where $\lambda_0$ can be determined from Eq. (1.6.12) or (1.6.13) taking into account the fact that the values of the physical-chemical parameters must be calculated on the basis of the mean temperature of the liquid at the examined cross-section of the tube. The exponent $\sigma$ is taken to be

$$\sigma = \begin{cases} 0.17 & \text{in heating } (\mu_s/\mu < 1), \\ 0.24 & \text{in cooling } (\mu_s/\mu > 1). \end{cases}$$

For nonisothermal flows of gases through a circular tube, the heat exchange rate can be estimated using the formula [267]

$$\text{Nu} = 0.023\,\text{Re}_d^{0.8}\,\text{Pr}^{0.4}\left(\frac{T}{T_s}\right)^{0.37}, \qquad (5.8.11)$$

where $T_s$ is the wall temperature; the values of all thermal-physical parameters must be determined on the basis of the mean temperature at the examined cross-section. This formula quite well agrees with experimental data for $8 \times 10^3 \le \text{Re}_d \le 8 \times 10^5$.

For liquid metals, the nonisothermal correction can, as a rule, be ignored [254].

More detailed information about heat transfer in turbulent nonisothermal flows through a circular tube or plane channel, as well as various relations for Nusselt numbers, can be found in the books [185, 254, 267, 406], which contain extensive literature surveys.

# 5.9. Thermogravitational and Thermocapillary Convection in a Fluid Layer

**Preliminary remarks.** In the preceding chapters it was assumed that the fluid velocity field is independent of the temperature and concentration distributions. However, there are a few phenomena in which the influence of these factors on the hydrodynamics is critical. This influence arises from the fact that various physical parameters of fluids, such as density, surface tension, etc. are temperature or concentration dependent.

For example, the convective motion of a liquid in a vessel whose opposite walls are maintained at different temperatures is due to the fact that the fluid density is normally a decreasing function of temperature. The lighter liquid near the heated wall tends to rise, whereas the heavier liquid near the opposite wall tends to lower. This is one of the examples in which the so-called gravitational (in this case, thermogravitational) convection manifests itself.

If the surface tension coefficient is not constant along the interface between two nonmixing fluids, then there arise additional tangential stresses on the interface; they are referred to as capillary stresses and can substantially affect the motion of the fluid or even solely determine it in the absence of gravitational and other forces. Phenomena due to surface tension gradients have a general name of Marangoni effects. In particular, if the dependence of surface tension on temperature is essential, then this effect is called thermocapillary, and if the dependence on the concentration is of importance, then we speak of a concentration-capillary effect.

The intense study of numerous problems related to the temperature or concentration gradient influence on the fluid motion is stimulated, apart from purely scientific interest, by a possibility of their wide applications in technology (first of all, chemical and space technology).

## 5.9-1. Thermogravitational Convection

*Statement of the problem.* Let us consider the motion of a viscous fluid in an infinite layer of constant thickness $2h$. The force of gravity is directed normally to the layer. The lower plane is a hard surface on which a constant temperature gradient is maintained. The nonuniformity of the temperature field results in two effects that can bring about the motion of the fluid, namely, the thermogravitational effect related to the heat expansion of the fluid and the appearance of Archimedes forces, and the thermocapillary effect (if the second surface is free) produced by tangential stresses on the interface due to the temperature dependence of the surface tension coefficient.

To describe the two-dimensional problem, we use the rectangular coordinate system $X, Y$, where the $X$-axis is directed oppositely to the temperature gradient on the lower surface and the $Y$-axis is directed vertically upward. The origin is chosen to be in the middle of the layer; therefore, $-h \leq Y \leq h$. The velocity and temperature fields are described by the equations [142, 143]

$$V_X \frac{\partial V_X}{\partial X} + V_Y \frac{\partial V_X}{\partial Y} = -\frac{1}{\rho} \frac{\partial P}{\partial X} + \nu \left( \frac{\partial^2 V_X}{\partial X^2} + \frac{\partial^2 V_X}{\partial Y^2} \right), \tag{5.9.1}$$

$$V_X \frac{\partial V_Y}{\partial X} + V_Y \frac{\partial V_Y}{\partial Y} = -\frac{1}{\rho} \frac{\partial P}{\partial Y} + \nu \left( \frac{\partial^2 V_Y}{\partial X^2} + \frac{\partial^2 V_Y}{\partial Y^2} \right) + \gamma g T_*, \tag{5.9.2}$$

$$\frac{\partial V_X}{\partial X} + \frac{\partial V_Y}{\partial Y} = 0, \tag{5.9.3}$$

$$V_X \frac{\partial T_*}{\partial X} + V_Y \frac{\partial T_*}{\partial Y} = \chi \left( \frac{\partial^2 T_*}{\partial X^2} + \frac{\partial^2 T_*}{\partial Y^2} \right). \tag{5.9.4}$$

Here $P$ is the pressure (taking into account the gravity potential), $\chi$ is the thermal diffusivity coefficient, $g$ is the gravitational acceleration, and $\gamma$ is the thermal expansion coefficient.

The thermogravitational motion is described in the Boussinesq approximation in which the variable density in the equations of motion (5.9.1)–(5.9.3) and in the convective heat conduction equation (5.9.4) is taken into account only in the Archimedes term (the last term in (5.9.2)). This term is proportional to the temperature deviation $T_*$ from the mean value. The thermocapillary motion is produced by surface forces, which are taken into account in the boundary condition on the free surface (see below).

For the one-dimensional flow of the fluid along the $X$-axis, the original equations (5.9.1)–(5.9.4) become [42]

$$\frac{1}{\rho}\frac{\partial P}{\partial X} = \nu\frac{\partial^2 V_X}{\partial Y^2}, \qquad \frac{1}{\rho}\frac{\partial P}{\partial Y} = \gamma g T_*,$$

$$V_X\frac{\partial T_*}{\partial X} = \chi\left(\frac{\partial^2 T_*}{\partial X^2} + \frac{\partial^2 T_*}{\partial Y^2}\right). \tag{5.9.5}$$

First, let us consider the case of purely thermogravitational convection. It corresponds to the case in which heat is supplied through both boundaries and a constant temperature gradient is maintained on both boundaries. The boundary conditions can be written in the form

$$T_* = -AX, \quad V_X = 0 \qquad \text{for} \quad Y = \pm h. \tag{5.9.6}$$

Here $A$ is the temperature gradient (for $A < 0$, the gradient is codirected with the $X$-axis). In (5.9.6), the temperature is measured from its value at $X = 0$.

We introduce the dimensionless variables and parameters

$$x = \frac{X}{h}, \quad y = \frac{Y}{h}, \quad v = \frac{h}{\nu}V_X, \quad p = \frac{P}{Ah^2\rho\gamma g}, \quad T = \frac{T_*}{Ah},$$

$$\mathrm{Pr} = \frac{\nu}{\chi}, \quad \mathrm{Gr} = \frac{Ah^4\gamma g}{\nu^2},$$

where $\mathrm{Pr}$ is the Prandtl number and $\mathrm{Gr}$ is the Grashof number. By substituting them into Eqs. (5.9.5) and the boundary conditions (5.9.6), we obtain

$$\mathrm{Gr}\,\frac{\partial p}{\partial x} = \frac{\partial^2 v}{\partial y^2}, \qquad \frac{\partial p}{\partial y} = T,$$

$$\mathrm{Pr}\,v\frac{\partial T}{\partial x} = \frac{\partial^2 T}{\partial x^2} + \frac{\partial^2 T}{\partial y^2}, \tag{5.9.7}$$

$$v = 0, \quad T = -x \qquad \text{for} \quad y = -1,$$
$$v = 0, \quad T = -x \qquad \text{for} \quad y = 1. \tag{5.9.8}$$

**Exact solution.** We seek a solution of problem (5.9.7), (5.9.8) in the form

$$v = v(y), \quad p = -(b + y)x + p_1(y), \quad T = -x + T_1(y). \tag{5.9.9}$$

As a result, we obtain the following velocity distribution for $v(y)$:

$$v(y) = \tfrac{1}{6}\,\mathrm{Gr}(y - y^3) + \tfrac{1}{2}b\,\mathrm{Gr}(1 - y^2). \tag{5.9.10}$$

There is an unknown constant $b$ in the solution (5.9.10). Let us calculate the rate of flow in the layer:

$$q \equiv \int_{-1}^{1} v(y)\,dy = \tfrac{2}{3}b\,\mathsf{Gr}. \tag{5.9.11}$$

By setting it equal to zero, we obtain $b = 0$. Note that in this section and in what follows, as well as in [42], we consider the case $q = 0$. But the problem on a flow with nonzero rate is also physically meaningful. For this case, the constant $b \neq 0$ is related to the rate of flow by formula (5.9.11).

Using Eqs. (5.9.7), formulas (5.9.9), and the boundary conditions (5.9.8), one can obtain the following expressions for $T_1(y)$ and $p_1(y)$:

$$\begin{aligned}
T_1(y) &= \tfrac{1}{360}\,\mathsf{Pr}\,\mathsf{Gr}(3y^5 - 10y^3 + 7y), \\
p_1(y) &= \tfrac{1}{720}\,\mathsf{Pr}\,\mathsf{Gr}(y^6 - 5y^4 + 7y^2) + \text{const}.
\end{aligned} \tag{5.9.12}$$

We see that the pressure is determined up to a constant additive term.

We note that the condition of zero rate of flow is based on the assumption that the flow "turns" as $x \to \pm\infty$. Our model can serve as an asymptotic description of the motion far from the ends of a plane gap closed from both sides.

---

### 5.9-2. Joint Thermocapillary and Thermogravitational Convection

***Statement of the problem.*** Let us consider a similar problem in which the upper boundary of the channel is free and the surface tension $\sigma$ on it depends linearly on temperature. The balance of tangential stresses on the free surface will then involve the thermocapillary stresses. The corresponding boundary condition has the form

$$\rho\nu\frac{\partial V_X}{\partial Y} = \sigma'\frac{\partial T_*}{\partial X} \qquad \text{for} \quad Y = h, \tag{5.9.13}$$

where $\sigma' = d\sigma/dT_* = \text{const}$. Here the left-hand side includes the viscous stress, and the right-hand side includes the thermocapillary stress.

The dimensionless equations and boundary conditions, as before, have the form (5.9.7), (5.9.8) except for the second boundary condition in (5.9.8), which must be replaced, according to (5.3.13), by

$$\frac{\partial v}{\partial y} = \mathsf{Ma}\,\frac{\partial T}{\partial x}, \qquad T = -x \qquad \text{for} \quad y = 1, \tag{5.9.14}$$

where $\mathsf{Ma} = Ah^2\sigma'/(\rho\nu^2)$ is the Marangoni number. (With this definition, it can be of either sign according to the signs of $A$ and $\sigma'$.*) One still can seek the general solution of problem (5.9.7), (5.9.14) in the form (5.9.9).

***Exact solution.*** The solution can be represented as the sum of three terms, each having a remarkably simple physical interpretation: the Poiseuille motion due to the constant dimensionless pressure gradient along the layer, the thermo-

---

* It is important to note that, for an overwhelming majority of fluids, surface tension decreases with temperature, and consequently, the inequality $\sigma' < 0$ is valid (in what follows, liquids for which $\sigma' > 0$ in a certain interval of the temperature variation will also be described).

gravitational motion, and the thermocapillary motion. The velocity distribution is of the form

$$v(y) = \tfrac{1}{2} b \, \mathrm{Gr}(-y^2 + 2y + 3) + \tfrac{1}{6} \, \mathrm{Gr}(-y^3 + 3y + 2) - \mathrm{Ma}(y + 1). \qquad (5.9.15)$$

The constant $b$ can be determined from the zero-rate-of-flow condition:

$$b = -\tfrac{1}{4} + \tfrac{3}{4} \, \mathrm{Ma} \, \mathrm{Gr}^{-1}. \qquad (5.9.16)$$

One can see from (5.9.15) that in the absence of thermogravitational forces and longitudinal pressure gradient ($b = \mathrm{Gr} = 0$), the velocity profile is linear. Then the rate of flow proves to be nonzero. At the same time, the expressions (5.9.15) and (5.9.16) show that a flow of zero rate can be produced by the Marangoni forces only if there is a nonzero longitudinal pressure gradient.

The velocity distribution for zero rate of flow is

$$v(y) = \tfrac{1}{24} \, \mathrm{Gr}(-4y^3 + 3y^2 + 6y - 1) - \tfrac{1}{8} \, \mathrm{Ma}(3y^2 + 2y - 1). \qquad (5.9.17)$$

Taking into account (5.9.17), for $T_1$ one can find

$$\begin{aligned}
T_1 = \; & \tfrac{1}{480} \, \mathrm{Pr} \, \mathrm{Gr}(4y^5 - 5y^4 - 20y^3 + 10y^2 + 16y - 5) \\
& + \tfrac{1}{96} \, \mathrm{Pr} \, \mathrm{Ma}(3y^4 + 4y^3 - 6y^2 - 4y + 3).
\end{aligned} \qquad (5.9.18)$$

We note that the velocity field (5.9.17) would not change if the linear temperature distribution were maintained only on the lower solid surface, whereas the free surface were thermally insulated. In this case, the second boundary condition (5.9.14) is replaced by the condition $\partial T / \partial y = 0$ for $y = 1$, and the solution still has the form (5.9.9).

In conclusion, we observe that to justify the suggested problem setting, one has to add the assumption of a flat free surface. Actually, the normal stress on the liquid surface prove to be variable in our cases, which will result in a distortion of the surface. But this effect is absent at large $g$, when any inner pressure is neutralized by an infinitesimal change of the surface shape.

### 5.9-3. Thermocapillary Motion. Nonlinear Problems

*Nonlinear dependence of the surface tension on temperature.* In the preceding, it was assumed that the dependence of the surface tension on temperature is linear. However, for a number of liquids, such as water solutions of high-molecular alcohols and some binary metallic alloys, it was experimentally proved that the function $\sigma = \sigma(T_*)$ is nonlinear and nonmonotone [264, 493, 494]. In Figure 5.3, experimental curves are presented [264] showing that $\sigma = \sigma(T_*)$ can have a pronounced minimum (the numbers on curves correspond to the number of carbon atoms in the alcohol molecule; the experiments were carried out at

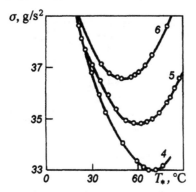

**Figure 5.3.** Experimental nonlinear curves of surface tension against temperature

low concentrations because high-molecule alcohols are poorly soluble in water.) This dependence can be approximated by the formula

$$\sigma = \sigma_0 + \tfrac{1}{2}\alpha(T_* - T_0)^2, \tag{5.9.19}$$

where $T_0$ is the temperature value corresponding the extremum of the surface tension coefficient.

**Statement of the problem.** Let us consider the problem of a steady-state thermocapillary motion in a liquid layer of thickness $h$. The motion is assumed to be two-dimensional. The dependence of the surface tension on temperature is assumed to be quadratic according to (5.9.19). The thermogravitational effect is not taken into account. It is assumed that the linear temperature distribution is maintained on the hard lower surface, and the plane surface of the layer is thermally insulated. The origin of the Cartesian coordinates $X, Y$ is placed on the solid surface at the point with temperature $T_0$. The velocity and temperature fields are described by Eqs. (5.9.1)–(5.9.4) with $\gamma g \equiv 0$.

The boundary conditions taking into account the quadratic dependence of the surface tension (5.9.19) on temperature have the form

$$V_X = 0, \quad V_Y = 0, \quad T_* = T_0 - AX \qquad \text{for} \quad Y = 0, \tag{5.9.20}$$

$$V_Y = 0, \quad \frac{\partial T_*}{\partial Y} = 0, \quad \rho\nu\frac{\partial V_X}{\partial Y} = \frac{\partial \sigma}{\partial X} \qquad \text{for} \quad Y = h. \tag{5.9.21}$$

By (5.9.20), the no-slip and no-flow conditions hold on the hard surface, and a linear temperature distribution is maintained. Condition (5.9.21) says that the no-flow condition on the free surface and the condition of zero heat flux through the free surface must hold, and the balance of tangential thermocapillary and viscous stresses must be provided. Taking into account the quadratic dependence (5.9.19) of the surface tension on temperature, we rewrite the right-hand side of the last condition in (5.9.21) using the relation

$$\frac{\partial \sigma}{\partial X} = \alpha(T_* - T_0)\frac{\partial T_*}{\partial X}.$$

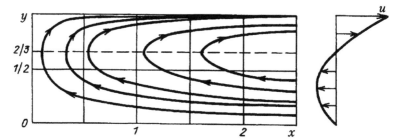

**Figure 5.4.** Streamlines and the profile of the longitudinal velocity component for thermocapillary flow in a liquid layer

**Solution of the problem.** The solution is sought [165] in the form

$$V_X = Ux\psi'(y), \quad V_Y = -U\psi(y),$$
$$T_* = T_0 - Ahx\Theta(y), \quad P = P_0 - \tfrac{1}{2}\rho U^2[\lambda x^2 + f(y)], \tag{5.9.22}$$

where $x = X/h$ and $y = Y/h$ are dimensionless coordinates, $U = \nu/h$ is the characteristic velocity, $P_0 = $ const is the pressure at the critical point on the hard surface (where $T_* = T_0$), and $\psi' = d\psi/dy$.

For the unknown functions $\psi(y)$, $\Theta(y)$, and $f(y)$ and the constant $\lambda$, treated as an eigenvalue, we obtain the following problem:

$$\psi''' + \psi\psi'' - (\psi')^2 + \lambda = 0, \quad f = \psi^2 + 2\psi',$$
$$\Theta'' - \mathsf{Pr}(\psi'\Theta - \psi\Theta') = 0;$$
$$\psi = 0, \quad \psi' = 0, \qquad \Theta = 1 \qquad \text{for} \quad y = 0; \tag{5.9.23}$$
$$\psi = 0, \quad \psi'' = \mathsf{Ma}\,\Theta^2, \quad \Theta' = 0 \qquad \text{for} \quad y = 1,$$

where $\mathsf{Ma} = \alpha A^2 h^3/(\rho\nu^2)$ is the Marangoni number modified for our special case.

The eigenvalue problem (5.9.23) was solved numerically in [44].

To study special features of the thermocapillary flow, we consider an approximate analytical solution of the problem at small Marangoni numbers under the assumption that the Prandtl number is of the order of 1.

For $\mathsf{Ma} = 0$, the problem has the solution $\psi = 0$, $f = 0$, $\lambda = 0$, $\Theta = 1$, which corresponds to a stagnant fluid with uniform temperature distribution across the layer.

For $|\mathsf{Ma}| \ll 1$, the solution is obtained by perturbation technique [44, 165]. We have the following expressions for the velocity, pressure, and temperature fields accurate to $O(\mathsf{Ma}^2)$ terms:

$$V_X = \tfrac{1}{4}U\,\mathsf{Ma}\,xy(3y - 2), \qquad V_Y = \tfrac{1}{4}U\,\mathsf{Ma}(1 - y)y^2,$$
$$T_* = T_0 - Ahx\left[1 - \tfrac{1}{48}\mathsf{Ma}\,\mathsf{Pr}(4 - 3y)y^3\right], \quad U = \nu/h, \tag{5.9.24}$$
$$P = P_0 - \tfrac{1}{4}\rho U^2\,\mathsf{Ma}\left[3(y^2 - x^2) - 2y\right].$$

In Figure 5.4, the streamlines of the thermocapillary flow (5.9.24) are shown together with the profile of the longitudinal dimensionless component $u = V_X/U$ of the flow velocity. The arrow directions correspond to the case $\mathsf{Ma} > 0$.

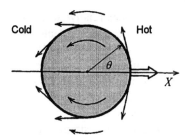

**Figure 5.5.** Drop motion due to temperature gradient. Thin arrows show the direction of thermocapillary stresses on the drop surface and of the flow induced by these stresses; thick arrow is the direction of the drop motion (it is assumed that surface tension is a decreasing function of temperature)

These results show that thermocapillary forces generate a complicated circulation liquid motion in the layer, and the flow changes its direction at the depth equal to 1/3 of the layer depth. Just as one can expect, the flow is symmetric with respect to the plane $X = 0$ with temperature $T_0$; the fluid flows out from the near-bottom layer along this plane.

# 5.10. Thermocapillary Drift of a Drop

### 5.10-1. Drift of a Drop in a Fluid With Temperature Gradient

*Statement of the problem.* Let us consider the thermocapillary effect for a drop in a temperature-nonuniform fluid medium [512]. Under an external temperature gradient, the temperature on the drop surface will not be constant along the surface, and therefore, one can expect the appearance of thermocapillary stresses directed from the hot pole of the drop to the cold pole if the surface tension coefficient is a decreasing function of temperature (Figure 5.5). When gravitational and other forces are absent, the induced flow makes the drop drift in the direction of temperature growth. This is the so-called thermocapillary drift of the drop. Here the Marangoni effect manifests itself in a pure form. If other forces are involved (say, the gravity force), then the Marangoni effect for this drop will be the change of its velocity.

Let us estimate the thermocapillary force applied to a drop and the velocity of the drop thermocapillary drift in the absence of gravitation. We assume the ambient fluid to be infinite and the nonuniform temperature field remote from the drop to be linear in $X$:

$$T_*^{(1)} \to AX + T_0 \qquad \text{as} \quad R \to \infty. \tag{5.10.1}$$

These assumptions are justified if the drop size is much less than both the characteristic length of the ambient fluid and the space scale of temperature variation.

Let us consider a steady-state motion of the drop at velocity $U_i$. Just as before, we assume that the surface tension depends linearly on temperature and that all other physical parameters of the liquids are constant. We also assume that the drop preserves a spherical shape by virtue of large capillary pressure preventing shape change.

We use the spherical coordinates relative to the drop center, in which the radial coordinate $R$ is measured from the drop center and the angle $\theta$ is measured from the positive direction of the $X$-axis. We supply all parameters and variables outside and inside the drop by subscripts 1 and 2, respectively.

We restrict ourselves to slow motions (small Reynolds numbers) described by the Stokes equations (2.1.2) and neglect the convective term in the heat equation (assuming that the Peclet number is small.)

First, we obtain the solution of the simpler, thermal part of the problem, which can be treated independently for $\mathsf{Pe} = 0$. The temperature outside and inside the drop satisfies the stationary heat equation

$$\Delta T_*^{(1)} = 0, \qquad \Delta T_*^{(2)} = 0. \tag{5.10.2}$$

Remote from the drop, the boundary condition (5.10.1) is used, and on the surface of the drop, the continuity conditions for the temperature and heat flux must be satisfied:

$$T_*^{(1)} = T_*^{(2)}, \quad \varkappa_1 \frac{\partial T_*^{(1)}}{\partial R} = \varkappa_2 \frac{\partial T_*^{(2)}}{\partial R} \qquad \text{for} \quad R = a, \tag{5.10.3}$$

where $\varkappa_1$ and $\varkappa_2$ are the thermal conductivity coefficients of the ambient fluid and of the drop, respectively.

**Exact solution.** Problem (5.10.1)–(5.10.3) can be solved by separation of variables, and the solution has the form

$$T_*^{(1)} = Aa\left( \frac{R}{a} + \frac{1-\delta}{2+\delta} \frac{a^2}{R^2} \right) \cos\theta + T_0, \quad T_*^{(2)} = \frac{3A}{2+\delta} R \cos\theta + T_0, \tag{5.10.4}$$

where $\delta = \varkappa_2/\varkappa_1$.

We now consider the hydrodynamic part of the problem, which is described by the Stokes equations (2.1.2). The fluid velocity components satisfy (2.2.2) remote from the drop, and the solution is bounded within the drop. On the interface, the no-flow condition (2.2.6) holds and condition (2.2.7) of continuity of the tangential velocity component must be satisfied. Moreover, the boundary condition of the tangential stress balance is to be used:

$$\mu_1 \left( \frac{\partial V_\theta^{(1)}}{\partial R} - \frac{V_\theta^{(1)}}{R} \right) - \mu_2 \left( \frac{\partial V_\theta^{(2)}}{\partial R} - \frac{V_\theta^{(2)}}{R} \right) = -\sigma' \frac{1}{R} \frac{\partial T_*^{(1)}}{\partial \theta} \qquad \text{for } R = a, \tag{5.10.5}$$

where the viscous stresses occur on the left-hand side of the equation, and the thermocapillary stresses, on the right-hand side. Here $\sigma' = d\sigma/dT_*^{(1)} < 0$.

We introduce the stream functions $\Psi^{(m)}$ by formulas (2.1.3) in each phase ($m = 1, 2$). Then we use the general solution of (2.1.6), in which we retain only terms with $n = 2$. Using the boundary conditions (2.2.2), (2.2.6), and (2.2.7) to determine the unknown constants, we find

$$\Psi^{(1)} = \frac{1}{2} a^2 U_i \left[ \frac{R^2}{a^2} + B\frac{R}{a} - (B+1)\frac{a}{R} \right] \sin^2\theta,$$

$$\Psi^{(2)} = \frac{1}{4} a^2 U_i (2B+3) \left( \frac{R^4}{a^4} - \frac{R^2}{a^2} \right) \sin^2\theta. \tag{5.10.6}$$

The constant $B$ is determined by condition (5.10.5) after the substitution of (5.10.4) and (5.10.6) as follows:

$$B = -\frac{1}{1+\beta}\left(1 + \frac{3}{2}\beta - \frac{\text{Ma}}{2+\delta}\right), \qquad \text{Ma} = \frac{Aa\sigma'}{\mu_1 U_i}, \tag{5.10.7}$$

where Ma is the Marangoni number.

**Drag force and drift velocity.** Formulas (5.10.6), (5.10.7) make it possible to calculate the force applied to the drop, $F = 4\pi\mu_1 aBU_i$. This can be rewritten as a sum

$$F = F_V + F_T,$$

$$F_V = -2\pi\mu_1 aU_i \frac{2+3\beta}{1+\beta}, \qquad F_T = -\frac{4\pi a^2 A\sigma'}{(2+\delta)(1+\beta)}. \tag{5.10.8}$$

The first term $F_V$ in (5.10.8) is just Hadamard–Rybczynski's result (2.2.15) for the drag of a drop in a translational flow. The second term $F_T$ is the thermocapillary force acting on the drop in the external temperature gradient due to the Marangoni effect.

We can find the drop velocity under a thermocapillary force in the absence of gravitation by setting the force $F$ in (5.10.8) equal to zero. As a result, we obtain

$$U_T = -\frac{2aA\sigma'}{\mu_1(2+\delta)(2+3\beta)}. \tag{5.10.9}$$

For $\sigma' = d\sigma/dT_*^{(1)} < 0$, this formula implies that signs of $U_T$ and $A$ coincide, and therefore, the drop will drift in the direction of temperature increase.

Note that for the thermocapillary drift velocity in the absence of gravitation, the result (5.10.9) is valid not only at small but at arbitrary Reynolds numbers. For $B = 0$, the flow (5.10.6) satisfies the complete equations of motion without dropping the convective terms (that is, the Navier–Stokes equations.) At the same time, we cannot omit the requirement of small Peclet numbers.

The results for a gas bubble can be obtained from these for a drop by setting $\delta = \beta = 0$.

In [58, 59, 466, 469], more precise results are cited taking into account higher-order approximations with respect to small Reynolds and Peclet numbers. For example, the following expression for thermocapillary drift velocity was found in [466]:

$$U_T = U_0\left(1 - \tfrac{301}{7200}\,\text{Pe}^2\right),$$

where $U_0 = -Aa\sigma'/(2\mu_1)$ and $\text{Pe} = a^2|A\sigma'|/(\mu_1\chi)$. Here $U_0$ is the bubble drift velocity in the zero approximation with respect to the Peclet number. The Peclet number Pe is calculated on the basis of the velocity $U_0$. As before, the Reynolds number is supposed to be zero (the Stokes approximation). One can see that taking account of convective mass transfer results in a decrease in the bubble drift velocity.

In [373], bubble thermocapillary drift was considered in the external temperature gradient at high Peclet numbers. The drift velocity was found to be

$$U_T = -\frac{Aa\sigma'}{3\mu_1}.$$

The analysis of the problem for arbitrary Reynolds and Peclet numbers can be performed only by numerical methods (e.g., see [403, 471].)

The nonstationary problem of drop acceleration by external temperature gradient was considered in [14, 373].

The results (5.10.8) for the thermocapillary force $F_T$ and (5.10.9) for the thermocapillary drift velocity, which were obtained under the assumption of a constant temperature gradient remote from the drop, prove to hold also for a varying gradient. These expressions can be rewritten in a vector form as follows [468]:

$$\mathbf{F}_T = -\frac{4\pi a^2 \sigma' \mathbf{A}}{(2+\delta)(1+\beta)}, \qquad \mathbf{U}_T = -\frac{2a\sigma' \mathbf{A}}{\mu_1(2+\delta)(2+3\beta)}.$$

Here $\mathbf{A}$ is the temperature field gradient of the ambient fluid in the absence of the drop, calculated at the position where the drop center is actually at the moment.

Some additional effects were considered for the thermocapillary motion of drops and bubbles in an external temperature gradient: interaction of a drop with a plane wall [285], and interaction of drops with bubbles or of bubbles with each other [12, 146]. In particular, it was shown in [12] that the interaction of drops of radius $a$ decreases with increasing distance $l$ between them as $(a/l)^3$ for thermocapillary drift, compared with $a/l$ for the motion in the gravitational field.

### 5.10-2. Drift of a Drop in Complicated Cases

*Drift for a nonlinear dependence of the surface tension on temperature.* In [173], a drop was considered in a constant external temperature gradient with a nonlinear dependence of the surface tension on temperature. In the case of a nonmonotone dependence and in the absence of gravitation, planes of drop equilibrium (stable and unstable) can occur. The appearance of such planes can hamper technological processes, for instance, the process of removing bubbles from the melt by applying a temperature gradient in the presence of microgravitation.

The analysis shows that if the function (5.9.19) has a minimum (i.e., $\alpha > 0$), then the equilibrium plane will be attractive and the equilibrium will be stable (the drift velocity outside the equilibrium plane always is directed to the plane.) For $\alpha < 0$, the equilibrium plane will be repulsive, and the equilibrium will be unstable (the drift velocity is directed from the plane.) Two drops in an equilibrium plane move towards each other if $\alpha > 0$; if $\alpha < 0$, then the drops diverge.

*Radiation-induced thermocapillary motion of a drop.* The temperature gradient is the simplest but not the unique method for bringing about the thermocapillary drift of a drop. If the drop is opaque and the fluid is transparent, one can move the drop by a light beam in a uniformly heated fluid. The radiation absorbed by the drop will heat it nonuniformly, thus producing thermocapillary stresses. For $d\sigma/dT_* < 0$, the drop will drift towards the warmer part, that is, towards the beam.

The corresponding problem was considered in [322, 389]. The radiation in [389] was assumed to have the form of a plane-parallel beam being absorbed on the drop surface as on a black body, but freely passing through the exterior fluid. The temperature remote from the drop is assumed to be constant. For the thermocapillary force and for the velocity of thermocapillary drift of the drop in the absence of gravitation, the following expressions were obtained ($J$ is the radiation flux power):

$$F_T = \frac{2\pi a^2 J \sigma'}{3\varkappa_1(2+\delta)(1+\beta)}, \quad U_T = \frac{aJ\sigma'}{3\mu_1\varkappa_1(2+\delta)(2+3\beta)}. \qquad (5.10.10)$$

Here the quantities $F_T$ and $U_T$ are positive if the corresponding vectors are codirected with the radiation beam and negative otherwise. As follows from the second formula in (5.10.10), if $\sigma' < 0$, then the drop drifts toward the beam.

## 5.10-2. General Formulas for Capillary Force and Drift Velocity

In all above cases, the distribution of the surface tension along the surface was obtained independently on the liquid motion as a consequence of neglecting the convective thermal conduction. To generalize this property, one can consider the Marangoni effect for a drop whose surface tension coefficient is a given function of the coordinates on the drop surface without specifying the origin of this nonuniformity. Then the following expression for the capillary force acting on a drop can be obtained [468]:

$$\mathbf{F}_T = -\frac{1}{2(1+\beta)} \int_S \nabla_s \sigma \, dS, \qquad (5.10.11)$$

where $\nabla_s$ is the gradient along the surface, and the integration is carried out over the entire drop surface.

By equating the sum of the drag force $F_V$ in (5.10.8) and the capillary force (5.10.11) with zero, one can find the velocity of the capillary drift of the drop (in the Stokes approximation):

$$\mathbf{U}_T = -\frac{1}{4\pi\mu_1 a(2+3\beta)} \int_S \nabla_s \sigma \, dS. \qquad (5.10.12)$$

These results, are rather general. They cover all cases in which the capillary stresses on the drop surface are stationary and do not depend on the motion of the fluid. Formulas (5.10.8) and (5.10.9), which were derived for Pe = 0, can be obtained from (5.10.11) and (5.10.12) for specific functions $\sigma$.

One can find more details on some achievements in this direction in the papers [85, 102, 145, 252, 391, 392, 467, 506].

# 5.11. Chemocapillary Effect in the Drop Motion

## 5.11-1. Preliminary remarks. Statement of the Problem

All the above cases of the Marangoni effect for a drop have one common feature, namely, the presence of an exterior asymmetry, which is not connected with motion. Essentially different situations are investigated in [270, 419], when the surface tension gradient is produced only in the motion of the liquid inside and outside the drop and, in turn, affects the motion.

For example, in [270], the influence of the surfactants on the drop motion is analyzed. When the drop is motionless, the surface film is uniform and the surface tension gradient is not produced. But if the drop moves, then the surfactants are redistributed on the surface, thus producing such a gradient. Since the surface tension is usually a decreasing function of the concentration, the capillary effect will manifest itself by surface retardation and by an increase in the drop drag. If the capillary effect is large, then it results in full cessation of movement on the drop or bubble surface, and the drag law proves to be the same as for a solid spherical particle. This conclusion is supported by numerous experimental studies [270]. From the point of view of hydrodynamics of the drop including membrane phase, this phenomenon was accounted in Section 2.2.

Further, following [419], we consider in detail another mechanism of the surface tension variability produced in the process of motion. We consider a drop moving at a constant velocity, on whose surface an exothermic or endothermal reaction is involved. It is assumed that a surfactants takes part in the reaction being dissolved in the surrounding liquid. Let the liquid temperature and the concentration of surfactants be constant remote from the drop, while on the interface, the concentration of the surfactants (the reactant) is zero (the diffusion regime of reaction). In this symmetric situation, the temperature variability, and, consequently, the thermocapillary stresses can be produced only in the process of motion of liquids.

To emphasize the role of a chemical reaction, such thermocapillary effects will be called chemo-thermocapillary. The problem on the steady-state translational Stokes flow past a drop is conventionally divided into three parts.

For the hydrodynamic part of the problem, we preserve all the most important assumptions, equations, and boundary conditions used above in Section 5.10.

The concentration part of the problem is described by the convective diffusion equation (4.4.3) (the subscript 1 is introduced for the stream function) and by the boundary conditions (4.4.4) and (4.4.5) specifying the concentration on the interface and remote from the drop. The diffusion Peclet number $Pe$ is assumed to be small.

As far as the thermal part of the problem is concerned, to describe the thermocapillary effect, it does not suffice to consider only the zero approximation with respect to low Peclet numbers, because in this case the temperature would be constant along the drop surface. Therefore, instead of Eqs. (5.10.2), we suggest the more general convective heat equations

$$
\begin{aligned}
\left(\mathbf{V}^{(1)} \cdot \nabla\right) T_*^{(1)} &= \chi_1 \Delta T_*^{(1)} && \text{outside of the drop,} \\
\left(\mathbf{V}^{(2)} \cdot \nabla\right) T_*^{(2)} &= \chi_2 \Delta T_*^{(2)} && \text{inside of the drop.}
\end{aligned}
\tag{5.11.1}
$$

Remote from the drop, the boundary condition (5.10.1) with $A = 0$ is used; the conditions of temperature continuity must hold on the interface (see the first boundary condition in (5.10.3)); the following heat flux balance condition, taking into account the heat production of the surface reaction, must be satisfied:

$$
\varkappa_1 \frac{\partial T_*^{(1)}}{\partial R} - \varkappa_2 \frac{\partial T_*^{(2)}}{\partial R} = Q D \frac{\partial C}{\partial R} \qquad \text{for} \quad R = a, \tag{5.11.2}
$$

where $Q$ is the thermal effect of the chemical reaction ($Q > 0$ for an exothermic reaction and $Q < 0$ for an endothermal reaction).

We note that the coupling between the hydrodynamic, diffusion, and thermal problems is provided by the convective terms in the equations of diffusion and heat conduction together with the two boundary conditions (5.10.5) and (5.11.2). To obtain the leading terms of the expansion with respect to low diffusion and thermal Peclet numbers, one can neglect the convective terms, so that the coupling between these problems will be provided only by the boundary conditions.

### 5.11-2. Drag Force and Velocity of Motion

The solution of the hydrodynamic part of the problem, as before, can be written in the form of the stream functions (5.10.6), where the constant $B$ remains undetermined as yet.

The approximate solution of the thermal and diffusion problem can be found by the method of matched asymptotic expansions (see Section 4.4) with the stream functions (5.10.6); one must retain only the zero and the first terms of the expansions with respect to low Peclet numbers and use the boundary conditions (5.10.5) and (5.11.2) to obtain the following values of the constant $B$ and the force acting on the drop:

$$
F = 4\pi \mu_1 a U_i B, \qquad B = -\frac{1 + \frac{3}{2}\beta + 3m}{1 + \beta + m}, \qquad m = -\frac{\mathsf{Ma}\,\mathsf{Pe}_T(1 - \mathsf{Le})}{12(2 + \delta)}. \tag{5.11.3}
$$

Here $\mathsf{Ma} = \dfrac{QC_i D\sigma'}{\varkappa_1 \mu_1 U_i}$ is the Marangoni number and $\mathsf{Le} = \dfrac{\chi_1}{D}$ is the Lewis number. In derivation of (5.11.3), it was assumed that the conditions $\mathsf{Pe} \approx \mathsf{Pe}_T$ and $\mathsf{Ma}\,\mathsf{Pe} \approx 1$ hold.

The drop velocity in the gravitation field can be found from the condition of the zero sum of the forces (5.11.3), the gravitational force, and the buoyancy force:

$$
U_i = \frac{(\rho_1 - \rho_2)a^2}{3\mu_1 B} g. \tag{5.11.4}
$$

For $B < 0$, the force (5.11.3) is, as usual, a drag force. For $B > 0$, the force (5.11.3) becomes a propulsion force, since it is codirected with the drop motion.

For $B > -\frac{3}{2}$, the flow pattern around the drop is similar to the Hadamard–Rybczynski flow (Figure 2.2). As $B$ decreases, the fluid circulation intensity decreases within the drop, and vanishes for $B = -\frac{3}{2}$. Under further decrease $(B < -\frac{3}{2})$, a circulation zone is produced around the drop. The direction of the inner circulation becomes opposite to the direction in the Hadamard–Rybczynski case. It follows from (5.11.3) that the drag force acting on the drop exceeds the Stokes force for a hard sphere.

In the limit case $\beta \to \infty$ (high viscosity of the drop substance), the thermocapillary effect does not influence the motion, $B \to -\frac{3}{2}$, the flow around the drop will be the same as for a hard sphere, and (5.11.3) implies the Stokes law (2.2.5). For $m = 0$ (no heat production or independence of the surface tension on temperature), the thermocapillary effect is absent, and (5.11.3) yields a usual drag force for a drop in the translational flow (2.2.15).

We note that a special regime of the so-called autonomous motion is possible, where the drop drifts at a constant nonzero velocity in the absence of any exterior factors [147, 148]. In this case the other possible regime (no motion) proves to be unstable. Effects similar to the ones considered in this section can be produced by the chemoconcentration-capillary mechanisms [149], as well as other factors different from surface chemical reactions, for example, by heat production within the drop [390].

*Remark.* The problem of mass transfer to a drop for the diffusion regime of reaction on its surface under the conditions of thermocapillary motion is stated in the same way as in its absence (see Section 4.4) taking into account the corresponding changes in the fluid velocity field. In [144], a more complicated problem is considered for the chemocapillary effect with the heat production, which was described in [147–149, 419]. It was assumed that a chemical reaction of finite rate occurs on the drop surface.

# Chapter 6
# Hydrodynamics and Mass and Heat Transfer in Non-Newtonian Fluids

So far we have considered motion and mass and heat transfer in Newtonian media, which are characterized by proportionality between the tangential stress and the corresponding rate of shear (note that there are no tangential stresses if the rate of shear is zero). Gases and single-phase low-molecular (i.e., simple) liquids obey this law closely. However, in practice one often deals with fluids of more complicated structure, such as polymer solutions and melts and disperse fluid systems (suspensions, emulsions, and pastes), which are characterized by a nonlinear relation between the tangential stress and the rate of shear. Such fluids are said to be non-Newtonian.

This chapter deals with the most commonly encountered (empirical and semi-empirical) rheological models of non-Newtonian fluids. Typical problems of hydrodynamics and heat and mass transfer are stated for power-law fluids, and the results of solutions are given for these problems.

## 6.1. Rheological Models of Non-Newtonian Incompressible Fluids

### 6.1-1. Newtonian Fluids

The classical hydrodynamics of viscous incompressible isotropic fluids is based on the generalized Newton law

$$\tau_{ij} = -P\delta_{ij} + 2\mu e_{ij} \qquad (i, j = 1, 2, 3),$$
$$\delta_{ij} = \begin{cases} 1 & \text{for } i = j, \\ 0 & \text{for } i \neq j, \end{cases} \qquad (6.1.1)$$

where the $\tau_{ij}$ are the stress tensor components, $P$ is the pressure, $\delta_{ij}$ is the Kronecker delta, $\mu$ is the dynamic viscosity of the fluid, and the $e_{ij}$ are the shear rate tensor components, which in the Cartesian coordinates $X_1$, $X_2$, $X_3$ are related to the fluid velocity components $V_1$, $V_2$, $V_3$ by the formula

$$e_{ij} = \frac{1}{2}\left( \frac{\partial V_i}{\partial X_j} + \frac{\partial V_j}{\partial X_i} \right). \qquad (6.1.2)$$

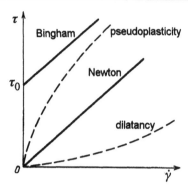

**Figure 6.1.** Typical flow curves for nonlinearly viscous fluids

Equation (6.1.1) contains only one rheological parameter $\mu$, which is independent of the kinematic (velocity, acceleration, and displacement) and dynamic (force and stress) characteristics of motion. The value of $\mu$, however, depends on temperature.

In the case of a one-dimensional simple shear Newtonian flow, Eq. (6.1.1) becomes

$$\tau = \mu \dot{\gamma}, \qquad (6.1.3)$$

where $\tau = \tau_{12}$, $\dot{\gamma} = \partial V_1/\partial X_2$, and $X_2$ is the coordinate perpendicular to the direction of the fluid velocity $V_1$.

For Newtonian fluids, the curve (6.1.3), called the flow curve, is a straight line passing through the origin (Figure 6.1).

Now let us briefly describe some models for the more complicated case of non-Newtonian fluids (a more detailed presentation of related problems can be found, for instance, in the books [19, 25, 47, 181, 206, 448]).

---

**6.1-2. Nonlinearly Viscous Fluids**

Under one-dimensional shear, many rheologically stable fluids of complex structure (whose rheological characteristics are time-independent) have a flow curve other than Newtonian. If the flow curve is curvilinear but still passes through the origin in the plane $\dot{\gamma}$, $\tau$, then the corresponding fluids are said to be nonlinearly viscous (often they are said to be purely viscous, anomalously viscous, or sometimes non-Newtonian).

Nonlinearly viscous fluids are further classified into pseudoplastic fluids, whose flow curve is convex, and dilatant fluids, whose flow curve is concave (both cases are shown by dashed lines in Figure 6.1).

Polymer solutions and melts, residual oils, rubber solutions, many petroleum products, paper pulps, biological fluids (blood, plasma), pharmaceutical compounds (emulsions, creams, and pastes), various food products (fats and sour cream) can serve as examples of pseudoplastic fluids. Dilatant properties are mainly exhibited by high-concentration or coarse-disperse systems (such as highly concentrated water suspensions of titanium dioxide, iron, mica, quartz powders or starch and wet river sand). Just as with Newtonian fluids, it is convenient to introduce the apparent (effective) viscosity $\mu_e$ by

$$\mu_e = \tau / \dot{\gamma}.$$

Pseudoplasticity is characterized by a decrease in viscosity with increasing shear rate. In this case, the medium behaves as if it were "thinning" and becomes more mobile. For dilatant fluids, the viscosity increases with increasing shear rate.

At present, there exist several dozens of rheological (mostly empirical) models of nonlinear viscous fluids. This is due to the fact that for the vast variety of fluid media of different physical nature, there is no rigorous general theory, similar to the molecular kinetic theory of gases, which would enable one to calculate the characteristics of molecular transport and the mechanical behavior of a medium on the basis of its interior microscopic structure.

Table 6.1 gives the most widespread rheological models of nonlinearly viscous fluids. Most of these models do not describe all aspects of the actual behavior of nonlinear viscous fluids in the entire range of the shear rate. Instead, they explain only some specific characteristic features of the flow. Table 6.1 contains quasi-Newtonian relations of two types, namely,

$$\tau = \mu_e(\dot{\gamma})\dot{\gamma}, \qquad \tau = \mu_e(\tau)\dot{\gamma}.$$

The coefficients of $\dot{\gamma}$ on the right-hand sides of these expressions can be treated as apparent non-Newtonian viscosities. These values allow one to assess to what extent the models are physically adequate to the behavior of specific fluid media.

It is known that the flow curve of any nonlinear viscous fluid has linear parts for very small and very large shear rates (Figure 6.1). By $\mu_0$ we denote the greatest "Newtonian viscosity" of pseudoplastic fluids that can be observed at "zero" shear rate, and by $\mu_\infty$, the least "Newtonian viscosity" corresponding to the "infinitely large" shear. Obviously, the model of power-law fluid (see the first row in Table 6.1) provides a good description of the actual behavior of nonlinear viscous media in the intermediate region between $\mu_0$ and $\mu_\infty$. However, in the limit cases $\dot{\gamma} \to 0$ and $\dot{\gamma} \to \infty$, this model leads to wrong results. The Ellis and Rabinovich models are valid for small and moderate stresses, but give zero viscosity as $\tau \to \infty$. The Sisko model leads to an infinitely large viscosity as $\dot{\gamma} \to 0$. The other models in Table 6.1 provide a good description of the qualitative structure of the entire flow curve. In the book [443], the numerical values of the most important parameters of some substances are given for the Ostwalde–de Waele, Ellis, and Reiner–Filippov rheological models.

## 6.1-3. Power-Law Fluids

At present, the model of power-law fluid, described in Table 6.1 for a one-dimensional flow, is used most commonly. The generalization of this model to

TABLE 6.1

Rheological models of nonlinear viscous fluids (according to
[443, 445, 452]); $\tau$ is the shear stress, and $\dot{\gamma} = \partial V_1 / \partial X_2$

| No | Fluid model, authors | Rheological equation |
|----|----------------------|----------------------|
| 1 | Power-law fluid, Ostwalde–de Waele | $\tau = k|\dot{\gamma}|^{n-1}\dot{\gamma}, \ n > 0$ |
| 2 | Sisko | $\tau = \left(A + B\mu_0|\dot{\gamma}|^{n-1}\right)\dot{\gamma}, \ n > 0$ |
| 3 | Prandtl | $\tau = A|\dot{\gamma}|\arcsin(\dot{\gamma}/B)$ |
| 4 | Williamson | $\tau = \left(\dfrac{A}{B+\dot{\gamma}} + \mu_0\right)\dot{\gamma}$ |
| 5 | Prandtl–Eyring | $\tau = \operatorname{arcsinh}(\dot{\gamma}/B)$ |
| 6 | Rabinovich | $\tau = \mu_0(1 + A\tau^2)^{-1}\dot{\gamma}$ |
| 7 | Ellis | $\tau = \dfrac{\dot{\gamma}}{A + B|\tau|^{m-1}}$ |
| 8 | Eyring | $\tau = A\dot{\gamma} + B\sin(C|\tau|)$ |
| 9 | Reiner–Filippov | $\tau = \left(\mu_\infty + \dfrac{\mu_0 - \mu_\infty}{A + B\tau^2}\right)\dot{\gamma}$ |

the three-dimensional case results in the equation of state (6.1.1), where

$$\mu = k(2I_2)^{\frac{n-1}{2}}. \tag{6.1.4}$$

(Here and in the following, for brevity, we denote the apparent viscosity $\mu_e$ of the fluid by $\mu$). The right-hand side of formula (6.1.4) contains two constants $k$ and $n$ and the quadratic invariant

$$I_2 = \sum_{i,j=1}^{3} e_{ij}e_{ij} = \frac{1}{4}\sum_{i,j=1}^{3}\left(\frac{\partial V_i}{\partial X_j} + \frac{\partial V_j}{\partial X_i}\right)^2 \tag{6.1.5}$$

of the shear rate tensor. The constant $k$ is called the consistence factor of the fluid. The larger $k$, the smaller is the fluidity. The exponent $n$ shows to what extent the behavior of the substance is non-Newtonian. The more $n$ differs from the unity (to either side), the more distinctly the viscosity is anomalous and the flow curve is nonlinear.

The values $0 < n < 1$ correspond to pseudoplastic fluids whose viscosity decreases with increasing shear rate. A Newtonian fluid is characterized by the

TABLE 6.2

Parameters of the power-law rheological equation for pseudoplastic substances

| Material | Concentration, % | Shear rate range, s$^{-1}$ | $n$ | $k$, N · s$^n$/m$^2$ |
|---|---|---|---|---|
| Starch glue | 1.54 | N/A | 0.952 | 0.003 |
| | 2.01 | N/A | 0.926 | 0.004 |
| | 2.89 | N/A | 0.794 | 0.035 |
| Water solution of carboxymethyl-cellulose | 0.09 | $10^3$ to $3 \times 10^4$ | 0.72 | 0.044 |
| | 0.22 | 10 to $10^3$ | 0.79 | 0.081 |
| | 0.22 | $10^3$ to $10^4$ | 0.63 | 0.302 |
| | 0.35 | $10^2$ to $10^3$ | 0.66 | 0.259 |
| | 0.35 | $10^3$ to $10^4$ | 0.58 | 0.429 |
| Paper pulp (water based) | 4.0 | N/A | 0.575 | 20.02 |
| Napalm in kerosene | 10.0 | N/A | 0.520 | 4.28 |
| Lime paste | 23.0 | N/A | 0.178 | 7.43 |
| Clay mortar | 33.0 | N/A | 0.171 | 7.2 |
| Cement stone in water solution | 54.3 | N/A | 0.153 | 2.51 |

parameter $n = 1$. The values $n > 1$ correspond to dilatant fluids whose viscosity increases with increasing shear rate.

The parameters $k$ and $n$ are assumed to be constant for a given fluid in some bounded range of shear rates. They can be determined from viscometric experiments and from the analysis of the so-called consistence curves. Table 6.2 presents the values of $k$ and $n$ for some substances [445].

It should be noted that for a sufficiently wide range of shear rates in real fluids, $k$ and $n$ are not constant. This does not prevent one from widely using the power-law rheological equation, since in practice one usually deals with a rather narrow range of shear rates.

In the following we often consider a rheological model more general than (6.1.4). In the three-dimensional case, this model is described by Eq. (6.1.1), where the apparent viscosity $\mu$ arbitrarily depends on the quadratic invariant of the shear rate tensor,

$$\mu = \mu(I_2). \qquad (6.1.6)$$

In Supplement 6 we present the equations of motion in various coordinate systems for non-Newtonian incompressible fluids governed by this law.

The first five models in Table 6.1 are special cases of (6.1.6).

## 6.1-4. Reiner–Rivlin Media

An important class of non-Newtonian fluids is formed by isotropic rheological stable media whose stress tensor $[\tau_{ij}]$ is a continuous function of the shear rate tensor $[e_{ij}]$ and is independent of the other kinematic and dynamic variables. One can rigorously prove that the most general rheological model satisfying these conditions is the following nonlinear model of a viscous non-Newtonian Stokes medium [19]:

$$\tau_{ij} = -P\delta_{ij} + 2\mu e_{ij} + 4\varepsilon \sum_{k=1}^{3} e_{ik} e_{kj}, \qquad (6.1.7)$$

where $\mu$ and $\varepsilon$ are scalar functions of the invariants

$$I_1 = e_{11} + e_{22} + e_{33}, \quad I_2 = \sum_{i,j=1}^{3} e_{ij} e_{ji}, \quad I_3 = \det[e_{ij}] \qquad (6.1.8)$$

of the shear rate tensor.

In the case of an incompressible fluid, the first invariant is zero, $I_1 = \operatorname{div} \mathbf{v} = 0$. For simple one- and two-dimensional flows (such as flows in thin films, longitudinal flow in a tube, and tangential flow between concentric cylinders), the third invariant $I_3$ is identically zero.

The scalar functions $\mu$ and $\varepsilon$ determine various rheological models of non-Newtonian media. For example, the case $\mu = \mathrm{const}$ and $\varepsilon = 0$ corresponds to the linear model (6.1.1) of a Newtonian fluid. By setting $\mu = k(2I_2)^{\frac{n-1}{2}}$ and $\varepsilon = 0$, we obtain the model (6.1.1) of a power-law nonlinear viscous fluid.

If we choose the coefficients $\mu$ and $\varepsilon$ in (6.1.7) to be nonzero constants, then we arrive at the Reiner–Rivlin model, which additively combines the Newton model with a tensor-quadratic component. In this case the constants $\mu$ and $\varepsilon$ are called, respectively, the shear and the dilatational (transverse) viscosity. Equation (6.1.7) permits one to give a qualitative description of specific features of the mechanical behavior of viscoelastic fluids, in particular, the Weissenberg effect (a fluid rises along a rotating shaft instead of flowing away under the action of the centrifugal force).

## 6.1-5. Viscoplastic Media

Besides of the media considered, there are media in which a finite yield stress $\tau_0$ is required to initiate flow. For such media, the intercept of the flow curve with the stress axis $\dot{\gamma} = 0$ is $(0, \tau_0)$, where $\tau_0$ is nonzero (Figure 6.1). The value of $\tau_0$ characterizes the plastic properties of a substance, and the slope of the flow curve to the $\dot{\gamma}$-axis characterizes its fluidity. Such media are said to be viscoplastic.

The combination of plasticity and viscosity, typical of these media, was discovered in 1889 by Shvedov for gelatin solutions and in 1919 by Bingham for

TABLE 6.3

Rheological models of viscoplastic fluids (according to [47, 443])

| No | Fluid model | Rheological equation |
|----|-------------|----------------------|
| 1 | Shvedov–Bingham | $\tau = \tau_0 \, \mathrm{sign}\,\dot{\gamma} + \mu_p\dot{\gamma}$ |
| 2 | Bulkley–Herschel | $\tau = \tau_0 \, \mathrm{sign}\,\dot{\gamma} + k|\dot{\gamma}|^{n-1}\dot{\gamma}$ |
| 3 | Casson | $\sqrt{\tau} = k_0 + k_1\sqrt{\dot{\gamma}}$ |
| 4 | Casson–Shul'man | $\tau^{1/n} = \tau_0^{1/n} + (\mu\dot{\gamma})^{1/n}$ |
| 5 | Shul'man | $\tau^{1/n} = \tau_0^{1/n} + (\mu\dot{\gamma})^{1/m}$ |

oil paints (viscous fluids applied to a smooth vertical surface necessarily drain off this surface sooner or later; therefore, a coat of paint remaining on the surface indicates that the paint possesses plastic properties).

Table 6.3 presents some models of viscoplastic media. The simplest and most commonly used model is the Shvedov–Bingham model (which is often referred to as Bingham model), corresponding to the upper straight line in Figure 6.1, where $\tau_0$ is the yield stress and $\mu_p$ is the plastic viscosity. This model is based on the assumption that a stagnant fluid has a rather rigid spatial structure which is capable of resisting any stress less than $\tau_0$. Beyond this value, the structure is broken down instantaneously, and the medium flows as a common Newtonian fluid under the shear stress $\tau - \tau_0$ (when the tangential stresses in the fluid become less than $\tau_0$, the structure is recovered). Quasisolid areas are formed wherever the shear stress is less than the yield stress.

The three-dimensional analog of the Shvedov–Bingham law has the form

$$e_{ij} = 0 \qquad \text{for} \quad |\tau| \le \tau_0,$$
$$\tau_{ij} = 2\left(\frac{\tau_0}{\sqrt{2I_2}} + \mu_p\right)e_{ij} \qquad \text{for} \quad |\tau| > \tau_0. \qquad (6.1.9)$$

The numerical values of the parameters $\tau_0$ and $\mu_p$ for various disperse systems containing sand, cement, and oil are given in the book [320].

Note that the Casson model (the third model in Table 6.3) fairly well describes various varnishes, paints, blood, food compositions like cocoa mass, and some other fluid disperse systems [443].

### 6.1-6. Viscoelastic Fluids

A long time ago, Maxwell pointed out that resin-like substances can be treated neither as solids nor as fluids. If the stress is applied slowly or acts for a sufficiently long time, then the resin behaves as a common viscous fluid. In this case the deformation continuously and irreversibly grows with time, and the shear rate is proportional to the applied stress according to the Newton law. On the other hand, if the stress is applied and released sufficiently rapidly, then the strain of the resin is proportional to the stress and is completely reversible.

On the basis of these observations, Maxwell suggested to combine Hooke's law (for elastic bodies) and Newton's law (for viscous fluids) additively into a single rheological equation of state, which has the following form in the one-dimensional case:

$$\tau + t_0 \frac{d\tau}{dt} = \mu \dot{\gamma}. \tag{6.1.10}$$

Here $t_0 = \mu/G$ is some characteristic time (the relaxation time), $G$ is the shear modulus, and $t$ is time.

Suppose that some constant deformation has been erected in a Maxwellian fluid and some effort is made to maintain this deformation in the course of time. Then the arising flow will gradually release the applied stress, and the effort needed to maintain the deformation will also decay. Under these conditions ($\tau = \tau_0$, $\dot{\gamma} = 0$ at $t = 0$, and $\gamma = \text{const}$ for $t > 0$), the solution of Eq. (6.1.10) has the form

$$\tau = \tau_0 \exp(-t/t_0),$$

so that we have exponential decay of stress (stress relaxation) in time. In time $t_0 = \mu/G$, the stress becomes approximately 2.7 times less than the initial value $\tau_0$.

Under very rapid mechanical actions or in observations with characteristic time $t < t_0$, the substance behaves as an ideal elastic medium. For $t \gg t_0$ the developing flow becomes stronger than the elastic deformation, and the substance can be treated as a simple Newtonian fluid. It is only if $t$ is of the same order of magnitude as $t_0$ that the elastic and viscous effects act simultaneously, and the complex nature of the deformation displays itself.

The three-dimensional analog of the Maxwell equation (6.1.10) has the form

$$\tau_{ij} + t_0 \frac{D \tau_{ij}}{D t} = -\delta_{ij} P + 2\mu e_{ij},$$

where

$$\frac{D}{D t} = \frac{\partial}{\partial t} + \sum_{i=1}^{3} V_i \frac{\partial}{\partial X_i}.$$

We must point out that all simple Newtonian substances, even such as air, water, and benzene, possess noticeable shear elasticity under very large loadings in the acoustic range of velocities. In this case, the characteristic deformation time must be of the order of $10^{-8}$ to $10^{-10}$ s (these are approximate relaxation times for simple fluids). Under these conditions, all simple fluids can be treated as viscoelastic media.

# 6.2. Motion of Non-Newtonian Fluid Films

### 6.2-1. Statement of the Problem. Formula for the Friction Stress

Let us consider steady-state laminar flow of a rheologically complex fluid along an inclined plane (Figure 1.3). We assume that the motion is sufficiently slow, so that the inertial forces (convective terms) can be neglected compared with the viscous friction and gravity. Let the film thickness $h$ be a constant much less than the film length. In this case, in the first approximation the normal velocity component $V_2$ is small compared with the tangential component $V = V_1$, and the derivatives along the film surface can be neglected compared with the normal derivatives.

Under these assumptions we arrive at a one-dimensional velocity profile $V = V(\xi)$ and pressure profile $P = P(\xi)$, where $\xi = h - Y$ is the coordinate measured from the wall along the normal. The corresponding equations of film flow have the form

$$\frac{\partial \tau}{\partial \xi} + \rho g \sin \alpha = 0, \tag{6.2.1}$$

$$\frac{\partial P}{\partial \xi} + \rho g \cos \alpha = 0. \tag{6.2.2}$$

One must supplement these equations with boundary conditions. On the free film surface, which is in contact with a gas, the tangential stress is zero, and the normal stress is equal to the atmosphere pressure $P_0$, that is,

$$\tau = 0, \quad P = P_0 \quad \text{at} \quad \xi = h. \tag{6.2.3}$$

On the impermeable wall, a no-slip condition is prescribed:

$$V = 0 \quad \text{at} \quad \xi = 0. \tag{6.2.4}$$

Equations (6.2.1) and (6.2.2) can be integrated independently. The solution satisfying (6.2.3) is given by the formulas

$$\tau = \rho g(h - \xi) \sin \alpha, \tag{6.2.5}$$
$$P = P_0 + \rho g(h - \xi) \cos \alpha. \tag{6.2.6}$$

One can see that the friction stress $\tau$ linearly increases from zero on the free surface to the maximum value $\tau_s = \rho g h \sin \alpha$ on the wall irrespective of the rheological properties of the medium.

### 6.2-2. Nonlinearly Viscous Fluids. Power-Law Fluids

*Nonlinearly viscous fluids. General formulas.* In the general case of nonlinearly viscous fluids, it is convenient to define the shear rate as a function of the stress:

$$\frac{dV}{d\xi} = f(\tau), \tag{6.2.7}$$

where the specific form of the function $f$ is determined by the rheological model of the fluid considered.

To obtain Eq. (6.2.7) for the rheological equations given in Table 6.1, one must express the shear rate $\dot{\gamma} = dV/d\xi$ via $\tau$.

We substitute $\tau$ from (6.2.5) into (6.2.7) and integrate with respect to $\xi$ taking account of the boundary condition (6.2.4) on the wall. As a result, we obtain the velocity profile

$$V = \int_0^\xi f\big(m(h-\zeta)\big)\, d\zeta = \frac{1}{m}\int_{m(h-\xi)}^{mh} f(\tau)\, d\tau, \quad \text{where } m = \rho g \sin\alpha. \quad (6.2.8)$$

The maximum velocity is attained on the free surface of the film at $\xi = h$:

$$U_{\max} = \frac{1}{m}\int_0^{mh} f(\tau)\, d\tau. \quad (6.2.9)$$

Let us calculate the mean flow rate velocity

$$\langle V \rangle = \frac{1}{h}\int_0^h V\, d\xi = \frac{1}{mh}\int_0^h \left[\int_{m(h-\xi)}^{mh} f(\tau)\, d\tau\right] d\xi$$

of the film flow. By using the formula

$$\int_0^h \left[\int_{m(h-\xi)}^{mh} f(\tau)\, d\tau\right] d\xi = \int_0^{mh} \left[\int_{h-\tau/m}^{h} f(\tau)\, d\xi\right] d\tau$$

to reverse the order of integration and then by integrating with respect to $\xi$, we obtain the following expression for the mean flow rate velocity of the fluid:

$$\langle V \rangle = \frac{1}{m^2 h}\int_0^{mh} \tau f(\tau)\, d\tau, \quad \text{where } m = \rho g \sin\alpha. \quad (6.2.10)$$

The amount $Q$ of the fluid passing per second through the cross-section of the film is called the volume rate of flow and is given by the integral

$$Q = \int_0^h V\, d\xi = h\langle V \rangle. \quad (6.2.11)$$

*Power-law fluids.* For a power-law fluid (the first model in Table 6.1), the dependence of the shear rate on the stress is given by (6.2.7) with

$$f(\tau) = \left(\frac{\tau}{k}\right)^{1/n}. \quad (6.2.12)$$

By substituting this function into (6.2.8)–(6.2.11), we can find the basic characteristics of the film flow of a power-law fluid along an inclined plane. The corresponding results are presented in Table 6.4.

One can see that the index $n$ of a power-law fluid substantially affects the velocity profile. With increasing pseudoplasticity the distribution of the velocity becomes more and more homogeneous, approaching a quasisolid distribution with profile $V = \langle V \rangle = $ const in the limit as $n \to 0$. On the contrary, dilatancy makes the flow field more and more nonuniform, and as $n \to \infty$ the velocity profile approaches the triangular shape given by

$$\frac{V}{\langle V \rangle} = 2\frac{\xi}{h}.$$

As above, the maximum velocity is attained on the free surface of the film,

$$U_{\max} = 2\langle V \rangle.$$

For $0 < n < \infty$, the possible range of the maximum velocity is given by the inequalities

$$\langle V \rangle < U_{\max} < 2\langle V \rangle.$$

In particular, for Newtonian fluids we have $U_{\max} = \frac{3}{2}\langle V \rangle$.

---

### 6.2-3. Viscoplastic Media. The Shvedov–Bingham Fluid

*Viscoplastic media. General formulas.* For viscoplastic media, the shear rate in general depends on the stress as follows:

$$\frac{dV}{d\xi} = \begin{cases} 0 & \text{for } 0 \le \tau \le \tau_0, \\ f(\tau) & \text{for } \tau_0 \le \tau \le \rho g h \sin \alpha. \end{cases} \tag{6.2.13}$$

To obtain the explicit form of the function $f(\tau)$ for the rheological models in question (see Table 6.3), one must express the shear rate $\dot\gamma = dV/d\xi$ via $\tau$.

The motion of plastic fluids with finite yield stress $\tau_0$ has some qualitative specific features not possessed by nonlinearly viscous fluids. Let us consider a layer of a viscoplastic fluid on an inclined plane whose slope is gradually varied. It follows from (6.2.5) that, irrespective of the rheological properties of the medium, the tangential stress decreases across the film from its maximum value $\tau_{\max} = \rho g h \sin \alpha$ on the solid wall to zero on the free surface. Therefore, a flow in a film of a viscoplastic fluid can be initiated only when the tangential stress on the wall becomes equal to or larger than the yield stress $\tau_0$:

$$\tau_0 = \rho g h_0 \sin \alpha_0. \tag{6.2.14}$$

The maximum angle of inclination for which the stagnant film is still held on the plane is

$$\alpha_0 = \arcsin\left(\frac{\tau_0}{\rho g h_0}\right).$$

TABLE 6.4.
Basic parameters of film flow of non-Newtonian fluids along an inclined plane ($m = \rho g \sin \alpha$)

| Film flow parameters | Fluid models | | |
|---|---|---|---|
| | Newtonian | Ostwalde–de Waele power-law fluid | Shvedov–Bingham |
| Rheological equation | $\tau = \mu \dot{\gamma}$ | $\tau = k|\dot{\gamma}|^{n-1}\dot{\gamma}$ | $\tau = \tau_0 \operatorname{sign}\dot{\gamma} + \mu_p \dot{\gamma}$ |
| Maximum velocity, $U_{max}$ | $\dfrac{mh^2}{2\mu}$ | $\dfrac{n}{n+1}\left(\dfrac{m}{k}\right)^{\frac{1}{n}} h^{\frac{n+1}{n}}$ | $\dfrac{(mh - \tau_0)^2}{2m\mu_p}$ |
| Mean velocity, $\langle V \rangle$ | $\dfrac{mh^2}{3\mu}$ | $\dfrac{n}{2n+1}\left(\dfrac{m}{k}\right)^{\frac{1}{n}} h^{\frac{n+1}{n}}$ | $\dfrac{2m^3h^3 - 3m^2h^2\tau_0 + \tau_0^3}{6m^2h\mu_p}$ |
| Shear rate on the wall, $\dot{\gamma}$ | $\dfrac{mh}{\mu}$ | $\left(\dfrac{mh}{k}\right)^{\frac{1}{n}}$ | $\dfrac{mh - \tau_0}{\mu_p}$ |
| Volume rate of flow, $Q$ | $\dfrac{mh^3}{3\mu}$ | $\dfrac{n}{2n+1}\left(\dfrac{m}{k}\right)^{\frac{1}{n}} h^{\frac{2n+1}{n}}$ | $\dfrac{2m^3h^3 - 3m^2h^2\tau_0 + \tau_0^3}{6m^2\mu_p}$ |
| Film thickness, $h$ | $\left(\dfrac{3\mu Q}{m}\right)^{1/3}$ | $\left(\dfrac{k}{m}\right)^{\frac{1}{2n+1}}\left(\dfrac{2n+1}{n}Q\right)^{\frac{n}{2n+1}}$ | Determined from the cubic equation $2m^3h^3 - 3m^2h^2\tau_0 + \tau_0^3 - 6m^2\mu_p Q = 0$ |
| Velocity profile, $V$ | $\dfrac{m}{2\mu}[h^2 - (h-\xi)^2]$ | $\dfrac{n}{n+1}\left(\dfrac{m}{k}\right)^{\frac{1}{n}}\left[h^{\frac{n+1}{n}} - (h-\xi)^{\frac{n+1}{n}}\right]$ | $\dfrac{m}{2\mu_p}[h^2 - (h-\xi)^2] - \dfrac{\tau_0}{\mu_p}\xi$ (for $0 \le \xi \le h - \tau_0/m$) |

The larger $\tau_0$ and the thinner the film, the larger is $\alpha_0$. For nonlinearly viscous media, the constant $\alpha_0$ is always zero.

To ensure that the film is held on a vertical plane ($\alpha_0 = \pi/2$), the equilibrium thickness of the film and the yield stress must satisfy the relation $h_0 = \tau_0/(\rho g)$, which follows from (6.2.14). This relation determines the thickness of the coating that remains on vertical surfaces.

Now let us choose a fixed angle $\alpha$ of inclination. Let the film thickness $h$ satisfy the inequality

$$h \geq \frac{\tau_0}{\rho g \sin \alpha}. \tag{6.2.15}$$

Then the entire flow region can be divided into the following two parts with distinct profiles:

(1) $0 \leq \xi \leq h - h_0$ is the shear near-wall region,
(2) $h - h_0 \leq \xi \leq h$ is the region of quasisolid motion.

Here

$$h_0 = \frac{\tau_0}{\rho g \sin \alpha}. \tag{6.2.16}$$

In the quasisolid region, adjacent to the free surface of the film, the fluid velocity is constant and is equal to the velocity on the boundary of the near-wall region at $\xi = h - h_0$. In the entire region of quasisolid motion, the velocity attains its maximum value $V = U_{max}$.

We substitute the expression (6.2.5) for $\tau$ into Eq. (6.2.13) and integrate with respect to $\xi$ taking account of the no-slip condition (6.2.4) on the wall. As a result, by simple transformations we obtain the velocity profile

$$V = \begin{cases} \dfrac{1}{m} \displaystyle\int_{m(h-\xi)}^{mh} f(\tau)\, d\tau & \text{for } 0 \leq \xi \leq h - h_0, \\[2ex] \dfrac{1}{m} \displaystyle\int_{\tau_0}^{mh} f(\tau)\, d\tau & \text{for } h - h_0 \leq \xi \leq h, \end{cases} \tag{6.2.17}$$

where $m = \rho g \sin \alpha$ and $h_0 = \tau_0/m$.

The maximum fluid velocity

$$U_{max} = \frac{1}{m} \int_{\tau_0}^{mh} f(\tau)\, d\tau \tag{6.2.18}$$

is attained in the entire region of quasisolid flow.

The mean flow rate velocity of the film flow of a viscoplastic fluid is given by the formula

$$\langle V \rangle = \frac{1}{m^2 h} \int_{\tau_0}^{mh} \tau f(\tau)\, d\tau. \tag{6.2.19}$$

The volume rate of flow can be found from (6.2.11), where $\langle V \rangle$ is determined by (6.2.19).

*The Shvedov–Bingham fluid.* For the Shvedov–Bingham fluid (the first model in Table 6.3), the dependence of the shear rate on the stress is given by (6.2.13), where

$$f(\tau) = \frac{\tau - \tau_0}{\mu_p}. \tag{6.2.20}$$

Substituting (6.2.20) into (6.2.17)–(6.2.19), we find the basic characteristics of film flow of a viscoplastic Shvedov–Bingham fluid along an inclined plane (the results of the corresponding calculations are presented in Table 6.4).

# 6.3. Mass Transfer in Films of Rheologically Complex Fluids

### 6.3-1. Mass Exchange Between a Film and a Gas

Following [47, 443, 444], let us consider absorption of weakly soluble gases on the surface of a fluid film flowing down an inclined plane. The steady-state velocity distribution inside the film is given by (6.2.8) for nonlinearly viscous fluids and by (6.2.17) for viscoplastic fluids.

Let us assume that the concentration of the absorbed substance on the film surface is constant and is equal to $C = C_s$ and that a pure fluid with zero concentration is supplied through the cross-section with the coordinate $X = 0$. We restrict our consideration to the case of high Peclet numbers, when the diffusion along the film can be neglected. In the diffusion boundary layer approximation (that is, when only the leading term $V \approx U_{max}$ of the expansion of the fluid velocity near the free boundary is considered), the concentration distribution inside the film is described under the above assumptions by the following equation and boundary conditions:

$$U_{max} \frac{\partial C}{\partial X} = D \frac{\partial^2 C}{\partial Y^2}; \tag{6.3.1}$$
$$X = 0, \quad C = 0; \quad Y = 0, \quad C = C_s,$$

where the coordinate $Y = 1 - \xi$ is measured inward the film along the normal to the surface.

The solution of problem (6.3.1) can be expressed via the complementary error function:

$$C = C_s \, \text{erfc} \left( \frac{Y}{2} \sqrt{\frac{U_{max}}{DX}} \right). \tag{6.3.2}$$

By differentiating this formula, we obtain the diffusion flux on the surface of the film:

$$j_* = -\rho D \frac{\partial C}{\partial Y} \bigg|_{Y=0} = \rho C_s \sqrt{\frac{U_{max} D}{\pi X}}. \tag{6.3.3}$$

For nonlinearly viscous and viscoplastic fluids, the maximum velocity $U_{max}$ in formula (6.3.3) can be calculated in the general case by using (6.2.9) and

(6.2.18), respectively. In particular, for a power-law fluid, one can take $U_{max}$ from Table 6.4, which implies

$$j_* = \rho C_s \left[ \frac{n}{n+1} \left( \frac{\rho g \sin \alpha}{k} \right)^{\frac{1}{n}} h^{\frac{n}{n+1}} \frac{D}{\pi X} \right]^{\frac{1}{2}}.$$

## 6.3-2. Dissolution of a Plate by a Fluid Film

Now let us consider mass transfer from a solid wall to a fluid film. We assume that the concentration on the surface of the plate is constant and is equal to $C_s$ and that a pure fluid is supplied through the input cross-section. In the diffusion boundary layer approximation, the velocity profile near the surface of the plate can be approximated by the expression

$$V \approx \left( \frac{dV}{d\xi} \right)_{\xi=0} \xi = f(mh)\xi, \qquad \text{where} \quad m = \rho g \sin \alpha.$$

Taking into account the preceding, we can write out the following problem for the concentration field in the diffusion boundary layer adjacent to the surface of the wall:

$$f(mh)\xi \frac{\partial C}{\partial X} = D \frac{\partial^2 C}{\partial \xi^2}; \tag{6.3.4}$$

$$X = 0, \quad C = 0; \qquad \xi = 0, \quad C = C_s.$$

The solution can be expressed via the incomplete gamma function as follows:

$$C = C_s \frac{1}{\Gamma(1/3)} \Gamma\left( \frac{1}{3}, \frac{f(mh)\xi^3}{9DX} \right). \tag{6.3.5}$$

By differentiating (6.3.5), we obtain the diffusion flux on the film surface:

$$j_* = -\rho D \frac{\partial C}{\partial \xi} \bigg|_{\xi=0} = 0.538 \, \rho C_s \left[ \frac{D^2 f(mh)}{X} \right]^{1/3}. \tag{6.3.6}$$

Using (6.2.12), we obtain the diffusion flux for a power-law fluid:

$$j_* = 0.538 \, \rho C_s \left[ \left( \frac{\rho g \sin \alpha}{k} \right)^{1/n} \frac{D^2}{X} \right]^{1/3}. \tag{6.3.7}$$

For a viscoplastic Shvedov–Bingham fluid, one must set $f(mh) = (\rho g h \sin \alpha - \tau_0)/\mu_p$ in the expression (6.3.6).

# 6.4. Motion of Non-Newtonian Fluids in Tubes and Channels

## 6.4-1. Circular Tube. Formula for the Friction Stress

Let us consider a steady-state axisymmetric flow of a non-Newtonian fluid in a straight horizontal circular tube of radius $a$. The coordinate $Z$ is measured along the tube axis and is directed downstream. We restrict our consideration to the hydrodynamically stabilized flow far from the input cross-section, where the streamlines are parallel to the tube axis. In this case, the pressure increment decreases with increasing $Z$, and the pressure gradient is negative and constant,

$$\frac{\partial P}{\partial Z} = -\frac{\Delta P}{L} = \text{const},$$

where $\Delta P$ is the pressure difference on the length $L$ of the tube.

In this problem, all derivatives of the velocity with respect to $t$, $Z$, and $\varphi$, as well as the velocity components $V_\varphi$ and $V_\mathcal{R}$, are zero. Thus, the equation of motion (see Supplement 6) is

$$\frac{1}{\mathcal{R}} \frac{d}{d\mathcal{R}} (\mathcal{R}\tau) + \frac{\Delta P}{L} = 0, \tag{6.4.1}$$

where $\tau = \tau_{\mathcal{R}Z}$.

The solution of Eq. (6.4.1) satisfying the boundedness condition ($|\tau| < \infty$) has the form

$$\tau = -\frac{\Delta P}{2L} \mathcal{R}. \tag{6.4.2}$$

One can see that the absolute value of the friction stress linearly increases from zero on the tube axis to its maximum value $\tau_s = a\Delta P/L$ on the tube wall irrespective of the type of the non-Newtonian fluid.

## 6.4-2. Circular Tube. Nonlinearly Viscous Fluids

*General formulas.* The shear rate in the tube is negative, that is, $\dot{\gamma} = dV/d\mathcal{R} < 0$, where $V = V_Z$. In the general case of nonlinearly viscous fluids, we represent the dependence of the shear rate on the stress as follows:

$$\frac{dV}{d\mathcal{R}} = -f(|\tau|), \tag{6.4.3}$$

where the form of the function $f(\tau) \geq 0$ is determined in accordance with the rheological model chosen.

By substituting the expression (6.4.2) into (6.4.3), we obtain an equation for the fluid velocity $V = V_Z$. The solution of this equation satisfying the no-slip condition on the tube wall ($V = 0$ at $\mathcal{R} = a$) has the form

$$V = \int_\mathcal{R}^a f\left(\frac{\Delta P}{2L} \mathcal{R}\right) d\mathcal{R}. \tag{6.4.4}$$

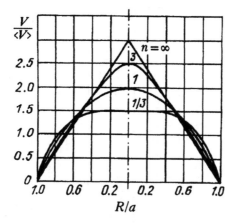

**Figure 6.2.** Characteristic velocity profiles for non-Newtonian fluids in a circular tube

The maximum velocity of the fluid is attained at the flow axis,

$$U_{max} = \int_0^a f\left(\frac{\Delta P}{2L}\mathcal{R}\right) d\mathcal{R}. \tag{6.4.5}$$

The volume rate of flow through the cross-section of the tube can be calculated by the formula

$$Q = \int_0^a 2\pi\mathcal{R}V\, d\mathcal{R} = \pi \int_0^a \mathcal{R}^2 f\left(\frac{\Delta P}{2L}\mathcal{R}\right) d\mathcal{R}, \tag{6.4.6}$$

and the mean flow rate velocity is given by

$$\langle V \rangle = \frac{Q}{\pi a^2}. \tag{6.4.7}$$

For the local coefficient of friction, we obtain

$$c_f = \frac{|\tau_s|}{\frac{1}{2}\rho\langle V \rangle^2} = \frac{a\Delta P}{\rho L\langle V \rangle^2},$$

where $\tau_s$ is the shear stress on the wall.

***Power-law fluids.*** For the special case of a power-law fluid described by the expressions (6.1.1) and (6.1.4), the function $f$ in (6.4.3) has the form

$$f = (\tau/k)^{1/n}. \tag{6.4.8}$$

By substituting (6.4.8) into (6.4.4)–(6.4.7), one can find the basic characteristics of motion of a power-law fluid in a circular tube. The results of the corresponding calculations [452, 508] are presented in Table 6.5 and are shown in Figure 6.2. One can see that the velocity profiles become more and more filled as the rheological parameter $n$ decreases. The limit case $n \to 0$ is characterized by a quasisolid motion of the fluid with the same velocity in the entire cross-section of the tube (it is only near the wall that the velocity rapidly decreases to zero). The parabolic Poiseuille profile corresponds to the Newtonian fluid ($n = 1$). The limit dilatant flow ($n \to \infty$) has a triangular profile, which is characterized by a linear law of velocity variation along the radius of the tube.

TABLE 6.5.

Basic parameters of flow of non-Newtonian fluids through circular tubes, $\dot\gamma = dV/dR$

| Fluid flow parameters | Fluid models | | |
|---|---|---|---|
| | Newtonian | Ostwalde–de Waele power-law fluid | Shvedov–Bingham |
| Rheological equation | $\tau = \mu\dot\gamma$ | $\tau = k\lvert\dot\gamma\rvert^{n-1}\dot\gamma$ | $\tau = \tau_0\,\mathrm{sign}\,\dot\gamma + \mu_p\dot\gamma$ |
| Maximum velocity, $U_{max}$ | $\dfrac{a^2\Delta P}{4\mu L}$ | $\dfrac{na}{n+1}\left(\dfrac{a\Delta P}{2kL}\right)^{\frac{1}{n}}$ | $\dfrac{L}{\mu_p\Delta P}\left(\dfrac{a\Delta P}{2L}-\tau_0\right)^2$ |
| Mean flow rate velocity, $\langle V\rangle$ | $\dfrac{a^2\Delta P}{8\mu L}$ | $\dfrac{na}{3n+1}\left(\dfrac{a\Delta P}{2kL}\right)^{\frac{1}{n}}$ | $\dfrac{1}{a^2\mu_p}\left[\dfrac{a^4\Delta P}{8L}-\dfrac{a^3\tau_0}{3}+\dfrac{2}{3}\tau_0^4\left(\dfrac{L}{\Delta P}\right)^3\right]$ |
| Shear rate on the wall, $\dot\gamma\rvert_{\mathcal{R}=a}$ | $-\dfrac{a\Delta P}{2\mu L}$ | $-\left(\dfrac{a\Delta P}{2kL}\right)^{\frac{1}{n}}$ | $-\dfrac{1}{\mu_p}\left(\dfrac{a\Delta P}{2L}-\tau_0\right)$ |
| Volume rate of flow, $Q$ | $\dfrac{\pi a^4\Delta P}{8\mu L}$ | $\dfrac{\pi na^3}{3n+1}\left(\dfrac{a\Delta P}{2kL}\right)^{\frac{1}{n}}$ | $\dfrac{\pi}{\mu_p}\left[\dfrac{a^4\Delta P}{8L}-\dfrac{a^3\tau_0}{3}+\dfrac{2}{3}\tau_0^4\left(\dfrac{L}{\Delta P}\right)^3\right]$ |
| Velocity profile, $V$ | $U_{max}\left[1-\left(\dfrac{\mathcal{R}}{a}\right)^2\right]$ | $U_{max}\left[1-\left(\dfrac{\mathcal{R}}{a}\right)^{\frac{n+1}{n}}\right]$ | $\dfrac{1}{\mu_p}\left[\dfrac{\Delta P}{4L}(a^2-\mathcal{R}^2)-\tau_0(a-\mathcal{R})\right]$ $\left(\text{for } \dfrac{2\tau_0 L}{\Delta P}\le\mathcal{R}\le a\right)$ |

---

| **6.4-3. Circular Tube.  Viscoplastic Media** |

*General formulas.* The dependence of the shear rate on the stress in flows of viscoplastic fluids in circular tubes can be represented as follows:

$$-\frac{dV}{dR} = \begin{cases} 0 & \text{for } 0 \le |\tau| \le \tau_0, \\ f(|\tau|) & \text{for } |\tau| \ge \tau_0. \end{cases} \tag{6.4.9}$$

Substituting this expression into (6.4.3), we obtain an equation for the velocity profile. One can see that for small pressure gradients satisfying the condition

$$\frac{a\Delta P}{2L} \le \tau_0,$$

the fluid in the tube is stagnant.

In the following we assume that $\frac{1}{2}a\Delta P/L > \tau_0$. For viscoplastic fluids (6.4.9), the solution of Eq. (6.4.3) satisfying the no-slip condition on the tube walls has the form

$$V = \begin{cases} \displaystyle\int_{\mathcal{R}}^{a} f\left(\frac{\Delta P}{2L}\mathcal{R}\right) d\mathcal{R} & \text{for } r_0 \le \mathcal{R} \le a, \\ \displaystyle\int_{r_0}^{a} f\left(\frac{\Delta P}{2L}\mathcal{R}\right) d\mathcal{R} & \text{for } 0 \le \mathcal{R} \le r_0, \end{cases} \tag{6.4.10}$$

where

$$r_0 = \frac{2L\tau_0}{\Delta P} \tag{6.4.11}$$

is the radius of the region of quasisolid motion. The maximum velocity of the fluid,

$$U_{\max} = \int_{r_0}^{a} f\left(\frac{\Delta P}{2L}\mathcal{R}\right) d\mathcal{R}, \tag{6.4.12}$$

is attained in the entire region $0 \le \mathcal{R} \le r_0$ of the quasisolid motion in a neighborhood of the tube axis, where the fluid moves as a whole.

The volume rate of flow through the tube cross-section can be calculated by the formula

$$Q = \pi \int_{r_0}^{a} \mathcal{R}^2 f\left(\frac{\Delta P}{2L}\mathcal{R}\right) d\mathcal{R}. \tag{6.4.13}$$

The mean flow rate velocity of the fluid is obtained by substituting (6.4.13) into (6.4.7).

*Shvedov–Bingham Fluids.* In the special case of a viscoplastic Shvedov–Bingham medium (the first model in Table 6.3), we have the following expression for the function $f$ in (6.4.9):

$$f(|\tau|) = \frac{|\tau| - \tau_0}{\mu_p}. \tag{6.4.14}$$

By substituting this dependence into the expressions (6.4.10)–(6.4.13), we can obtain all basic characteristics of this flow. The results of the corresponding calculations [276, 508] are presented in Table 6.5.

The velocity profile is shown in Figure 6.3, where the hatched area corresponds to the region of quasisolid motion.

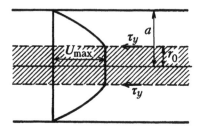

**Figure 6.3.** Velocity profile in the flow of a viscoplastic Shvedov–Bingham medium

### 6.4-4. Plane Channel

Now let us consider a steady-state hydrodynamically stabilized flow of a non-Newtonian fluid through a plane channel of width $2h$. Let us introduce Cartesian coordinates $X$, $\xi$ with $X$-axis directed downstream along the lower wall and with coordinate $\xi$ measured inward the channel along the normal to this wall $(0 \le \xi \le 2h)$. Since the problem is symmetric about the midline $\xi = h$, it suffices to consider the lower half of the region, $0 \le \xi \le h$.

The steady-state flow far from the input cross-section is characterized by a constant negative pressure gradient $\partial P/\partial X = -\Delta P/L = \mathrm{const}$ ($\Delta P$ is the total pressure difference on the length $L$ of the channel); moreover, the transverse velocity of the fluid is zero. The tangential velocity $V = V_X$ depends only on the coordinate $\xi$ and satisfies the equation $\tau'_\xi = -\Delta P/L$. By integrating this equation and by taking into account the symmetry condition ($\tau = 0$ at $\xi = h$), we obtain

$$\tau = \frac{\Delta P}{L}(h - \xi). \qquad (6.4.15)$$

Up to the different notation ($\Delta P/L \rightarrow \rho g \sin \alpha$), formula (6.4.15) coincides with the expression (6.2.5) for shear stresses, which was obtained earlier for film flows. Therefore, we can calculate the velocity profile $V$ in a plane channel (in the region $0 \le \xi \le h$), the maximum velocity $U_{\max}$, and the mean flow rate velocity $\langle V \rangle$ for nonlinear viscous fluids by formulas (6.2.8)–(6.2.11) and for viscoplastic fluids by formulas (6.2.17)–(6.2.19) if we formally replace $\rho g \sin \alpha$ by $\Delta P/L$ in these formulas.

For power-law and Shvedov–Bingham fluids, the basic characteristics of flow in a plane channel can be found from Table 6.4, where one must set $m = \Delta P/L$.

# 6.5. Heat Transfer in Channels and Tubes. Account of Dissipation

### 6.5-1. Plane Channel

Let us consider the problem of dissipative heating of a non-Newtonian fluid in a plane channel with isothermal walls on which the same constant temperature is maintained:

$$\xi = 0, \quad T = T_s; \qquad \xi = 2h, \quad T = T_s. \tag{6.5.1}$$

(Here we use the coordinate system introduced in Section 7.6, and $T$ denotes the dimensional temperature). If the fluid temperature at the input cross-section is equal to the wall temperature $T_s$, then along some initial part of the tube the fluid is gradually heated by internal friction until a balance is achieved between the heat withdrawal through the walls and the dissipative heat release. In the region where such an equilibrium is established, the fluid temperature does not vary along the channel, that is, the temperature field is stabilized (provided that the velocity profile is also stabilized). In what follows we just study this thermally and hydrodynamically stabilized flow.

Following [208, 276], we assume that the heat release does not affect the physical properties of the fluid (that is, viscosity, density, and thermal conductivity coefficient are temperature independent). In this case, the velocity profile can be found independently of the heat problem (see Section 6.4).

In the general case, the equation for the temperature distribution in the region of heat stabilization can be obtained from the equation of heat transfer in Supplement 6 (where the convective terms are equal to zero and the temperature depends only on the transverse coordinate). This equation has the form

$$\varkappa T''_{\xi\xi} + \tau V'_\xi = 0, \tag{6.5.2}$$

where $\varkappa$ is the thermal conductivity coefficient, $\tau = \mu V'_\xi$ is the shear stress, $\mu$ is the apparent viscosity, and the primes stand for $\xi$-derivatives.

Since the problem is symmetric about the midplane $\xi = h$ of the channel, it suffices to consider only the lower half $0 \le \xi \le h$ of the region with the symmetry condition

$$\xi = h, \qquad T'_\xi = 0 \tag{6.5.3}$$

imposed on the boundary.

Let us consider the case of an arbitrary viscoplastic fluid with yield stress $\tau_0$ (similar results for nonlinear viscous fluids correspond to $\tau_0 = 0$). To obtain the temperature profile, we proceed as follows. First, in the near-wall shear region $0 \le \xi \le h - h_0$, where $h_0 = \tau_0 L / \Delta P$, we solve Eq. (6.5.2) with the boundary conditions (6.5.1). Then in the quasisolid region $h - h_0 \le \xi \le h$, we solve Eq. (6.5.2) with $V'_\xi = 0$ under the boundary condition (6.5.3). Finally, we match the two solutions on the common boundary $\xi = h_0$. This procedure results in the following temperature distribution in the channel:

$$T - T_s = \begin{cases} \dfrac{1}{\varkappa} \displaystyle\int_0^\xi \left( \int_\xi^{h-h_0} \tau V'_\xi \, d\xi \right) d\xi & \text{for } 0 \le \xi \le h - h_0, \\[2mm] T_{max} - T_s & \text{for } h - h_0 \le \xi \le h, \end{cases} \tag{6.5.4}$$

where the maximum temperature $T_{max}$ is determined by

$$T_{max} - T_s = \frac{1}{\varkappa} \int_0^{h-h_0} \left( \int_\xi^{h-h_0} \tau V'_\xi \, d\xi \right) d\xi, \qquad h_0 = L \frac{\tau_0}{\Delta P}. \tag{6.5.5}$$

The heat flux to the wall is given by

$$q_T = \varkappa \left( \frac{dT}{d\xi} \right)_{\xi=0} = \int_0^{h-h_0} \tau V_\xi' \, d\xi. \tag{6.5.6}$$

For a power-law fluid, one must set $h_0 = 0$ and $\tau = k(V_\xi')^n$ in (6.5.4)–(6.5.6). The results of the corresponding calculations are presented in Table 6.6. This table also contains the basic parameters of heat exchange in viscoplastic Shvedov–Bingham fluids (the values $\tau_0 = 0$ and $\mu_p = \mu$ correspond to a Newtonian fluid).

## 6.5-2. Circular Tube

Under the same assumptions (the tube temperature is constant and is equal to $T_s$, and the physical properties of the medium are temperature independent), the temperature distribution in the region of heat stabilization in a circular tube of radius $a$ satisfies the following equation and boundary conditions (with allowance for dissipative heating):

$$\frac{\varkappa}{\mathcal{R}} \frac{d}{d\mathcal{R}} \left( \mathcal{R} \frac{dT}{d\mathcal{R}} \right) = -\tau V_\mathcal{R}'; \tag{6.5.7}$$

$$\mathcal{R} = 0, \quad \mathcal{R} \frac{dT}{d\mathcal{R}} = 0; \qquad \mathcal{R} = a, \quad T = T_s, \tag{6.5.8}$$

where $\tau$ is the shear stress and $V$ is the fluid velocity.

For viscoplastic media with yield stress $\tau_0$, the solution of problem (6.5.7), (6.5.8) has the form

$$T - T_s = \begin{cases} \dfrac{1}{\varkappa} \displaystyle\int_\mathcal{R}^a \left( \int_{r_0}^\mathcal{R} \tau V_\mathcal{R}' \mathcal{R} \, d\mathcal{R} \right) \dfrac{d\mathcal{R}}{\mathcal{R}} & \text{for } r_0 \leq \mathcal{R} \leq a, \\[3mm] T_{max} - T_s & \text{for } 0 \leq \mathcal{R} \leq r_0, \end{cases} \tag{6.5.9}$$

where $r_0 = 2L\tau_0/\Delta P$ and the maximum temperature $T_{max}$ is given by

$$T_{max} - T_s = \frac{1}{\varkappa} \int_{r_0}^a \left( \int_{r_0}^\mathcal{R} \tau V_\mathcal{R}' \mathcal{R} \, d\mathcal{R} \right) \frac{d\mathcal{R}}{\mathcal{R}}. \tag{6.5.10}$$

The heat flux to the tube wall can be found by using the expression

$$q_T = \frac{1}{a} \int_{r_0}^a \tau V_\mathcal{R}' \mathcal{R} \, d\mathcal{R}. \tag{6.5.11}$$

For nonlinear viscous fluids, the basic parameters of heat exchange are given by formulas (6.5.9)–(6.5.11) with $r_0 = 0$.

TABLE 6.6.
Basic parameters of heat exchange of non-Newtonian fluids in plane channels and circular tubes (notation: $m = \Delta P/L$)

| Heat exchange parameters | Fluid models | |
| --- | --- | --- |
| | Ostwalde–de Waele power-law fluid | Shvedov–Bingham |
| Rheological equation | $\tau = k|\dot\gamma|^{n-1}\dot\gamma$ | $\tau = \tau_0\,\mathrm{sign}\,\dot\gamma + \mu_p\dot\gamma$ |
| **Plane channel** | | |
| Temperature difference, $\Delta T = T_{\max} - T_s$ | $\dfrac{kn^2 h^2}{\varkappa(2n+1)(3n+1)}\left(\dfrac{mh}{k}\right)^{\frac{n+1}{n}}$ | $\dfrac{m^2}{12\varkappa\mu_p}(h-h_0)^4$ |
| Heat flux, $q_T$ | $\dfrac{knh}{2n+1}\left(\dfrac{mh}{k}\right)^{\frac{n+1}{n}}$ | $\dfrac{m^2}{3\mu_p}(h-h_0)^3$ |
| Temperature profile, $T - T_s$ | $\dfrac{kn^2}{\varkappa(2n+1)(3n+1)}\left(\dfrac{m}{k}\right)^{\frac{n+1}{n}}\left[h^{\frac{3n+1}{n}} - (h-\xi)^{\frac{3n+1}{n}}\right]$ | $\dfrac{m^2}{12\varkappa\mu_p}\left[(h-h_0)^4 - (h-h_0-\xi)^4\right]$ (for $0 \le \xi \le h - h_0$, where $h_0 = \tau_0/m$) |
| **Circular tube** | | |
| Temperature difference, $\Delta T = T_{\max} - T_s$ | $\dfrac{kn^2 a^2}{\varkappa(3n+1)^2}\left(\dfrac{am}{2k}\right)^{\frac{n+1}{n}}$ | $\dfrac{m^2}{4\varkappa\mu_p}\left[\dfrac{1}{16}(a^4-r_0^4) - \dfrac{2}{9}r_0(a^3-r_0^3) + \dfrac{1}{4}r_0^2(a^2-r_0^2) + \dfrac{1}{12}r_0^4\ln\dfrac{r_0}{a}\right]$ |
| Heat flux, $q_T$ | $\dfrac{kna}{3n+1}\left(\dfrac{am}{2k}\right)^{\frac{n+1}{n}}$ | $\dfrac{m^2}{16a\mu_p}\left[(a-r_0)^4 + \dfrac{4}{3}r_0(a-r_0)^3\right]$ |
| Temperature profile, $T - T_s$ | $\dfrac{kn^2 a^2}{\varkappa(3n+1)^2}\left(\dfrac{am}{2k}\right)^{\frac{n+1}{n}}\left[1 - \left(\dfrac{\mathcal{R}}{a}\right)^{\frac{3n+1}{n}}\right]$ | $\dfrac{m^2}{4\varkappa\mu_p}\left[\dfrac{1}{16}(a^4-\mathcal{R}^4) - \dfrac{2}{9}r_0(a^3-\mathcal{R}^3) + \dfrac{1}{4}r_0^2(a^2-\mathcal{R}^2) + \dfrac{1}{12}r_0^4\ln\dfrac{\mathcal{R}}{a}\right]$ (for $r_0 \le \mathcal{R} \le a$, where $r_0 = 2\tau_0/m$) |

For power-law and Shvedov–Bingham fluids, the results of the corresponding calculations are presented in Table 6.6 (the value $\tau_0 = 0$ corresponds to Newtonian fluids).

It follows from Tables 6.5 and 6.6 that the maximum temperature difference for a power-law fluid in a circular tube can be expressed via the mean flow rate velocity $\langle V \rangle$ of the flow as follows:

$$T_{max} - T_s = \frac{k}{\varkappa} \left( \frac{3n+1}{na} \right)^{n-1} \langle V \rangle^{n+1}.$$

The corresponding result for a Newtonian fluid ($n = 1$, $k = \mu$) is independent of the tube radius,

$$T_{max} - T_s = \frac{\mu}{\varkappa} \langle V \rangle^2. \tag{6.5.12}$$

The formulas obtained in this section can be used for a majority of common fluids. Flows of extremely viscous fluids possess specific qualitative features, which are described in the next section.

# 6.6. Hydrodynamic Thermal Explosion in Non-Newtonian Fluids

## 6.6-1. Nonisothermal Flows. Temperature Equation

*Rheological equation in the nonisothermal case.* The most general one-dimensional nonisothermal equation of state for non-Newtonian fluids can be written in the form $F(\tau, \dot{\gamma}, T) = 0$, where $\tau$ is the tangential stress, $\dot{\gamma}$ is the shear rate, and $T$ is the dimensional temperature. Some special types of the equation of state are presented in Tables 6.1 and 6.3, where the rheological parameters $n$, $A$, $B$, $C$, $\mu_0$, $\mu_\infty$, and $\tau_0$ must be treated as functions of temperature $T$.

Let us consider power-law fluids in more detail. Experiments [141] show that the index $n$ of non-Newtonian behavior of a substance may be treated as a constant if the temperature differences in the flow region do not exceed 30 to 50 K. The medium consistence $k = k(T)$ is much more sensitive to temperature inhomogeneities and decreases with increasing $T$. Therefore, the rheological equation of state for a power-law fluid in the nonisothermal case can be written as follows:

$$\tau = k(T)|\dot{\gamma}|^{n-1}\dot{\gamma}. \tag{6.6.1}$$

Now let us show that in the nonisothermal motion of fluid in tubes and channels some critical phenomena may occur related to the existence of a maximum admissible pressure gradient. Once this value is exceeded, the steady-state flow pattern is violated. This is accompanied by an accelerated decrease in the apparent viscosity and increase in the fluid velocity. This phenomenon is known as the hydrodynamic thermal explosion [52] and is caused by the nonlinear dependence of the apparent viscosity on temperature. Specifically, under certain external conditions, the dissipative heat release in the bulk of the fluid exceeds the heat transfer to the tube walls.

In what follows we assume that the thermal conductivity coefficient of the medium is independent of temperature.

***Equation for the temperature distribution.*** In the region of hydrodynamic and thermal stabilization, nonisothermal rectilinear steady-state flow of a power-law fluid in a circular tube of radius $a$ at constant temperature on the tube surface is described by Eqs. (6.4.1), (6.5.7), and (6.6.1). We assume that the no-slip condition is satisfied on the tube wall, and the boundary conditions for temperature are given in (6.5.8).

The solution of Eq. (6.4.1) bounded as $\mathcal{R} \to 0$ is given by (6.4.2). By eliminating $\tau$ from (6.5.7) with the use of (6.4.2), we obtain

$$V_{\mathcal{R}}' = \frac{2\varkappa}{A\mathcal{R}^2}(\mathcal{R}T_{\mathcal{R}}')_{\mathcal{R}}', \qquad \tau = -\frac{A\mathcal{R}}{2}, \qquad (6.6.2)$$

where $A = \Delta P/L$ is the pressure gradient.

On substituting these expressions into (6.6.1) and performing some elementary transformations, we arrive at the equation

$$T_{\mathcal{R}\mathcal{R}}'' + \frac{1}{\mathcal{R}}T_{\mathcal{R}}' + \frac{1}{\varkappa}\left(\frac{\Delta P}{2L}\right)^{\frac{n+1}{n}} \mathcal{R}^{\frac{n+1}{n}} [k(T)]^{-\frac{1}{n}} = 0 \qquad (6.6.3)$$

for temperature, which must be supplemented by the boundary conditions (6.5.8).

From (6.6.1) we obtain the expression

$$V = \left(\frac{\Delta P}{2L}\right)^{1/n} \int_{\mathcal{R}}^{a} \left[\frac{\mathcal{R}}{k(T)}\right]^{1/n} d\mathcal{R} \qquad (6.6.4)$$

for the fluid velocity, which can be used as soon as we determine the relation $T = T(\mathcal{R})$ by solving problem (6.6.3), (6.5.8).

## 6.6-2. Exact Solutions. Critical conditions

***Exponential dependence of the consistence factor on temperature.*** For highly viscous Newtonian fluids (such as glycerin), an exponential dependence of viscosity on temperature [133] is usually assumed. Extending this law to the consistence factor of power-law fluids, we can write [52, 253, 300, 443]

$$k = k_0 \exp[-\alpha(T - T_0)], \qquad (6.6.5)$$

where $k_0$, $\alpha$, and $T_0$ are empirical constants.

Let us substitute the expression (6.6.5) into (6.6.3) and introduce the new variables

$$y = \left(\frac{\mathcal{R}}{a}\right)^{\frac{3n+1}{2n}}, \qquad w = \frac{\alpha}{n}(T - T_s). \qquad (6.6.6)$$

As a result, we obtain the problem

$$yw''_{yy} + w'_y + \varepsilon_n y e^w = 0;$$
$$y = 0, \quad (yw'_y) = 0; \qquad y = 1, \quad w = 0,$$

(6.6.7)

where the parameter $\varepsilon_n$ is given by

$$\varepsilon_n = \frac{4n\alpha a^2 k_0}{\varkappa(3n+1)^2} \left(\frac{a\Delta P}{2k_0 L}\right)^{\frac{n+1}{n}} \exp\left[\frac{\alpha}{n}(T_s - T_0)\right].$$

(6.6.8)

Problem (6.6.7) up to notation coincides with the classical thermal explosion problem [133]. This fact together with formula (6.6.6) allows one to find the temperature distribution in the tube for a nonisothermal flow of a power-law fluid [50]. Namely,

$$T = T_s + \frac{n}{\alpha} \ln \frac{8}{\varepsilon_n} - \frac{2n}{\alpha} \ln\left[b\left(\frac{\mathcal{R}}{a}\right)^{\frac{3n+1}{n}} + \frac{1}{b}\right].$$

(6.6.9)

Here the integration constant $b$ satisfies a quadratic equation with the roots

$$b_1 = \left(\frac{2}{\varepsilon_n}\right)^{1/2} - \left(\frac{2}{\varepsilon_n} - 1\right)^{1/2}, \quad b_2 = \left(\frac{2}{\varepsilon_n}\right)^{1/2} + \left(\frac{2}{\varepsilon_n} - 1\right)^{1/2},$$

(6.6.10)

where the parameter $\varepsilon_n$ is defined in (6.6.8). These two roots determine two different temperature profiles; the stable solution is determined by $b_1$ and the unstable solution, by $b_2$. The thermal explosion occurs under the critical condition $b_1 = b_2$ (the constants in (6.6.10) are equal to each other), which corresponds to $\varepsilon_n = 2$. For $\varepsilon_n > 2$, problem (6.6.7) does not have a solution; therefore, there cannot be a steady-state rectilinear flow in the tube. In this case, the heat due to viscous friction cannot be completely released through the tube walls and results in an accelerated increase in temperature (that is, a thermal explosion).

Now let us write $b = b_1$ and consider the case $\varepsilon_n < 2$.

By substituting (6.6.9) into (6.6.4) and by taking account of (6.6.5) and (6.6.8), we find the velocity of the power-law fluid in the form

$$V = \frac{4(3n+1)^2 \varkappa L}{n\alpha a^2 b^2 \Delta P} \int_{\mathcal{R}/a}^{1} \zeta^{\frac{1}{n}} \left(\zeta^{\frac{3n+1}{n}} + b^{-2}\right)^{-2} d\zeta.$$

(6.6.11)

For $n = 1$, which corresponds to a Newtonian fluid, we obtain the following profile [52] from (6.6.11):

$$V = \frac{16\varkappa bL}{\alpha a^2 \Delta P} \left[\frac{b}{1+b^2} + \arctan b - \frac{by}{1+b^2y^2} - \arctan(by)\right], \quad y = \left(\frac{\mathcal{R}}{a}\right)^2.$$

(6.6.12)

By setting $y = 0$ in (6.6.12), we calculate the velocity on the axis of the tube:

$$U_{max} = \frac{16 \varkappa b L}{\alpha a^2 \Delta P} \left( \frac{b}{1 + b^2} + \arctan b \right).$$

For a power-law fluid, the volume rate of flow, $Q$, through the cross-section of the tube is equal to

$$Q = 2\pi \int_0^a V \mathcal{R} \, d\mathcal{R} = \frac{4\pi(3n + 1)\varkappa b^2 L}{\alpha(1 + b^2)\Delta P}. \tag{6.6.13}$$

By passing to the limit as $\alpha \to 0$, we obtain an isothermal flow of fluid. It follows from formula (6.6.10) that $b \to (\varepsilon_n/8)^{1/2}$ as $\varepsilon_n \to 0$. Thus, we can rewrite Eq. (6.6.13) as follows:

$$Q = \frac{8b^2}{\varepsilon_n(1 + b^2)} Q_{is}, \qquad Q_{is} = \tfrac{1}{2}\pi(3n + 1)\varkappa \frac{\varepsilon_n L}{\alpha \Delta P}, \tag{6.6.14}$$

where $Q_{is}$ is the volume rate of flow in the isothermal case at $T \equiv T_s$.

Since the critical conditions of hydrodynamic thermal explosion are characterized by the limit values $\varepsilon_n = 2$ and $b = 1$, it follows from (6.6.14) that

$$Q_* = 2Q_{is}. \tag{6.6.15}$$

Obviously, for a nonisothermal flow in a circular tube of a power-law fluid with exponential dependence of the consistence factor on temperature and arbitrary index $n$, the critical value of the volume rate of flow is twice that in the isothermal case.

Note that in [51] a nonisothermal flow of a power-law fluid between two parallel planes one of which moves at a constant velocity (a Couette flow) was studied, as well as a gravity flow in an annular gap and a flow between two rotating cylinders of a fluid with exponential consistence factor (6.6.5) under constant wall temperature.

***Power-law dependence of consistence factor on temperature.*** The experiments described [286] dealt with water solutions of carboxymethylcellulose, whose flow curve is well described by the power-law Ostwalde–de Waele equation. It was shown that the index $n$ is virtually invariable in the temperature interval 15 to 60°C, whereas the medium consistence factor can be approximated by the expression

$$k = k_0 \left( 1 + B_n \frac{T - T_0}{T_0} \right)^{-n}, \tag{6.6.16}$$

where $B_n$ is a characteristic constant of the material. For the solutions in question, $n$ varies in the interval 0.33 through 1.0.

For the power-law dependence (6.6.16) of the consistence factor on temperature, the solution of problem (6.6.3), (6.5.8) can be expressed via the Bessel function $J_0(x)$ as follows [515]:

$$\frac{B_n T + (1 - B_n)T_0}{B_n T_s + (1 - B_n)T_0} = \frac{J_0\left( \sigma \mathcal{R}^{\frac{3n+1}{2n}} \right)}{J_0\left( \sigma a^{\frac{3n+1}{2n}} \right)}, \tag{6.6.17}$$

where the coefficient $\sigma$ is given by the formula

$$\sigma = \frac{2n}{3n+1} \left( \frac{B_n}{\varkappa T_0} \right)^{\frac{1}{2}} \left( \frac{\Delta P}{2L} \right)^{\frac{n+1}{2n}} k_0^{-\frac{1}{2n}}.$$

For Newtonian fluids $n = 1$, formula (6.6.17) was derived in [208].

Let $x_1 \approx 2.405$ be the first zero of the Bessel function $J_0(x_1)$. It follows from (6.6.17) that by increasing the pressure gradient $\Delta P/L$ according to the law $\sigma a^{(3n+1)/(2n)} \to x_1$, one can obtain arbitrarily large temperature on the flow axis. For $\sigma a^{(3n+1)/(2n)} \geq x_1$, there is no bounded solution of problem (6.6.3), (6.5.8), (6.6.16) at all. In this case, the dissipative heat released cannot be completely withdrawn through the walls, and hence, rapid transient heating of the medium is initiated.

The temperature distribution for the flow of a fluid with power-law dependence (6.6.16) of consistence on temperature in a plane channel, as well as some other solutions, was obtained in [299, 302].

A similar problem on a nonisothermal rectilinear flow of a viscoplastic Shvedov–Bingham fluid in a circular tube for the case in which the yield stress and the plastic viscosity are inversely proportional to temperature was studied in [298].

So far we have considered nonisothermal flows of non-Newtonian fluids with allowance for dissipative heating and the dependence of the apparent viscosity on temperature. It has been assumed that the wall temperature is constant and convective heat transfer is absent.

In [300–302], thermohydrodynamic problems for non-Newtonian fluids were studied under the assumption that temperature varies along the walls of the tube (or channel); in this case, convective heat transfer plays an important role. It was assumed that the dependence of the apparent viscosity of the medium on temperature is exponential or power-law; dissipative heat release was neglected. In one-dimensional steady-state flows of this type, the pressure gradient varies along the tube. It was shown that in some cases a situation typical of thermal explosion may arise. In this situation, heat supply due to fluid convection exceeds heat withdrawal through the walls. It was also discovered that there exists another mechanism for crisis phenomena to arise. If there is a constant heat withdrawal from the tube walls and the fluid velocity is sufficiently small, then the intensive cooling of the fluid may result in an accelerated increase of the fluid viscosity, which, in turn, results in flow choking.

# 6.7. Hydrodynamic and Diffusion Boundary Layers in Power-Law Fluids

## 6.7-1. Hydrodynamic Boundary Layer on a Flat Plate

*Statement of the problem.* Numerous practical applications stimulated the appearance of a vast variety of papers dealing with the boundary layer theory for nonlinearly viscous fluids obeying the power-law rheological law (e.g., see [8, 47, 181, 445, 448]). The main attention is paid to the investigation of self-similar problems, since their solutions allow one to study the characteristic properties of the boundary layer and to develop and justify approximate methods for numerical calculations. A detailed analysis of dilatant fluid flows shows that the region in which the tangential velocity varies is strictly localized in space [330, 520, 521].

Let us consider a steady-state isothermal flow of a power-law fluid past a thin flat plate. The velocity of the incoming flow is $U_i$. We assume that the coordinates $X$ and $Y$ are directed along the plate and transverse to the plate, respectively, and the origin is placed at the front edge of the plate. We denote the tangential and transverse velocities by $V_X$ and $V_Y$, respectively.

The main dimensionless parameter for power-law fluids is the generalized Reynolds number introduced by the formula

$$\mathsf{Re} = \frac{\rho L^n U_i^{2-n}}{k} \sim \frac{\text{inertial force}}{\text{friction force}}, \tag{6.7.1}$$

where $L$ is a dimensional quantity taken to be the characteristic length.

At high Reynolds numbers, the terms in the equation of motion (see Supplement 6) and in the continuity equation with regard to the expressions (6.1.1) and (6.1.4) can be estimated by the same scheme as for Newtonian fluids. As a result, after isolating the leading terms of the corresponding asymptotic expansions, we obtain

$$V_X \frac{\partial V_X}{\partial X} + V_Y \frac{\partial V_X}{\partial Y} = \frac{k}{\rho} \frac{\partial}{\partial Y} \left( \left| \frac{\partial V_X}{\partial Y} \right|^{n-1} \frac{\partial V_X}{\partial Y} \right), \tag{6.7.2}$$

$$\frac{\partial V_X}{\partial X} + \frac{\partial V_Y}{\partial Y} = 0. \tag{6.7.3}$$

These equations, which are considered in the domain $X \geq 0$, $Y \geq 0$, must be supplemented by the boundary conditions

$$V_X(X,0) = V_Y(X,0) = 0, \quad V_X(0,Y) = U_i, \quad V_X(X,\infty) = U_i. \tag{6.7.4}$$

**Exact solutions.** The solution of problem (6.7.2)–(6.7.4) is reduced to integrating the ordinary third-order differential equation

$$|f_{\zeta\zeta}''|^{n-1} f_{\zeta\zeta\zeta}''' + f f_{\zeta\zeta}'' = 0 \tag{6.7.5}$$

with the boundary conditions

$$f(0) = 0, \quad f_\zeta'(0) = 0, \quad f_\zeta'(\infty) = 1. \tag{6.7.6}$$

The fluid velocities $V_X$ and $V_Y$ and the self-similar variable $\zeta$ can be expressed via the coordinates $X$ and $Y$ and the function $f(\zeta)$ as follows:

$$V_X = U_i f'_\zeta, \quad V_Y = \frac{1}{n+1}\left[\frac{n(n+1)kU_i^{2n-1}}{\rho X^n}\right]^{\frac{1}{n+1}}(\zeta f'_\zeta - f),$$

$$\zeta = \left[\frac{\rho U_i^{2-n}}{n(n+1)kX}\right]^{\frac{1}{n+1}} Y. \tag{6.7.7}$$

In [514, 515], exact closed-form solutions of problem (6.7.5), (6.7.6) were obtained for pseudoplastic fluids with $n = \frac{1}{5}, \frac{1}{4}, \frac{1}{2}, \frac{3}{5}, \frac{5}{7}$. Let us write out two of these solutions in a parametric form.

For $n = \frac{1}{5}$, we have

$$f = at^2, \quad \zeta = b\int_0^t (1+t^3)^{1/3}dt, \tag{6.7.8}$$

where $a = 2^{-1/6} \cdot 5^{5/6}$, $b = 10^{5/6}$, and $t$ is a parameter, $0 \le t < \infty$).
For $n = \frac{3}{5}$, we have

$$f = at^2(1-t^3)^{-1/2}, \quad \zeta = b\int_0^t (1-t^3)^{-3/2}dt, \tag{6.7.9}$$

where $a = 2^{-3/4} \cdot 3^{1/2} \cdot 5^{5/8}$, $b = 2^{-7/4} \cdot 3^{3/2} \cdot 5^{5/8}$, and $0 \le t < 1$.

It was proved in [330] that for dilatant fluids ($n > 1$) the velocity varies only in a bounded region $0 \le \zeta \le \zeta_*$ near the plate. (Outside this region, for $\zeta \ge \zeta_*$, the fluid velocity is constant and is equal to $U_i$.) The function $f$ and the boundary $\zeta = \zeta_*$ of the localization region can be found by solving Eq. (6.7.5) with two boundary conditions (6.7.6) on the plate surface and two additional conditions $f'_\zeta(\zeta_*) = 1$ and $f''_{\zeta\zeta}(\zeta_*) = 0$. Outside the localization region, for $\zeta \ge \zeta_*$, the function $f$ is linear, $f = \zeta - \zeta_* + f(\zeta_*)$.

For the special case $n = 2$, the value $\zeta_* \approx 1.849$ can be found from the transcendental equation $2\cos\left(\frac{1}{2}\sqrt{3}\,\zeta_*\right) = -\exp\left(\frac{3}{2}\zeta_*\right)$, and in the localization region $0 \le \zeta \le \zeta_*$, the solution has the form [330]

$$f(\zeta) = \frac{\exp(-\zeta) + 2\exp(\frac{1}{2}\zeta)\sin(\frac{1}{2}\sqrt{3}\,\zeta - \frac{1}{6}\pi)}{-\exp(-\zeta_*) + 2\exp(\frac{1}{2}\zeta_*)\sin(\frac{1}{2}\sqrt{3}\,\zeta_* + \frac{1}{6}\pi)}.$$

Numerical solutions of problem (6.7.5), (6.7.6) for various values of the index $n$ ($0.1 \le n \le 2.0$) can be found in the book [445].

The second derivative $f''_{\zeta\zeta}(0)$ on the plate surface is nicely approximated by the formula

$$f''_{\zeta\zeta}(0) = 0.062 + 0.43\,n - 0.0245\,n^3, \tag{6.7.10}$$

whose error does not exceed 1% for $0.2 \le n \le 2.0$.

***Formulas for friction coefficients.*** Let us introduce the following dimensionless friction coefficients: the local friction coefficient

$$c_f = \frac{\tau_s}{\frac{1}{2}\rho U_i^2} = 2(n^2 + n)^{-\frac{n}{n+1}} \operatorname{Re}_X^{-\frac{1}{n+1}} \left[ f_{\zeta\zeta}''(0) \right]^n \tag{6.7.11}$$

and the mean friction coefficient

$$\langle c_f \rangle = \frac{2}{\rho U_i^2} \frac{1}{L} \int_0^L \tau_s \, dX = 2(n+1)(n^2+n)^{-\frac{n}{n+1}} \operatorname{Re}^{-\frac{1}{n+1}} \left[ f_{\zeta\zeta}''(0) \right]^n; \tag{6.7.12}$$

here $\operatorname{Re}_X = \rho X^n U_i^{2-n}/k$ is the local Reynolds number and $\operatorname{Re}$ is given by (6.7.1), where $L$ is the plate length.

Numerical solutions of problem (6.7.5), (6.7.6) show that the friction coefficients are nicely approximated by the expressions

$$c_f = \frac{2.266 - 1.22\,n + 0.28\,n^2}{n+1} \operatorname{Re}_X^{-\frac{1}{n+1}},$$

$$\langle c_f \rangle = (2.266 - 1.22\,n + 0.28\,n^2)\operatorname{Re}^{-\frac{1}{n+1}}, \tag{6.7.13}$$

whose maximum error does not exceed 0.5% for $0.1 \leq n \leq 2.0$.

For weakly nonlinearly viscous fluids that obey the viscous friction law $\tau = \tau(\dot\gamma)$, the local friction coefficient can be found by using the approximate formulas

$$c_f = \frac{2\tau(w)}{\rho U_i^2}, \qquad X = 0.22\,\rho U_i^3 \int_w^\infty \frac{dw}{w^2 \tau(w)}, \tag{6.7.14}$$

where $w = (\dot\gamma)_{Y=0}$ is the shear rate on the plate surface. To find the dependence of $c_f$ on $X$, one has to calculate the integral and then eliminate $w$ from (6.7.14).

Formulas (6.7.14) can be derived by the integral method in which the profile corresponding to a Newtonian fluid is taken as the tangential velocity profile.

Test calculations for power-law fluids show that the maximum error of (6.7.14) is 5% for $0.8 \leq n \leq 1.3$ and 9% for the wider interval $0.5 \leq n \leq 1.8$.

### 6.7-2. Hydrodynamic Boundary Layer on a V-Shaped Body

We shall now investigate the plane problem involving flow of a power-law fluid past a V-shaped body (a wedge). In the process of potential flow of an ideal fluid past the front critical point of the V-shaped body with an angle $\omega$ of taper (see Figure 1.6), the velocity near the vertex is

$$U(X) = AX^m. \tag{6.7.15}$$

Here the $X$-axis is directed along the wedge surface, $A$ is a constant, and the exponent $m$ and angle $\omega$ are related by [427]

$$m = \frac{\omega}{2\pi - \omega}.$$

The steady-state flow of a power-law fluid in a plane boundary layer near the surface of a V-shaped body is described elsewhere by the system of equations [445]

$$V_X \frac{\partial V_X}{\partial X} + V_Y \frac{\partial V_X}{\partial Y} = U \frac{dU}{dX} + \frac{k}{\rho} \frac{\partial}{\partial Y} \left( \left| \frac{\partial V_X}{\partial Y} \right|^{n-1} \frac{\partial V_X}{\partial Y} \right),$$

$$\frac{\partial V_X}{\partial X} + \frac{\partial V_Y}{\partial Y} = 0. \tag{6.7.16}$$

Here the $Y$-axis is directed along the normal to the wedge surface (given by $Y = 0$), $\rho$ is the fluid density, $V_X$ and $V_Y$ are, respectively, longitudinal and transverse liquid velocity components, and $U = U(X)$ is defined by Eq. (6.7.15).

To complete the statement of the problem, Eqs. (6.7.16) must be supplemented by the no-slip boundary conditions for fluid adhering to the wedge surface

$$V_X = V_Y = 0 \quad \text{at} \quad Y = 0, \tag{6.7.17}$$

and also by the condition

$$V_X \rightarrow U(X) \quad \text{as} \quad Y \rightarrow \infty, \tag{6.7.18}$$

for asymptotic matching of the longitudinal velocity component on the outer surface of the boundary layer with the fluid velocity in the flow core (6.7.15).

In the special case $m = 0$, problem (6.7.16)–(6.7.18) transforms into problem (6.7.2)–(6.7.4) for a steady-state flow of a power-law fluid past a flat plate.

The solution to the problem (6.7.16)–(6.7.18) is found form [445]

$$V_X = AX^m \Phi'(\xi), \quad \xi = BX^{\frac{2m-nm-1}{n+1}} Y,$$

$$V_Y = \frac{A}{B} X^{\frac{2nm-n-m}{n+1}} \left[ \frac{m-2nm-1}{n+1} \Phi(\xi) + \frac{nm-2m-1}{n+1} \xi \Phi'(\xi) \right], \tag{6.7.19}$$

where the primes stand for the derivative with respect to $\xi$ and the parameter $B$ is determined by

$$B = \left( A^{2-n} \frac{2nm - m + 1}{n+1} \frac{\rho}{k} \right)^{\frac{1}{n+1}}. \tag{6.7.20}$$

Substituting the expressions (6.7.19) into system (6.7.16) and the boundary conditions (6.7.17) and (6.7.18) and using (6.7.15) to determine the unknown function $\Phi = \Phi(\xi)$, we obtain the problem involving the third-order ordinary differential equation

$$n(\Phi'')^{n-1} \Phi''' + \Phi \Phi'' = b[(\Phi')^2 - 1],$$

$$\Phi = \Phi' = 0 \quad \text{at} \quad \xi = 0, \tag{6.7.21}$$

$$\Phi' \rightarrow 1 \quad \text{as} \quad \xi \rightarrow \infty,$$

where the constant $b$ depends on both geometric and rheological parameters of the problem in the following manner:

$$b = \frac{m(n+1)}{2nm - m + 1}.$$  (6.7.22)

From now on we assume that $0 \leq b < \infty$.

The viscous friction on the wedge surface is determined by

$$\tau_s = k\left(\left|\frac{\partial V_X}{\partial Y}\right|_{Y=0}\right)^n = k\left(ABX^{\frac{3m-1}{n+1}}|\Phi''(0)|\right)^n,$$  (6.7.23)

and the local friction coefficient $c_f$ is

$$c_f = \frac{\tau_s}{\frac{1}{2}\rho U^2} = \frac{2k}{\rho}A^{n-2}B^n X^{\frac{nm-n-2m}{n+1}}\left(|\Phi''(0)|\right)^n.$$  (6.7.24)

To use the expressions (6.7.23) and (6.7.24), we must know the value of the second derivative of $\Phi$ on the surface $\xi = 0$. For the special case of a Newtonian fluid, which corresponds to $n = 1$, see the approximate formula (1.8.10) for $\Phi''(0)$.

To estimate the second derivative, the following approximate formula can be used [359]:

$$\Phi''(0) = \left[(0.07 + 0.4n)^{n+1} + \frac{2}{3}\frac{n+1}{n}b\right]^{\frac{1}{n+1}}.$$  (6.7.25)

A comparison was made of the results based on formula (6.7.25) with the data of [445] obtained using a numerical solution of problem (6.7.21) with $m = 1$ (this value corresponds to a plane flow near the critical line at $\omega = \pi$). The maximum error of relation (6.7.25) for $0.2 \leq n \leq 2.2$ is 2%.

### 6.7-3. Diffusion Boundary Layer on a Flat Plate

Convective mass and heat transfer to a plate in a longitudinal flow of a non-Newtonian fluid was considered in [443]. By solving the corresponding problem in the diffusion boundary layer approximation (at high Peclet numbers), we arrive at the following expression for the dimensionless diffusion flux:

$$j = \frac{1}{\Gamma(\frac{1}{3})}\left[\frac{3}{2}\frac{(2n+1)}{n+1}f''_{\zeta\zeta}(0)\right]^{\frac{1}{2}}\left[\frac{\mathsf{Re}}{n(n+1)}\right]^{\frac{1}{3(n+1)}}\mathsf{Pe}^{\frac{1}{3}}\left(\frac{X}{L}\right)^{-\frac{n+2}{3(n+1)}},$$

where $\mathsf{Re} = \rho L^n U_i^{2-n}/k$; $\mathsf{Pe} = LU_i/D$, $L$ is the characteristic length, and $f$ is the solution of problem (6.7.5), (6.7.6). The second derivative $f''_{\zeta\zeta}(0)$ can be calculated by the approximate formula (6.7.10).

# 6.8. Submerged Jet of a Power-Law Fluid

## 6.8-1. Statement of the Problem

Let us consider the plane problem about the outflow of an incompressible power-law fluid from a narrow horizontal opening into an infinite space filled with the same medium. We introduce Cartesian coordinates $X, Y$ with the $X$-coordinate measured from the opening along the jet.

We assume that the opening is infinitely narrow and the outflow velocity of the fluid is so large that the longitudinal momentum of the jet remains finite,

$$J_0 = \int_{-\infty}^{+\infty} \rho V_X^2 \, dY = \text{const}. \tag{6.8.1}$$

By viscous friction, the jet spurting from the opening at a large velocity partly entrains the surrounding fluid and simultaneously slows down, so that a thin boundary layer is formed. This boundary layer is symmetric with respect to the $X$-axis, and its thickness grows downstream. The pressure across the jet is constant. Since the fluid remote from the opening is stagnant, it follows that the pressure gradient is zero in the entire flow region.

The velocity profile in the jet is described by the boundary layer equations (6.7.2), (6.7.3) together with the boundary conditions stating that the velocity profile is symmetric about the flow axis,

$$V_Y = 0, \quad \frac{\partial V_X}{\partial Y} = 0 \quad \text{at} \quad Y = 0, \tag{6.8.2}$$

the condition that the velocity decays remote from the opening,

$$V_X \to 0 \quad \text{as} \quad Y \to \pm\infty, \tag{6.8.3}$$

and the integral condition (6.8.1).

## 6.8-2. Exact Solutions

By using the self-similar variable $\eta$ and an auxiliary function $F = F(\eta)$, we can represent the velocity components as follows [443, 445]:

$$\begin{aligned}
\eta &= A(k/\rho)^{-\frac{1}{n+1}} X^{-\frac{2}{3n}} Y, \\
V_X &= [3n(n+1)]^{\frac{1}{2-n}} A^{\frac{n+1}{2-n}} X^{-\frac{1}{3n}} F'_\eta, \\
V_Y &= \frac{1}{3n}[3n(n+1)]^{\frac{1}{2-n}} A^{\frac{2n-1}{2-n}} (k/\rho)^{\frac{1}{n+1}} X^{\frac{1-3n}{3n}} (\eta F'_\eta - F),
\end{aligned} \tag{6.8.4}$$

where the constant $A$ is determined by condition (6.8.1) of conservation of momentum and the prime stands for the derivative with respect to $\eta$.

By substituting the expressions (6.8.4) into (6.7.2), (6.7.3), (6.8.2), and (6.8.3), we arrive at the ordinary third-order differential equation

$$n|F''_{\eta\eta}|^{n-1} F'''_{\eta\eta\eta} + (n+1)[FF''_{\eta\eta} + (F'_\eta)^2] = 0 \qquad (6.8.5)$$

with the boundary conditions

$$F = F''_{\eta\eta} = 0 \quad \text{at} \quad \eta = 0; \qquad F'_\eta \to 0 \quad \text{as} \quad \eta \to \pm\infty. \qquad (6.8.6)$$

The solution of problem (6.8.5), (6.8.6) satisfying the additional normalization condition $F'_\eta(0) = 1$ can be represented in the implicit form

$$\eta = \begin{cases} \displaystyle\int_0^F \left[1 - (2n-1)(n+1)^{\frac{1-n}{n}} F^{\frac{n+1}{n}}\right]^{\frac{n}{1-2n}} dF & \text{for } n \ne \tfrac{1}{2}, \\[2ex] \displaystyle\int_0^F \exp\left(\tfrac{3}{4} F^3\right) dF & \text{for } n = \tfrac{1}{2}. \end{cases} \qquad (6.8.7)$$

From (6.8.4) and (6.8.1) we obtain the constant $A$:

$$A = [3n(n+1)]^{-\frac{2}{3n}} \left[2 \int_0^\infty (F'_\eta)^2 \, d\eta\right]^{\frac{n-2}{3n}} \left[\frac{J_0}{\rho} \left(\frac{\rho}{k}\right)^{\frac{1}{n+1}}\right]^{\frac{2-n}{3n}}. \qquad (6.8.8)$$

The improper integral can be calculated by passing from the variable $\eta$ to the function $F$ according to (6.8.7). In particular, for $\tfrac{1}{2} < n < 2$ we obtain

$$\int_0^\infty (F'_\eta)^2 \, d\eta = n(n+1)^{-\frac{2}{n+1}} (2n-1)^{-\frac{n}{n+1}} \frac{\Gamma\left(\dfrac{n}{n+1}\right)\Gamma\left(\dfrac{3n-1}{2n-1}\right)}{\Gamma\left(\dfrac{n}{n+1} + \dfrac{3n-1}{2n-1}\right)}, \qquad (6.8.9)$$

where $\Gamma(x)$ is the gamma function. For $n = \tfrac{1}{2}$ we have

$$\int_0^\infty (F'_\eta)^2 \, d\eta = 4^{1/3} \cdot 3^{-4/3} \, \Gamma(1/3) \approx 0.983.$$

Let us first consider the case of a Newtonian fluid. We calculate the integral (6.8.7) for $n = 1$ and then express $F$ via $\eta$. As a result, we obtain

$$F = \tanh\eta \qquad (n = 1). \qquad (6.8.10)$$

Formulas (6.8.8) and (6.8.9) determine the constant $A$:

$$A = \left(\frac{J_0}{48 \, \rho\sqrt{\nu}}\right)^{1/3} \approx 0.275 \left(\frac{J_0}{\rho\sqrt{\nu}}\right)^{1/3}, \qquad (6.8.11)$$

where $\nu$ is the kinematic viscosity of the fluid.

By substituting (6.8.10) and (6.8.11) into (6.8.4), we arrive at the velocity distribution in a plane jet of a Newtonian fluid (see Subsection 1.4-2).

Now let us consider the dependence of qualitative characteristics of a jet flow of a power-law fluid on the rheological parameter $n$.

It follows from (6.8.7) that for $0 < n \le \tfrac{1}{2}$ the function $F$ increases infinitely as $\eta \to \infty$, and for $n > \tfrac{1}{2}$ the function $F$ tends as $\eta \to \infty$ to the finite limit

$$F(\infty) = (n+1)^{\frac{n-1}{n+1}} (2n-1)^{-\frac{n}{n+1}}. \qquad (6.8.12)$$

---

## 6.8-3. Jet Width and Volume Rate of Flow

The fluid velocity attains its maximum on the flow axis and decreases according to the law $U_{max} \sim X^{-1/(3n)}$. Thus, the smaller $n$, the more rapid the velocity decreases.

We define the 1%-width $\delta(X)$ of the jet as twice the distance from the jet axis to a point with coordinate $y^0$ at which the longitudinal velocity differs from its limit value by 1% :

$$\delta(X) = 2y^0 = 2\frac{\eta^0}{A}\left(\frac{k}{\rho}\right)^{\frac{1}{n+1}} X^{\frac{2}{3n}}, \qquad (6.8.13)$$

where $\eta^0$ is the value of the self-similar variable for which $V_X/U_{max} = F'_\eta = 0.01$. It follows from (6.8.13) that for $n > \frac{2}{3}$ the jet is convex outward, for $n = \frac{2}{3}$ the jet boundaries are rectilinear, and for $n < \frac{2}{3}$ the jet boundaries are diverging parabolas with a cuspidal singular point on the flow axis.

Now we calculate the volume rate of flow per unit length of the opening as follows:

$$Q = \int_{-\infty}^{+\infty} V_X\, dY = 2F(\infty)\left[3n(n+1)A^{2n-1}\right]^{\frac{1}{2-n}} \left(\frac{k}{\rho}\right)^{\frac{1}{n+1}} X^{\frac{1}{3n}}. \qquad (6.8.14)$$

For $\frac{1}{2} < n < 2$, we must substitute $F(\infty)$ from (6.8.12) into (6.8.14). Obviously, the rate of flow increases with the distance from the opening, since the jet entrains the stagnant fluid on its sides. The rate of flow also increases with increasing momentum. As the index $n$ decreases from 1 to $\frac{1}{2}$, the rate of flow grows infinitely, $\lim_{n \to 1/2} Q = \infty$. For $0 < n < \frac{1}{2}$, the volume rate of flow is infinite.

# 6.9. Motion and Mass Exchange of Particles, Drops, and Bubbles in Non-Newtonian Fluids

## 6.9-1. Drag Coefficients

***Power-law fluids.*** Here we briefly discuss the motion of spherical bubbles, drops, and particles at a constant velocity $U_i$ in a power-law non-Newtonian.

In the case of an inertia-free flow (at low Reynolds numbers) of a quasi-Newtonian power-law fluid with rheological index $n$ close to 1 past a gas bubble, the drag coefficient can be calculated by the formula [190]

$$c_f = \frac{2|F|}{\pi a^2 \rho U_i} = 3^{\frac{n-1}{2}}\frac{13 + 4n - 8n^2}{(n+2)(2n+1)}\frac{8}{\mathsf{Re}}, \qquad (6.9.1)$$

where $\mathsf{Re} = \rho a^n U_i^{2-n}/k$ is the Reynolds number and $a$ is the bubble radius.

Once can readily see that for pseudoplastic fluids the drag coefficient is larger and for dilatant fluids the drag coefficient is less than for a Newtonian fluid.

For a uniform Stokes flow of a Newtonian fluid past a drop, the drag coefficient can be represented in the form [303, 304]

$$c_f = \frac{24\,\Phi_n(\beta)}{2^n\,\mathrm{Re}},\tag{6.9.2}$$

where the function $\Phi_n$ characterizes the rheological properties of the flow and is a function of the dimensionless parameters $n$, $\beta = \mu a^{n-1}/(kU_i^{n-1})$, and $\mu$ is the viscosity of the drop.

For a gas bubble and a solid particle, which corresponds to the limit values $\beta = 0$ and $\beta = \infty$, the function $\Phi_n$ is nicely approximated by the formula

$$\Phi_n(0) = 0.81 + 0.46\,n - 0.6\,n^2 \qquad \text{(bubble)},\tag{6.9.3}$$

$$\Phi_n(\infty) = 1.65 + 0.1\,n - 0.75\,n^2 \qquad \text{(solid particle)},\tag{6.9.4}$$

whose error does not exceed 1.5% for $0.6 \le n \le 1.0$ (compared with the results provided by numerical solutions [303, 304]).

For the case of a drop, which corresponds to finite values $0 < \beta < \infty$, the function $\Phi_n(\beta)$ can be calculated by using the approximate formula

$$\Phi_n(\beta) = \frac{1}{\beta+1}\Phi_n(0) + \frac{\beta}{\beta+1}\Phi_n(\infty)\tag{6.9.5}$$

whose maximum error is about 3%.

The expressions (6.9.2)–(6.9.5) allow one to calculate the drag coefficients for particles, drops, and bubbles in a Stokes flow of a power-law fluid.

The velocity of a drop falling by gravity in a power-law fluid at low Reynolds numbers is given by the formula [69]

$$U_i = 2a\left\{\frac{ag|\rho_1 - \rho_2|}{9k\Phi_n(\beta)}\right\}^{1/n},\tag{6.9.6}$$

where $\rho_1$ and $\rho_2$ are the densities of the drop and the continuous phase, respectively.

In [294] the rheological behavior of concentrated suspensions with a non-Newtonian continuous phase described by the power-law model was studied in the framework of the cellular model.

*Viscoplastic fluids.* In the case of a spherical bubble in a translational Stokes flow of a viscoplastic Shvedov–Bingham fluid with a small yield stress, the following two-term asymptotic expansion is valid for the drag coefficient [37]:

$$c_f = 8\,(1 + 3.22\,\mathrm{Bi})\,\mathrm{Re}^{-1}, \qquad \mathrm{Re} = a\rho U_i/\mu_p,\tag{6.9.7}$$

where $\mathrm{Bi} = a\tau_0/(\mu_p U_i) \ll 1$ is the Bingham number, $\tau_0$ is the Bingham yield stress, and $\mu_p$ is the Bingham plastic viscosity.

The yield stress of viscoplastic fluids may be estimated by observing the motion/no motion of a sphere. For example, the yield stress for carbopol solutions was evaluated in [182].

The motion and sedimentation of particles in non-Newtonian fluids was considered by direct numerical simulation in [129, 130, 193, 194, 207, 484].

The books [92, 112, 272] present a detailed review of investigations related to the motion of particles, drops, and bubbles in a non-Newtonian fluid, as well as numerous formulas and curves determining the drag force.

---

### 6.9-2. Sherwood Numbers

In the case of mass exchange between a bubble and a translational Stokes flow of a quasi-Newtonian power-law fluid ($n$ is close to unity), one can use the following simple approximate formula for calculating the mean Sherwood number at high Peclet numbers:

$$\mathsf{Sh} = \left[ (0.5 - 0.3\,n)\,\mathsf{Pe} \right]^{1/2}. \tag{6.9.8}$$

The results predicted by formula (6.9.8) were compared with the experimental data of [92, 292] for the liquid phase mass transfer coefficients from the absorption of nitrogen bubbles in aqueous solutions of carboxymetil cellulose and carbopol at low Reynolds numbers. The maximum error of the formula is about 5% for $0.7 \le n \le 1.0$.

In the case of a spherical bubble in a translational flow of a viscoplastic fluid with low Bingham numbers, one can use the following formula for the mean Sherwood number [37]:

$$\mathsf{Sh} = 0.461\,(1 + 0.5\,\mathsf{Bi})^{1/2}\,\mathsf{Pe}^{1/2}.$$

In the case of mass exchange between a particle and a translational flow of a quasi-Newtonian power-law fluid, the mean Sherwood number can be estimated using the relation

$$\mathsf{Sh} = \left[ (4.95 - 4.7\,n)\,\mathsf{Pe} \right]^{1/3}. \tag{6.9.9}$$

The results predicted by formula (6.9.9) well agree with the experimental data of [92, 292] at low Reynolds numbers and high Peclet numbers for $0.7 \le n \le 1.0$.

The book [92] present a detailed review of investigations related to the mass exchange of particles, drops, and bubbles in non-Newtonian fluids.

# 6.10. Transient and Oscillatory Motion of Non-Newtonian Fluids

### 6.10-1. Transient Motion of an Infinite Flat Plate

*Power-law fluids.* Let us consider a stagnant semi-infinite fluid bounded by a rigid plane $-\infty < X < \infty$, $Y \ge 0$. The problem of flow evolution for $t > 0$

resulting from an abrupt start-up of motion of the plane at a constant velocity has the following mathematical statement. The equation of motion in terms of stresses has the form

$$\rho \frac{\partial V_X}{\partial t} = \frac{\partial \tau_{XY}}{\partial Y}. \tag{6.10.1}$$

The initial and boundary conditions are

$$V_X = 0 \qquad \text{at} \quad t = 0, \tag{6.10.2}$$
$$V_X = U = \text{const} \quad \text{at} \quad Y = 0. \tag{6.10.3}$$

For the power-law fluid (see the first model in Table 6.1, $0 < n \leq 1$)

$$\tau_{XY} = k \left| \frac{\partial V_X}{\partial Y} \right|^{n-1} \frac{\partial V_X}{\partial Y}, \tag{6.10.4}$$

substituting (6.10.4) into (6.10.1) and taking into account the fact that $\partial V_X / \partial Y \leq 0$ for Stokes' first problem, we obtain the following equation for $V_X(t, Y)$:

$$\rho \frac{\partial V_X}{\partial t} = -k \frac{\partial}{\partial Y} \left( -\frac{\partial V_X}{\partial Y} \right)^n. \tag{6.10.5}$$

This equation was solved in [39] by introducing the self-similar variable

$$\eta = \frac{Y}{n+1} \left( \frac{\rho}{kU^{n-1}t} \right)^{\frac{1}{n+1}}. \tag{6.10.6}$$

As a result, the problem for the dimensionless velocity field

$$\Phi_n(\eta) = V_n(t, Y)/U \tag{6.10.7}$$

can be represented as the boundary value problem for the ordinary differential equation

$$\frac{d^2 \Phi_n}{d\eta^2} \left( -\frac{d\Phi_n}{d\eta} \right)^{n-1} + \frac{(n+1)^n}{n} \eta \frac{d\Phi_n}{d\eta} = 0 \tag{6.10.8}$$

with the boundary conditions

$$\Phi_n = 1 \qquad \text{at} \quad \eta = 0, \tag{6.10.9}$$
$$\Phi_n \to 0 \qquad \text{as} \quad \eta \to \infty. \tag{6.10.10}$$

For Newtonian fluids ($n = 1$), Eq. (6.10.8) becomes linear and has the well-known solution (1.9.3).

For an arbitrary $n$, the solution of (6.10.8) with regard to (6.10.10) can be written in the form of the integral [39]

$$\Phi_n(\eta) = \left[ \frac{(1+n)^n(1-n)}{2n} \right]^{-\frac{1}{1-n}} \int_\eta^\infty (C_n + \eta^2)^{-\frac{1}{1-n}} \, d\eta, \tag{6.10.11}$$

TABLE 6.7

Analytical expressions for dimensionless self-similar
profiles of transient velocity for various $n$

| $n$ | $C_n$ | $\Phi_n$ | $\eta_1$ |
|-----|-------|----------|----------|
| $\frac{1}{3}$ | 0.866 | $1 - f_n$ | 6.57 |
| $\frac{1}{2}$ | 1.637 | $1 - \frac{2}{\pi}(g_n f_n + \arcsin f_n)$ | 4.30 |
| $\frac{2}{3}$ | 2.838 | $1 - \frac{2}{\pi}\left(\frac{2}{3}g_n^3 f_n + g_n f_n + \arcsin f_n\right)$ | 3.05 |
| $\frac{5}{6}$ | 5.978 | $1 - \frac{2}{\pi}\left(\frac{128}{315}g_n^9 f_n + \frac{16}{35}g_n^7 f_n + \frac{8}{15}g_n^5 f_n \right.$ $\left. + \frac{2}{3}g_n^3 f_n + g_n f_n + \arcsin f_n\right)$ | 2.29 |

where the constant $C_n$ is determined from condition (6.10.9).

Note that solutions of the form (6.10.11) are physically meaningful only for
pseudoplastic fluids ($n < 1$).

In [39], for some values of $n < 1$, the constants $C_n$ are calculated and
closed-form expressions are obtained for the dimensionless self-similar profiles
of velocity fields in terms of elementary functions in the new variables

$$f_n = \sqrt{\frac{\eta^2}{C_n + \eta^2}} \quad \text{and} \quad g_n = \sqrt{\frac{C_n}{C_n + \eta^2}}. \tag{6.10.12}$$

These results are presented in Table 6.7.

The numbers in the last column of the table (for $\eta_1$) are the values of the
self-similar variable $\eta$ at which the medium velocity is equal to 1% of the wall
velocity.

The curves representing the velocity profiles given in the table can be found
in [39] and [41].

*Viscoelastic fluids.* In the monograph [181], the exact solution of Stokes'
first problem (6.10.1)–(6.10.3) was obtained for Maxwellian fluids with the rheological law (6.1.10), which has the following form for the problem in question:

$$\tau_{XY} - t_0 \frac{\partial \tau_{XY}}{\partial t} = \mu \frac{\partial V_X}{\partial Y}, \tag{6.10.13}$$

where $t_0$ is the stress relaxation time.

The exact solution of problem (6.10.1)–(6.10.3), (6.10.13) obtained in [181]
by the Laplace transform method has the form

$$V_X(t, Y) = U\left\{1 - \frac{1}{\pi}\int_0^\infty \frac{1}{\alpha}\exp\left(-\alpha\frac{t}{t_0}\right)\sin\left[\frac{Y}{\sqrt{\nu t_0}}\sqrt{\alpha(1-\alpha)}\right]d\alpha\right\}. \tag{6.10.14}$$

Note that the spectrum of eigenvalues of the problem (the integration variable $\alpha$
in (6.10.14)) is continuous.

### 6.10-2. Oscillating Flat-Plate Flow for Maxwellian Fluids

Stokes' second problem deals with the behavior of a semi-infinite fluid if the wall bounding the fluid performs harmonic oscillations in its plane. This problem is stated as a problem without initial data [181, 482]: only the boundary condition

$$V_X = U \cos \omega t \qquad \text{at} \quad Y = 0 \tag{6.10.15}$$

is added to the equation of motion (6.10.1) and the rheological law (6.10.13).

The problem is reduced to solving a linear partial differential equation of hyperbolic type (the telegraph equation [482]) in the region $t > 0$, $0 \le Y < \infty$. Specifically, we have the equation

$$\frac{\partial V_X}{\partial t} + t_0 \frac{\partial^2 V_X}{\partial t^2} = \nu \frac{\partial^2 V_X}{\partial Y^2} \tag{6.10.16}$$

with the boundary condition (6.10.15).

Strictly speaking, to obtain a unique solution in an infinite region, condition (6.10.17) must be supplemented by the Sommerfeld radiation condition [234, 317], so as to eliminate waves coming from infinity. The solution of this problem has the form

$$V_X(t, Y) = U \exp\left[-\sqrt{(\Lambda - \omega^2 t_0^2)y}\right] \cos\left[\omega t - \sqrt{(\Lambda + \omega^2 t_0^2)y}\right],$$

$$\Lambda = \omega t_0 \sqrt{1 + \omega^2 t_0^2}, \quad y = \frac{Y}{2\sqrt{\nu t_0}}.$$

Obviously, the oscillation frequency is the same and is equal to $\omega$ at any point of the half-space, whereas the amplitude and the phase of the oscillations varies with the spatial coordinate $Y$.

In the case of high frequencies ($\omega t_0 \gg 1$), we have the asymptotics

$$V_X(t, Y) = U \exp\left(-\tfrac{1}{2}\sqrt{2y}\right) \cos\left(\omega t - \omega t_0 \sqrt{2y}\right);$$

that is, the solution is a decaying wave propagating from the oscillating wall into the bulk of fluid at the velocity $\sqrt{\nu/t_0}$.

### 6.10-3. Transient Simple Shear Flow of Shvedov–Bingham Fluids

*Statement of the problem.* The problem of a transient flow of a Shvedov–Bingham fluid was exactly solved in [434]. Suppose that in the half-space $Y > 0$ we originally have a steady-state flow of a Shvedov–Bingham fluid with a uniform transverse velocity gradient under the action of the tangential stress $\tau_1 > \tau_0$ on the boundary, where $\tau_0$ is the yield stress. At time $t = 0$, the tangential stress on the boundary abruptly drops to a value $\tau_2$ such that $|\tau_2| < \tau_0$. In this case, the front $Y = Y_*(t)$ of the solidified fluid (moving at a uniform velocity) starts to propagate into the bulk of the layer.

The only nonzero velocity component $V_X$ is directed along the $X$-axis coinciding with the boundary of the half-space. This component depends on $t$ and $Y$ and satisfies equation (6.10.1), where $\rho$ is the fluid density and $\tau_{XY}$ is the corresponding component of the stress tensor.

In this case, the rheological equation for a Shvedov–Bingham fluid has the form

$$\tau_{XY} = \mu_p \frac{\partial V_X}{\partial Y} = \begin{cases} 0 & \text{for } |\tau| < \tau_0, \\ \tau - \text{sign}(\tau)\tau_0 & \text{for } |\tau| > \tau_0, \end{cases} \tag{6.10.17}$$

where $\mu_p$ is the Bingham (plastic) viscosity.

The problem in question involves the initial conditions

$$V_X = GY, \quad \tau = \tau_2, \quad Y_* = 0 \quad \text{at} \quad t = 0, \tag{6.10.18}$$

where $G = \tau_1/\mu_p$.

The boundary conditions have the form

$$\tau = \begin{cases} \tau_1 & \text{for } t < 0 \text{ with } |\tau_1| > \tau_0 \\ \tau_2 & \text{for } t > 0 \text{ with } |\tau_2| < \tau_0 \end{cases} \quad \text{at} \quad Y = 0,$$

$$V_X|_{Y=Y_*-0} = V_X|_{Y=Y_*+0} \quad \text{at} \quad Y = Y_*(t), \tag{6.10.19}$$

$$\tau|_{Y=Y_*-0} = \tau|_{Y=Y_*+0} \quad \text{at} \quad Y = Y_*(t),$$

$$V_X \to GY \quad \text{as} \quad Y \to \infty,$$

**Exact solution.**   The exact self-similar solution of problem (6.10.1), (6.10.17)–(6.10.19) is given by the formulas

$$V_X(t, Y) = \begin{cases} \lambda \dfrac{\tau_1}{\mu_p} \sqrt{\nu t} \left[ 1 + \dfrac{2\exp(-\lambda^2/4) - \sqrt{\pi}\,\lambda\,\text{erfc}(\lambda/2)}{\sqrt{\pi}\,\lambda\,\text{erfc}(\lambda/2)} \right] \\ \qquad\qquad\qquad\qquad \text{if } 0 \le Y/\sqrt{\nu t} \le \lambda, \\[2mm] \dfrac{\tau_1}{\mu_p} Y \left\{ 1 + \dfrac{2\sqrt{\nu t}\exp[-Y^2/(4\nu t)] - \sqrt{\pi}\,Y\,\text{erfc}(Y/\sqrt{4\nu t})}{\sqrt{\pi}\,Y\,\text{erfc}(\lambda/2)} \right\} \\ \qquad\qquad\qquad\qquad \text{if } \lambda \le Y/\sqrt{\nu t}; \end{cases} \tag{6.10.20}$$

$$\frac{\tau(t, Y) - \tau_0}{\tau_0 - \tau_2} = \begin{cases} -1 + \dfrac{Y}{\lambda\sqrt{\nu t}} & \text{if } 0 \le Y/\sqrt{\nu t} \le \lambda, \\[2mm] \dfrac{\tau_1}{\tau_0 - \tau_2} \left\{ 1 + \dfrac{Y\exp[-Y^2/(4\nu t)] - \sqrt{\pi \nu t}\,\text{erfc}(Y/\sqrt{\nu t})}{\sqrt{\pi \nu t}\,\text{erfc}(\lambda/2)} \right\} \\ \qquad\qquad\qquad\qquad \text{if } \lambda \le Y/\sqrt{\nu t}; \end{cases} \tag{6.10.21}$$

where $\nu = \mu_p/\rho$ and the dimensionless parameter $\lambda$, which is an eigenvalue of problem (6.10.1), (6.10.17)–(6.10.19), satisfies the equation

$$\frac{\lambda\exp(-\lambda^2/4)}{\sqrt{\pi}\,\text{erfc}(\lambda/2)} = \frac{\tau_0 - \tau_2}{\tau_1}. \tag{6.10.22}$$

The motion of the front of the solidified fluid obeys the equation $Y_*(t) = \lambda\sqrt{\nu t}$.

The solution of Eq. (6.10.22) has the asymptotics $\lambda = \sqrt{\pi}\,(\tau_0 - \tau_2)/\tau_1$ for $(\tau_0 - \tau_2)/\tau_1 \ll 1$.

# Chapter 7
# Foams: Structure and Some Properties

Foam is a technological medium commonly encountered in various branches of industry such as chemical, biotechnological, mining, and food industries. In column bubblers, which comprise an important class of apparatuses for exchange processes in gas-liquid systems, lavish foam generation is virtually inevitable, since the fluids used there almost invariably contain surfactants of chemical, biological, or mineral origin. These substances stabilize the interface [125, 297, 371, 413], so that a foam layer becomes a homogeneous technological working medium with unique physico-chemical and hydrodynamic properties.

The presence of excess surface energy in such gas-liquid disperse system predetermines its nonequilibrium. However, by virtue of the stabilizing effect of surfactants, foam possesses a metastable structure and has a certain lifetime [118]. Its properties slowly relax under the action of external factors provided that the latter do not exceed some threshold values beyond which the foam structure is destroyed.

It has been recently discovered [273] that foam can also exist in steady state, where the phase volume fractions and the bubble size distribution are independent of time. This phenomenon occurs if the foam is a closed (in the terminology of [375]) but not isolated system, that is, there is energy exchange but no mass exchange between the foam and the ambient medium. Such conditions arise in closed vibrating containers under withdrawal of the dissipative heat due to vibrational motion. If a certain threshold of specific kinetic vibrational energy is exceeded, then, after some relaxation time, the foam becomes steady-state (and forms dissipative structures [311]) with Gaussian distribution of bubble sizes.

Since the specific area of the foam interface is rather large, one can effectively use foam in numerous chemico-technological processes such as desorption, cleaner flotation, separation, gas exchange, control of gas filtration through porous media, and transportation of loose materials and aggressive fluids [32, 383, 429]. At the same time, the role of foam in chemico-physical processes is not unambiguous. Progressing gravitational dehydration, excessive strength of foam structures, and retardation of processes on actual foam interfaces decrease the efficiency of exchange processes so that, at a certain stage of the evolution of a foam structure, a considerable part of the foam volume becomes ineffective from the technological viewpoint. For this reason, the problems of forced destruction of foam structures [480] comprise an important part of the general problem of foam control.

# 7.1. Fundamental Parameters. Models of Foams

Foam is a special gas-liquid medium with gaseous disperse phase and liquid continuous phase.* The continuous phase is the binding phase continuously filling all space between the disperse gaseous inclusions. In monodisperse foam containing at most 74% of gas, the gas bubbles are spherical. Such foam, in the nomenclature of [280], is called "Kugelschaum" (spherical foam). With increasing gas content, bubbles begin to deform, and plane interfaces with rounded-off transition parts between them are formed. In the limit, the foam cells become polyhedral (we have "Polyedrschaum" according to [280]).

Actual foam contains bubbles whose shape is intermediate between spheres and polyhedra. Such foam is said to be cellular [214, 280]. The distinction between the cellular and polyhedral kinds of foam is rather conventional and is determined by very low moisture contents (of the order of some tenth of per cent). Nevertheless, the polyhedral model of foam cells is used rather frequently [38, 125, 244, 438, 480].

## 7.1-1. Multiplicity, Dispersity, and Polydispersity of Foams

*Multiplicity.* One of the most important quantitative characteristics of foam is the multiplicity $K$. This is a conventional name for the dimensionless reciprocal of the volume fraction of fluid in the foam [38, 125, 280, 480]. If $\Phi$ is the volume fraction of gas in the foam, then

$$K = (1 - \Phi)^{-1}. \qquad (7.1.1)$$

This quantity is sufficiently large (usually, of the order of $10^2$ to $10^3$) and more convenient than the moisture content, which, accordingly, is $10^{-2}$ to $10^{-3}$. In terms of multiplicity, the conventional range of cellular foam is $4 \leq K \leq 170$ [214].

*Dispersity.* Foam cells usually have the shape of rounded polyhedra. Therefore, it is convenient to choose the radius $a$ of the volume-equivalent bubble, that is, of a spherical bubble of the same volume as the cell as the single linear dimension characterizing the interior scale of foam. Foam consisting of cells of the same size is said to be monodisperse. This kind of foam is extremely rare. Usually, there exists a spectrum of radii $a_1, \ldots, a_n$; in this case, the foam dispersity characterizes the mean linear dimension of the cell [214]:

$$\bar{a} = \frac{1}{n} \sum_{i=1}^{n} a_i. \qquad (7.1.2)$$

---

* Sometimes, the disperse phase is another liquid immiscible with the first one. In this case, we have a biliquid foam.

Sometimes the foam dispersity is characterized by the specific interfacial area $\varepsilon$ equal to the total surface of gaseous bubbles per unit volume of foam,

$$\varepsilon = 3\frac{K-1}{K} \sum_{i=1}^{n} n_i a_i^2 \Big/ \sum_{i=1}^{n} n_i a_i^3, \qquad (7.1.3)$$

where $n_i$ is the number of bubbles of size $a_i$ per unit volume of foam. For monodisperse foam,

$$\varepsilon = \frac{3}{a} \frac{K-1}{K}. \qquad (7.1.4)$$

**Polydispersity.** If a liquid is injected into polydisperse cellular foam (a process which decreases the foam multiplicity) until the foam becomes spherical with the same size distribution of bubbles, then the obtained spherical foam is said to be equivalent. The equivalent foam is characterized by the multiplicity called the minimum multiplicity $K_{\min}$. Obviously, the minimum multiplicity of polydisperse foam is larger than that of monodisperse foam, since in gaps between densely packed spheres of the same largest size, spheres of smaller sizes can be located. Thus the value $K_{\min}$ can be used as a quantitative measure of polydispersity [214]. While $K_{\min} = 3.86$ for monodisperse foam, for actual polydisperse foam we have $K_{\min} \sim 10$ to 15; in practice $K_{\min}$ never exceeds 20 [480].

In [118] the following size distribution of bubbles is proposed, which closely agrees with the size distribution in actual foam:

$$f(a) = \frac{6\alpha a}{(1+\alpha a^2)^4}, \qquad (7.1.5)$$

where $\alpha$ is the parameter of the distribution function. One can readily verify that this function is normalized and the parameter $\alpha$ is related to the modal value $\bar{a}$ of the radius as follows:

$$\alpha = \tfrac{1}{7}(\bar{a})^{-2}. \qquad (7.1.6)$$

We point out that in steady-state foam (typical of closed but nonisolated systems) the distribution function is Gaussian [273].

Generalizing (7.1.5), one can propose the multiparameter distribution function

$$f(a) = \sum_{i=1}^{n} \gamma_i \frac{2(\beta_i - 1)\alpha_i a}{(1+\alpha_i a^2)^{\beta_i}} \qquad (7.1.7)$$

with the normalization condition

$$\sum_{i=1}^{n} \gamma_i = 1, \qquad (7.1.8)$$

where $\alpha_i$, $\beta_i$, and $\gamma_i$ are constants chosen so as to achieve the best possible agreement with experimental data.

The surface-average (Sauter mean [383]) radius of cells is also introduced by using (7.1.4) as follows:

$$a_s = \frac{3}{\varepsilon} \frac{K-1}{K}. \qquad (7.1.9)$$

This quantity is used for the description of gas exchange in polydisperse foams.

| **7.1-2. Capillary Pressure and Capillary Rarefaction** |

*Capillary pressure.* This term denotes the pressure difference between foam bubbles and the atmosphere:

$$\Delta P_g = P_g - P_a, \tag{7.1.10}$$

where $P_g$ is the pressure in the gaseous phase of the foam and $P_a$ is the atmosphere pressure. This variable is determined by Laplace's additional term due to the curvature of the interface between the atmosphere and the upper surface of bubbles in the foam layer adjacent to the atmosphere. This is an integral characteristic of foam, since the pressure in lower layers of bubbles is the same in view of plane interfaces between these layers [38]. The capillary pressure was calculated in the early thirties by Derjaguin [115]. This pressure depends on the surface tension of the foaming solution $\sigma$ and the specific interfacial area $\varepsilon$ in the foam as follows [125]:

$$\Delta P_g = \tfrac{2}{3}\sigma\varepsilon. \tag{7.1.11}$$

For monodisperse foam it follows from (7.1.4) that

$$\Delta P_g = \frac{2\sigma}{a}\frac{K-1}{K}. \tag{7.1.12}$$

We note that in actual foam the capillary pressure is not large and usually does not exceed hundreds of Pascals (is of the order of 0.001 atm.).

*Capillary rarefaction.* The continuous fluid phase filling the porous space between deformed bubbles has a common concave boundary with these bubbles. It follows that the local pressure $P_l$ in the liquid phase is less than the pressure $P_g$ in the gaseous phase and these variables are related by

$$P_l = P_g - \sigma\kappa_s, \tag{7.1.13}$$

where $\kappa_s$ is the local mean curvature of the interface.

The excess of the atmosphere pressure over the pressure in the liquid phase,

$$\Delta P_l = P_a - P_l, \tag{7.1.14}$$

will be called the capillary rarefaction $\Delta P_l$. By comparing (7.1.10) with (7.1.13) and (7.1.14), one can see that

$$\Delta P_l = \sigma\kappa_s - \Delta P_g. \tag{7.1.15}$$

In contrast with the capillary pressure $\Delta P_g$, the capillary rarefaction $\Delta P_l$ is a local variable. The distribution of $\Delta P_l$ in the continuous phase determines all interior flows of liquid in foam. At the same time, the averaged integral capillary rarefaction determines the capability of foam to absorb liquid, and it presents certain strength to the foam body.

Sometimes, the term osmotic dispersion pressure [378, 383] is used instead of capillary rarefaction. The osmotic pressure is defined as the excessive external pressure that must be applied to the semipermeable membrane interface between foam and fluid to stop the flux of the fluid sucked into the foam from the free volume. In this case, it is assumed that foam cell faces are flat, and therefore, the capillary pressure in foam bubbles is zero.

According to (7.1.15), the capillary rarefaction depends on the mean curvature $\kappa_s$ of the internal foam surface at the nodes of the foam structure. This variable was calculated in [378] for a monodisperse foam with cells modeled by pentagonal dodecahedra:

$$\kappa_s = \frac{0.584}{a} \frac{\Phi^{1/3}}{(1 - \Phi)^{1/2}}. \qquad (7.1.16)$$

It follows from [378, 383] that the formula for the osmotic pressure differs from the corresponding formula for the capillary rarefaction only by the coefficient $g(\Phi) = (1 - 1.83\sqrt{1 - \Phi})^2$ which characterizes the fraction of the membrane area adjacent to flat faces of foam cells. In the case of polydisperse foam, it is also expedient to use formula (7.1.16) with $a$ replaced by the Sauter mean radius (7.1.9).

In the case of monodisperse foam, another approximation formula for the interface mean curvature was obtained in [156] by using the deformation theory [430] and the rounded dodecahedron model; namely,

$$\kappa_s = \frac{9.33}{a}[1 - \exp(-0.00825\,K)]. \qquad (7.1.17)$$

The error of this formula is 3% in the multiplicity interval $30 \le K \le 300$.

### 7.1-3. The Polyhedral Model of Foams

*Preliminary remarks. Models of the foam cell.* The polyhedral shape of foam cells is the limit shape as the foam multiplicity grows infinitely. At the same time, this is a rather convenient structural model for actual foam with finite multiplicity. A polyhedron constructed of liquid films must satisfy the following two rules, stated by Plateau [9, 379, 407]:

1. Three films must meet along one edge making equal dihedral angles (of $120°$).

2. Three edges must meet at one node making equal angles (of $109°26'16''$) between each other.

It is also natural to require that the number of faces (films) $F$, edges (the Plateau borders) $B$, and nodes $N$ shall satisfy the fundamental topological Euler theorem about polyhedra [101, 479]:

$$N - B + F = 2. \qquad (7.1.18)$$

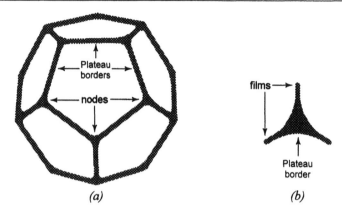

**Figure 7.1.** A foam cell model: (a) pentagonal dodecahedron, (b) section of the Plateau border

The simplest space figure (regular polyhedron) which closely satisfies the Plateau conditions is a pentagonal dodecahedron whose twelve faces are equal regular pentagons [280]. This model has the following drawbacks: it is impossible to fill the space with dodecahedra, and their parameters are inconsistent with the statistics of actual foam. Statistical experiments show [282] that in the mean a polyhedral cell of foam must have 13.7 faces and each face must have 5.1 edges. For this reason, it was suggested in [101] that a random filling of identical bodies (that is, statistical honeycomb) is a spatial tessellation with parameters corresponding the above-described statistics.

It was repeatedly proposed to use Kelvin's tetrakaidecahedron (that is, minimal truncated octahedron) [381, 407, 479] with eight hexagonal and six quadrangular faces as the polyhedral model of a foam cell and of a cell of any three-dimensional biological tissue. Note, however, that it was statistically shown [195] that Kelvin's tetrakaidecahedron is encountered in biological tissues among other tetrakaidecahedral cells only in 10% of the cases.

It was shown in [244, 245] that the polyhedral foam structure is better modeled by "compact" tetrakaidecahedral cells. The point is that according to the Euler theorem (7.1.18), the number of faces is fixed, and so the specific interfacial area and hence the free energy are the less, the closer in shape are the faces of the polyhedra and the closer is the shape of these faces to circular. Elongated faces of polyhedra are of low probability; therefore, the shape of the polyhedron must be nearly spherical. Such a hypothetical tetrakaidecahedron was said to be compact. It is very difficult to calculate the geometric parameters of such a hypothetical polyhedron (the surface area, the edge length, etc.); therefore, one usually takes the well-known results for the corresponding characteristics of the pentagonal dodecahedron and makes the corresponding corrections. For example, the total length of edges is increased $\sqrt{14/12}$ times. We note that such corrections are close to unity (the difference is 5 to 10%).

Thus, in approximate modeling it is possible to use the following polyhedral model of the foam structure [214] (see Figure 7.1):

1. Foam cells have the shape of equal pentagonal dodecahedra with rounded edges and vertices.

2. Deformed bubbles are divided by thin plane-parallel films of regular pentagonal shape.

3. At the junctions between three films, there are some thicker areas (called Plateau borders) whose cross-section is constant along the edge and has the shape of a curvilinear Plateau triangle formed by arcs of three adjacent circles.

4. Four Plateau borders meet at a node, which is one more structural element of foam. It is a spherical tetrahedron formed by four concave spherical surfaces. Up to the multiplicity $K \approx 200$, nodes contain the main part of continuous phase, but for $K > 600$ most part of the moisture is contained in Plateau borders [214].

*Formulas for geometric parameters.* The radius of curvature $R_n$ of nodal menisci is one of the most important quantitative characteristics of foam. The other quantity connected with the former is the radius $R_b$ of Plateau borders.

Although, according to (7.1.15), the capillary rarefaction is a local variable, its values at nodes and in Plateau borders are equalized rather rapidly due to the flow of liquid, which allows one to assume that the capillary rarefaction at nodes and in Plateau borders is a unified integral characteristic of foam. This means that the mean curvature of menisci at nodes and borders is the same. However, since nodes possess a spherical curvature and Plateau borders a cylindrical curvature, the radius of curvature of the latter must be two times less than that of the nodal menisci,

$$R_b = \tfrac{1}{2}R_n. \tag{7.1.19}$$

The quantities $R_n$ and $R_b$ are related to a unified quantity of the mean curvature of the nodes and borders interface by the formulas

$$\kappa_s = \frac{1}{R_b} = \frac{2}{R_n}. \tag{7.1.20}$$

Therefore, for an estimation of their dependence of the multiplicity and dispersity, the above-mentioned correlations (7.1.16) and (7.1.17) can be used. For an express estimation, in the literature, there are correlations such as

$$\frac{R_b}{a} = \frac{C}{\sqrt{K}}, \tag{7.1.21}$$

which differ from each other by the value of the coefficient $C$. For example, we have $C = 1.782$ in [266], $C = 1.73$ in [244], and $C = 1.628$ in [383].

As regards such a geometric parameter as the film thickness $h$, in actual foam, the film thickness gradually decreases due to slow pressing-out of liquid from films into Plateau borders. This process is virtually terminating when the pressure difference between the phases (equal to the sum of the capillary pressure and the capillary rarefaction) is compensated for by the disjoining pressure $\Pi(h)$, which arises in thin films under the interaction of adsorption layers [116]:

$$\Delta P_l + \Delta P_g = \sigma \kappa_s = \frac{\sigma}{R_b}, \tag{7.1.22}$$

$$\frac{\sigma}{R_b} = \Pi(h). \tag{7.1.23}$$

In practice, the disjoining pressure $\Pi(h)$ must be taken into account only for films of thickness $10^{-9} \leq h \leq 10^{-7}$m. The disjoining pressure comprises some components of different physical nature [383]: Van der Waals attraction, electrostatic repulsion, steric elastic interaction, etc. The approximation formulas for these components in various regions of $h$ acquire different forms [116, 122, 383].

As was already noted, since there is no definite borderline between cellular and polyhedral foam, the following relation can be used as a criterion for the foam to be polyhedral [244]:

$$h \ll R_b \ll a. \qquad (7.1.24)$$

The most important geometric parameters of polyhedral foam are the length of a Plateau border and its cross-section area. These quantities can be found in [214, 244]. For example, for the pentagonal dodecahedron model, the length of the Plateau border is

$$b = 0.818a. \qquad (7.1.25)$$

For monodisperse foam, the area $S_b$ of the cross-section of the Plateau border is

$$S_b = 0.161 R_b^2. \qquad (7.1.26)$$

For polydisperse foam, this value is substantially less.

## 7.2. Envelope of Foam Cells

### 7.2-1. Capsulated Structure of Foam Cells

A complicated many-layer structure of foam cells is formed when gas bubbles are sparged into solutions of surfactants. According to [429], each bubble has a two-sided envelope which is a layer of the solvent containing hydrophilic polar parts of surfactant molecules (see Figure 7.2). Nonpolar hydrophobic parts of molecules on the inner surface of the envelope are oriented toward the bubble, and on the outer surface, outward the envelope. Between two cells, each of which is a capsule with envelope, there is a lamella, that is, an interlayer of a complicated structure. In the middle of the lamella, there is a liquid layer that is a continuous phase. On each of two surfaces of this layer, there is a monolayer of the surfactant. The hydrophobic parts ("tails") of surfactants molecules in each monolayer and the "tails" of the envelope form two direct plate micellae [413], which separate the envelope and the continuous liquid film at the center. Thus, gas bubbles in foam are separated at least by five distinct layers. The multilayer structure of a foam lamella is well seen in photographs (e.g., see [429], p. 54). This fact is also confirmed by the ladder-type shape of the disjoining pressure

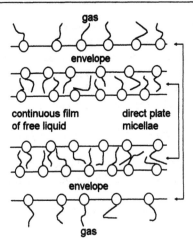

**Figure 7.2.** The lamella structure

isotherm [125, 213] for a black foam lamella (a film from which virtually all free liquid is pressed out).

All this makes foam cells stable and is an essential difference between foams and emulsions. This fact was not taken into account in many papers, e.g., in [214, 438]. The foam cell does not loose its individuality even if it remains alone on the liquid surface [429]. The two-sided envelope provides sufficient rigidity and elasticity of foams [249].

## 7.2-2. Elasticity of the Solution Surface

***Surfactants on an interface.*** It is well-known [480] that pure liquids cannot form stable foams. Liquids gain the capacity to form a foam if they contain dissolved or mixed surfactants.

Surfactant molecules consist of a nonpolar hydrophobic part of rather large size ("tail") and a hydrophilic part ("head") [9, 297, 371]. As a rule, the hydrophobic part is a hydrocarbon chain, and the hydrophilic part is a carboxyl or a sulphate group, or its analogs. Since dipole forces or hydrogen bonds are absent, the cohesive forces in hydrocarbons are considerably less than those in dipole solvents such as water. Therefore, the surfactant molecules concentrate in the near-surface adsorption layer, since it is thermodynamically the most beneficial. The adsorption layer is structured so that the hydrophilic parts of surfactant molecules are tightly bound to the solvent, while the hydrophobic "tails" are considerably less bound to water and are disposed outside the water phase in the adjacent boundary layer of gas or some nonpolar liquid. As long as the adsorption layer of the surfactant is strongly rarefied, the surfactant molecules lie parallel to the surface. With increasing concentration, the hydrophobic parts of molecules leave the surface and become "inclined" in the mean. When the adsorption layer is filled, they become completely "upright" [9].

***Specific adsorption and surface pressure.*** If we introduce the notions of specific adsorption $\Gamma$ (the mass of surfactants per unit area of the surface or interface) and of the saturation adsorption $\Gamma_\infty$ (specific adsorption of surfactants for the limit filling of the adsorption layer), then it is suggested to consider the collection of surfactant molecules in the adsorption layer as a peculiar two-dimensional gas governed by the Van Laar equation [45, 240, 250]

$$\Delta\sigma = \sigma_0 - \sigma = \tilde{R}T_*\Gamma_\infty \ln(1 - \Gamma/\Gamma_\infty)^{-1}. \tag{7.2.1}$$

Here $\Delta\sigma$ is the surface pressure, $\sigma_0$ is the surface tension of the pure solvent, $\sigma$ is the surface tension of the solution, $\tilde{R}$ is the gas constant, and $T_*$ is the absolute temperature.

For low filling of the adsorption layer ($\Gamma \ll \Gamma_\infty$), Eq. (7.2.1) is equivalent to the equation of state for an ideal gas,

$$\Delta\sigma = \tilde{R}T_*\Gamma. \tag{7.2.2}$$

***Elastic properties of interface.*** The surface tension of the solution interface is less than the surface tension of the pure solvent interface. The difference is equal to the surface pressure of surfactant molecules [9, 109, 414]. This does not contradict the fact that the films forming the skeleton of the foam possess increased strength and elasticity. The equilibrium surface layer of a pure liquid is ideally inelastic. Under the action of external forces, the free surface increases not because of extension (an increase in the distance between the molecules in the near-surface layer) but because new molecules are coming from the bulk. A decrease in the equilibrium tension as some amount of surfactant is added does not mean that the elasticity of the surface decreases, since this surface does not possess elastic properties under slow external actions. Nevertheless, we point out that even surfaces of pure liquids possess elastic properties [465] (dynamic surface tension [232]) under very rapid external actions whose characteristic time is less than the time of self-adsorption relaxation of the surface layer. This property must not depend on the existence of an adsorption layer of surfactant. At the same time, surfactants impart additional elastic properties to the surface both at low and high strain rates.

The work of external forces in the deformation of the surface of a surfactant solution changes the free energy of the system because of changing both the surface area and the surface tension [248, 414]. Under the assumption that temperature and volume are constant, the free energy is one of the basic notions of thermodynamics. In our case, it coincides with the interface energy $\Psi = \sigma S$. The rate of change of the free energy $\Psi$ for varying surface area $S$ can be written as

$$\left(\frac{\partial\Psi}{\partial S}\right)_{T,V} = \sigma + S\left(\frac{\partial\sigma}{\partial S}\right)_{T,V}. \tag{7.2.3}$$

where $V$ is the volume of the system. Note that here and henceforth, $S$ is a dimensional quantity.

The first term on the right-hand side determines the surface (equilibrium or dynamic) tension $\sigma$, and the second term is the modulus of elasticity of the adsorption layer

$$E = \left( \frac{\partial \sigma}{\partial \ln S} \right)_{T,V}, \qquad (7.2.4)$$

where the modulus of elasticity may be either equilibrium (Gibbs elasticity [248]) or dynamic (Marangoni elasticity [240]).

*Marangoni and Gibbs elasticity.* The mechanism of elastic action of the adsorption layer can be represented as follows. Any deformation of the surface accompanying, for example, an increase in its area decreases the quantity of adsorbed surfactants per unit area. This decreases the surface pressure of surfactant molecules and hence increases the surface tension that counteracts further elongation of the surface. If the concentration of surfactants in the adsorption layer is small, then the two-dimensional gas of surfactant molecules is governed by the equation of state

$$\sigma_0 - \sigma = \frac{N_S M}{S} \tilde{R} T_*, \qquad (7.2.5)$$

which follows from (7.2.2). Here $N_S$ is the number of moles of the surfactant in the adsorption layer and $M$ is the molecular mass of surfactants.

By assuming that $N_S = $ const, which takes place either for insoluble surfactants or for the case in which the characteristic time of external action is substantially less than the relaxation time of the adsorption layer, and by differentiating (7.2.5) with respect to $S$, we obtain the following estimate for the Marangoni modulus of elasticity:

$$E_M = S \left( \frac{\partial \sigma}{\partial S} \right)_{T,V,N_S} = \sigma_0 - \sigma. \qquad (7.2.6)$$

Let us represent the total number $N$ of moles in the system as

$$N = \frac{CV}{M} + N_S = \text{const}, \qquad (7.2.7)$$

where $C$ is the partial density of surfactants in the solution. Then we can estimate the Gibbs modulus of elasticity as follows:

$$E_G = S \left( \frac{\partial \sigma}{\partial S} \right)_{T,V,N} = E_M + \tilde{R} T_* V \left( \frac{\partial C}{\partial S} \right)_{T,V,N}. \qquad (7.2.8)$$

Obviously, $E_G < E_M$, since the equilibrium value of the derivative satisfies the inequality $\left( \frac{\partial C}{\partial S} \right)_{T,V,N} < 0$. For a finite rate of deformation, the actual value of the modulus of elasticity lies between $E_M$ and $E_G$. For sufficiently thin films, when $N_s \gg CV/M$, the condition $N_s \simeq$ const is satisfied approximately.

### 7.2-3. Elasticity of Foam Cell Elements

Since the boundary of a foam cell is a multilayer film [429], the elasticity of this shell is many times larger than the elasticity of the adsorption layer.

For higher concentrations of surfactants in the adsorption layer, surface pressure, as well as some cohesive forces, may act between the adsorbed molecules. The adsorption layer acquires some additional strength, whose maximum value corresponds to the state of an incompletely filled layer with inclined position of adsorbed molecules whose hydrophobic "tails" interlace each other [9]. Moreover, the elasticity of a strongly developed surface may decrease if the system contains a limited amount of surfactants [125].

Films obtained by using colloid rather than foaming agents possess the highest strength properties. In this case, micellae may be adsorbed near the surface [413], and the colloid becomes a gel exhibiting the strength properties of a solid body [9, 386]. For colloid solutions, the rigidity of the envelope grows with concentration and attains the limit value corresponding to the saturated adsorption layer. In some cases, one can observe an irreversible transition of the substance on the surface into a solid insoluble state (denaturation).

# 7.3. Kinetics of Surfactant Adsorption in Liquid Solutions

### 7.3-1. Mass Transfer Problems for Surfactants

*Equations and a condition on the surface.* Nonsteady-state distribution of surfactants in the volume $V$ and on the surface $S$ is described by the equation of convective diffusion in the bulk and on the surface, respectively [250]:

$$\frac{\partial C}{\partial t} + (\mathbf{V} \cdot \nabla)C = D\Delta C, \tag{7.3.1}$$

$$\frac{\partial \Gamma}{\partial t} + (\mathbf{V}_s \cdot \nabla_s)\Gamma = D_s\Delta_s\Gamma - \Gamma(\nabla_s \cdot \mathbf{V}_s) + j_*. \tag{7.3.2}$$

The subscripts "s" indicate the corresponding variables on the surface $S$.

On the surface we write out the following boundary condition for Eq. (7.3.1):

$$-D\frac{\partial C}{\partial n}\bigg|_S = j_*. \tag{7.3.3}$$

*Kinetics equation of the reversible adsorption.* The flux $j_*$ of surfactants from the bulk to the surface is given by the kinetic equation

$$j_* = \beta\left(1 - \frac{\Gamma}{\Gamma_\infty}\right)C_s - \alpha\Gamma, \tag{7.3.4}$$

which is the law of acting surfaces [250, 511] and is an analog of the Langmuir equation [21, 415]. Here $\beta$ and $\alpha$ are the kinetic adsorption and desorption coefficients, and $\Gamma_\infty$ is the specific adsorption corresponding to the limit saturation of the adsorption layer. Formula (7.3.4) describes the kinetics of the pure adsorption process, that is, the exchange of surfactant molecules between the surface monolayer and the adjacent solution provided that there is a potential barrier between these layers.

*Kinetic coefficients.* The kinetic adsorption and desorption coefficients can be estimated [82, 219, 412] if the form of the potential $\Phi(Z)$ of interaction between a particle (a surfactant molecule) and the solution surface is known; here $Z$ is the coordinate measured from the surface into the bulk of liquid. If the function $\Phi(Z)$ has the form of a potential barrier with a potential well, then the saddle-point method [261] implies

$$\beta = D\sqrt{-\frac{\Phi_{max}^{(2)}}{2\pi kT_*}}\,\exp\left(-\frac{\Phi_{max}}{kT_*}\right),\tag{7.3.5}$$

$$\alpha = \beta(\Phi_{max} - \Phi_{min})\left[\sum_{n=2}^{\infty}\frac{(-1)^{n+1}}{(n+1)!}\Phi_{max}^{(n)}(Z_{max} - Z_{min})^{n+1}\right]^{-1},\tag{7.3.6}$$

where $k$ is the Boltzmann constant, $\Phi_{max}$ and $\Phi_{max}^{(n)}$ are the value of the potential and of its $n$th derivative, respectively, at the point of maximum (that is, at $Z = Z_{max}$), and $\Phi_{min}$ is the value of the potential at the point of minimum (at $Z = Z_{min}$).

## 7.3-2. Kinetics of Surfactant Adsorption in Foam Films

*Equation of dynamics of the adsorption layer.* In the case of foams, the main interface consists of films in which the liquid is virtually stagnant. The film surface is even more constrained. Therefore, Eq. (7.3.1) describes the molecular (or Brownian) diffusion in the bulk of liquid, and Eq. (7.3.2), under the additional assumption that the specific adsorption is rapidly smoothed along the interface (i.e., $\Gamma = \Gamma(t)$), describes the dynamics of a localized (or ideal) adsorption layer [119, 250, 511]:

$$\frac{d\Gamma}{dt} = \beta\left(1 - \frac{\Gamma}{\Gamma_\infty}\right)C_s - \alpha\Gamma - \Gamma\frac{d}{dt}\ln S(t).\tag{7.3.7}$$

The value $C_s$ is the instantaneous local concentration of surfactants on the boundary of the foam film. Note that the last term in Eq. (7.3.7) originates [250, 415] from the term containing the divergence of the surface velocity in Eq. (7.3.2).

*Dynamics of the adsorption layer for the foam film.* Taking into account the fact that in actual foam of high multiplicity the film thickness $h$ usually does not exceed 10 micrometers, the time $t_d = h^2/D$ of diffusion relaxation in the film is a fraction of a second, and hence the concentration is equalized throughout the film thickness almost instantaneously. For this reason, we can omit the subscripts $S$ on $C_s$ indicating the corresponding surface and assume that

$$C_s = C(t)\qquad\text{for}\qquad t > t_d.\tag{7.3.8}$$

One can assume that $C = $ const and $S = $ const at the boundary of the surfactant solution. In this case, we can find the solution of Eqs. (7.3.7), (7.3.8) in the form

$$\Gamma(t) = \Gamma(0)\exp(-t/t_a) + Ct_a\beta[1 - \exp(-t/t_a)], \qquad (7.3.9)$$

where $\Gamma(0)$ are the initial values of $\Gamma$, $t_a = (\alpha + \beta C/\Gamma_\infty)^{-1}$ is the adsorption relaxation time. Usually, the adsorption relaxation time $t_a$ is much larger than $t_d$. For example, according to [201], $t_a \approx 10^3$ to $10^4$ s for protein solutions, which are also surfactants. Thus, the filling of the adsorption layer is governed by a kinetic mechanism.

We must point out that if the adsorption layer contacts with a sufficiently deep liquid, then the diffusion relaxation time can be comparable with the adsorption relaxation time. In this case, the kinetics of the adsorption layer filling, which is determined by Eqs. (7.3.3) and (7.3.4), can be diffusion-controllable for small volume concentrations of surfactants in the solution or be governed by a diffusion-kinetic mechanism for higher concentrations [274]. A pure kinetic region of the adsorption layer filling is possible only in thin layers of surfactant solutions, for example, in liquid elements of foam structures.

By passing to the limit as $t \to \infty$ in (7.3.9), we obtain the equation

$$\Gamma = \beta C \left( \alpha + \frac{\beta C}{\Gamma_\infty} \right)^{-1} \qquad (7.3.10)$$

for the isotherm of the surfactant adsorption, which has a purely Langmuir form [415].

---

### 7.3-3. Kinetics of Surfactant Adsorption in a Transient Foam Body

The continuous liquid phase in a foam layer is a surfactant solution whose volume and interface vary in the course of evolution. The system evolves, its volume varies because of syneresis, and the interface area varies because of the diffusion gas exchange between the system and the ambient medium. For a system with variable volume $V(t)$ and surface area $S(t)$, we can write out the balance equation for the surfactant mass in the system in the form

$$\frac{d}{dt}(VC + S\Gamma) = C\frac{dV}{dt}. \qquad (7.3.11)$$

Taking into account relations (7.3.7) and using (7.3.11), we obtain the following equation describing the variation of the surfactant concentration:

$$\frac{dC}{dt} = -\frac{S(t)}{V(t)}\left[\beta\left(1 - \frac{\Gamma}{\Gamma_\infty}\right)C - \alpha\Gamma\right]. \qquad (7.3.12)$$

For given external actions $V(t)$ and $S(t)$, one must solve Eqs. (7.3.7), (7.3.8), and (7.3.12) for $C = C(t)$ and $\Gamma = \Gamma(t)$ simultaneously.

In this process, the modulus of elasticity of the adsorption layer is also a function of time and is given by

$$E = S \left( \frac{\partial \sigma}{\partial S} \right)_T.$$

Taking into account (7.2.1), we obtain

$$E(t) = \frac{\tilde{R} T_* S}{1 - \Gamma / \Gamma_\infty} \left( \frac{d\Gamma}{dt} \right) \left( \frac{dS}{dt} \right)^{-1},$$

where $\Gamma(t)$ is determined by the solution (7.3.8), (7.3.12).

We point out that the elasticity of the lamella, which is the basic topological element of the foam structure separating the disperse gaseous inclusions, must be substantially larger than the elasticity of the adsorption layer.

The point is that the lamella [429] has a complex multilayer structure (see Figure 7.2). It consists of two boundaries of foam cells, two direct plate micellae [413], and a liquid film, which is a part of the continuous phase of the foam. All in all, the lamella has six parallel adsorption layers. Hence, its modulus of elasticity must be many times larger than that of a simple adsorption layer.

# 7.4. Internal Hydrodynamics of Foams. Syneresis and Stability

*Preliminary remarks.* Foam is a fluid multiphase continuous medium possessing an internal structure. The disperse phase consists of gaseous bubbles capsulated into multilayer elastic envelopes consisting of some adsorption layers of surfactants and submersed into a continuous phase. The capsules are in a close contact with one another so that they are locally deformed and form a structure with an anisotropic distribution of the continuous phase around each cell.

In contrast with two-phase bubble-containing fluids, aerosols, and emulsions, foam has a least three phases. Along with gas and the free continuous liquid phase, foam contains the so-called "skeleton" phase, which includes adsorption layers of surfactants and the liquid between these layers inside the capsule envelope. The volume fraction of the "skeleton" phase is extremely small even compared with the volume fraction of the free liquid. Nevertheless, this phase determines the foam individuality and its structure and rheological properties. It is the frame of reference with respect to which the diffusion motion of gas and the hydrodynamic motion of the free liquid can occur under the action of external forces and internal inhomogeneities. At the same time, the elements of the "skeleton" phase themselves can undergo strain and relative displacements as well as mass exchange with the other phases (solvent evaporation and condensation and surfactant adsorption and desorption).

The evolution of a foam system, that is, the spatial redistribution of the substance takes place in regions of very complicated geometric and topological structure. If we were to consider problems of transfer in phases with regard to all the corresponding boundary conditions on the interfaces, it would be extremely difficult to implement this approach and in the end we would obtain a great body of excessive information about the fields of microparameters of the system. In practice, it suffices to know the averaged parameters and fluxes described in the framework of mechanics of heterogeneous media [312, 313].

### 7.4-1. Internal Hydrodynamics of Foams

*Syneresis.* The most actual problem in the study of foam systems is the spatial redistribution of liquid phase under the action of external fields and internal inhomogeneities.

The outflow of a liquid from foam under the action of gravity field was considered in [15] and termed "syneresis." Later on, syneresis was attributed to capillary effects, primarily due to the gradients of capillary rarefaction. The importance of these effects was already mentioned in [266], but the gradient of capillary rarefaction was rightly set to be zero, since only steady-state flows of liquids through foam layer were considered there. The fundamental role of the gradient of capillary rarefaction in the process of evolution of a foam layer in syneresis was also pointed out in [215, 335].

Indeed, as fluid flows, foam channels closed above grow in thickness at the bottom, thus creating an increasing counteraction to the gravitational force, which slows down the outflow until equilibrium is attained [214]. It should be noted that this effect is possible only in closed deformable channels with negative curvature, which are typical of foam. According to [324], the capillary rarefaction is a characteristic of the foam compressibility and determines its elastic resistance to the strain caused by the liquid redistribution.

*Hydroconductivity.* A phenomenological theory of syneresis was proposed in [239, 243–247]. There, in accordance with the linear theory of syneresis, for the local density $\mathbf{q}$ of flux of the extensive variable $V$, which is the volume fraction of the liquid phase (the reciprocal of the foam multiplicity $K$), the following expression was proposed [245]:

$$\mathbf{q} = H\left[(\rho_l - \rho_g)\mathbf{g} + \nabla(\Delta P_l)\right].\tag{7.4.1}$$

Here $\rho_l$ and $\rho_g$ are the liquid and gas densities, respectively, $\mathbf{g}$ is the vector of the gravitational acceleration, and $\Delta P_l$ is the capillary rarefaction given by (7.1.10) and (7.1.15). The kinetic coefficient $H$ was called the coefficient of hydroconductivity and calculated for polyhedral foam models [245, 246]. Generally speaking, the variable $H$ is a tensor, but usually the isotropic approximation is used, where this parameter is a scalar. Various expressions for the coefficient $H$ were proposed and made more precise in [125, 214, 245]. Thus, different approaches used to calculate the coefficient of hydroconductivity were analyzed in [488]. For example, the structure of spherical and cellular foam was studied under the assumption that liquid flows through a porous layer according to the Kozeni–Karman model, and the Lemlich–Poiseuille model of channel hydroconductivity was used for polyhedral foam.

At present, the following relations are recommended as the basic experimentally verified formulas for calculating the coefficients of hydroconductivity [257]: for cellular polydisperse foam [214],

$$H = \frac{4.64 \times 10^{-2} a^2}{\mu (K-1)^2 K_{\min}}; \tag{7.4.2}$$

for polyhedral foam [488],

$$H = \frac{7.44 \times 10^{-3} a^2 (K-1)^2}{\mu K^4} \sqrt{\frac{4}{K_{\min}}}. \tag{7.4.3}$$

In these formulas, $K = 1/V$. The coefficient $\sqrt{4/K_{\min}}$ is introduced for approximate consideration of polydispersity, since it is well known [481] that the hydroconductivity of polydisperse polyhedral foam is 1.5 to 2 times less than that of monodisperse foam. At the same time, substantiation of formula (7.4.3) for polydisperse foams is doubtful, since polydisperse foam can hardly be polyhedral.

| 7.4-2. Generalized Equation of Syneresis |

*Equation of syneresis. Syneresis coefficient.* Together with the liquid flux (7.4.1) relative to the "skeleton" phase, there also exists translational transfer $V\mathbf{U}$ determined by the local velocity field $\mathbf{U}$ of the entire foam (or of its "skeleton" phase). In this case, the law of conservation of liquid mass has the form

$$\frac{\partial V}{\partial t} + \nabla \cdot (\mathbf{q} + V\mathbf{U}) = 0$$

and allows us to obtain the following generalized equation of syneresis in the laboratory frame of reference:

$$\begin{aligned} \frac{\partial V}{\partial t} + (\mathbf{U} \cdot \nabla)V &= \nabla \cdot [\mathfrak{D}(V)\nabla V] \\ &- (\rho_1 - \rho_g)\frac{dH}{dV}(\mathbf{g} \cdot \nabla V) - (\rho_1 - \rho_g)H(V)\nabla \cdot \mathbf{g} - V\nabla \cdot \mathbf{U}, \end{aligned} \tag{7.4.4}$$

where the newly introduced variable

$$\mathfrak{D}(V) = -H\frac{d\Delta P_1}{dV} \tag{7.4.5}$$

is called the syneresis coefficient [239].

The left-hand side of (7.4.4) is the substantial derivative of the volume fraction of liquid in foam. The last term on the right-hand side is important only if we take into account the volume compressibility of foam. The penultimate term does not vanish if the mass force depends on the coordinates (for example, if syneresis is considered in centrifugal fields). For gravitational syneresis, this term is zero.

In [246] the hydroconductivity and syneresis coefficients were calculated for the channel version of hydroconductivity of polyhedral foam:

$$H = 3.3 \times 10^{-3} \frac{a^2}{\mu} V^2, \tag{7.4.6}$$

$$\mathfrak{D} = 9.5 \times 10^{-4} \frac{\sigma a}{\mu} \sqrt{V}. \tag{7.4.7}$$

We point out that the accuracy of formula (7.4.3) is two times less than that of formula (7.4.6), since the last formula was derived under the assumption that the motion in the walls of Plateau borders is retarded.

**Boundary conditions.** The conventional boundary conditions for Eq. (7.4.4) at the boundary of the foam region prescribe the normal component $q_n$ of the volume rate of flow per unit volume. This flow characteristic and the variables $V_s$ and $\left. \frac{dV}{dn} \right|_s$ at the boundary of the region are nonlinearly related by Eqs. (7.4.1) and (7.4.5)–(7.4.7).

On the interface between the foam and the ambient liquid medium, one can set the local multiplicity equal to the multiplicity $K_{min}$ of the spherical foam. On the interface between the foam and a porous filter, one can set the value of the volume moisture content provided by this filter [246],

$$V_s = 0.33 \frac{\sigma^2}{a^2 (P_f - P_g)^2}, \tag{7.4.8}$$

where $P_f - P_g$ is the pressure difference on the filter. In contrast with the case in which $q_n$ is given, these conditions are linear.

### 7.4-3. Gravitational and Centrifugal Syneresis

*Gravitational syneresis. Distribution of the foam multiplicity.* In [247], Eq. (7.4.4) was solved for a vertical steady-state ($\partial V / \partial t = 0$) stagnant ($\mathbf{U} = \mathbf{0}$) nonirrigated ($\mathbf{q} = \mathbf{0}$) foam column under the condition that the foam multiplicity attains its minimum ($K_{min}$) at the interface between the foam and the liquid (in the cross-section $Z = 0$). The following quadratic dependence of the foam multiplicity on the height was obtained:

$$K(Z) = \left[ \sqrt{K_{min}} + 1.73 \frac{(\rho_l - \rho_g) g a}{\sigma} Z \right]^2. \tag{7.4.9}$$

Gravitational syneresis problems in a somewhat different setting were also studied in [377, 383].

*Centrifugal syneresis. Distribution of the foam moisture.* By using the syneresis equation in centrifugal fields, the theory of centrifugal plate foam suppressor, which is a set of conical plates rotating around a common axis, was developed in [489, 490]. The following distribution along the generatrix of the plate was obtained for the moisture content averaged over the width of the gap between the plates:

$$V = \left[ \frac{1}{V_0^2} + \frac{4}{3} A(\xi_0^3 - \xi^3) \right]^{-1/2}, \tag{7.4.10}$$

where $V_0$ is the moisture content at the inlet of the foam suppressor (at $\xi = \xi_0$) and $\xi$ is the ratio of the coordinate along the generatrix of the plate to the inlet cross-section radius $r_1$. The dimensionless parameter $A$ characterizes the mode of operation and the plate geometry and can be calculated by the formula

$$A = \frac{\pi}{18} \frac{a^2 \omega^2 r_1^3}{\nu Q} \sin^2 \alpha \cos \alpha, \tag{7.4.11}$$

where $\omega$ is the angular speed of rotation, $\alpha$ is the angle formed by the inclined generatrix with the axis of rotation, $Q$ is volume rate of foam flow through the gap between the plates, $a$ is the radius of the equivalent cell of foam, and $\nu$ is the kinematic viscosity of the liquid.

## 7.4-4. Barosyneresis

Transient problems of syneresis are of great interest. For example, the transient syneresis in a stagnant foam layer ($\mathbf{U} = \mathbf{0}$) under the action of constant mass forces is governed by a complex nonlinear parabolic equation. Some self-similar solutions and "traveling wave" type solutions were found in [152] for some special forms of this equation. For one-dimensional barosyneresis ($\mathbf{g} = \mathbf{0}$), Eq. (7.4.4) has the form

$$\frac{\partial V}{\partial t} = 10^{-3} \frac{\sigma a}{\mu} \frac{\partial}{\partial Z} \left( V^{1/2} \frac{\partial V}{\partial Z} \right). \tag{7.4.12}$$

Some well-known exact solutions of this equation, including wave-like and self-similar solutions, as well as the solutions corresponding to the peaking mode of operation [425], are presented in the reference books [197, 516]. However, it should be noted that these exact solutions exist only for some special initial and boundary conditions, which follow from the form of the solution itself, and hence, the interpretation of these solutions has always been a problem.

For natural boundary conditions, as a rule, one must use numerical methods or approximate analysis. For example, a numerical solution describing the wave of capillary "suction" was obtained in [152], and a solution describing centrifugal syneresis was obtained in [490] by using expansions with respect to a small parameter.

For the steady-state ($\partial V / \partial t = 0$) one-dimensional barosyneresis ($\mathbf{g} = \mathbf{0}$) in a stagnant ($\mathbf{U} = \mathbf{0}$) irrigated ($\mathbf{q} \neq \mathbf{0}$) foam column adjacent to a filter ($V_s = \text{const}$), the solution was also obtained in [247]. The distribution of multiplicity in the foam layer in a centrifuge was presented in the same monograph.

| 7.4-5. Stability, Evolution, and Rupture of Foams |

*Disjoining pressure. Critical thickness of the foam film.* Since a foam structure is metastable, it continuously evolves under the action of external and internal factors. The main processes determining the evolution of spherical foam are syneresis caused by mass forces and diffusion redistribution of the gas phase. For mature cellular and polyhedral foams, the determining process is the thinning of films as the liquid flows out to Plateau borders owing to a difference in capillary rarefaction. The thinning process is quite long but not infinitely long; it is just this process that determines the life time of foam, which may range from few seconds to several days [384].

There is a factor that opposes the thinning of the foam film. It is called the disjoining pressure and denoted by $\Pi(h)$, where $h$ is the film thickness. The disjoining pressure manifests itself for $h \lesssim 10^{-7}$ m. Usually, three components are distinguished in the disjoining pressure [116]:

$$\Pi(h) = \Pi_m(h) + \Pi_e(h) + \Pi_s(h), \qquad (7.4.13)$$

The molecular component of the disjoining pressure, $\Pi_m(h)$, is negative (repulsive). It is caused by the London–van der Waals dispersion forces. The ion-electrostatic component, $\Pi_e(h)$, is positive (attractive). It arises from overlapping of double layers at the surface of charge-dipole interaction. At last, the structural component, $\Pi_s(h)$, is also positive (attractive). It arises from the short-range elastic interaction of closed adsorption layers.

The disjoining pressure $\Pi(h)$, defined by Eq. (7.4.13) as the sum of the three components, is a nonmonotone function of the film thickness $h$ for $h < 10^{-6}$ m. In this range of $h$, there is one or two intervals (depending on the type and concentration of the surfactant and electrolyte in the dispersive liquid) where

$$\frac{d\Pi}{dh} < 0. \qquad (7.4.14)$$

This inequality is the most severe condition of the hydrodynamic stability of the liquid film [241]. Therefore, an instability and rupture of the film are considered to occur when the film thickness attains a critical value $h_{cr}$, as the liquid flows out of the film. The critical thickness is determined by the condition

$$\left.\frac{d\Pi}{dh}\right|_{h=h_{cr}} = 0. \qquad (7.4.15)$$

*Mechanism of rupture. Black films.* The mechanism of hydrodynamic instability of thin foam films was analyzed in [278, 279, 411]. The stability of ultrathin films is governed by a competition between capillary forces and the molecular component of the disjoining pressure. An instability can arise when $d\Pi/dh > 0$ and the capillary pressure is not too large. This is possible if

$$h < h_{cr} \ll \Lambda, \qquad (7.4.16)$$

where $\Lambda$ is perturbation wave length. Linear time [278, 411] and space-time [442] stability analysis established that an instability takes place when inequalities (7.4.16) are satisfied. The linear and weakly nonlinear [124, 437] consideration allows one to determine the interval of "unstable" wave numbers, the most amplified wave numbers, and the maximum time growth rate of the perturbation amplitude.

In some cases, the growth of perturbations leads to the formation of spots of thinner metastable films (with thickness about 10 nm). The film at the spots is so thin that it appears black in reflected light. Such films are often referred to as "black" films. These objects are obliged by their origin to a sufficiently large value of the structural component of the disjoining pressure, which determines the existence of the second interval where $d\Pi/dh < 0$ on the $\Pi(h)$ curve. A rupture of the black films can also take place, but this mechanism is connected with a display of the "vacancy" instability [125].

*Rupture of foams.* In [242], the concept of critical thickness is suggested as a criterion of the rupture of foams. This means that polyhedral foam will rupture as soon as the thickness of some films making the faces of the foam polyhedron cells attains a critical value. Adopting the channel model of the foam, i.e., assuming that the liquid completely resides in Plateau borders, and using Eqs. (7.1.21) and (7.1.23), one can represent the volume fraction of the liquid phase, $V$ (the reciprocal of the foam multiplicity $K$) in the form

$$V = 0.33 \frac{\sigma^2}{a^2 \Pi^2(h)},$$

where $\sigma$ is the coefficient of surface tension, $a$ the radius of the equivalent cell, and $\Pi(h)$ the disjoining pressure. For $h = h_{cr}$ we have

$$V_{cr} = 0.33 \frac{\sigma^2}{a^2 \Pi^2(h_{cr})}.$$

A foam bed may rupture at any part of its volume, but the most probable place of rupture is the surface of contact with atmosphere.

*Rupture of a vertical foam bed.* Consider the stationary problem of the rupture of a vertical flowing foam column under the action of gravitational syneresis [242]. Let the $Z$-axis be directed vertically upward along the axis of the foam column. At the section $Z = 0$ (at the interface with the solution), the column receives fresh spherical foam ($V = 1/4$). The foam moves upward at a translational speed $U$. Moving downward along the Plateau borders is a flow of the liquid phase; this motion occurs due to the action of gravity forces (according to the mechanism of hydroconduction). Since the problem is stationary, the total flux is zero. In this particular case, the general equation of syneresis (7.4.4) becomes

$$\frac{\partial V}{\partial Z} = \frac{UV - \Delta \rho g H(V)}{\mathfrak{D}(V)}, \qquad (7.4.17)$$

where $\Delta\rho = \rho_l - \rho_g$ is the difference of the liquid and gas phase densities, $H(V)$ the local coefficient of hydroconduction of the foam (see Eq. (7.4.6)), and $\mathfrak{D}(V)$ the local coefficient of syneresis (see Eq. (7.4.7)). The boundary conditions are

$$V = \tfrac{1}{4} \quad \text{at} \quad Z = 0, \qquad V = V_{cr} \quad \text{at} \quad Z = Z_0. \qquad (7.4.18)$$

The first boundary condition is used for determining the constant of integration and the second, for determining the height $Z_0$ of the stationary flowing column. The exact solution of problem (7.4.17), (7.4.18) has the form [242]

$$V(Z) = \frac{U}{\alpha\Delta\rho\, g} \tanh^2\left( \frac{2\beta}{\sqrt{\alpha\Delta\rho\, gU}\, Z} \right),$$

where, according to relations (7.4.6) and (7.4.7), $\alpha = 3.3 \times 10^{-3}\, a^2/\mu$ and $\beta = 9.5 \times 10^{-4}\, \sigma a/\mu$. The height of the stationary foam column is then defined as

$$Z_0 = \frac{2\beta}{\sqrt{\alpha\Delta\rho\, gU}} \operatorname{Artanh} \sqrt{\frac{U}{\alpha\Delta\rho\, gV_{cr}}}$$

In [242], an additional boundary condition at the upper boundary of the foam column is also discussed for the case of nonstationary syneresis.

# 7.5. Rheological Properties of Foams

### 7.5-1. Macrorheological Models of Foams

Foam is a complex rheological body possessing all three basic macrorheological properties (elasticity, plasticity, and viscosity) to a variable extent.

It is convenient to describe these properties in terms of the following mechanical models [396]: the Hooke body (an elastic spring), the Saint-Venant body modeling dry friction (a bar on a solid surface), and the Newton body (a piston in a vessel filled with a viscous fluid). By using various combinations of these elementary models (connected in parallel and/or in series), one can describe situations which are rather complex from the rheological viewpoint.

The key point in the rheological classification of substances is the question as to whether the substance has a preferred shape or a natural state or not [19]. If the answer is "yes," then this substance is said to be "solid-shaped"; otherwise it is referred to as "fluid-shaped" [508]. The simplest model of a viscoelastic "solid-shaped" substance is the Kelvin body [396] or the Voigt body [508], which consists of a Hooke and a Newton body connected in parallel. This model describes deformations with time-lag and elastic aftereffects. A classical model of viscoplastic "fluid-shaped" substance is the Maxwell body [396], which consists of a Hooke and a Newton body connected in series and describes stress relaxation.

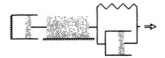

**Figure 7.3.** Macrorheological model of foam behavior (foam body)

It is well known [38, 118, 125, 280, 379] that for foam there exists a yield stress $\tau_0$ that classifies the types of rheological behavior of foam as follows: for $\tau < \tau_0$, the foam is a "solid-shaped" substance, and for $\tau \geq \tau_0$, it is "fluid-shaped." For this reason, mechanical models of foam must include the Saint-Venant body. One of the simplest macrorheological models of the foam body is shown in Figure 7.3.

### 7.5-2. Shear Modulus, Effective Viscosity, and Yield Stress

*Shear modulus.* One of the most important characteristics of the foam elasticity is the shear modulus $G$, which is the coefficient of proportionality between the tangential stress $\tau$ in "solid-shaped" foam ($\tau < \tau_0$) and the shear strain $\gamma$. This variable was theoretically calculated by Derjagin [115]

$$G = \frac{4}{15}\sigma\varepsilon. \tag{7.5.1}$$

If we estimate the specific interfacial area $\varepsilon$ for a monodisperse foam by formula (7.1.4), then we can rewrite this relation as

$$G = 0.8\frac{\sigma}{a}\frac{K-1}{K}. \tag{7.5.2}$$

Later on Khan [225] also theoretically obtained the following estimate for the shear modulus of dry two-dimensional foam under the assumption that the foam cells are regular hexagonal cells and the deformation is not small:

$$G = \frac{2\sigma}{a\sqrt{3\gamma^2 + 12}}. \tag{7.5.3}$$

We believe that the following semiempirical Princen–Kiss formula [380] is most reliable for practical computations:

$$G = 1.769\left(0.288 - \frac{1}{K}\right)\left(\frac{K-1}{K}\right)^{1/3}\frac{\sigma}{a}. \tag{7.5.4}$$

*Effective viscosity of viscoelastic foam.* Foam exhibits viscous properties even if it is in the "solid-shaped" state. The point is that fluid elements of a foam cell also contribute to the shear deformation. The rate of energy dissipation in these elements depends on the frequency of the applied action, or, which is the same, on the shear rate $\dot{\gamma}$. The effective viscosity, which determines dissipative properties of such media, was theoretically calculated in [428] for the two-dimensional foam model:

$$\mu_e = 6.7\mu\,\text{Ca}^{-1/3}, \tag{7.5.5}$$

where $\mu$ is the fluid viscosity and $\text{Ca} = \mu a\dot{\gamma}/\sigma$ is the capillarity factor. The rate of energy dissipation turns out to be proportional to $\omega^{5/3}$, where $\omega$ is the action frequency. Thus, for high frequencies of action, the foam viscosity is essential even in the "solid-shaped" state.

*Yield stress and plastic viscosity.* The most important rheological characteristic determining the foam behavior ("solid-shaped" or "fluid-shaped") is the yield stress $\tau_0$. This variable was calculated in [379] for a two-dimensional foam model:

$$\tau_0 = 1.28\frac{\sigma}{a}\left(\frac{1-K}{K}\right)^{4/3}\left(0.288 - \frac{1}{K}\right). \tag{7.5.6}$$

In [295] the following correlation was obtained for $\tau_0$ by using an experimental plant containing a tube of diameter of 14 mm with a helical thread on the internal surface (to prevent the slip of the foam column in the piston mode of operation):

$$\frac{4\tau_0\rho a^2}{\mu^2} = 0.61K^{0.18}\left(\frac{2\sigma\rho a}{\mu^2}\right)^{0.49}\left(\frac{8g\rho^2 a^3}{\mu^2}\right)^{0.35}, \tag{7.5.7}$$

where $\rho$ is the fluid density. The accuracy of this formula is $\pm 10\%$. In these experiments, the foam multiplicity $K$ varied from 36 to 322, the foam dispersity $a$ from 0.175 mm to 0.5 mm, and the solution viscosity $\mu$ from 1.5 Pa $\cdot$ s to 10.5 Pa $\cdot$ s. In all cases, the surfactant solution was the 0.4% solution of sulphonol in distilled water containing either 5.2 or 30 mass percent of glycerin (to change the viscosity), while the variables $\rho$ and $\sigma$ varied slightly. In [295], an approximate formula was also obtained for the plastic (Bingham) viscosity $\mu_p$,

$$\frac{\mu_p}{\mu} = 8.8 \times 10^{-5}K^{0.99}\left(\frac{2\sigma\rho a}{\mu^2}\right)^{2}\left(\frac{8g\rho^2 a^3}{\mu^2}\right)^{-0.98}, \tag{7.5.8}$$

whose error does not exceed 17%.

Note that $\tau_0$ and $\mu_p$ grow with increasing multiplicity $K$ and decrease with increasing dispersity $\bar{a}$. At the same time, $\tau_0$ and $\mu_p$ depend on the viscosity $\mu$ of the solution in a qualitatively different way. While $\tau_0$ grows, $\mu_p$ decreases with increasing $\mu$. This means that foam with a larger fraction of the liquid phase is closer to an ideally plastic fluid with a larger yield strength.

Some other empirical formulas are known for calculating viscoplastic rheological parameters [383]. For example, the formula for the yield stress has the form

$$\tau_0 = \frac{\sigma}{a}\left(\frac{K-1}{K}\right)^{1/3}Y(K), \tag{7.5.9}$$

where the Sauter (surface area average) radius of foam cells must be taken as $a$. The function $Y(K)$ is also plotted in [383] for the narrow multiplicity region $5 \leq K \leq 25$. However, in this region the function $Y(K)$ can be approximated with accuracy up to 12% by the formula

$$Y(K) = 0.095 \left\{ 1 - \exp\left[ -0.1(K-4)^{4/3} \right] \right\}. \qquad (7.5.10)$$

The corresponding correlation formula for the plastic viscosity has the form

$$\mu_p = \frac{\tau_0}{\dot{\gamma}} + 32 \left( \frac{K-1}{K} - 0.73 \right) \mu \, \text{Ca}^{-1/2}. \qquad (7.5.11)$$

These results were obtained in experiments with biliquid foam (emulsions such as "oil in water") and polymer surfactants for drop diameters $8.3 \times 10^{-6}$ to $26.3 \times 10^{-6}$ m and for the disperse phase viscosity $2.65 \times 10^{-3}$ to $6.6 \times 10^{-1}$ Pa $\cdot$ s.

### 7.5-3. Other Approaches and Problems

The recent papers [155, 156] demonstrate a new approach to the study of rheological properties of foam by the example of enforced flow of structurized foam through channels. In particular, some hypothetical models are proposed for shear strain and stress states in foam structures. It is also pointed out that because of the elastic component inherent in foam of high multiplicity, its rheology is of thixotropic (time-dependent) character, and such foam cannot be described by a single flow curve. Shear stresses lead a redistribution of multiplicity over the cross-section of the channel because of the Weissenberg effect. The central core contains less moisture, elastic forces dominate, and the core moves as a piston. At the same time, at the periphery the moisture is greater, the capillary rarefaction disappears and the shear slip becomes possible. In this case, it is possible to calculate such important characteristics of moving foam as the multiplicity distribution over the cross-section and the yield stress depending on the initial homogeneous multiplicity $K_0$.

Over few last decades, applied problems have arisen which are related to the behavior of foams in porous media. It turns out that aqueous surfactant-stabilized foams can drastically reduce the gas mobility in porous media [334]. This fact is of great applied significance in petroleum and gas industry. In [106, 189, 233], mechanisms of foam control of gas migration through porous media are presented.

# Supplements

## S.1. Exact Solutions of Linear Heat and Mass Transfer Equations

This section contains some solutions presented in the books [73, 79, 231, 370, 482, 516]. The dimensionless time is denoted by $\tau = \chi t / a^2$.

### S.1-1. Heat Equation

Let us consider the simplest heat equation with one spatial variable:

$$\frac{\partial T}{\partial \tau} = \frac{\partial^2 T}{\partial x^2}.$$

*1. Some particular solutions* ($A, B, \lambda$ are arbitrary constants):

1. $T = Ax + B$,

2. $T = A \exp(\lambda^2 \tau \pm \lambda x) + B$,

3. $T = A \exp(-\lambda^2 \tau) \cos(\lambda x) + B$,

4. $T = A \exp(-\lambda^2 \tau) \sin(\lambda x) + B$,

5. $T = A \dfrac{1}{\sqrt{\tau}} \exp\left(-\dfrac{x^2}{4\tau}\right) + B$,

6. $T = A \dfrac{x}{\tau^{3/2}} \exp\left(-\dfrac{x^2}{4\tau}\right) + B$,

7. $T = A \operatorname{erf}\left(\dfrac{x}{2\sqrt{\tau}}\right) + B$,

where $\operatorname{erf} z \equiv \dfrac{2}{\sqrt{\pi}} \displaystyle\int_0^z \exp(-\xi^2)\, d\xi$ is the error function.

*2. Region* $-\infty < x < +\infty$

$$T = f(x) \qquad \text{at} \quad \tau = 0 \qquad \text{(initial condition)}.$$

The solution is

$$T = \frac{1}{2\sqrt{\pi\tau}} \int_{-\infty}^{+\infty} \exp\left[-\frac{(x-\xi)^2}{4\tau}\right] f(\xi)\, d\xi.$$

*The special case* $f(x) = \begin{cases} A & \text{for } |x| < x_0, \\ B & \text{for } |x| > x_0. \end{cases}$

*The solution is*

$$T = \frac{1}{2}(A - B)\left[\operatorname{erf}\left(\frac{x_0 - x}{2\sqrt{\tau}}\right) + \operatorname{erf}\left(\frac{x_0 + x}{2\sqrt{\tau}}\right)\right] + B.$$

### 3. Region $0 \le x < +\infty$. The first boundary value problem.

3.1.      $T = f(x)$    at   $\tau = 0$     (initial condition),

         $T = 0$      at   $x = 0$     (boundary condition).

The solution is

$$T = \frac{1}{2\sqrt{\pi\tau}}\int_0^{+\infty}\left\{\exp\left[-\frac{(x-\xi)^2}{4\tau}\right] - \exp\left[-\frac{(x+\xi)^2}{4\tau}\right]\right\} f(\xi)\,d\xi.$$

*The special case $f(x) = A$.*
*The solution is*

$$T = A\operatorname{erf}\left(\frac{x}{2\sqrt{\tau}}\right).$$

3.2.      $T = 0$      at   $\tau = 0$     (initial condition),

         $T = g(\tau)$   at   $x = 0$     (boundary condition).

The solution is

$$T = \frac{x}{2\sqrt{\pi}}\int_0^{\tau}\exp\left[-\frac{x^2}{4(\tau - \zeta)}\right]\frac{g(\zeta)\,d\zeta}{(\tau - \zeta)^{3/2}}.$$

*The special case $g(\tau) = \begin{cases} A & \text{for } 0 < \tau < \tau_0, \\ B & \text{for } \tau_0 < \tau. \end{cases}$*
*The solution is*

$$T = \begin{cases} A\operatorname{erfc}\left(\dfrac{x}{2\sqrt{\tau}}\right) & \text{for } 0 < \tau < \tau_0, \\[2ex] A\operatorname{erfc}\left(\dfrac{x}{2\sqrt{\tau}}\right) + (B - A)\operatorname{erfc}\left(\dfrac{x}{2\sqrt{\tau - \tau_0}}\right) & \text{for } \tau_0 < \tau, \end{cases}$$

*where* $\operatorname{erfc} x \equiv 1 - \operatorname{erf} x$.

3.3.      $T = f(x)$    at   $\tau = 0$     (initial condition),

         $T = g(\tau)$   at   $x = 0$     (boundary condition).

The solution is

$$T = \frac{1}{2\sqrt{\pi\tau}}\int_0^{+\infty}\left\{\exp\left[-\frac{(x-\xi)^2}{4\tau}\right] - \exp\left[-\frac{(x+\xi)^2}{4\tau}\right]\right\} f(\xi)\,d\xi$$
$$+ \frac{x}{2\sqrt{\pi}}\int_0^{\tau}\exp\left[-\frac{x^2}{4(\tau - \zeta)}\right]\frac{g(\zeta)\,d\zeta}{(\tau - \zeta)^{3/2}}.$$

### 4. Region $0 \le x < +\infty$. The second boundary value problem.*

4.1.      $T = f(x)$    at   $\tau = 0$     (initial condition),

         $\partial_x T = 0$    at   $x = 0$     (boundary condition).

---

\* In what follows, along with the conventional notation, the following brief notation for partial derivatives is used: $\partial_\tau T \equiv \dfrac{\partial T}{\partial \tau}$, $\partial_x T \equiv \dfrac{\partial T}{\partial x}$, $\partial_{xx} T \equiv \dfrac{\partial^2 T}{\partial x^2}$.

The solution is

$$T = \frac{1}{2\sqrt{\pi\tau}} \int_0^{+\infty} \left\{ \exp\left[-\frac{(x-\xi)^2}{4\tau}\right] + \exp\left[-\frac{(x+\xi)^2}{4\tau}\right] \right\} f(\xi)\, d\xi.$$

4.2.      $T = 0$      at   $\tau = 0$      (initial condition),

           $\partial_x T = g(\tau)$   at   $x = 0$      (boundary condition).

The solution is

$$T = -\frac{1}{\sqrt{\pi}} \int_0^{\tau} \exp\left[-\frac{x^2}{4(\tau-\zeta)}\right] \frac{g(\zeta)}{\sqrt{\tau-\zeta}}\, d\zeta.$$

*The special case $g(\tau) = -A$.*
*The solution is*

$$T = 2A\sqrt{\frac{\tau}{\pi}} \exp\left(-\frac{x^2}{4\tau}\right) - Ax\operatorname{erfc}\left(\frac{x}{2\sqrt{\tau}}\right).$$

4.3.      $T = f(x)$      at   $\tau = 0$      (initial condition),

           $\partial_x T = g(\tau)$   at   $x = 0$      (boundary condition).

The solution is

$$T = \frac{1}{2\sqrt{\pi\tau}} \int_0^{+\infty} \left\{ \exp\left[-\frac{(x-\xi)^2}{4\tau}\right] + \exp\left[-\frac{(x+\xi)^2}{4\tau}\right] \right\} f(\xi)\, d\xi$$
$$- \frac{1}{\sqrt{\pi}} \int_0^{\tau} \exp\left[-\frac{x^2}{4(\tau-\zeta)}\right] \frac{g(\zeta)}{\sqrt{\tau-\zeta}}\, d\zeta.$$

## 5. Region $0 \le x < +\infty$. *The third boundary value problem.*

5.1.      $T = f(x)$      at   $\tau = 0$      (initial condition),

           $\partial_x T - kT = 0$   at   $x = 0$      (boundary condition).

The solution is

$$T = \frac{1}{2\sqrt{\pi\tau}} \int_0^{+\infty} G(x,\xi,\tau) f(\xi)\, d\xi,$$

where

$$G(x,\xi,\tau) = \exp\left[-\frac{(x-\xi)^2}{4\tau}\right] + \exp\left[-\frac{(x+\xi)^2}{4\tau}\right]$$
$$- 2k \int_0^{+\infty} \exp\left[-\frac{(x+\xi+\eta)^2}{4\tau} - k\eta\right] d\eta.$$

5.2.      $T = 0$      at   $\tau = 0$   (initial condition),

           $\partial_x T - kT = kg(\tau)$   at   $x = 0$   (boundary condition).

The solution is

$$T = -\frac{k}{\sqrt{\pi}} \int_0^{\tau} \frac{g(\zeta)}{\sqrt{\tau-\zeta}} H(x,\tau-\zeta)\, d\zeta,$$

where

$$H(x,\tau) = \exp\left(-\frac{x^2}{4\tau}\right) - k\int_0^{+\infty} \exp\left[-\frac{(x+\eta)^2}{4\tau} - k\eta\right] d\eta.$$

*The special case* $g(\tau) = \begin{cases} -A & \text{for } 0 < \tau < \tau_0, \\ -B & \text{for } \tau_0 < \tau. \end{cases}$

*The solution is*

$$T = \begin{cases} AW(x,\tau) & \text{for } 0 < \tau < \tau_0, \\ AW(x,\tau) + (B-A)W(x,\tau-\tau_0) & \text{for } \tau_0 < \tau, \end{cases}$$

*where*

$$W(x,\tau) = \operatorname{erfc}\left(\frac{x}{2\sqrt{\tau}}\right) - \exp(kx + k^2\tau)\operatorname{erfc}\left(\frac{x}{2\sqrt{\tau}} + k\sqrt{\tau}\right).$$

5.3.        $T = f(x)$        at   $\tau = 0$    (initial condition),

           $\partial_x T - kT = kg(\tau)$    at   $x = 0$    (boundary condition).

The solution is

$$T = \frac{1}{2\sqrt{\pi\tau}} \int_0^{+\infty} G(x,\xi,\tau)f(\xi)\,d\xi - \frac{k}{\sqrt{\pi}} \int_0^{\tau} \frac{g(\zeta)}{\sqrt{\tau-\zeta}} H(x,\tau-\zeta)\,d\zeta,$$

where $G(x,\xi,\tau)$ and $H(x,\tau)$ can be found in items 5.1 and 5.2.

### 6. Region $0 \le x < l$. The first boundary value problem.

6.1. Homogeneous boundary conditions:

          $T = f(x)$      at   $\tau = 0$     (initial condition),

          $T = 0$         at   $x = 0$     (boundary condition),

          $T = 0$         at   $x = l$     (boundary condition).

The solution is

$$T = \sum_{n=1}^{\infty} b_n \exp\left(-\frac{n^2\pi^2\tau}{l^2}\right) \sin\left(\frac{n\pi x}{l}\right), \quad b_n = \frac{2}{l}\int_0^l f(x)\sin\left(\frac{n\pi x}{l}\right)\,dx.$$

*The special case* $f(x) = A$.
*The solution is*

$$T = \frac{4A}{\pi} \sum_{n=0}^{\infty} \frac{1}{(2n+1)} \exp\left[-\frac{(2n+1)^2\pi^2\tau}{l^2}\right] \sin\left[\frac{(2n+1)\pi x}{l}\right].$$

*The special case* $f(x) = Ax$.
*The solution is*

$$T = \frac{2Al}{\pi} \sum_{n=1}^{\infty} \frac{(-1)^{n-1}}{n} \exp\left(-\frac{n^2\pi^2\tau}{l^2}\right) \sin\left(\frac{n\pi x}{l}\right).$$

6.2. Nonhomogeneous boundary conditions:

          $T = f(x)$      at   $\tau = 0$     (initial condition),

          $T = g(\tau)$     at   $x = 0$     (boundary condition),

          $T = h(\tau)$     at   $x = l$     (boundary condition).

The solution is

$$T = g(\tau) + \frac{x}{l}\left[h(\tau) - g(\tau)\right] + \frac{2}{l}\sum_{n=1}^{\infty} M_n(\tau)\exp\left(-\frac{n^2\pi^2\tau}{l^2}\right)\sin\left(\frac{n\pi x}{l}\right),$$

where

$$M_n(\tau) = \int_0^l f(x)\sin\left(\frac{n\pi x}{l}\right)dx + \frac{n\pi}{l}\int_0^\tau \exp\left(\frac{n^2\pi^2\tau}{l^2}\right)g(\tau)\,d\tau$$

$$- (-1)^n\frac{n\pi}{l}\int_0^\tau \exp\left(\frac{n^2\pi^2\tau}{l^2}\right)h(\tau)\,d\tau - \frac{l}{\pi n}\exp\left(\frac{n^2\pi^2\tau}{l^2}\right)\left[g(\tau) - (-1)^n h(\tau)\right].$$

### 7. Region $0 \leq x \leq l$. The second boundary value problem.

$$
\begin{array}{llll}
T = f(x) & \text{at} & \tau = 0 & \text{(initial condition)}, \\
\partial_x T = 0 & \text{at} & x = 0 & \text{(boundary condition)}, \\
\partial_x T = 0 & \text{at} & x = l & \text{(boundary condition)}.
\end{array}
$$

The solution is

$$T = \sum_{n=0}^{\infty} b_n \exp\left(-\frac{n^2\pi^2\tau}{l^2}\right)\cos\left(\frac{n\pi x}{l}\right),$$

where

$$b_0 = \frac{1}{l}\int_0^l f(x)\,dx, \quad b_n = \frac{2}{l}\int_0^l f(x)\cos\left(\frac{n\pi x}{l}\right)dx; \quad n = 1,\,2,\,\ldots$$

### 8. Region $0 \leq x \leq l$. The third boundary value problem.

$$
\begin{array}{llll}
T = f(x) & \text{at} & \tau = 0 & \text{(initial condition)}, \\
\partial_x T - kT = 0 & \text{at} & x = 0 & \text{(boundary condition)}, \\
\partial_x T + kT = 0 & \text{at} & x = l & \text{(boundary condition)},
\end{array}
$$

where $k > 0$. The solution is

$$T = \sum_{n=1}^{\infty}\int_0^l \frac{y_n(x)y_n(\xi)}{\|y_n\|^2}\exp(-\lambda_n^2\tau)f(\xi)\,d\xi,$$

where

$$y_n(x) = \cos(\lambda_n x) + \frac{k}{\lambda_n}\sin(\lambda_n x), \quad \|y_n\|^2 = \frac{k}{\lambda_n^2} + \frac{l}{2}\left(1 + \frac{k^2}{\lambda_n^2}\right).$$

Here the $\lambda_n$ are the positive roots of the transcendental equation

$$\frac{\tan(\lambda l)}{\lambda} = \frac{2k}{\lambda^2 - k^2}.$$

**9. Region $0 \leq x \leq l$. Mixed boundary value problem.**

$$
\begin{array}{llll}
T = f(x) & \text{at} & \tau = 0 & \text{(initial condition)}, \\
T = 0 & \text{at} & x = 0 & \text{(boundary condition)}, \\
\partial_x T = 0 & \text{at} & x = l & \text{(boundary condition)}.
\end{array}
$$

The solution is

$$
T = \int_0^l G(x, \xi, \tau) f(\xi) \, d\xi,
$$

where

$$
G(x, \xi, \tau) = \vartheta \left( \frac{x - \xi}{4l}, \frac{\tau}{l^2} \right) - \vartheta \left( \frac{x + \xi}{4l}, \frac{\tau}{l^2} \right)
$$
$$
+ \vartheta \left( \frac{x + \xi - 2l}{4l}, \frac{\tau}{l^2} \right) - \vartheta \left( \frac{x - \xi - 2l}{4l}, \frac{\tau}{l^2} \right).
$$

Here $\vartheta(x, \tau)$ is the Jacobi function

$$
\vartheta(x, \tau) = 1 + 2 \sum_{n=1}^{\infty} \exp(-\pi^2 n^2 \tau) \cos(2\pi n x) = \frac{1}{\sqrt{\pi \tau}} \sum_{n=-\infty}^{+\infty} \exp \left[ -\frac{(x - n)^2}{4\tau} \right].
$$

The first series rapidly converges for large $\tau$, and the second for small $\tau$.

## S.1-2. Heat Equation With a Source

Let us consider a one-dimensional heat equation with a linear source:

$$
\frac{\partial T}{\partial \tau} = \frac{\partial^2 T}{\partial x^2} + bT.
$$

This equation with $T$ replaced by $C$ and $b > 0$ is encountered in mass transfer problems with a first-order volume reaction.

*1. Some particular solutions* ($A$, $B$, and $\lambda$ are arbitrary constants):

1.  $T = (Ax + B)e^{b\tau}$,

2.  $T = A \exp\left[(\lambda^2 + b)\tau \pm \lambda x\right] + Be^{b\tau}$,

3.  $T = A \exp\left[(b - \lambda^2)\tau\right] \cos(\lambda x) + Be^{b\tau}$,

4.  $T = A \exp\left[(b - \lambda^2)\tau\right] \sin(\lambda x) + Be^{b\tau}$,

5.  $T = A \dfrac{1}{\sqrt{\tau}} \exp\left(-\dfrac{x^2}{4\tau} + b\tau\right) + Be^{b\tau}$,

6.  $T = A \dfrac{x}{\tau^{3/2}} \exp\left(-\dfrac{x^2}{4\tau} + b\tau\right) + Be^{b\tau}$,

7.  $T = Ae^{b\tau} \operatorname{erf}\left(\dfrac{x}{2\sqrt{\tau}}\right) + Be^{b\tau}$,

where erf $z$ is the error function.

**2. Simplifying transformation.** The change $T = e^{b\tau}u$ leads to the equation $\partial_\tau u = \partial_{xx}u$ considered in Subsection S.1-1. The initial condition for the new variable $u$ does not change, and the nonhomogeneous part in the boundary conditions is multiplied by $e^{-b\tau}$. Taking this into account, we can readily obtain the solution of the original equation with initial and boundary conditions considered in Subsection S.1-1. By way of example, we write out solutions of two typical problems.

**3. Region $-\infty < x < +\infty$.**

$$T = f(x) \qquad \text{at} \qquad \tau = 0 \qquad \text{(initial condition).}$$

The solution is

$$T = \frac{1}{2\sqrt{\pi\tau}} \int_{-\infty}^{+\infty} \exp\left[-\frac{(x-\xi)^2}{4\tau} + b\tau\right] f(\xi)\,d\xi.$$

**4. Region $0 \le x < +\infty$. The first boundary value problem.**

$$T = f(x) \qquad \text{at} \qquad \tau = 0 \qquad \text{(initial condition),}$$
$$T = g(\tau) \qquad \text{at} \qquad x = 0 \qquad \text{(boundary condition).}$$

The solution is

$$T = \frac{1}{2\sqrt{\pi\tau}} \int_0^{+\infty} \left\{ \exp\left[-\frac{(x-\xi)^2}{4\tau}\right] - \exp\left[-\frac{(x+\xi)^2}{4\tau}\right] \right\} e^{b\tau} f(\xi)\,d\xi$$
$$+ \frac{x}{2\sqrt{\pi}} \int_0^\tau \exp\left[-\frac{x^2}{4(\tau-\zeta)}\right] \exp[b(\tau-\zeta)] \frac{g(\zeta)\,d\zeta}{(\tau-\zeta)^{3/2}}.$$

## S.1-3. Heat Equation in the Cylindrical Coordinates

Let us consider the equation

$$\frac{\partial T}{\partial \tau} = \frac{\partial^2 T}{\partial r^2} + \frac{1}{r}\frac{\partial T}{\partial r}.$$

This equation is encountered in heat transfer problems with cylindrical symmetry (e.g., heat exchange between a circular cylinder and the ambient medium, with $r$ being the dimensionless radial coordinate).

**1. Some particular solutions** ($A$, $B$, and $\lambda$ are arbitrary constants):

1.  $T = A + B \ln r$,

2.  $T = A + 4B\tau + Br^2$,

3.  $T = A + \dfrac{B}{\tau} \exp\left(-\dfrac{r^2}{4\tau}\right)$,

4.  $T = A + B \displaystyle\int_1^z e^{-\zeta}\frac{d\zeta}{\zeta}, \qquad z = \dfrac{r^2}{4\tau}$,

5.  $T = \exp(-\lambda^2\tau)J_0(\lambda r)$,

where $J_0(z)$ is the Bessel function.

### 2. Region $0 \leq r \leq L$. The first boundary value problem.

2.1.

$$T = T_0 \quad \text{at} \quad \tau = 0 \quad \text{(initial condition)},$$
$$T = T_L \quad \text{at} \quad r = L \quad \text{(boundary condition)},$$
$$T \neq \infty \quad \text{at} \quad r = 0 \quad \text{(boundary condition)}.$$

where $T_0 = \text{const}$ and $T_L = \text{const}$.

The solution is

$$\frac{T(r,\tau) - T_L}{T_0 - T_L} = \sum_{n=1}^{\infty} \frac{2}{\mu_n J_1(\mu_n)} \exp\left(-\mu_n^2 \frac{\tau}{L^2}\right) J_0\left(\mu_n \frac{r}{L}\right),$$

where the $\mu_n$ are the roots of the Bessel function, $J_0(\mu_n) = 0$. We give numerical values of the first five roots (to the fourth decimal place): $\mu_1 = 2.4048$, $\mu_2 = 5.5201$, $\mu_3 = 8.6537$, $\mu_4 = 11.7915$, and $\mu_5 = 14.9309$. We have $\mu_{n+1} - \mu_n \to \pi$ as $n \to \infty$.

2.2.

$$T = f(r) \quad \text{at} \quad \tau = 0 \quad \text{(initial condition)},$$
$$T = 0 \quad \text{at} \quad r = L \quad \text{(boundary condition)},$$
$$T \neq \infty \quad \text{at} \quad r = 0 \quad \text{(boundary condition)}.$$

The solution is

$$T(r,\tau) = \sum_{n=1}^{\infty} A_n \exp\left(-\frac{\mu_n^2 \tau}{L^2}\right) J_0\left(\mu_n \frac{r}{L}\right),$$

where

$$A_n = \frac{2}{L^2 J_1^2(\mu_n)} \int_0^L r f(r) J_0\left(\mu_n \frac{r}{L}\right) dr.$$

Here the $\mu_n$ are the roots of the Bessel function, $J_0(\mu_n) = 0$.

### 3. Region $0 \leq r \leq L$. The second boundary value problem.

3.1.

$$T = T_0 \quad \text{at} \quad \tau = 0 \quad \text{(initial condition)},$$
$$\partial_r T = g_L \quad \text{at} \quad r = L \quad \text{(boundary condition)},$$
$$T \neq \infty \quad \text{at} \quad r = 0 \quad \text{(boundary condition)},$$

where $T_0 = \text{const}$ and $g_L = \text{const}$.

The solution is

$$T(r,\tau) = T_0 + g_L L \left[\frac{2\tau}{L^2} - \frac{1}{4} + \frac{r^2}{2L^2} - \sum_{n=1}^{\infty} \frac{2}{\mu_n^2 J_0(\mu_n)} \exp\left(-\mu_n^2 \frac{\tau}{L^2}\right) J_0\left(\mu_n \frac{r}{L}\right)\right],$$

where the $\mu_n$ are the roots of the Bessel function. $J_1(\mu_n) = 0$. We give numerical values of the first five roots (to the fourth decimal place): $\mu_1 = 3.8317$, $\mu_2 = 7.0156$, $\mu_3 = 10.1735$, $\mu_4 = 13.3237$, and $\mu_5 = 16.4706$. We have $\mu_{n+1} - \mu_n \to \pi$ as $n \to \infty$.

3.2.

$$T = f(r) \quad \text{at} \quad \tau = 0 \quad \text{(initial condition)},$$
$$\partial_r T = g(\tau) \quad \text{at} \quad r = L \quad \text{(boundary condition)},$$
$$T \neq \infty \quad \text{at} \quad r = 0 \quad \text{(boundary condition)}.$$

The solution is

$$T(r,\tau) = \frac{2}{L^2} \int_0^L r f(r)\, dr + \frac{2}{L} \int_0^\tau g(\zeta)\, d\zeta + \sum_{n=1}^{\infty} \exp\left(-\frac{\mu_n^2 \tau}{L^2}\right) J_0\left(\frac{\mu_n r}{L}\right) H_n(\tau),$$

where

$$H_n(\tau) = \frac{1}{J_0^2(\mu_n)} \left[ \frac{2}{L^2} \int_0^L r f(r) J_0\left(\frac{\mu_n r}{L}\right) dr + \frac{2}{L} \int_0^\tau g(\zeta) \exp\left(\frac{\mu_n^2 \zeta}{L^2}\right) d\zeta \right].$$

Here the $\mu_n$ are the roots of the Bessel function: $J_1(\mu_n) = 0$.

#### 4. Region $0 \le r \le L$. The third boundary value problem.

$$\begin{array}{llll}
T = T_0 & \text{at} & \tau = 0 & \text{(initial condition)}, \\
\partial_r T = k(T_L - T) & \text{at} & r = L & \text{(boundary condition)}, \\
T \ne \infty & \text{at} & r = 0 & \text{(boundary condition)},
\end{array}$$

where $k = \text{const}$, $T_0 = \text{const}$, and $T_L = \text{const}$.

The solution is

$$\frac{T(r,\tau) - T_0}{T_L - T_0} = 1 - \sum_{n=1}^{\infty} A_n \exp\left(-\frac{\mu_n^2 \tau}{L^2}\right) J_0\left(\frac{\mu_n r}{L}\right),$$

where

$$A_n = \frac{2 J_1(\mu_n)}{\mu_n [J_0^2(\mu_n) + J_1^2(\mu_n)]}.$$

Here the $\mu_n$ are the positive roots of the equation $\mu_n J_1(\mu_n) - kL J_0(\mu_n) = 0$.

### S.1-4. Heat Equation in Spherical Coordinates

Let us consider the equation

$$\frac{\partial T}{\partial \tau} = \frac{\partial^2 T}{\partial r^2} + \frac{2}{r} \frac{\partial T}{\partial r}.$$

This equation is encountered in heat transfer problems with spherical symmetry (e.g., heat exchange between a ball and the ambient medium, with $r$ being the dimensionless radial coordinate). Note that the change of variables $u(r,\tau) = rT(r,\tau)$ leads to the equation $\partial_\tau u = \partial_{rr} u$ with constant coefficients, considered in Subsection S.1-1.

#### 1. Some particular solutions ($A$, $B$, and $\lambda$ are arbitrary constants):

1.  $T = A + Br^{-1}$,

2.  $T = A + 6B\tau + Br^2$,

3.  $T = A + \dfrac{B}{\tau^{3/2}} \exp\left(-\dfrac{r^2}{4\tau}\right)$,

4.  $T = A + \dfrac{B}{r\sqrt{\tau}} \exp\left(-\dfrac{r^2}{4\tau}\right)$,

5.  $T = Ar^{-1} \exp(\lambda^2 \tau \pm \lambda r) + B$,

6.  $T = Ar^{-1} \exp(-\lambda^2 \tau) \cos(\lambda r) + B$,

7.  $T = Ar^{-1} \exp(-\lambda^2 \tau) \sin(\lambda r) + B$.

### 2. Region $0 \leq r \leq L$. The first boundary value problem.

2.1.

$$T = T_0 \quad \text{at} \quad \tau = 0 \quad \text{(initial condition)},$$
$$T = T_L \quad \text{at} \quad r = L \quad \text{(boundary condition)},$$
$$T \neq \infty \quad \text{at} \quad r = 0 \quad \text{(boundary condition)},$$

where $T_0 = \text{const}$ and $T_L = \text{const}$.
The solution is

$$\frac{T(r,\tau) - T_L}{T_0 - T_L} = 2 \sum_{n=1}^{\infty} \frac{(-1)^{n+1} L}{\pi n r} \sin\left(\frac{\pi n r}{L}\right) \exp\left(-\frac{\pi^2 n^2 \tau}{L^2}\right).$$

2.2.

$$T = f(r) \quad \text{at} \quad \tau = 0 \quad \text{(initial condition)},$$
$$T = T_L \quad \text{at} \quad r = L \quad \text{(boundary condition)},$$
$$T \neq \infty \quad \text{at} \quad r = 0 \quad \text{(boundary condition)}.$$

The solution is

$$T(r,\tau) = T_L + \sum_{n=1}^{\infty} \frac{A_n}{r} \sin\left(\frac{\pi n r}{L}\right) \exp\left(-\frac{\pi^2 n^2 \tau}{L^2}\right),$$

where

$$A_n = \frac{2}{L} \int_0^L r f(r) \sin\left(\frac{\pi n r}{L}\right) dr.$$

### 3. Region $0 \leq r \leq L$. The second boundary value problem.

3.1.

$$T = T_0 \quad \text{at} \quad \tau = 0 \quad \text{(initial condition)},$$
$$\partial_r T = g_L \quad \text{at} \quad r = L \quad \text{(boundary condition)},$$
$$T \neq \infty \quad \text{at} \quad r = 0 \quad \text{(boundary condition)},$$

where $T_0 = \text{const}$ and $g_L = \text{const}$.
The solution is

$$T(r,\tau) = T_0 + g_L \left[ \frac{3\tau}{L} + \frac{5r^2 - 3L^2}{10L} - \sum_{n=1}^{\infty} \frac{2L^2}{\mu_n^3 \cos(\mu_n) r} \sin\left(\frac{\mu_n r}{L}\right) \exp\left(-\frac{\mu_n^2 \tau}{L^2}\right) \right],$$

where the $\mu_n$ are the positive roots of the transcendental equation $\tan(\mu_n) - \mu_n = 0$. We give numerical values of the first five roots (to the fourth decimal place): $\mu_1 = 4.4934$, $\mu_2 = 7.7253$, $\mu_3 = 10.9041$, $\mu_4 = 14.0662$, and $\mu_5 = 17.2208$.

3.2.

$$T = f(r) \quad \text{at} \quad \tau = 0 \quad \text{(initial condition)},$$
$$\partial_r T = g(\tau) \quad \text{at} \quad r = L \quad \text{(boundary condition)},$$
$$T \neq \infty \quad \text{at} \quad r = 0 \quad \text{(boundary condition)}.$$

The solution is

$$T(r,\tau) = \frac{3}{L^3} \int_0^L r^2 f(r) \, dr + \frac{3}{L} \int_0^\tau g(\zeta) \, d\zeta + \sum_{n=1}^{\infty} \exp\left(-\frac{\mu_n^2 \tau}{L^2}\right) \sin\left(\frac{\mu_n r}{L}\right) H_n(\tau),$$

where

$$H_n(\tau) = \frac{2}{\mu_n^2 \cos(\mu_n)r} \left[ \frac{1}{L\mu_n} \int_0^L r f(r) \sin\left(\frac{\mu_n r}{L}\right) dr + \int_0^\tau g(\zeta) \exp\left(\frac{\mu_n^2 \zeta}{L^2}\right) d\zeta \right].$$

Here the $\mu_n$ are the positive roots of the transcendental equation $\tan(\mu_n) - \mu_n = 0$.

**4. Region $0 \leq r \leq L$. The third boundary value problem.**

$$\begin{array}{llll} T = T_0 & \text{at} & \tau = 0 & \text{(initial condition),} \\ \partial_r T = k(T_L - T) & \text{at} & r = L & \text{(boundary condition),} \\ T \neq \infty & \text{at} & r = 0 & \text{(boundary condition),} \end{array}$$

where $k = \text{const}$, $T_0 = \text{const}$, and $T_L = \text{const}$.
The solution is

$$\frac{T(r,\tau) - T_0}{T_L - T_0} = 1 - \sum_{n=1}^\infty A_n \frac{L}{r} \sin\left(\frac{\mu_n r}{L}\right) \exp\left(-\frac{\mu_n^2 \tau}{L^2}\right),$$

where

$$A_n = \frac{2}{\mu_n} \frac{\sin \mu_n - \mu_n \cos \mu_n}{\mu_n - \sin \mu_n \cos \mu_n}.$$

Here the $\mu_n$ are the positive roots of the transcendental equation $(kL-1)\tan(\mu_n) + \mu_n = 0$.

# S.2. Formulas For Constructing Exact Solutions

## S.2-1. Duhamel Integrals

**1. Homogeneous equations with nonhomogeneous boundary conditions.** The solution of the boundary value problem

$$\frac{\partial T}{\partial \tau} = f(x)\frac{\partial^2 T}{\partial x^2} + g(x)\frac{\partial T}{\partial x} + h(x)T, \tag{1}$$

$$\begin{array}{llll} T = 0 & \text{at} & \tau = 0 & \text{(initial condition),} & (2) \\ T = g(\tau) & \text{at} & x = 0 & \text{(boundary condition),} & (3) \\ T = 0 & \text{at} & x = l & \text{(boundary condition)} & (4) \end{array}$$

with nonstationary boundary condition at $x = 0$ can be expressed by the formula (Duhamel integral)

$$T(x,\tau) = \int_0^\tau \frac{\partial W}{\partial \tau}(x, \tau - \zeta)g(\zeta)\,d\zeta$$

in terms of the solution $W(x,\tau)$ of the auxiliary problem for Eq. (1) with the initial condition (2) and boundary condition (4) (in the equation and the initial and boundary conditions, $T$ must be replaced by $W$) and the simpler stationary boundary condition at $x = 0$

$$W = 1 \quad \text{at} \quad x = 0 \quad \text{(boundary condition).} \tag{5}$$

This formula can also be used for $l = \infty$.

A similar formula holds for a homogeneous boundary condition at $x = a$ and a nonhomogeneous nonstationary boundary condition at $x = b$.

**2. Nonhomogeneous equations with homogeneous boundary conditions.** The solution of the boundary value problem for the nonhomogeneous equation

$$\frac{\partial T}{\partial \tau} = f(x)\frac{\partial^2 T}{\partial x^2} + g(x)\frac{\partial T}{\partial x} + h(x)T + \Phi(x,\tau),\tag{6}$$

$$T = 0 \quad \text{at} \quad \tau = 0 \quad \text{(initial condition)},\tag{7}$$

$$T = 0 \quad \text{at} \quad x = 0 \quad \text{(boundary condition)},\tag{8}$$

$$T = 0 \quad \text{at} \quad x = l \quad \text{(boundary condition)},\tag{9}$$

can be expressed by the formula (Duhamel integral)

$$T(x,\tau) = \int_0^\tau U(x,\,\tau - \zeta;\,\zeta)\,d\zeta$$

via the solution $U(x,\tau;\zeta)$ of the auxiliary problem for the homogeneous equation

$$\frac{\partial U}{\partial \tau} = f(x)\frac{\partial^2 U}{\partial x^2} + g(x)\frac{\partial U}{\partial x} + h(x)U\tag{10}$$

with the boundary conditions (8) and (9) (in which $T$ must be replaced by $U$) and with the following nonhomogeneous initial condition depending on the parameter $\zeta$:

$$U = \Phi(x,\zeta) \quad \text{at} \quad \tau = 0 \quad \text{(initial condition)}.\tag{11}$$

This formula can also be used for $l = \infty$.

---

### S.2-2. Problems With Volume Reaction

Let us consider the boundary value problem

$$\frac{\partial T}{\partial \tau} = f(x)\frac{\partial^2 T}{\partial x^2} + g(x)\frac{\partial T}{\partial x} - kT,\tag{12}$$

$$T = 0 \quad \text{at} \quad \tau = 0 \quad \text{(initial condition)},\tag{13}$$

$$T = T_0 \quad \text{at} \quad x = 0 \quad \text{(boundary condition)},\tag{14}$$

$$T = T_l \quad \text{at} \quad x = l \quad \text{(boundary condition)},\tag{15}$$

where $k$, $T_0$, and $T_l$ are some constants.

Equation (12) is commonly encountered in problems of chemical technology in which the function $T$ plays the role of concentration and the parameter $k$, of the constant rate of volume chemical reaction (see Section 3.1).

The solution of the problem with volume reaction (12)–(15) can be expressed by the formula

$$T(x,\tau) = k\int_0^\tau e^{-k\zeta}\widetilde{T}(x,\zeta)\,d\zeta + e^{-k\tau}\widetilde{T}(x,\tau)$$

in terms of the solution $\widetilde{T}(x,\tau)$ of the simpler auxiliary equation without volume chemical reaction

$$\frac{\partial \widetilde{T}}{\partial \tau} = f(x)\frac{\partial^2 \widetilde{T}}{\partial x^2} + g(x)\frac{\partial \widetilde{T}}{\partial x}\tag{16}$$

with the same initial and boundary conditions (13)–(15) (in which $T$ must be replaced by $\widetilde{T}$). This formula can also be used for $l = \infty$.

# S.3. Orthogonal Curvilinear Coordinates

This section contains some material borrowed from the books [179, 231, 255, 276].

## S.3-1. Arbitrary Orthogonal Coordinates

The curvilinear coordinates $x^1$, $x^2$, $x^3$ are defined as functions of the rectangular Cartesian coordinates $x$, $y$, $z$:

$$x^1 = x^1(x, y, z), \quad x^2 = x^2(x, y, z), \quad x^3 = x^3(x, y, z).$$

Using these formulas, one can express $x$, $y$, $z$ in terms of the curvilinear coordinates $x^1$, $x^2$, $x^3$ as follows:

$$x = x(x^1, x^2, x^3), \quad y = y(x^1, x^2, x^3), \quad z = z(x^1, x^2, x^3).$$

The metric tensor components $g_{ij}$ are determined by the formulas

$$g_{ij}(x^1, x^2, x^3) = \frac{\partial x}{\partial x^i}\frac{\partial x}{\partial x^j} + \frac{\partial y}{\partial x^i}\frac{\partial y}{\partial x^j} + \frac{\partial z}{\partial x^i}\frac{\partial z}{\partial x^j};$$
$$g_{ij}(x^1, x^2, x^3) = g_{ji}(x^1, x^2, x^3); \qquad i, j = 1, 2, 3.$$

A system of coordinates is orthogonal if

$$g_{ij}(x^1, x^2, x^3) = 0 \quad \text{for} \quad i \ne j.$$

In this case the third invariant of the metric tensor is given by

$$g = g_{11}g_{22}g_{33}.$$

In what follows, we present the basic differential operators in the orthogonal curvilinear coordinates $x^1$, $x^2$, $x^3$. The corresponding unit vectors are denoted by $i_1$, $i_2$, and $i_3$.
The gradient of a scalar $P$ is

$$\nabla P = \frac{1}{\sqrt{g_{11}}}\frac{\partial P}{\partial x^1}i_1 + \frac{1}{\sqrt{g_{22}}}\frac{\partial P}{\partial x^2}i_2 + \frac{1}{\sqrt{g_{33}}}\frac{\partial P}{\partial x^3}i_3.$$

The divergence of a vector $\mathbf{V} = i_1 V_1 + i_2 V_2 + i_3 V_3$ is

$$\nabla \cdot \mathbf{V} = \frac{1}{\sqrt{g}}\left[\frac{\partial}{\partial x^1}\left(V_1\sqrt{\frac{g}{g_{11}}}\right) + \frac{\partial}{\partial x^2}\left(V_2\sqrt{\frac{g}{g_{22}}}\right) + \frac{\partial}{\partial x^3}\left(V_3\sqrt{\frac{g}{g_{33}}}\right)\right].$$

The gradient of a scalar $C$ with respect to a vector $\mathbf{V}$ is

$$(\mathbf{V} \cdot \nabla)C = \frac{V_1}{\sqrt{g_{11}}}\frac{\partial C}{\partial x^1} + \frac{V_2}{\sqrt{g_{22}}}\frac{\partial C}{\partial x^2} + \frac{V_3}{\sqrt{g_{33}}}\frac{\partial C}{\partial x^3}.$$

The gradient of a vector $\mathbf{W}$ with respect to a vector $\mathbf{V}$ is

$$(\mathbf{V} \cdot \nabla)\mathbf{W} = i_1(\mathbf{V} \cdot \nabla)W_1 + i_2(\mathbf{V} \cdot \nabla)W_2 + i_3(\mathbf{V} \cdot \nabla)W_3.$$

The rotor of a vector $\mathbf{V}$ is

$$\nabla \times \mathbf{V} = \mathbf{i}_1 \frac{\sqrt{g_{11}}}{\sqrt{g}} \left[ \frac{\partial}{\partial x^2} \left( V_3 \sqrt{g_{33}} \right) - \frac{\partial}{\partial x^3} \left( V_2 \sqrt{g_{22}} \right) \right]$$

$$+ \mathbf{i}_2 \frac{\sqrt{g_{22}}}{\sqrt{g}} \left[ \frac{\partial}{\partial x^3} \left( V_1 \sqrt{g_{11}} \right) - \frac{\partial}{\partial x^1} \left( V_3 \sqrt{g_{33}} \right) \right]$$

$$+ \mathbf{i}_3 \frac{\sqrt{g_{33}}}{\sqrt{g}} \left[ \frac{\partial}{\partial x^1} \left( V_2 \sqrt{g_{22}} \right) - \frac{\partial}{\partial x^2} \left( V_1 \sqrt{g_{11}} \right) \right].$$

The Laplace operator of a scalar $C$:

$$\Delta C \equiv \frac{1}{\sqrt{g}} \left[ \frac{\partial}{\partial x^1} \left( \frac{\sqrt{g}}{g_{11}} \frac{\partial C}{\partial x^1} \right) + \frac{\partial}{\partial x^2} \left( \frac{\sqrt{g}}{g_{22}} \frac{\partial C}{\partial x^2} \right) + \frac{\partial}{\partial x^3} \left( \frac{\sqrt{g}}{g_{33}} \frac{\partial C}{\partial x^3} \right) \right].$$

The Laplacian of a vector $\mathbf{V}$:

$$\Delta \mathbf{V} = \nabla (\nabla \cdot \mathbf{V}) - \nabla \times (\nabla \times \mathbf{V}).$$

## S.3-2. Cylindrical Coordinates $\mathcal{R}$, $\varphi$, $Z$

The transformations of coordinates $(0 \le \varphi \le 2\pi)$ are

$$\mathcal{R} = \sqrt{X^2 + Y^2}, \quad \tan \varphi = Y/X, \quad Z = Z \quad (\sin \varphi = Y/\mathcal{R}),$$
$$X = \mathcal{R} \cos \varphi, \quad Y = \mathcal{R} \sin \varphi, \quad Z = Z.$$

The metric tensor components are

$$g_{\mathcal{R}\mathcal{R}} = 1, \quad g_{\varphi\varphi} = \mathcal{R}^2, \quad g_{ZZ} = 1, \quad \sqrt{g} = \mathcal{R}.$$

The gradient of a scalar $P$ is

$$\nabla P = \frac{\partial P}{\partial \mathcal{R}} \mathbf{i}_{\mathcal{R}} + \frac{1}{\mathcal{R}} \frac{\partial P}{\partial \varphi} \mathbf{i}_{\varphi} + \frac{\partial P}{\partial Z} \mathbf{i}_Z.$$

The divergence of a vector $\mathbf{V}$ is

$$\nabla \cdot \mathbf{V} = \frac{1}{\mathcal{R}} \frac{\partial (\mathcal{R} V_{\mathcal{R}})}{\partial \mathcal{R}} + \frac{1}{\mathcal{R}} \frac{\partial V_{\varphi}}{\partial \varphi} + \frac{\partial V_Z}{\partial Z}.$$

The gradient of a scalar $C$ with respect to a vector $\mathbf{V}$ is

$$(\mathbf{V} \cdot \nabla)C = V_{\mathcal{R}} \frac{\partial C}{\partial \mathcal{R}} + \frac{V_{\varphi}}{\mathcal{R}} \frac{\partial C}{\partial \varphi} + V_Z \frac{\partial C}{\partial Z}.$$

The gradient of a vector $\mathbf{W}$ with respect to a vector $\mathbf{V}$ is

$$(\mathbf{V} \cdot \nabla)\mathbf{W} = (\mathbf{V} \cdot \nabla)W_{\mathcal{R}} \mathbf{i}_{\mathcal{R}} + (\mathbf{V} \cdot \nabla)W_{\varphi} \mathbf{i}_{\varphi} + (\mathbf{V} \cdot \nabla)W_Z \mathbf{i}_Z.$$

The rotor of a vector $\mathbf{V}$ is

$$\nabla \times \mathbf{V} = \left( \frac{1}{\mathcal{R}} \frac{\partial V_Z}{\partial \varphi} - \frac{\partial V_{\varphi}}{\partial Z} \right) \mathbf{i}_{\mathcal{R}} + \left( \frac{\partial V_{\mathcal{R}}}{\partial Z} - \frac{\partial V_Z}{\partial \mathcal{R}} \right) \mathbf{i}_{\varphi} + \frac{1}{\mathcal{R}} \left[ \frac{\partial (\mathcal{R} V_{\varphi})}{\partial \mathcal{R}} - \frac{\partial V_{\mathcal{R}}}{\partial \varphi} \right] \mathbf{i}_Z.$$

The Laplacian of a scalar $C$ is

$$\Delta C = \frac{1}{\mathcal{R}} \frac{\partial}{\partial \mathcal{R}} \left( \mathcal{R} \frac{\partial C}{\partial \mathcal{R}} \right) + \frac{1}{\mathcal{R}^2} \frac{\partial^2 C}{\partial \varphi^2} + \frac{\partial^2 C}{\partial Z^2}.$$

*Remark.* Cylindrical coordinates $\mathcal{R}$, $\varphi$, $Z$ also used as polar coordinates on the plane $XY$.

## S.3-3. Spherical Coordinates $R$, $\theta$, $\varphi$

The transformations of coordinates ($0 \leq \theta \leq \pi$, $0 \leq \varphi \leq 2\pi$) are

$$R = \sqrt{X^2 + Y^2 + Z^2}, \quad \theta = \arccos \frac{Z}{R}, \quad \tan \varphi = \frac{Y}{X} \quad \left( \sin \varphi = \frac{Y}{\sqrt{X^2 + Y^2}} \right),$$

$$X = R \sin \theta \cos \varphi, \quad Y = R \sin \theta \sin \varphi, \quad Z = R \cos \theta.$$

The metric tensor components are

$$g_{RR} = 1, \quad g_{\theta\theta} = R^2, \quad g_{\varphi\varphi} = R^2 \sin^2 \theta, \quad \sqrt{g} = R^2 \sin \theta.$$

The gradient of a scalar $P$ is

$$\nabla P = \frac{\partial P}{\partial R} \mathbf{i}_R + \frac{1}{R} \frac{\partial P}{\partial \vartheta} \mathbf{i}_\theta + \frac{1}{R \sin \theta} \frac{\partial P}{\partial \varphi} \mathbf{i}_\varphi.$$

The divergence of a vector $\mathbf{V}$ is

$$\nabla \cdot \mathbf{V} = \frac{1}{R^2} \frac{\partial}{\partial R} \left( R^2 V_R \right) + \frac{1}{R \sin \theta} \frac{\partial}{\partial \theta} \left( \sin \theta \, V_\theta \right) + \frac{1}{R \sin \varphi} \frac{\partial V_\varphi}{\partial \varphi}.$$

The gradient of a scalar $C$ with respect to a vector $\mathbf{V}$ is

$$(\mathbf{V} \cdot \nabla) C = V_R \frac{\partial C}{\partial R} + \frac{V_\theta}{R} \frac{\partial C}{\partial \theta} + \frac{V_\varphi}{R \sin \theta} \frac{\partial C}{\partial \varphi}.$$

The gradient of a vector $\mathbf{W}$ with respect to a vector $\mathbf{V}$ is

$$(\mathbf{V} \cdot \nabla) \mathbf{W} = (\mathbf{V} \cdot \nabla) W_R \mathbf{i}_R + (\mathbf{V} \cdot \nabla) W_\theta \mathbf{i}_\theta + (\mathbf{V} \cdot \nabla) W_\varphi \mathbf{i}_\varphi.$$

The rotor of a vector $\mathbf{V}$ is

$$\nabla \times \mathbf{V} = \frac{1}{R \sin \theta} \left[ \frac{\partial (\sin \theta \, V_\varphi)}{\partial \theta} - \frac{\partial V_\theta}{\partial \varphi} \right] \mathbf{i}_R$$

$$+ \frac{1}{R} \left[ \frac{1}{\sin \theta} \frac{\partial V_R}{\partial \varphi} - \frac{\partial (RV_\varphi)}{\partial R} \right] \mathbf{i}_\theta + \frac{1}{R} \left[ \frac{\partial (RV_\theta)}{\partial R} - \frac{\partial V_R}{\partial \theta} \right] \mathbf{i}_\varphi.$$

The Laplacian of a scalar $C$ is

$$\Delta C = \frac{1}{R^2} \frac{\partial}{\partial R} \left( R^2 \frac{\partial C}{\partial R} \right) + \frac{1}{R^2 \sin \theta} \frac{\partial}{\partial \theta} \left( \sin \theta \frac{\partial C}{\partial \theta} \right) + \frac{1}{R^2 \sin^2 \theta} \frac{\partial^2 C}{\partial \varphi^2}.$$

## S.3-4. Coordinates of a Prolate Ellipsoid of Revolution $\sigma$, $\tau$, $\varphi$

The transformations of coordinates ($\sigma \geq 1 \geq \tau \geq -1$) are

$$X^2 = m^2(\sigma^2 - 1)(1 - \tau^2) \cos^2 \varphi, \quad Y^2 = m^2(\sigma^2 - 1)(1 - \tau^2) \sin^2 \varphi, \quad Z = m\sigma\tau.$$

The special system of coordinates $u$, $v$, $\varphi$ ($0 \leq u < \infty$, $0 \leq v \leq \pi$, $0 \leq \varphi \leq 2\pi$) is

$$\sigma = \cosh u, \quad \tau = \cos v, \quad \varphi = \varphi,$$

$$X = m \sinh u \sin v \cos \varphi, \quad Y = m \sinh u \sin v \sin \varphi, \quad Z = m \cosh u \cos v.$$

The metric tensor components are

$$g_{\sigma\sigma} = m^2\frac{\sigma^2-\tau^2}{\sigma^2-1}, \quad g_{\tau\tau} = m^2\frac{\sigma^2-\tau^2}{1-\tau^2}, \quad g_{\varphi\varphi} = m^2(\sigma^2-1)(1-\tau^2),$$

$$\sqrt{g} = m^3(\sigma^2-\tau^2), \quad g_{uu} = g_{vv} = m^2(\sinh^2 u + \sin^2 v), \quad g_{\varphi\varphi} = m^2\sinh^2 u \sin^2 v.$$

The gradient of a scalar $P$ is

$$\nabla P = \frac{1}{m}\sqrt{\frac{\sigma^2-1}{\sigma^2-\tau^2}}\frac{\partial P}{\partial \sigma}\mathbf{i}_\sigma + \frac{1}{m}\sqrt{\frac{1-\tau^2}{\sigma^2-\tau^2}}\frac{\partial P}{\partial \tau}\mathbf{i}_\tau + \frac{1}{m\sqrt{(1-\tau^2)(\sigma^2-1)}}\frac{\partial P}{\partial \varphi}\mathbf{i}_\varphi.$$

The divergence of a vector $\mathbf{V}$ is

$$\nabla \cdot \mathbf{V} = \frac{1}{m(\sigma^2-\tau^2)}\left\{\frac{\partial}{\partial \sigma}\left[V_\sigma\sqrt{(\sigma^2-\tau^2)(\sigma^2-1)}\right]\right.$$

$$\left.+ \frac{\partial}{\partial \tau}\left[V_\tau\sqrt{(\sigma^2-\tau^2)(1-\tau^2)}\right] + \frac{\partial}{\partial \varphi}\left[V_\varphi\frac{\sigma^2-\tau^2}{\sqrt{(\sigma^2-1)(1-\tau^2)}}\right]\right\}.$$

The gradient of a scalar $C$ with respect to a vector $\mathbf{V}$ is

$$(\mathbf{V}\cdot\nabla)C = \frac{V_\sigma}{m}\sqrt{\frac{\sigma^2-1}{\sigma^2-\tau^2}}\frac{\partial C}{\partial \sigma} + \frac{V_\tau}{m}\sqrt{\frac{1-\tau^2}{\sigma^2-\tau^2}}\frac{\partial C}{\partial \tau} + \frac{V_\varphi}{m\sqrt{(\sigma^2-1)(1-\tau^2)}}\frac{\partial C}{\partial \varphi}.$$

The gradient of a vector $\mathbf{W}$ with respect to a vector $\mathbf{V}$ is

$$(\mathbf{V}\cdot\nabla)\mathbf{W} = (\mathbf{V}\cdot\nabla)W_\sigma\mathbf{i}_\sigma + (\mathbf{V}\cdot\nabla)W_\tau\mathbf{i}_\tau + (\mathbf{V}\cdot\nabla)W_\varphi\mathbf{i}_\varphi.$$

The Laplacian of a scalar $C$ is

$$\Delta C = \frac{1}{m^2(\sigma^2-\tau^2)}\left\{\frac{\partial}{\partial \sigma}\left[(\sigma^2-1)\frac{\partial C}{\partial \sigma}\right] + \frac{\partial}{\partial \tau}\left[(1-\tau^2)\frac{\partial C}{\partial \tau}\right] + \frac{\sigma^2-\tau^2}{(\sigma^2-1)(1-\tau^2)}\frac{\partial^2 C}{\partial \varphi^2}\right\}.$$

## S.3-5. Coordinates of an Oblate Ellipsoid of Revolution $\sigma, \tau, \varphi$

The transformations of coordinates ($\sigma \geq 0, -1 \leq \tau \leq 1$) are

$$X^2 = m^2(1+\sigma^2)(1-\tau^2)\cos^2\varphi, \quad Y^2 = m^2(1+\sigma^2)(1-\tau^2)\sin^2\varphi, \quad Z = m\sigma\tau.$$

The special system of coordinates $u, v, \varphi$ ($0 \leq u < \infty, 0 \leq v \leq \pi, 0 \leq \varphi \leq 2\pi$) is

$$\sigma = \sinh u, \quad \tau = \cos v, \quad \varphi = \varphi,$$

$$X = m\cosh u \sin v \cos\varphi, \quad Y = m\cosh u \sin v \sin\varphi, \quad Z = m\sinh u \cos v.$$

The components of the metric tensor are

$$g_{\sigma\sigma} = m^2\frac{\sigma^2+\tau^2}{1+\sigma^2}, \quad g_{\tau\tau} = m^2\frac{\sigma^2+\tau^2}{1-\tau^2}, \quad g_{\varphi\varphi} = m^2(1+\sigma^2)(1-\tau^2),$$

$$\sqrt{g} = m^3(\sigma^2+\tau^2), \quad g_{uu} = g_{vv} = m^2(\sinh^2 u + \cos^2 v), \quad g_{\varphi\varphi} = m^2\cosh^2 u \sin^2 v.$$

The gradient of a scalar $P$ is

$$\nabla P = \frac{1}{m}\sqrt{\frac{\sigma^2+1}{\sigma^2+\tau^2}}\frac{\partial P}{\partial\sigma}\mathbf{i}_\sigma + \frac{1}{m}\sqrt{\frac{1-\tau^2}{\sigma^2+\tau^2}}\frac{\partial P}{\partial\tau}\mathbf{i}_\tau + \frac{1}{m\sqrt{(1-\tau^2)(\sigma^2+1)}}\frac{\partial P}{\partial\varphi}\mathbf{i}_\varphi.$$

The divergence of a vector $\mathbf{V}$ is

$$\nabla\cdot\mathbf{V} = \frac{1}{m(\sigma^2+\tau^2)}\left\{\frac{\partial}{\partial\sigma}\left[V_\sigma\sqrt{(\sigma^2+\tau^2)(\sigma^2+1)}\right]\right.$$
$$\left.+\frac{\partial}{\partial\tau}\left[V_\tau\sqrt{(\sigma^2+\tau^2)(1-\tau^2)}\right]+\frac{\partial}{\partial\varphi}\left[V_\varphi\frac{\sigma^2+\tau^2}{\sqrt{(\sigma^2+1)(1-\tau^2)}}\right]\right\}.$$

The gradient of a scalar $C$ with respect to a vector $\mathbf{V}$ is

$$(\mathbf{V}\cdot\nabla)C = \frac{V_\sigma}{m}\sqrt{\frac{\sigma^2+1}{\sigma^2+\tau^2}}\frac{\partial C}{\partial\sigma} + \frac{V_\tau}{m}\sqrt{\frac{1-\tau^2}{\sigma^2+\tau^2}}\frac{\partial C}{\partial\tau} + \frac{V_\varphi}{m\sqrt{(\sigma^2+1)(1-\tau^2)}}\frac{\partial C}{\partial\varphi}.$$

The gradient of a vector $\mathbf{W}$ with respect to a vector $\mathbf{V}$ is

$$(\mathbf{V}\cdot\nabla)\mathbf{W} = (\mathbf{V}\cdot\nabla)W_\sigma\mathbf{i}_\sigma + (\mathbf{V}\cdot\nabla)W_\tau\mathbf{i}_\tau + (\mathbf{V}\cdot\nabla)W_\varphi\mathbf{i}_\varphi.$$

The Laplacian of a scalar $C$ is

$$\Delta C = \frac{1}{m^2(\sigma^2+\tau^2)}\left\{\frac{\partial}{\partial\sigma}\left[(1+\sigma^2)\frac{\partial C}{\partial\sigma}\right]+\frac{\partial}{\partial\tau}\left[(1-\tau^2)\frac{\partial C}{\partial\tau}\right]+\frac{\sigma^2+\tau^2}{(1+\sigma^2)(1-\tau^2)}\frac{\partial^2 C}{\partial\varphi^2}\right\}.$$

## S.3-6. Coordinates of an Elliptic Cylinder $\sigma$, $\tau$, $Z$

The transformations of coordinates ($\sigma\geq 0, -1\leq\tau\leq 1$) are

$$X = b\sigma\tau, \quad Y^2 = b^2(\sigma^2-1)(1-\tau^2), \quad Z = Z.$$

The special system of coordinates $u$, $v$, $Z$ ($0\leq u<\infty, 0\leq v\leq\pi$) is

$$\sigma = \cosh u, \quad \tau = \cos v, \quad Z = Z,$$
$$X = b\cosh u\cos v, \quad Y = b\sinh u\sin v, \quad Z = Z.$$

The components of the metric tensor are

$$g_{\sigma\sigma} = b^2\frac{\sigma^2-\tau^2}{\sigma^2-1}, \quad g_{\tau\tau} = b^2\frac{\sigma^2-\tau^2}{1-\tau^2}, \quad g_{ZZ} = 1,$$
$$g_{uu} = g_{vv} = b^2(\sinh^2 u + \sin^2 v), \quad g_{ZZ} = 1.$$

The Laplacian is

$$\Delta C = \frac{1}{b^2(\sinh^2 u + \sin^2 v)}\left(\frac{\partial^2 C}{\partial u^2} + \frac{\partial^2 C}{\partial v^2}\right) + \frac{\partial^2 C}{\partial Z^2}$$
$$= \frac{\sqrt{\sigma^2-1}}{b^2(\sigma^2-\tau^2)}\frac{\partial}{\partial\sigma}\left(\sqrt{\sigma^2-1}\frac{\partial C}{\partial\sigma}\right) + \frac{\sqrt{1-\tau^2}}{b^2(\sigma^2-\tau^2)}\frac{\partial}{\partial\tau}\left(\sqrt{1-\tau^2}\frac{\partial C}{\partial\tau}\right) + \frac{\partial^2 C}{\partial Z^2}.$$

*Remark.* Coordinates of an elliptic cylinder $\sigma$, $\tau$, $Z$ also used as elliptic coordinates on the plane $XY$.

# S.4. Convective Diffusion Equation in Miscellaneous Coordinate Systems

The convective diffusion equation in Cartesian coordinates has the form of Eq. (3.1.1).

### S.4-1. Diffusion Equation in Cylindrical and Spherical Coordinates

The convective diffusion equation in cylindrical coordinates has the form

$$\frac{\partial C}{\partial t} + V_R \frac{\partial C}{\partial R} + \frac{V_\theta}{R} \frac{\partial C}{\partial \theta} + V_Z \frac{\partial C}{\partial Z} = D \left[ \frac{1}{R} \frac{\partial}{\partial R} \left( R \frac{\partial C}{\partial R} \right) + \frac{1}{R^2} \frac{\partial^2 C}{\partial \theta^2} + \frac{\partial^2 C}{\partial Z^2} \right].$$

The convective diffusion equation in spherical coordinates has the form

$$\frac{\partial C}{\partial t} + V_R \frac{\partial C}{\partial R} + \frac{V_\theta}{R} \frac{\partial C}{\partial \theta} + \frac{V_\varphi}{R \sin \theta} \frac{\partial C}{\partial \varphi}$$

$$= D \left[ \frac{1}{R^2} \frac{\partial}{\partial R} \left( R^2 \frac{\partial C}{\partial R} \right) + \frac{1}{R^2 \sin \theta} \frac{\partial}{\partial \theta} \left( \sin \theta \frac{\partial C}{\partial \theta} \right) + \frac{1}{R^2 \sin^2 \theta} \frac{\partial^2 C}{\partial \varphi^2} \right].$$

### S.4-2. Diffusion Equation in Arbitrary Orthogonal Coordinates

The convective diffusion equation in arbitrary orthogonal coordinates has the form

$$\frac{\partial C}{\partial t} + \frac{V_1}{\sqrt{g_{11}}} \frac{\partial C}{\partial x^1} + \frac{V_2}{\sqrt{g_{22}}} \frac{\partial C}{\partial x^2} + \frac{V_3}{\sqrt{g_{33}}} \frac{\partial C}{\partial x^3}$$

$$= \frac{D}{\sqrt{g}} \left[ \frac{\partial}{\partial x^1} \left( \frac{\sqrt{g}}{g_{11}} \frac{\partial C}{\partial x^1} \right) + \frac{\partial}{\partial x^2} \left( \frac{\sqrt{g}}{g_{22}} \frac{\partial C}{\partial x^2} \right) + \frac{\partial}{\partial x^3} \left( \frac{\sqrt{g}}{g_{33}} \frac{\partial C}{\partial x^3} \right) \right],$$

where $g_{11}$, $g_{22}$, and $g_{33}$ are the metric tensor components and $g = g_{11} g_{22} g_{33}$.

# S.5. Equations of Fluid Motion in Miscellaneous Coordinate Systems

Here we use the model of a viscous incompressible fluid, which is described by Navier–Stokes equations.

The Navier–Stokes equations in Cartesian coordinates have the form of Eqs. (1.1.2). These are considered in conjunction with the continuity equation (1.1.1).

### S.5-1. Navier–Stokes Equations in Cylindrical Coordinates

The continuity equation is

$$\frac{\partial V_R}{\partial R} + \frac{1}{R} \frac{\partial V_\varphi}{\partial \varphi} + \frac{\partial V_Z}{\partial Z} + \frac{V_R}{R} = 0.$$

The equations of motion ($F_R$, $F_\varphi$, and $F_Z$ are the components of the external volume force) are

$$\mathbb{H} V_R - \frac{V_\varphi^2}{R} = -\frac{1}{\rho} \frac{\partial P}{\partial R} + \nu \left( \Delta V_R - \frac{V_R}{R^2} - \frac{2}{R^2} \frac{\partial V_\varphi}{\partial \varphi} \right) + F_R,$$

$$\mathbb{H} V_\varphi + \frac{V_R V_\varphi}{R} = -\frac{1}{\rho R} \frac{\partial P}{\partial \varphi} + \nu \left( \Delta V_\varphi - \frac{V_\varphi}{R^2} + \frac{2}{R^2} \frac{\partial V_R}{\partial \varphi} \right) + F_\varphi,$$

$$\mathbb{H} V_Z = -\frac{1}{\rho} \frac{\partial P}{\partial Z} + \nu \Delta V_Z + F_Z,$$

where the differential operators $\mathbb{H}$ and $\Delta$ are given by the formulas

$$\mathbb{H} \equiv \frac{\partial}{\partial t} + V_R \frac{\partial}{\partial \mathcal{R}} + \frac{V_\varphi}{\mathcal{R}} \frac{\partial}{\partial \varphi} + V_Z \frac{\partial}{\partial Z},$$

$$\Delta \equiv \frac{\partial^2}{\partial \mathcal{R}^2} + \frac{1}{\mathcal{R}} \frac{\partial}{\partial \mathcal{R}} + \frac{1}{\mathcal{R}^2} \frac{\partial^2}{\partial \varphi^2} + \frac{\partial^2}{\partial Z^2}.$$

---

### S.5-2. Navier–Stokes Equations in Spherical Coordinates

The continuity equation is

$$\frac{\partial}{\partial R} \left( R^2 \sin\theta V_R \right) + \frac{\partial}{\partial \theta} (R \sin\theta V_\theta) + \frac{\partial}{\partial \varphi} \left( R V_\varphi \right) = 0.$$

The equations of motion are

$$\mathrm{M}\, V_R - \frac{V_\theta^2 + V_\varphi^2}{R} = -\frac{1}{\rho} \frac{\partial P}{\partial R} + \nu \Delta V_R$$

$$- \frac{2\nu}{R^2} \left( \frac{\partial V_\theta}{\partial \theta} + \frac{1}{\sin\theta} \frac{\partial V_\varphi}{\partial \varphi} + V_R + \cot\theta\, V_\theta \right) + F_R,$$

$$\mathrm{M}\, V_\theta + \frac{V_R V_\theta - V_\varphi^2 \cot\theta}{R} = -\frac{1}{\rho} R \frac{\partial P}{\partial \theta} + \nu \Delta V_\theta$$

$$+ \frac{\nu}{R^2 \sin^2\theta} \left( 2 \sin^2\theta \frac{\partial V_R}{\partial \theta} - 2 \cos\theta \frac{\partial V_\varphi}{\partial \varphi} - V_\theta \right) + F_\theta,$$

$$\mathrm{M}\, V_\varphi + \frac{V_R V_\varphi + V_\theta V_\varphi \cot\theta}{R} = -\frac{1}{\rho R \sin\theta} \frac{\partial P}{\partial \varphi} + \nu \Delta V_\varphi$$

$$+ \frac{\nu}{R^2 \sin^2\theta} \left( 2 \sin\theta \frac{\partial V_R}{\partial \varphi} + 2 \cos\theta \frac{\partial V_\theta}{\partial \varphi} - V_\varphi \right) + F_\varphi,$$

where the differential operators $\mathrm{M}$ and $\Delta$ are given by the formulas

$$\mathrm{M} \equiv \frac{\partial}{\partial t} + V_R \frac{\partial}{\partial R} + \frac{V_\theta}{R} \frac{\partial}{\partial \theta} + \frac{V_\varphi}{R \sin\theta} \frac{\partial}{\partial \varphi},$$

$$\Delta \equiv \frac{1}{R^2} \frac{\partial}{\partial R} \left( R^2 \frac{\partial}{\partial R} \right) + \frac{1}{R^2 \sin\theta} \frac{\partial}{\partial \theta} \left( \sin\theta \frac{\partial}{\partial \theta} \right) + \frac{1}{R^2 \sin^2\theta} \frac{\partial^2}{\partial \varphi^2}.$$

# S.6. Equations of Motion and Heat Transfer of Non-Newtonian Fluids

In this section we present the equations of motion and heat transfer for incompressible non-Newtonian fluids governed by the rheological equation of state (7.1.1) when the apparent viscosity $\mu = \mu(I_2, T)$ arbitrarily depends on the second invariant $I_2$ of the shear rate tensor and on the temperature $T$. This section contains some material from the books [47, 320, 443]. For the continuity equation in cylindrical and spherical coordinates, see Supplement 5.3.

S.6-1. Equations in Rectangular Cartesian Coordinates

The equations of motion are

$$\rho\left(\frac{\partial V_i}{\partial t} + V_j \frac{\partial V_i}{\partial X_j}\right) = -\frac{\partial P}{\partial X_i} + \frac{\partial}{\partial X_j}\left(\mu \frac{\partial V_i}{\partial X_j}\right) + \frac{\partial \mu}{\partial X_j}\frac{\partial V_j}{\partial X_i} + \rho F_i,$$

where $\rho$ is the fluid density, $i, j = 1, 2, 3$, and the summation is taken over the index $j$.

The heat transfer equation is

$$\rho c_p\left(\frac{\partial T}{\partial t} + V_j \frac{\partial T}{\partial X_j}\right) = \varkappa \frac{\partial^2 T}{\partial X_j^2} + 2\mu I_2,$$

where $\varkappa$ and $c_p$ are the thermal conductivity coefficient and specific heat of the fluid (these parameters are assumed to be constant) and the summation is taken over the index $j = 1, 2, 3$. The last summand on the right-hand side of the heat transfer equation takes into account the dissipative heating of fluid, and the invariant $I_2$ is calculated by formula (6.1.5).

S.6-2. Equations in Cylindrical Coordinates

The equations of motion are

$$\rho\left(\frac{\partial V_R}{\partial t} + V_R \frac{\partial V_R}{\partial R} + \frac{V_\varphi}{R}\frac{\partial V_R}{\partial \varphi} + V_Z \frac{\partial V_R}{\partial Z} - \frac{V_\varphi^2}{R}\right)$$
$$= \rho F_R + \frac{\partial \tau_{RR}}{\partial R} + \frac{1}{R}\frac{\partial \tau_{R\varphi}}{\partial \varphi} + \frac{\partial \tau_{RZ}}{\partial Z} + \frac{\tau_{RR} - \tau_{\varphi\varphi}}{R},$$

$$\rho\left(\frac{\partial V_\varphi}{\partial t} + V_R \frac{\partial V_\varphi}{\partial R} + \frac{V_\varphi}{R}\frac{\partial V_\varphi}{\partial \varphi} + V_Z \frac{\partial V_\varphi}{\partial Z} + \frac{V_R V_\varphi}{R}\right)$$
$$= \rho F_\varphi + \frac{\partial \tau_{R\varphi}}{\partial R} + \frac{1}{R}\frac{\partial \tau_{\varphi\varphi}}{\partial \varphi} + \frac{\partial \tau_{\varphi Z}}{\partial Z} + \frac{2\tau_{R\varphi}}{R},$$

$$\rho\left(\frac{\partial V_Z}{\partial t} + V_R \frac{\partial V_Z}{\partial R} + \frac{V_\varphi}{R}\frac{\partial V_Z}{\partial \varphi} + V_Z \frac{\partial V_Z}{\partial Z}\right)$$
$$= \rho F_Z + \frac{\partial \tau_{RZ}}{\partial R} + \frac{1}{R}\frac{\partial \tau_{\varphi Z}}{\partial \varphi} + \frac{\partial \tau_{ZZ}}{\partial Z} + \frac{\tau_{RZ}}{R},$$

where the stress tensor components are given by the formulas

$$\tau_{RR} = -P + 2\mu\frac{\partial V_R}{\partial R}, \qquad \tau_{\varphi\varphi} = -P + 2\mu\left(\frac{1}{R}\frac{\partial V_\varphi}{\partial \varphi} + \frac{V_R}{R}\right),$$

$$\tau_{ZZ} = -P + 2\mu\frac{\partial V_Z}{\partial Z}, \qquad \tau_{R\varphi} = \mu\left(\frac{1}{R}\frac{\partial V_R}{\partial \varphi} + \frac{\partial V_\varphi}{\partial R} - \frac{V_\varphi}{R}\right),$$

$$\tau_{\varphi Z} = \mu\left(\frac{\partial V_\varphi}{\partial Z} + \frac{1}{R}\frac{\partial V_Z}{\partial \varphi}\right), \qquad \tau_{RZ} = \mu\left(\frac{\partial V_Z}{\partial R} + \frac{\partial V_R}{\partial Z}\right).$$

The equation of heat transfer is

$$\rho c_p\left(\frac{\partial T}{\partial t} + V_R \frac{\partial T}{\partial R} + \frac{V_\varphi}{R}\frac{\partial T}{\partial \varphi} + V_Z \frac{\partial T}{\partial Z}\right)$$
$$= \varkappa\left[\frac{1}{R}\frac{\partial}{\partial R}\left(R\frac{\partial T}{\partial R}\right) + \frac{1}{R^2}\frac{\partial^2 T}{\partial \varphi^2} + \frac{\partial^2 T}{\partial Z^2}\right] + 2\mu I_2.$$

## S.6-3. Equations in Spherical Coordinates

The equation of motion is

$$
\rho\left(\frac{\partial V_R}{\partial t} + V_R\frac{\partial V_R}{\partial R} + \frac{V_\theta}{R}\frac{\partial V_R}{\partial \theta} + \frac{V_\varphi}{R\sin\theta}\frac{\partial V_R}{\partial \varphi} - \frac{V_\theta^2 + V_\varphi^2}{R}\right) = \rho F_R
$$
$$
+ \frac{1}{R^2}\frac{\partial}{\partial R}\left(R^2\tau_{RR}\right) + \frac{1}{R\sin\theta}\frac{\partial}{\partial \theta}\left(\sin\theta\,\tau_{R\theta}\right) + \frac{1}{R\sin\theta}\frac{\partial\tau_{R\varphi}}{\partial \varphi} - \frac{\tau_{\theta\theta} + \tau_{\varphi\varphi}}{R},
$$
$$
\rho\left(\frac{\partial V_\theta}{\partial t} + V_R\frac{\partial V_\theta}{\partial R} + \frac{V_\theta}{R}\frac{\partial V_\theta}{\partial \theta} + \frac{V_\varphi}{R\sin\theta}\frac{\partial V_\theta}{\partial \varphi} + \frac{V_R V_\theta - V_\varphi^2\cot\theta}{R}\right) = \rho F_\theta
$$
$$
+ \frac{1}{R^2}\frac{\partial}{\partial R}\left(R^2\tau_{R\theta}\right) + \frac{1}{R\sin\theta}\frac{\partial}{\partial \theta}\left(\sin\theta\,\tau_{\theta\theta}\right) + \frac{1}{R\sin\theta}\frac{\partial\tau_{\theta\varphi}}{\partial \varphi} + \frac{\tau_{R\theta} - \tau_{\varphi\varphi}\cot\theta}{R},
$$
$$
\rho\left(\frac{\partial V_\varphi}{\partial t} + V_R\frac{\partial V_\varphi}{\partial R} + \frac{V_\theta}{R}\frac{\partial V_\varphi}{\partial \theta} + \frac{V_\varphi}{R\sin\theta}\frac{\partial V_\varphi}{\partial \varphi} + \frac{V_R V_\varphi + V_\theta V_\varphi\cot\theta}{R}\right) = \rho F_\varphi
$$
$$
+ \frac{1}{R^2}\frac{\partial}{\partial R}\left(R^2\tau_{R\varphi}\right) + \frac{1}{R\sin\theta}\frac{\partial}{\partial \theta}\left(\sin\theta\,\tau_{\theta\varphi}\right) + \frac{1}{R\sin\theta}\frac{\partial\tau_{\varphi\varphi}}{\partial \varphi} + \frac{\tau_{R\varphi} + \tau_{\theta\varphi}\cot\theta}{R},
$$

where the stress tensor components are given by the formulas

$$
\tau_{RR} = -P + 2\mu\frac{\partial V_R}{\partial R}, \quad \tau_{\theta\theta} = -P + 2\mu\left(\frac{1}{R}\frac{\partial V_\theta}{\partial \theta} + \frac{V_R}{R}\right),
$$
$$
\tau_{\varphi\varphi} = -P + 2\mu\left(\frac{1}{R\sin\theta}\frac{\partial V_\varphi}{\partial \varphi} + \frac{V_R}{R} + \frac{V_\theta\cot\theta}{R}\right),
$$
$$
\tau_{R\theta} = \mu\left(\frac{1}{R}\frac{\partial V_R}{\partial \theta} + \frac{\partial V_\theta}{\partial R} - \frac{V_\theta}{R}\right), \quad \tau_{R\varphi} = \mu\left(\frac{1}{R\sin\theta}\frac{\partial V_R}{\partial \varphi} + \frac{\partial V_\varphi}{\partial R} - \frac{V_\varphi}{R}\right),
$$
$$
\tau_{\theta\varphi} = \mu\left(\frac{1}{R\sin\theta}\frac{\partial V_\theta}{\partial \varphi} + \frac{1}{R}\frac{\partial V_\varphi}{\partial \theta} - \frac{V_\varphi\cot\theta}{R}\right).
$$

The equation of heat transfer is

$$
\rho c_p\left(\frac{\partial T}{\partial t} + V_R\frac{\partial T}{\partial R} + \frac{V_\theta}{R}\frac{\partial T}{\partial \theta} + \frac{V_\varphi}{R\sin\theta}\frac{\partial T}{\partial \varphi}\right)
$$
$$
= \frac{\varkappa}{R^2}\left[\frac{\partial}{\partial R}\left(R^2\frac{\partial T}{\partial R}\right) + \frac{1}{\sin\theta}\frac{\partial}{\partial \theta}\left(\sin\theta\frac{\partial T}{\partial \theta}\right) + \frac{1}{\sin^2\theta}\frac{\partial^2 T}{\partial \varphi^2}\right] + 2\mu I_2.
$$

The second invariant of the shear rate tensor in the spherical coordinates has the form

$$
I_2 = \left(\frac{\partial V_R}{\partial R}\right)^2 + \left(\frac{1}{R}\frac{\partial V_\theta}{\partial \theta} + \frac{V_R}{R}\right)^2 + \left(\frac{1}{R\sin\theta}\frac{\partial V_\varphi}{\partial \varphi} + \frac{V_R}{R} + \frac{V_\theta\cot\theta}{R}\right)^2
$$
$$
+ \frac{1}{2}\left(\frac{1}{R}\frac{\partial V_R}{\partial \theta} - \frac{V_\theta}{R}\right)^2 + \frac{1}{2}\left(\frac{1}{R\sin\theta}\frac{\partial V_R}{\partial \varphi} + \frac{\partial V_\varphi}{\partial R} - \frac{V_\varphi}{R}\right)^2
$$
$$
+ \frac{1}{2}\left(\frac{1}{R\sin\theta}\frac{\partial V_\theta}{\partial \varphi} + \frac{1}{R}\frac{\partial V_\varphi}{\partial \theta} - \frac{V_\varphi\cot\theta}{R}\right)^2.
$$

For power-law fluids (7.1.4), the dissipative term in heat transfer equations can be calculated by the formula $2\mu I_2 = k(2I_2)^{\frac{n+1}{2}}$.

# References

1. **Abramson, V. I. and Fishbein, G. A.**, *Some problems of convective diffusion toward a spherical particle at* Re $\geq$ 1000, J. Eng. Phys. Thermophys., Vol. 32, No. 6, 1977.
2. **Abramson, V. I., Rivkind, V. Ya., and Fishbein, G. A.**, *Nonstationary mass exchange with a heterogeneous chemical reaction in laminar flow past a sphere*, J. Eng. Phys. Thermophys., Vol. 30, No. 1, 1976.
3. **Abramovich, G. N., Girshovich, T. A., Krashennikov, S. Yu., Sekundov, A. N., and Smirnova, I. P.**, *Theory of Turbulent Jets*, Nauka, Moscow, 1984 [in Russian].
4. **Abramowitz, M. and Stegun, I. A.** (editors), *Handbook of Mathematical Functions*, National Bureau of Standards, 1964.
5. **Acrivos, A.**, *A note of the rate of heat or mass transfer from a small sphere freely suspended in linear shear field*, J. Fluid Mech., Vol. 98, No. 2, pp. 299–304, 1980.
6. **Acrivos, A. and Goddard, J. D.**, *Asymptotic expansions for laminar forced-convection heat and mass transfer. Part 1. Low speed flows*, J. Fluid Mech., Vol. 23, No. 2, pp. 273–291, 1965.
7. **Acrivos, A. and Taylor, T. D.**, *Heat and mass transfer from single sphere in Stokes flow*, Phys. Fluids, Vol. 5, No. 4, pp. 387–394, 1962.
8. **Acrivos, A., Shah, M. J., and Petersen, E. E.**, *Momentum and heat transfer in laminar boundary-layer flows of non-Newtonian fluids past external surfaces*, AIChE J., Vol. 6, No. 2, pp. 312–317, 1960.
9. **Adamson, A. W. and Gast, A.P.**, *Physical Chemistry of Surfaces*, Wiley, New York, 1997.
10. **Akselrud, G. A. and Molchanov, A. D.**, *Dissolution of Solid Substances*, Khimiya, Moscow, 1977 [in Russian].
11. **Alekseenko, S. V., Nakoryakov, V. E., and Pokusaev, B. G.**, *Wave Flow of Liquid Films*, Begell House Inc., New York, 1994.
12. **Anderson, J. L.**, *Droplet interactions in thermocapillary motion*, Int. J. Mult. Flow, Vol. 11, No. 6, pp. 813–824, 1985.
13. **Anderson, J. L.**, *Prediction of the concentration dependence of macromolecular diffusion coefficients*, Ind. Eng. Chem. Fundam., Vol. 12, No. 4, pp. 488–490, 1973.
14. **Antanovskii, L. K. and Kopbosynov, B. K.**, *Nonstationary thermocapillary drift of a viscous fluid drop*, J. Appl. Mech. Techn. Phys., No. 2, 1986.
15. **Arbuzov, K. N. and Grebenshchikov, B. N.**, *On a study of the foam stability. I. Kinetics of foam syneresis*, Rus. J. Phys. Chem., Vol. 10, No. 1, 1937.
16. **Ariel, P. D.**, *The flow near a rotating disk: an approximate solution*, Trans. ASME, J. Appl. Mech., Vol. 63, No. 2, pp. 436–438, 1996.
17. **Ascoli, E. P., Dandy, D. S., and Leal, L. G.**, *Buoyancy-driven motion of a deformable drop toward a plane wall at low Reynolds number*, J. Fluid Mech., Vol. 213, No. 2, pp. 287–311, 1990.
18. **Astarita, G.**, *Mass Transfer With Chemical Reaction*, Elsevier Publ. Comp., Amsterdam, 1967.
19. **Astarita, G. and Marrucci, G.**, *Principles of Non-Newtonian Fluid Mechanics*, McGraw-Hill, London, 1974.

20. **Astavin, V. S., Korolev, I. O., and Ryazantsev, Yu. S.,** *Heat flow in a channel with discontinuity of the temperature on the wall,* Fluid Dynamics, Vol. 14, No. 5, pp. 802–805, 1979.

21. **Aveyard, R. and Haydon, D. A.,** *An Introduction to the Principles of Surface Chemistry,* Pergamon Press, Cambrige, 1973.

22. **Baeva, M., Baev, P., and Kaplan, A.,** *An analysis of the heat transfer from a moving elliptical cylinder,* J. Phys. D, Vol. 30, No. 8, pp. 1190–1196, 1997.

23. **Barenblatt, G. I.,** *Dimensional Analysis,* Gordon and Breach Publ., New York, 1989.

24. **Barenblatt, G. I. and Chernyi, G. G.,** *On moment relations on surface of discontinuity in dissipative media,* J. Appl. Math. Mech. (PMM), Vol. 27, No. 5, pp. 1205–1218, 1963.

25. **Barnes, H. A., Hutton, J. F., and Walters, K.,** *An Introduction to Rheology,* Elsevier Publ. Comp., Amsterdam, 1989.

26. **Batchelor, G. K.,** *An Introduction to Fluid Dynamics,* Cambridge Univ. Press, Cambridge, 1967.

27. **Batchelor, G. K.,** *Mass transfer from a particle suspended in fluid with a steady linear ambient velocity distribution,* J. Fluid Mech., Vol. 95, No. 2, pp. 369–400, 1979.

28. **Bateman, H. and Erdélyi, A.,** *Higher Transcendental Functions, Vol. 1,* McGraw-Hill Book Corp., New York, 1953.

29. **Bateman, H. and Erdélyi, A.,** *Higher Transcendental Functions, Vol. 2,* McGraw-Hill Book Corp., New York, 1953.

30. **Bateman, H. and Erdélyi, A.,** *Higher Transcendental Functions, Vol. 3,* McGraw-Hill Book Corp., New York, 1955.

31. **Bauer, H. F.,** *Diffusion, convection and chemical reaction in a channel,* Int. J. Heat Mass Transfer, Vol. 19, No. 5, pp. 479–486, 1976.

32. **Bazilevsky, A., Kornev, K., and Rozhkov, A.,** *Convective phenomena in foam motion through channels of variable cross section and porous media,* Proc. ASME Symp. Rheology and Fluid Mechanics of Nonlinear Materials, pp. 123–128, Marcel Dekker Inc., New York, 1996.

33. **Beavers, G. S. and Joseph, D. D.,** *Boundary conditions at a naturally permeable wall,* J. Fluid Mech., Vol. 30, No. 1, pp. 197–207, 1967.

34. **Beavers, G. S., Sparrow, E. M., and Magnuson, R. A.,** *Experiments on coupled parallel flows in a channel and boundary porous medium,* Trans. ASME, J. Basic Eng., Vol. 92, No. 4, pp. 843–848, 1970.

35. **Bejan, A.,** *Heat Transfer,* Wiley, Chichester, 1995.

36. **Berdichevski, V., Fridlyand, A., and Sutyrin V.,** *Prediction of turbulent velocity profile in Couette and Poiseuille flows from first principles,* Phys. Rev. Letters, Vol. 76, No. 21, pp. 3967–3970, 1996.

37. **Bhavaraju, S. M., Mashelkar, R. A., and Blanch, H. W.,** *Bubble motion and mass transfer in non-Newtonian fluids,* AIChE J., Vol. 24, No. 6, pp. 1063–1076, 1978.

38. **Bikerman, J. J.,** *Foams,* Springer-Verlag, New York, 1973.

39. **Bird, R. B.,** *Unsteady pseudoplastic flow near a moving wall,* AIChE J., Vol. 5, No. 4, pp. 565–625, 1959.

40. **Bird, R. B. and Curtiss, C. F.,** *Tangential Newtonian flow in annuli. 1. Unsteady state velocity profiles,* Chem. Eng. Sci., Vol. 11, No. 2, pp. 108–113, 1959.

41. **Bird, R. B., Stewart, W. E., and Lightfoot, E. N.,** *Transport Phenomena,* Wiley, New York, 1965.

42. **Birikh, R. V.,** *On temperature convection in a horizontal fluid layer,* J. Appl. Mech. Techn. Phys., No. 3, pp. 67–72, 1966.

43. **Blasius, H.**, *Crenzschichten in Flussigkeiten mit Kleiner Reibung*, Zeitschr. für Math. und Phys., Bd. 56, Ht. 1, S. 1–37, 1908.

44. **Bobkov, N. N. and Gupalo, Yu. P.**, *The flow pattern in a liquid layer and the spectrum of the boundary-value problem when the surface tension depends nonlinearly on the temperature*, J. Appl. Math. Mech. (PMM), Vol. 60, No. 6, pp. 999–1005, 1996.

45. **de Boer, J. H.**, *The Dynamical Character of Adsorption*, Clarendon Press, Oxford, 1953.

46. **Bogatykh, I. S.**, *On obtaining the drag coefficient of fluid particles or solid phases dispersed in a gaseous flow*, Rus. J. Appl. Chem., Vol. 60, No. 12, 1987.

47. **Böhme, G.**, *Non-Newtonian Fluid Mechanics*, Elsevier Science Publ., Amsterdam, 1987.

48. **Borishanskii, V. M., Kutateladze, S. S., Novikov, I. I., et al.**, *Liquid Metallic Heat Transfer Media*, Atomizdat, Moscow, 1976 [in Russian].

49. **Borzykh, A. A. and Cherepanov, G. P.**, *The plane problem of the theory of convective heat transfer and mass exchange*, J. Appl. Math. Mech. (PMM), Vol. 42, No. 5, 1978.

50. **Bostandzhiyan, S. A. and Chernyaeva, S. M.**, *On hydrodynamical heat "explosion" of a non-Newtonian fluid*, Doklady AN SSSR, Vol. 170, No. 2, pp. 301–304, 1966 [in Russian].

51. **Bostandzhiyan, S. A. and Chernyaeva, S. M.**, *Some problems of nonisothermal steady flow of non-Newtonian fluid*, Fluid Dynamics, Vol. 1, No. 3, pp. 55–57, 1966.

52. **Bostandzhiyan, S. A., Merzhanov, A. G., and Khudyaev, S. I.**, *On hydrodynamical thermal "explosion,"* Doklady AN SSSR, Vol. 163, No. 1, pp. 133–136, 1965 [in Russian].

53. **Boussinesq, M. I.**, *Calcul du pouvoir refroidissant des courants fluids*, J. de Math. Pures et Appliques, Bd. 1, Ht. 6, S. 285–332, 1905.

54. **Boyadjiev, Ch. and Beschkov, V.**, *Mass Transfer in Liquid Film Flows*, Publ. House Bulgar. Acad. Sci., Sofia, 1984.

55. **Bozzi, L. A., Feng, J. Q., Scott, T. C., and Pearlstein A. J.**, *Steady axisymmetric motion of deformable drops falling or rising through a homoviscous fluid in a tube at intermediate Reynolds number*, J. Fluid Mech., Vol. 336, pp. 1–32, 1997.

56. **Bradshaw, P. (editor)**, *Turbulence*, Springer-Verlag, Berlin, 1978.

57. **Bradshaw, P., Launder, B., and Limley, J.**, *Collaborative testing of turbulence models*, AIAA 91-0215, 1991.

58. **Bratukhin, Yu. K.**, *Flow of an inhomogeneously heated fluid past a gaseous bubble at small Marangoni numbers*, J. Eng. Phys. Thermophys., Vol. 32, No. 2, 1977.

59. **Bratukhin, Yu. K.**, *Thermocapillary drift of a droplet of viscous fluid*, Fluid Dynamics, Vol. 10, No. 5, pp. 833–837, 1975.

60. **Brauer, H. and Schmidt-Traub, H.**, *Kopplung von Stofftransport und chemischer Reaction und Platten und Kugeln sowie in Poren*, Chemic Ingenieur Technik, Bd. 45, Ht. 5, S. 341–344, 1973.

61. **Brenner, H.**, *Effect of finite boundaries on the Stokes resistance on arbitrary particle*, J. Fluid Mech., Vol. 12, No. 1, pp. 35–48, 1962.

62. **Brenner, H.**, *Forced convection-heat and mass transfer at small Peclet numbers from particle of arbitrary shape*, Chem. Eng. Sci., Vol. 18, No. 2, pp. 109–122, 1963.

63. **Brenner, H.**, *On the invariance of the heat transfer coefficient to flow reversal in Stokes and potential streaming flows past particles of arbitrary shape*, J. Math. Phys. Sci., Vol. 1, p. 173, 1967.

64. **Bretshnaider, S.,** *Properties of Fluids.* *(Calculational Methods for Engineers),* Khimiya, Leningrad, 1966 [in Russian].

65. **Bridgman, P. W.,** *Dimensional Analysis,* Yale Univ. Press, New-Haven, 1931.

66. **Brignell, A. S.,** *Solute extraction from an internally circulating spherical liquid drop,* Int. J. Heat Mass Transfer, Vol. 18, No. 1, pp. 61–68, 1975.

67. **Brounshtein, B. I. and Rivkind, V. Ya.,** *Mass and heat transfer with closed stream lines for large Peclet numbers,* Physics Doklady, Vol. 26, No. 10, 1981.

68. **Brounshtein, B. I. and Fishbein, G. A.,** *Hydrodynamics, Mass and Heat Exchange in Disperse Systems,* Khimiya, Leningrad, 1977 [in Russian].

69. **Brounshtein, B. I. and Shchegolev, V. V.,** *Hydrodynamics, Mass and Heat Exchange in Column Devices,* Khimiya, Leningrad, 1988 [in Russian].

70. **Brown, G. M.,** *Heat or mass transfer in a fluid in laminar flow in circular or flat conduit,* AIChE J., Vol. 6, No. 2, pp. 179–183, 1960.

71. **Brunn, P. O.,** *Absorption by bacterial cell: Interaction between receptor sites and the effect of fluid motion,* Trans. ASME, J. Biomechan. Eng., Vol. 103, No. 1, pp. 32–37, 1981.

72. **Bubnov, M. M., Dianov, E. M., Kazenin, D. A., et al.,** *On the problem of metallic protective facing of the optical fiber,* Doklady Chem. Techn., Vol. 337, No. 5, 1994.

73. **Butkovskii, A. G.,** *Characteristics of Systems With Distributed Parameters,* Nauka, Moscow, 1979 [in Russian].

74. **Buyevich, Yu. A.,** *On convective diffusion toward particles of a condensed polydisperse cloud of solid spheres,* J. Eng. Phys. Thermophys., Vol. 23, No. 4, 1972.

75. **Buyevich, Yu. A. and Kazenin, D. A.,** *Limit problems of heat and mass transfer toward a cylinder and a sphere submerged into an infiltrating granular layer,* J. Appl. Mech. Techn. Phys., No. 5, 1977.

76. **Buyevich, Yu. A. and Korneev, Yu. A.,** *On mass and heat exchange between phases in a concentrated disperse system,* J. Eng. Phys. Thermophys., Vol. 25, No. 4, 1973.

77. **Buyevich, Yu. A. and Shchelchkova, I. N.,** *Flow of dense suspensions,* Progr. Aerospace Sci., Vol. 18, No. 2-A, pp. 121–150, 1978.

78. **Buevich, Yu. A. and Shchelchkova, I. N.,** *Rheological properties of homogeneous finely divided suspensions. Steady-state flows,* J. Eng. Phys. Thermophys., Vol. 33, No. 5, 1977.

79. **Carslow, H. S. and Jaeger, J. C.,** *Conduction of Heat in Solids,* Pergamon Press, New York, 1959.

80. **Cebeci, T. and Bradshaw, P.,** *Physical and Computional Aspects of Convective Heat Transfer,* Springer-Verlag, New York, 1984.

81. **Chambré, P. L. and Acrivos, A.,** *On chemical surface reactions in laminar boundary layer flows,* J. Appl. Phys., Vol. 27, No. 11, pp. 1322–1328, 1956.

82. **Chandrasekhar, S.,** *Stochastic problems in physics and astronomy,* Rev. Modern Phys., Vol. 15, No. 1, pp. 1–89, 1943.

83. **Chanson, H.,** *Air Bubble Entainment in Free-Surface Turbulent Shear Flows,* Acad. Press, New York, 1996.

84. **Chao, B. T.,** *Transient heat and mass transfer to translating droplet,* Trans. ASME, J. Heat Transfer, Vol. 91, No. 2, pp. 273–291, 1969.

85. **Chen, J., Stebe, K. J.,** *Surfactant-induced retardation of the thermocapillary migration of a droplet,* J. Fluid Mech., Vol. 340, pp. 35–60, 1997.

86. **Chen, X.-J., Chen, T.-K., and Zhou, F.-De. (editors),** *Multiphase Flow and Heat Transfer,* Begell House Inc., New York, 1994.

87. **Cheng, H. and Papanicolaou, G.,** *Flow past periodic arrays of spheres at low Reynolds number,* J. Fluid Mech., Vol. 335, pp. 189–212, 1997.

88. **Chernyakov, A. V. and Kazenin, D. A.,** *Thermodynamic criterion for nonbreaking flow-down of a film,* Rus. Chem. Industry, No. 8, 1997.

89. **Chervenivanova, E. and Zapryanov, Z.,** *On the deformation of compound multiphase drops at low Reynolds Numbers,* Physicochemical Hydrodynamics, Vol. 11, pp. 243–259, 1989.

90. **Chervenivanova, E. and Zapryanov, Z.,** *On the deformation of two droplets in a quasisteady Stokes flow,* Int. J. Mult. Flow, Vol. 11, No. 5, pp. 721–738, 1985.

91. **Chervenivanova, E. and Zapryanov, Z.,** *The slow motion of droplets perpendicular to a deformable flat fluid interface,* Quart. J. Mech. Appl. Math, Vol. 41, pp. 419–444, 1988.

92. **Chhabra, R. P.,** *Bubbles, Drops, and Particles in Non-Newtonian Fluids,* CRC Press, Boca Raton, 1993.

93. **Chwang, A. T. and Wu, T. Y.,** *Hydrodynamics of low Reynolds number flow. Part 2. Singularity method for Stokes flows,* J. Fluid Mech., Vol. 67, No. 4, pp. 787–815, 1975.

94. **Clift, R., Grace, J. R., and Weber, M. E.,** *Bubbles, Drops and Particles,* Acad. Press, New York, 1978.

95. **Cochran, W. G.,** *The flow due to a rotating disk,* Proc. Cambr. Phil. Soc., Vol. 30, pp. 365–375, 1934.

96. **Cole, G. D.,** *Perturbation Methods in Applied Mathematics,* Blaisdell Publ. Comp., Waltham, 1968.

97. **Collins, R. E.,** *Flow of Fluids Through Porous Materials,* Reinhold Publ. Corp., New York, 1961.

98. **Comer, J. L. and Kleinstreuer, C.,** *A numerical investigation of laminar flow past nonspherical solids and droplets,* Trans. ASME, J. Fluid Eng., Vol. 117, No. 1, pp. 170–175, 1995.

99. **Cotta, R. M.,** *Integral Transforms in Computational Heat and Fluid Flow,* CRC Press, Boca Raton, 1993.

100. **Cox, R. G., Zia, I. Y. Z., and Mason, S. G.,** *Particle motion in sheared suspensions. XXV. Streamlines around cylinders and spheres,* J. Colloid Interface Sci., Vol. 27, No. 1, pp. 7–18, 1968.

101. **Coxeter, H. S. M.,** *Introduction to Geometry,* Wiley, New York, 1969.

102. **Cuenot, B., Magnaudet, J., and Spennate, B.,** *The effect of slightly soluble surfactant on the flow around spherical bubble,* J. Fluid Mech., Vol. 339, pp. 25–54, 1997.

103. **Danckwerts, P. V.,** *Gas-Liquid Reactions,* McGraw-Hill, New York, 1970.

104. **Danckwerts, P. V.,** *Absorption by simultaneous diffusion and chemical reaction into particles of various shapes and into falling drops,* Trans. Faradey Soc., Vol. 47, No. 2, pp. 1014–1023, 1951.

105. **Darcy, H.,** *Recherches experimentals relatives aux mouvement de leáu dans les tayaux,* Meḿ. preś. Academ. Sci., Inst. Imperial de France, 1858, t. 15.

106. **Dautov, R., Kornev, K., and Mourzenko, V.,** *Foam patterning in porous media,* Phys. Rev. E, Vol. 55, No. 6, pp. 6929–6944, 1997.

107. **Davidson, J. F., Clift, R., and Harrison, D. (editors),** *Fluidization,* Wiley, New York, 1985.

108. **Davis, E. J.,** *Exact solutions for a class of heat and mass transfer problems,* Can. J. Chem. Eng., Vol. 51, No. 5, pp. 562–572, 1973.

109. **Davis, H. T.,** *Mechanics of Phases, Interfaces, and Thin Films,* VCH, New York, 1996.

110. **Deavours, C. A.,** *An exact solution for the temperature distribution in parallel plate Poiseuille flow,* Trans. ASME, J. Heat Transfer, Vol. 96, No. 4, 1974.

111. **Decker, W. D.,** *Bubble Column Reactors,* Wiley, New York, 1992.

112. **De Kee, D. and Chhabra, R. P.,** *Transport Processes in Bubbles, Drops and Particles,* Hemisphere, New York, 1992.

113. **Dennis, S. C. R. and Walker, J. D. A.,** *Calculation of the steady flow past a sphere at low and moderate Reynolds number,* J. Fluid Mech., Vol. 48, No. 4, pp. 771–778, 1971.

114. **Dennis, S. C. R., Walker, J. D. A., and Hudson, J. D.,** *Heat transfer from a sphere at low Reynolds numbers,* J. Fluid Mech., Vol. 60, No. 2, pp. 273–283, 1973.

115. **Derjaguin, B. V.,** *Elastic properties of foams,* Rus. J. Phys. Chem., Vol. 2, No. 6, 1931.

116. **Derjaguin, B. V., Churaev, N. V., and Muller, V. M.,** *Surface Forces,* Consultants Bureau, New York, 1987.

117. **Devnin, S. I.,** *Aeromechanics of Constructions of Bad Streamlined Shapes: Reference Book,* Sudostroenie, Leningrad, 1983 [in Russian].

118. **De Vris, K.,** *Foam Stability,* Center, Amsterdam, 1957.

119. **Dukhin, S. S., Kretzschmar, G., and Miller, R.,** *Dynamics of Adsorption at Liquid Interfaces: Theory, Experiment, Application,* Elsevier Science Ltd., Oxford, 1995.

120. **Dullien, F. A. L.,** *Statistical test of Vigners correlation of liquid-phase diffusion coefficients,* Ind. Eng. Chem. Fundam., Vol. 10, No. 1, pp. 41–49, 1971.

121. **Eckert, E. R. G. and Drake, R. M.,** *Analysis of Heat and Mass Transfer,* Hemisphere Publ., New York, 1987.

122. **Elimelech, M., Gregory, J., Jia, X., and Willions, R. A.,** *Particle Deposition and Aggregation,* Butterworth Heinemann, Oxford, 1995.

123. **Emanuel, G.,** *Analytical Fluid Dynamics,* CRC Press, Boca Raton, 1993.

124. **Erneux, T. and Davis, S. H.,** *Nonlinear rupture of free films,* Phys. Fluids A, Vol. 5, pp. 1117–1122, 1993.

125. **Exerowa, D. and Kruglyakov, P. M.,** *Foam and Foam Films,* Elsevier Science Ltd., Oxford, 1997.

126. **Faber, T. E.,** *Fluid Dynamics for Physicists,* Cambridge Univ. Press, Cambridge, 1995.

127. **Falkner, V. M.,** *The resistance of a smooth flat plate with turbulent boundary layer,* Aircraft Eng., Vol. 15, pp. 65–69, 1943.

128. **Falkner, V. M. and Skan, S. W.,** *Some approximate solutions of the boundary layer equations,* Phil. Mag., Vol. 12, 1931.

129. **Feng, J., Huang, P. Y., and Joseph, D. D.,** *Dynamic simulation of sedimentation of solid particles in an Oldroyd-B fluids,* J. Non-Newtonian Fluid Mech., Vol. 63, pp. 63–88, 1996.

130. **Feng, J., Joseph, D. D., Glowinski, R., and Pan, T. W.,** *A tree-dimensional computation of the force and torque on an ellipsoid setting slowly through a viscoelastic fluid,* J. Fluid Mech., Vol. 283, pp. 1–16, 1995.

131. **Feng, Zhi-Gang and Michaelides, E. E.,** *Unsteady heat transfer from a sphere at small Peclet numbers,* Trans. ASME, J. Fluids Eng., Vol. 118, No. 1, pp. 96–102, 1996.

132. **Frankel, N. A. and Acrivos, A.,** *Heat and mass transfer from small spheres and cylinders freely suspended in shear flow,* Phys. Fluids, Vol. 11, No. 9, pp. 1913–1918, 1968.

133. **Frank-Kamenetskii, D. A.,** *Diffusion and Heat Transfer in Chemical Kinetics,* Nauka, Moscow, 1987 [in Russian].

134. **Friedlander, S. K.,** *Mass and heat transfer to single spheres and cylinders at low Reynolds numbers,* AIChE J., Vol. 3, No. 1, pp. 43–48, 1957.

135. **Frisch, Ur.,** *Turbulence (The Legacy of A. N. Kolmogorov)*, Cambridge Univ. Press, Cambridge, 1995.

136. **Froindorfer, B., Muller, H., Vetoshkin, A. G., et al.,** *Mathematical description of rheological models of foams*, Rus. J. Appl. Chem., Vol. 59, No. 12, 1986.

137. **Froment, G. F. and Bischoff, K. B.,** *Chemical Reactor Analysis and Design*, Wiley, New York, 1991.

138. **Frost, W. and Moulden, T. H.,** *Handbook of Turbulence*, Plenum Press, New York, 1977.

139. **Fuks, N. A.,** *Mechanics of Aerosols*, Izd. AN SSSR, Moscow, 1955 [in Russian].

140. **Eglit, E. E. and Hodges, D. H. (editors),** *Continuum Mechanics via Problems and Exercises, Vol. 1*, World Scientific, Singapore, 1996.

141. **Galperin, D. I., Moshev, V. V., and Stepanova, V. G.,** *Thermodynamical properties of stratified ethyl cellulose*, Colloid Journal, Vol. 23, No. 1, 1961.

142. **Gershuni, G. Z. and Zhukhovitskii, E. M.,** *Convective Stability of Incompressible Fluid*, Nauka, Moscow, 1972 [in Russian].

143. **Gershuni, G. Z., Zhukhovitskii, E. M., and Nepomnyashchii, A. A.,** *Stability of Convective Flows*, Nauka, Moscow, 1989 [in Russian].

144. **Golovin, A. A.,** *Influence of Maragoni Effects on Hydrodynamics and Mass Transfer in Fluid Extraction*, PhD Thesis in Engineering, Moscow, Karpov Institute of Physical Chemistry, 1989 [in Russian].

145. **Golovin, A. A.,** *Thermocapillary interaction between a solid particle and a gas bubble*, Int. J. Mult. Flow, Vol. 21, No. 4, pp. 715–719, 1995.

146. **Golovin, A. A., Nir, A., and Pismen, L. M.,** *Spontaneous motion of two droplets caused by mass transfer*, Ind. Eng. Chem. Res., Vol. 34, No. 10, pp. 3278–3288, 1995.

147. **Golovin, A. A. and Ryazantsev, Yu. S.,** *Drift of a reacting droplet due to the chemoconcentration capillary effect*, Fluid Dynamics, Vol. 25, No. 3, pp. 370–378, 1990.

148. **Golovin, A. A., Gupalo, Yu. P., and Ryazantsev, Yu. S.,** *On chemothermocapillary effect caused by the motion of a drop in liquid*, Physics Doklady, Vol. 31, No. 9, 1986.

149. **Golovin, A. A., Gupalo, Yu. P., and Ryazantsev, Yu. S.,** *Chemoconcentration capillary effect associated with the motion of drop in a liquid*, Fluid Dynamics, Vol. 23, No. 1, pp. 122–128, 1988.

150. **Golovin, A. M. and Zhivotyagin, A. F.,** *Influence of a volume chemical reaction on mass transfer inside a drop at large Peclet numbers*, Moscow Univ. Math. Mech. Bull., No. 4, 1979.

151. **Golovin, A. M. and Zhivotiagin, A. F.,** *Unsteady convective mass transfer inside a drop at high Peclet numbers*, J. Appl. Math. Mech. (PMM), Vol. 47, No. 5, pp. 625–632, 1983.

152. **Gol'dfarb, I. I., Kann K. B., and Shreiber, I. R.,** *Liquid flow in foams*, Fluid Dynamics, Vol. 23, No. 2, pp. 244–249, 1988.

153. **Goldshtik, M. A.,** *Processes of Transfer in Granular Layer*, Institute of Thermophysics, Novosibirsk, 1984 [in Russian].

154. **Gonor, A. L. and Rivkind, V. Ya.,** Dynamics of drops, Itogi Nauki i Tekhniki (Mekhanika Zhidkosti i Gaza), Vol. 17, 1982 [in Russian].

155. **Gotovtsev, V. M.,** *Viscoelastic model for the plug flow of a foam in a cylindrical channel*, Theor. Found. Chem. Eng., Vol. 30, No. 6, pp. 523–529, 1996.

156. **Gotovtsev, V. M.,** *Viscoelastoplastic flow of a foam in a cylindrical channel*, Theor. Found. Chem. Eng., Vol. 31, No. 4, pp. 306–311, 1997.

157. **Graetz, L.**, *Über die Warmeleitungsfähigkeit von Flüssigkeiten*, Annln. Phys., Bd. 18, S. 79–84, 1883.

158. **Grigull, U.**, *Wärmeübertragung in laminarer Strömung mit Reibungswärme*, Chemie-Ingenieur-Technik, pp. 480–483, 1955.

159. **Grigull, U. and Sandner, H.**, *Heat Conduction*, Hemisphere Publ., New York, 1984.

160. **Grossmann S.**, *Asymptotic dissipation rate in turbulence*, Phys. Rev. E, Vol. 51, pp. 6275–6277, 1995.

161. **Gukhman, A. A.**, *Introduction to the Theory of Similarity*, Acad. Press, London, 1965.

162. **Gukhman, A. A. and Zaitsev, A. A.**, *Generalized Analysis*, Faktorial, Moscow, 1998 [in Russian].

163. **Gulyaev, A. N., Kozlov, V. E., and Sekundov, A. N.**, *A universal one-parameter model for turbulent viscosity*, Fluid Dynamics, Vol. 28, No. 4, pp. 485–494, 1993.

164. **Gupalo, Yu. P. and Ryazantsev, Yu. S.**, *Diffusion on a particle in the shear flow of a viscous fluid. Approximation of the diffusion boundary layer*, J. Appl. Math. Mech. (PMM), Vol. 36, No. 3, pp. 447–451, 1972.

165. **Gupalo, Yu. P. and Ryazantsev, Yu. S.**, *Thermocapillary motion of a liquid with a free surface with nonlinear dependence of the surface tension on the temperature*, Fluid Dynamics, Vol. 23, No. 5, pp. 752–757, 1988.

166. **Gupalo, Yu. P., Polyanin, A. D., and Ryazantsev, Yu. S.**, *Mass and Heat Exchange Between Reacting Particles and Flow*, Nauka, Moscow, 1985 [in Russian].

167. **Gupalo, Yu. P., Polyanin, A. D., and Ryazantsev, Yu. S.**, *Mass transfer in a diffusional trail of a drop in Stokes flow*, J. Appl. Math. Mech. (PMM), Vol. 41, No. 2, pp. 298–302, 1977.

168. **Gupalo, Yu. P., Polyanin, A. D., and Ryazantsev, Yu. S.**, *Mass transfer interaction of moving particles in a reactive dispersive system*, Acta Astronautica, Vol. 5, pp. 1213–1219, 1978.

169. **Gupalo, Yu. P., Polyanin, A. D., and Ryazantsev, Yu. S.**, *Some general invariance relations in problems of convective heat and mass transfer at large Peclet numbers*, Fluid Dynamics, Vol. 16, No. 6, pp. 877–881, 1981.

170. **Gupalo, Yu. P., Polyanin, A. D., Priadkin, P. A., and Ryazantsev, Yu. S.**, *On the unsteady mass transfer on a drop in a viscous fluid stream*, J. Appl. Math. Mech. (PMM), Vol. 42, No. 3, pp. 462–471, 1978.

171. **Gupalo, Yu. P., Polyanin, A. D., Ryazantsev, Yu. S., and Sergeev, Yu. A.**, *Convective diffusion to a drop under arbitrary conditions of absorption. Diffusion boundary layer approximation*, Fluid Dynamics, Vol. 14, No. 6, pp. 862–866, 1979.

172. **Gupalo, Yu. P., Polyanin, A. D., Ryazantsev, Yu. S., and Sergeev, Yu. A.**, *On convective mass transfer in a system of periodically situated spheres*, J. Appl. Mech. Techn. Phys., No. 4, 1979.

173. **Gupalo, Yu. P., Rednikov, A. E., and Ryazantsev, Yu. S.**, *Thermocapillary drift of a drop in the case when the surface tension depends non-linearly on the temperature*, J. Appl. Math. Mech. (PMM), Vol. 53, No. 3, pp. 332–339, 1989.

174. **Gupalo, Yu. P., Ryazantsev, Yu. S., and Sergeev, Yu. A.**, *Diffusion flux to a distorted gas bubble at large Reynolds numbers*, Fluid Dynamics, Vol. 11, No. 4, pp. 548–553, 1976.

175. **Gupalo, Yu. P., Ryazantsev, Yu. S., and Ulin, V. I.**, *Diffusion in a particle in a homogeneous translational-shear flow*, J. Appl. Math. Mech. (PMM), Vol. 39, No. 3, pp. 472–479, 1975.

176. **Haase, R.**, *Thermodynamik der Irreversiblen Prozesse*, Dr. Dietrich Stainkopff Verlag, Darmstadt, 1963.

177. **Hadamard, J. S.**, *Mouvement permanent lent d'une sphere liquide et visqueuse dans un liquide visqueux*, Comp. Rend. Acad. Sci. Paris, Vol. 152, No. 25, pp. 1735–1739, 1911 and Vol. 154, No. 3, p. 109, 1912.

178. **Hamba, F.**, *Estimate of constants in the $K - \varepsilon$ model of turbulence by using large eddy simulation*, J. Phys. Soc. Japan, Vol. 56, No. 10, pp. 3405–3408, 1987.

179. **Happel, J. and Brenner, H.**, *Low Reynolds Number Hydrodynamics*, Prentice-Hall, Englewood Cliffs, 1965.

180. **Harper, J. F. and Moore, D. W.**, *The motion of a spherical liquid drop at high Reynolds number*, J. Fluid Mech., Vol. 32, No. 2, pp. 367–391, 1968.

181. **Harris, J.**, *Rheology and Non-Newtonian Flow*, Longman, London, 1977.

182. **Hartnett, J. P. and Hu, R. Y. Z.**, *The yield stress: an engineering reality*, J. Rheol., Vol. 33, p. 671, 1989.

183. **Hetsroni, G. (editor)**, *Handbook of Multiphase Systems*, Hemisphere Publ. Corp., Washington, 1982.

184. **Hewitt, G. F., Shires, G. L., and Bott, T. R.**, *Process Heat Transfer*, Begell House, New York, 1994.

185. **Hewitt, G. F. and Spalding, D. B.**, *Encyclopedia of Heat and Mass Transfer*, Hemisphere, New York, 1986.

186. **Hieber, C. A. and Gebhart, B.**, *Low Reynolds number heat transfer from a circular cylinder*, J. Fluid Mech., Vol. 32, No. 1, pp. 21–28, 1968.

187. **Hill, R. and Power, G.**, *Extremum principles for slow viscous flow and approximate calculation of drag*, Quarterly J. Mech. Appl. Math., Vol. 9, No. 3, pp. 313–319, 1956.

188. **Hinze, J. O.**, *Turbulence*, McGraw-Hill, New York, 1975.

189. **Hirasaki, G. J. and Lawson, J.**, *Mechanism of foam flow in porous media: apparent viscosity in smooth capillaries*, Soc. Petr. Eng. J., Vol. 25, pp. 176–190, 1985.

190. **Hirose, T. and Moo-Young, M.**, *Bubble drag and mass transfer in non-Newtonian fluids: creeping flow with power law fluids*, Can. J. Chem. Eng., Vol. 47, No. 3, pp. 265–267, 1969.

191. **Hobler, T.**, *Minimun Zraszania Powierznchi*, Chimia Stosowana, Bd. 2B, S. 145–159, 1964.

192. **Holms, P., Lumley, J. L., and Berkooz, G.**, *Turbulence, Coherent Structures, Dynamical Systems and Symmetry*, Cambridge Univ. Press, Cambridge, 1996.

193. **Huang, P. Y., Feng, J., Hu H. H., and Joseph, D. D.**, *Direct simulation of the motion of solid particles in Couette and Poiseuille flows of viscoelastic fluids*, J. Fluid Mech., Vol. 343, pp. 73–96, 1997.

194. **Huang, P. Y., Hu, H. H., and Joseph D. D.**, *Direct simulation of the sedimentation of elliptic particles in Oldroyd-B fluids*, J. Fluid Mech., Vol. 362, pp. 297–325, 1998.

195. **Hulbary, R. L.**, *Three-dimensional cells shape in the tuberous roots of Asparagus and in the leaf Rhoeo*, Amer. J. Botany, Vol. 35, No. 5, pp. 558–566, 1948.

196. **Ibragimov, M. Kh., Subbotin, V. I., Bobkov, V. P., Sabelev, G. I., and Taranov, G. S.**, *Structure of Turbulent Stream and Mechanisms of Heat Transfer in Channels*, Atomizdat, Moscow, 1978 [in Russian].

197. **Ibragimov, N. H. (editor)**, *CRC Handbook of Lie Group to Differential Equations, Vol. 1*, Boca Raton, CRC Press, 1994.

198. **Idelchik, I. E.**, *Handbook of Hydraulic Resistance*, Begell House, New York, 1994.

199. **Incropera, F. P. and Dewitt, D. P.**, *Fundamentals of Heat and Mass Transfer*, Wiley, New York, 1996.

200. **Isachenko, V. P., Osipove, V. A., and Sukomel, A. S.**, *Heat Transfer*, Energoizdat, Moscow, 1981 [in Russian].

201. **Izmailova, V. N., Yampolskaya, G. P., and Summ, B. D.**, *Surface Phenomena in Protein Systems*, Khimiya, Moscow, 1988 [in Russian].

202. **Janke, E., Emde, F., and Lösch, F.**, *Tafeln Höherer Funktionen*, Teubner Verlogsgesellschaft, Stuttgart, 1960.

203. **Johnk, R. E. and Hanratty, T. J.**, *Temperature profiles for turbulent flow of air in a pipe. The fully developed heat-transfer region*, Chem. Eng. Sci., Vol. 17, No. 11, pp. 867–879, 1962.

204. **Jones, A. S.**, *Extensions to the solution of the Graets problem*, Int. J. Heat Mass Transfer, Vol. 14, No. 4, pp. 619–623, 1971.

205. **Jones, O. C. and Michivoshi, I.** (editors), *Dynamics of Two-Phase Flows*, Begell House Inc., New York, 1992.

206. **Joseph, D. D.**, *Fluid Dynamics of Viscoelastic Liquids*, Springer-Verlag, 1990.

207. **Joseph, D. D. and Feng J.**, *A note on the forces that move particles in a second-order fluid*, J. Non-Newtonian Fluid Mech., Vol. 64, pp. 299–302, 1996.

208. **Kaganov, S. A.**, *On steady-state laminar flow of incompressible fluid in a plane channel and in a circular cylindrical tube with regard to heat friction and dependence of viscosity on temperature*, J. Appl. Mech. Techn. Phys., No. 3, 1962.

209. **Kader, B. A.**, *Temperature and concentration profiles in fully turbulent boundary layers*, Int. J. Heat Mass Transfer, V. 24, No. 9, pp. 1541–1544, 1981.

210. **Kader, B. A. and Yaglom, A. M.**, *Heat and mass transfer laws for fully turbulent wall flows*, Int. J. Heat Mass Transfer, Vol. 15, No. 12, pp. 2329–2351, 1972.

211. **Kader, B. A. and Yaglom, A. M.**, *Influence of roughness and longitudinal pressure gradient on turbulent boundary layers*, Itogi Nauki i Tekhniki (Mekhanika Zhidkosti i Gaza), Vol. 18, pp. 3–111, 1984 [in Russian].

212. **Kader, B. A. and Yaglom, A.M.**, *Similarity laws for turbulent flows*, Itogi Nauki i Techniki (Mekhanika Zhidkosti i Gasa), Vol. 15, pp. 81–155, 1980 [in Russian].

213. **Kajiyama, T., Aizawa, M.**, *New Developments in Construction and Functions of Organic Thin Films*, Elsevier Science Ltd., Oxford, 1996.

214. **Kann, K. B.**, *Capillary Hydrodynamics of Foams*, Nauka, Novosibirsk, 1989 [in Russian].

215. **Kann, K. B.**, *Some laws of foam syneresis. Outflow*, Colloid Journal, Vol. 40, No. 5, 1978.

216. **Kaplun, S. and Lagerstrom, P. A.**, *Asymptotic expansions of Navier–Stokes solutions for small Reynolds numbers*, J. Math. Mech., Vol. 6, pp. 585–593, 1957.

217. **Kaŕmań, Th., von**, *Mechanische Ähnlichkeit und Turbulenz*, Verhandlg. d. III Intern. Kongress fuŕ Techn. Mechanik, Stockholm, Bd. 1, S. 85, 1930.

218. **Kassoy, D. R.**, *Heat transfer from circular cylinders at low Reynolds number*, Phys. Fluids, Vol. 10, No. 5, pp. 938–946, 1967.

219. **Kazenin, D. A. and Makeyev, A. A.**, *On the determination of depth filter colloidal particle size separation properties*, Proc. 5-th World Cong. Chem. Eng., Vol. 5, pp. 534–540, 1996.

220. **Keil, F., Mackens, W., Voß, H., and Werther, J.**, *Scientific Computation in Chemical Engineering*, Springer-Verlag, 1996.

221. **Keller, L. V. and Fridman, A. A.**, *Differentialgleichuny für die turbulente bewegung einer kompressiblen flussighit*, Proc. 1st Intern. Congr. Appl. Mech., pp. 395–405, Delft, 1924.

222. **Kendoush, A. A.**, *Theory of convective heat and mass transfer to spherical-cap bubbles*, AIChE J., Vol. 40, No. 9, pp. 1440–1448, 1994.

223. **Kestin, J. and Richardson, P. D.**, *Heat transfer across turbulent incompressible boundary layers*, Int. J. Heat Mass Transfer, Vol. 6, No. 2, pp. 147–189, 1963.

224. **Kevorkian, J. and Cole, J. D.**, *Perturbation Methods in Applied Mathematics*, Springer-Verlag, 1981.

225. **Khan, S. A.**, *Foam rheology: Relation between extensional and shear deformations in high gas fraction foams*, Rheological Acta, Vol. 26, No. 1, pp. 78–84, 1987.

226. **Kholpanov, L. P. and Shkadov, V. Ya.**, *Hydrodynamics and Heat Transfer with the Interface*, Nauka, Moscow, 1990 [in Russian].

227. **Kholpanov, L. P., Zaporozhets, V. P., Zibert, G. K., and Kashchitskii, Ya. A.**, *Mathematical Modeling of Nonlinear Thermohydrogasodynamical Processes*, Nauka, Moscow, 1998 [in Russian].

228. **Kim, J. and Hussain, F.**, *Propagation velocity of perturbations in turbulent channel flow*, Phys. Fluids A, Vol. 5, No. 3, pp. 695–706, 1993.

229. **Kim, J., Moin, P., and Moser, R.**, *Turbulence statistics in fully developed channel flow at low Reynolds-number*, Fluid Mech., Vol. 177, pp. 133–166, 1987.

230. **Kolmogorov, A. N.**, *A refinement of previous hypotheses concerning the local structure of turbulence in a viscous incompressible fluid at high Reynolds number*, J. Fluid Mech., Vol. 13, No. 1, pp. 82–85, 1962.

231. **Korn, G. A. and Korn, T. M.**, *Mathematical Handbook for Scientists and Engineers*, McDraw-Hill Book Comp., New York, 1961.

232. **Kochurova, N. N. and Rusanov, A. I.**, *On nonequilibrium thermodynamics of the dynamic surface tension*, Colloid Journal, Vol. 46, No. 1, 1984.

233. **Kornev, K. and Shugai, G. A.**, *Thermodynamic and hydrodynamic peculiarities of a lamella confined in a cylindrical pore*, Phys. Rev. E, Vol. 58, No. 6, 1998.

234. **Koshlyakov, N. S., Gliner, E. B., and Smirnov, M. M.**, *Partial Differential Equations of Mathematical Physics*, Vysshaya Shkola, Moscow, 1970 [in Russian].

235. **Kovatcheva, N. P., Polyanin, A. D., and Kurdyumov, V. N.**, *Mass transfer from a particle in a shear flow with surface reactions*, Acta Mech. (Springer-Verlag), Vol. 101, pp. 155–160, 1993.

236. **Kovatcheva, N. T., Polyanin, A. D., and Zapryanov, Z. D.**, *The change of the diffusivity with the change of the concentration of the solvent in a solution*, Acta Mech. (Springer-Verlag), Vol. 80, pp. 259–272, 1989.

237. **Knyazeva, E. N. and Kurdyumov, S. P.**, *Laws of Evolution and Self-organization for Complex Systems*, Nauka, Moscow, 1994 [in Russian].

238. **Kronig, R. and Brink, J. C.**, *On the theory of extraction from falling droplets*, Appl. Sci. Res., Vol. A2, No. 2, pp. 142–154, 1950.

239. **Krotov, V. V.**, *Generalized syneresis equation*, Colloid Journal, Vol. 46, No. 1, 1984.

240. **Krotov, V. V.**, *Rheological analysis of the Marangoni effect for an ideal interphase layer*, Colloid Journal, Vol. 48, No. 1, 1986.

241. **Krotov, V. V.**, *The hydrodynamic stability of polyhedral disperse systems and their kinetics under the conditions of spontaneous breakdown. 1. Aspects of the hydrodynamic stability*, Colloid Journal, Vol. 48, No. 4, 1986.

242. **Krotov, V. V.**, *The hydrodynamic stability of polyhedral disperse systems and their kinetics under the conditions of spontaneous breakdown. 3. Kinetics associated with the instability of films*, Colloid Journal, Vol. 48, No. 6, 1986.

243. **Krotov, V. V.**, *The structure, syneresis, and kinetics of destruction of polyhedral disperse systems*. In: Problems of Thermodynamics of Heterogeneous Systems and the Theory of Surface Phenomena. Vol. 6, pp. 110–191; Izd. Leningrad. Univ., Leningrad, 1982 [in Russian].

244. **Krotov, V. V.**, *The theory of syneresis of foams and concentrated emulsions. 1. Local multiplicity of polyhedral disperse systems*, Colloid Journal, Vol. 42, No. 6, 1980.

245. **Krotov, V. V.,** *The theory of syneresis of foams and concentrated emulsions. 2. Local hydroconduction of concentrated disperse systems,* Colloid Journal, Vol. 42, No. 6, 1980.

246. **Krotov, V. V.,** *The theory of syneresis of foams and concentrated emulsions. 3. Local syneresis equation and setting of boundary conditions,* Colloid Journal, Vol. 43, No. 1, 1981.

247. **Krotov, V. V.,** *The theory of syneresis of foams and concentrated emulsions. 4. Some analytical solutions of the one-dimensional syneresis equation,* Colloid Journal, Vol. 43, No. 2, 1981.

248. **Krotov, V. V. and Rusanov, A. I.,** *Gibbs elasticity and stability of fluid objects.* In: Problems of the Thermodynamics of Heterogeneous Systems and of the Theory of Surface Phenomena. Vol. 1, pp. 157–198; Izd. Leningrad. Univ., Leningrad, 1971 [in Russian].

249. **Krotov, V. V. and Rusanov, A. I.,** *Quasistatic processes in liquid films.* In: Problems of the Thermodynamics of Heterogeneous Systems and of the Theory of Surface Phenomena. Vol. 2, pp. 147–178; Izd. Leningrad. Univ., Leningrad, 1973 [in Russian].

250. **Krotov, V. V. and Rusanov, A. I.,** *To the kinetics of adsorption of surface-active substances in liquid solutions,* Colloid Journal, Vol. 39, No. 1, 1977.

251. **Kurdyumov, V. N. and Polyanin, A. D.,** *Mass transfer problem for particles, drops and bubbles in a shear flow,* Fluid Dynamics, Vol. 25, No. 4, pp. 611–615, 1990.

252. **Kurdyumov, V. N., Rednikov, A. Ye., and Ryazantsev, Yu. S.,** *Thermocapillary motion of a bubble with heat generation at the interface,* Micrograv. Q, Vol. 4, No. 1, p. 5, 1994.

253. **Kutateladze, S. S.,** *Foundations of the Theory of Heat Transfer,* Atomizdat, Moscow, 1979 [in Russian].

254. **Kutateladze, S. S.,** *Heat Transfer and the Hydrodynamical Resistance. A Reference Book,* Energoatomizdat, Moscow, 1990 [in Russian].

255. **Kutepov, A. M., Polyanin, A. D., Zapryanov, Z. D., Vyazmin, A. V., and Kazenin, D. A.,** *Chemical Hydrodynamics,* Kvantum, Moscow, 1996 [in Russian].

256. **Kutepov, A. M., Sterman, L. S., and Styushin, N. G.,** *Hydrodynamics and Heat Exchange in Vapor Generation,* Vysshaya Shkola, Moscow, 1977 [in Russian].

257. **Kuznetsova, L. L. and Kruglyakov, P. M.,** *Study of the regularities of surfactant solutions flow along Plateau–Gibbs channels in foam,* Doklady Phys. Chem., Vol. 260, No. 4, 1981.

258. **Lagerstrom, P. A.,** *Perturbation Methods,* Springer-Verlag, New York, 1988.

259. **Lamb, H.,** *Hydrodynamics,* Dover Publ., New York, 1945.

260. **Landau, L. D. and Lifschitz, E. M.,** *Fluid Mechanics. Course of Theoretical Physics,* Pergamon Press, Oxford, 1987.

261. **Lavrent'ev, M. A. and Shabat, B. V.,** *Methods of the Theory of Functions of a Complex Variable,* Nauka, Moscow, 1973 [in Russian].

262. **Le Clair, B. P. and Hamielec, A. E.,** *A theoretical and experimental study of the internal circulation in water drops falling at terminal velocity in air,* J. Atmosph. Sci., Vol. 29, No. 4, 1972.

263. **Le Clair, B. P. and Hamielec, A. E.,** *Viscous flow through particle assemblies at intermediate Reynolds numbers. A cell model for transport in bubble swarms,* Can. J. Chem. Eng., Vol. 49, No. 6, pp. 713–720, 1971.

264. **Legros, J. C., Limbourg, M. C., and Petre, G.,** *Influence of a surface tension minimum as a function of temperature on the Marangoni convection,* Acta Astronautica, Vol. 11, No. 2, pp. 143–147, 1984.

265. **Lekhtmakher, S. O.**, *Settling of particles from laminar flow in dependence on the Peclet number*, J. Eng. Phys. Thermophys., Vol. 20, No. 3, pp. 546–549, 1971.

266. **Leonard, R. A. and Lemlich, R.**, *A study of interstitial liquid flow in foam*, AIChE J., Vol. 11, No. 1, pp. 18–29, 1965.

267. **Leont'ev, A. I.** (editor), *Theory of Heat and Mass Transfer*, Vysshaya Shkola, Moscow, 1979 [in Russian].

268. **Lesieur, M.**, *Turbulence in Fluids*, Kluwer Acad. Publ., Dordrecht, 1997.

269. **Leveque, M. A.**, *Les lois de la transmission de chaleur par convection*, Ann. Mines, Bd. 13, S. 527–532, 1928.

270. **Levich, V. G.**, *Physicochemical Hydrodynamics*, Prentice-Hall, Englewood Cliffs, New Jersey, 1962.

271. **Levich, V. G., Krylov, V. S., and Vorotilin, V. P.**, *To the theory of unstable diffusion from a moving drop*, Doklady AN SSSR, Vol. 161, No. 3, pp. 648–652, 1965 [in Russian].

272. **Levitskii, S. P. and Shulman, Z. P.**, *Dynamics and Heat and Mass Transfer Between Bubbles and Polymer Fluids*, Nauka i Tekhnika, Minsk, 1990 [in Russian].

273. **Licinio, P. and Figneiredo, J. M. A.**, *Steady foam states*, Europhys. Lett., Vol. 36, No. 3, pp. 173–178, 1996.

274. **Lin, S. Y., Chang, H. Ch., and Chen, E. M.**, *The effect of bulk concentration on surfactant adsorption processes: the shift from diffusion-control to mixed kinetic-diffusion control with bulk concentration*, J. Chem. Eng. Japan, Vol. 29, No. 4, pp. 634–641, 1996.

275. **Loewenberg, M. and Hinch, E. J.**, *Collision of two deformable drops in shear flow*, J. Fluid Mech., Vol. 338, pp. 299–316, 1997.

276. **Loitsyanskiy, L. G.**, *Mechanics of Liquids and Gases*, Begell House, New York, 1996.

277. **Lykov, A. V.**, *Theory of Heat Conduction*, Vysshaya Shkola, Moscow, 1967 [in Russian].

278. **Maldarelli, C. and Jain, R. K.**, *The hydrodynamic stability of thin films*. In: Thin Liquid Films (editor Ivanov, I. B.), Marcell Dekkert Inc., New York, 1988.

279. **Maldarelli, C., Jain, R. K., Ivanov, I. B., and Ruckenstein, E.**, *Stability of symmetric and asymmetric thin liquid films to short and long wavelength perturbations*, J. Colloid Interface Sci., Vol. 78, pp. 118–143, 1980.

280. **Manegold, E.**, *Schaum, Strassenbahn*, Chemic and Technic, Heidelberg, 1953.

281. **Masliyah, J. H. and Epstein, N.**, *Numerical solution of heat and mass transfer from spheroids in steady axisymmetric flow*, Progress Heat Mass Transfer, Vol. 6, pp. 613–632, 1972.

282. **Matzke, E. B.**, *The three-dimensional shape of bubbles in foam–an analysis of the role of surface forces in three-dimensional cell shape determination*, Amer. J. Botany, Vol. 33, No. 1, pp. 58–80, 1946.

283. **Maxworthy, T., Qnann, C., Kürten, M., and Durst, F.**, *Experiments on the rise of air bubbles in clean viscous liquids*, J. Fluid Mech., Vol. 321, pp. 421–441, 1996.

284. **McLachlan, N. W.**, *Theory and Application of Mathieu Functions*, Clarendon Press, Oxford, 1947.

285. **Meyyappan, M., Wilcox, W. R., and Subramanian, R. S.**, *Thermocapillary migration of a bubble normal to a plane surface*, J. Colloid Interface Sci., Vol. 94, pp. 243–257, 1981.

286. **Mizushina, T. and Kurivaki, Yu.**, *Heat transfer in laminar flow of pseudoplastic fluids in a circular tube*. In: Heat and Mass Transfer, Vol. 3, Minsk, 1968 [in Russian].

287. **Mobius, D. and Miller, R.,** *Drops and Bubbles in Interfacial Research,* Elsevier Science Ltd., Oxford, 1997.

288. **Moin, P. and Mahesh, K.,** *Direct numerical simulation: A tool in turbulence research,* Annual Rev. Fluid Mech., Vol. 30, 1988.

289. **Monin, A. S. and Yaglom, A. M.,** *Statistical Fluid Mechanics: Theory of Turbulence, Vol. 1,* Gidrometeoizdat, St. Petersburg, 1992 [in Russian].

290. **Monin, A. S. and Yaglom, A. M.,** *Statistical Fluid Mechanics: Theory of Turbulence, Vol. 2,* Gidrometeoizdat, St. Petersburg, 1996 [in Russian].

291. **Moore, D. M.,** *The velocity of rise of distorted gas bubbles in a liquid of small viscosity,* J. Fluid Mech., Vol. 23, No. 4, pp. 749–766, 1965.

292. **Moo-Young, M., Hirose, T., and Ali, S.,** *Rheological effects on liquid phase mass transfer in two phase dispersions: results for creeping flow,* Proc. 5th Int. Congr. Rheol., p. 233, Kyoto, 1970.

293. **Morrison, F. A.,** *Transient heat and mass transfer to a drop in a electric field,* Trans. ASME, J. Heat Transfer, Vol. 99, No. 2, pp. 269–274, 1977.

294. **Moshev, V. V. and Ivanov, V. A.,** *Rheological Behavior of Concentrated Non-Newtonian Suspensions,* Nauka, Moscow, 1990 [in Russian].

295. **Muller, H., Vetoshkin, A. G., Kazenin, D. A., et al.,** *Rheological behavior of gas-liquid foams,* Rus. J. Appl. Chem., Vol. 62, No. 3, 1989.

296. **Müller, W.,** *Zum Problem der Anlanfströmung einer Flussigkeit im geraden Rohr mit Kreisring- und Kreisquerschnitt,* Zs. angew Math. Mech. (ZAMM), Bd. 16, Ht. 4, S. 227–238, 1936.

297. **Myers, D.,** *Surfactant Science and Technology,* VCH Publ. Inc., New York, 1992.

298. **Naidenov, V. I.,** *Non-isothermic instability of flow of viscoplastic fluids in tubes,* High Temperature, Vol. 28, No. 3, 1990.

299. **Naidenov, V. I.,** *On integral equations describing the temperature distribution in plane flow of non-Newtonian media,* J. Appl. Mech. Techn. Phys., No. 5, 1983.

300. **Naidenov, V. I.,** *On nonlinear equations of self-similar non-isothermic motion of viscous fluid,* Comput. Math. and Math. Phys., Vol. 28, No. 12, 1988.

301. **Naidenov, V. I. and Polyanin, A. D.,** *On some nonlinear convective–heat effects in the theory of filtration and hydrodynamics,* Physics Doklady, Vol. 29, No. 11, 1984.

302. **Naidenov, V. I. and Polyanin, A. D.,** *On some nonisothermic fluid flows,* J. Appl. Mech. Techn. Phys., No. 3, 1990.

303. **Nakano, Y. and Tien, C.,** *Creeping flow a power-low fluid over a Newtonian fluid drop,* AIChE J., Vol. 14, pp. 145–151, 1968.

304. **Nakano, Y. and Tien, C.,** *Viscous incompressible non-Newtonian flow around fluid sphere at intermediate Reynolds number,* AIChE J., Vol. 16, No. 4, pp. 554–569, 1970.

305. **Nakayama, W. and Yang, K.-T.,** *Computers and Computing in Heat Transfer Science and Engineering,* CRC Press, Boca Raton, 1993.

306. **Nakoryakov, V. E., Pokusaev, B. G., and Shreiber, I. R.,** *Wave Propagation in Gas-Liquid Media,* CRC Press – Begell House, Boca Raton, 1993.

307. **Natanson, G. L.,** *Diffusion settling of aerosols on a cylinder in flow with small entrainment factor,* Doklady AN SSSR, Vol. 112, No. 1, pp. 100–103, 1957 [in Russian].

308. National Bureau of Standarts, *Tables Relating to Mathieu Function,* Columbia Univ. Press, New York, 1951.

309. **Newman, J.,** *Mass transfer to the rear of a cylinder at high Schmidt numbers,* J. Ind. Eng. Chem. Fundamentals, Vol. 8, No. 3, pp. 82–86, 1969.

310. **Nicodemus, R., Grossmann, S., and Holthaus, M.,** *The background flow method. Part 1. Constructive approach to bounds on energy dissipation,* J. Fluid Mech., Vol. 363, pp. 281–300, 1998.

311. **Nicolis, G. and Prigogine, I.,** *Self-Organization in Nonequilibrium Systems,* Wiley, New York, 1977.

312. **Nigmatulin, R. I.,** *Dynamics of Multiphase Media. Part. 1,* Nauka, Moscow, 1987 [in Russian].

313. **Nigmatulin, R. I.,** *Foundations of Mechanics of Heterogeneous Media,* Nauka, Moscow, 1978 [in Russian].

314. **Nikitin, N. V.,** *Direct numerical modeling of three-dimensional turbulent flows in pipes of circular cross-section,* Fluid Dynamics, Vol. 29, No. 6, pp. 749–758, 1994.

315. **Nikitin, N. V.,** *Spectral finite-element method for the analysis of turbulent flows of incompressible fluid through tubes and channels,* Comput. Math. and Math. Phys., Vol. 34, No. 6, pp. 785–798, 1994.

316. **Nikitin N. V.,** *Statistical characteristics of wall turbulence,* Fluid Dynamics, Vol. 31, No. 3, pp. 361–370, 1996.

317. **Nikolskii, V. V. and Nikolskaya, T. I.,** *Electrodynamics and Propagation of Radio Waves,* Nauka, Moscow, 1989 [in Russian].

318. **Nikuradse, J.,** *Gesetzmässigkeiten der turbulenten Strömung in glatten Rohren,* VDI-Forschungsheft, No. 356, 1932.

319. **Nusselt, W.,** *Abhängigkeit der Wärmeübergangzahl con der Rohrlänge,* VDI Zeitschrift, Bd. 54, Ht. 28, S. 1154–1158, 1910.

320. **Ogibalov, P. M. and Mirzadzhanzade, A. Kh.,** *Unsteady Motions of Viscoplastic Media,* Izd. Moskov. Univ., Moscow, 1970 [in Russian].

321. **Oellrich, L., Schmidt-Traub, H., and Brauer, H.,** *Theoretische Berechnung des Stofftransport in der Umgebung einer Einzelblase,* Chem. Eng. Sci., Vol. 28, No. 3, pp. 711–721, 1973.

322. **Oliver, D. L. R. and DeWitt, K. J.,** *Surface tension driven flows for a droplet in a microgravity environment,* Int. J. Heat Mass Transfer, Vol. 31, No. 7, pp. 1534–1537, 1988.

323. **O'Neill, M. E. and Stewartson, K.,** *On the slow motion of a sphere parallel to a nearly plane wall,* J. Fluid Mech., Vol. 27, pp. 705–724, 1967.

324. **Ostrovskii, G. M. and Nekrasov, V. A.,** *A mathematical model for outflow of liquid from foam,* Theor. Found. Chem. Eng., Vol. 30, No. 6, pp. 599–603, 1996.

325. **Oseen, C. W.,** *Über die Stokes'sche Formel, und über eine verwandte Aufgabe in der Hydrodynamik,* Ark. Math. Astronom. Fys., Bd. 6, Ht. 29, 1910.

326. **Paneli, D. and Gutfinger, C.,** *Fluid Mechanics,* Cambridge Univ. Press, Cambridge, 1997.

327. **Paskonov, V. M., Polezhaev, V. I., and Chudov, L. A.,** *Numerical Modeling of Heat and Mass Exchange Processes,* Nauka, Moscow, 1984 [in Russian].

328. **Patankar, S. V.,** *Numerical Heat Transfer and Fluid Flow,* Hemisphere, New York, 1980.

329. **Patankar, S. V. and Spalding, D. B.,** *Heat and Mass Transfer in Boundary Layers,* Morgan-Grampian, London, 1967.

330. **Pavlov, K. B.,** *Boundary-layer theory in non-Newtonian nonlinearly viscous media,* Fluid Dynamics, Vol. 13, No. 3, pp. 360–366, 1978.

331. **Perlmutter, D. D.,** *Stability of Chemical Reactors,* Prentice-Hall, Englewood Cliffs, 1972.

332. **Perry, J. H.,** *Chemical Engineers Handbook,* McGraw-Hill, New York, 1950.

333. **Persillon, H. and Braza, M.,** *Physical analysis of the transition to turbulence in the wake of a circular cylinder by three-dimensional Navier-Stokes simulation,* J. Fluid Mech., Vol. 365, pp. 23–88, 1998.

334. **Persoff, P., Pruess, K., Bonson, S. M., and Wu, Y. S.,** *Aqueous foams for control of gas migration and water coning in aquiter gas storage,* Energy Source, Vol. 12, pp. 479–497, 1990.

335. **Pertsov, A. V., Chernin, V. N., Chistyakov, B. E., and Shchukin, E. D.,** *Capillary effects and hydrostatic stability of foams,* Doklady Phys. Chem., Vol. 238, No. 6, 1978.

336. **Petrov, A. G.,** *Circulation inside viscous deformed drops moving in gas with constant velocity,* J. Appl. Mech. Techn. Phys., No. 6, 1989.

337. **Petrov, A. G.,** *Curvilinear motion of an ellipsoidal bubble,* J. Appl. Mech. Techn. Phys., No. 3, 1972.

338. **Petrov, A. G.,** *Inner flow of viscous drop,* Proc. Third Int. Aeros. Conf., Kyoto, Japan, pp. 339–342, 1990.

339. **Petrov, A. G.,** *Internal flow and deformation of viscous drops,* Moscow Univ. Math. Mech. Bull., No. 3, 1988.

340. **Petrov, A. G.,** *Energy dissipation rate in a viscous fluid with condition for shear stress ot the boundary stream line,* Physics Doklady, Vol. 34, No. 2, 1989.

341. **Petukhov, B. S.,** *Heat Transfer and Drag in Laminar Flow of Liquids in Tubes,* Energiya, Moscow, 1967 [in Russian].

342. **Poe, G. G.,** *Closed streamline flows past rotating particles: inertial effects, lateral migration, heat transfer,* Ph. D. dissertation, Stanford Univ., 1975.

343. **Poe, G. G. and Acrivos, A.,** *Closed streamline flows past small rotating particles; heat transfer at high Peclet numbers,* Int. J. Mult. Flow, Vol. 2, No. 4, pp. 365–377, 1976.

344. **Polezhaev, V. I., Bune, A. V., Verezub, N. A., et al.,** *Mathematical Modeling of Convective Heat and Mass Transfer on the Base of the Navier–Stokes Equations,* Nauka, Moscow, 1987 [in Russian].

345. **Pohlhausen, E.,** *Warmeaustausch zwischen festen Koïpern und Flüssigkeiten mit kleiner Reibung und kleiner Warmeleitung,* Zs. angew Math. Mech. (ZAMM), Bd. 1, S. 15, 1921.

346. **Polubarinova-Kochina, P. Ya.,** *Theory of Underground Water Motion,* Nauka, Moscow, 1977 [in Russian].

347. **Polyanin, A. D.,** *An asymptotic analysis of some nonlinear boundary-value problems of convective mass and heat transfer of reacting particles with the flow,* Int. J. Heat Mass Transfer, Vol. 27, No. 2, pp. 163–189, 1984.

348. **Polyanin, A. D.,** *Diffusional interaction of drops in a liquid,* Fluid Dynamics, Vol. 13, No. 2, pp. 192–204, 1978.

349. **Polyanin, A. D.,** *Method for solution of some non-linear boundary value problems of a non-stationary diffusion-controlled (thermal) boundary layer,* Int. J. Heat Mass Transfer, Vol. 25, No. 4, pp. 471–485, 1982.

350. **Polyanin, A. D.,** *On diffusion interaction of solid particles at high Peclet numbers,* J. Appl. Math. Mech. (PMM), Vol. 42, No. 2, pp. 315–326, 1978.

351. **Polyanin, A. D.,** *On nonisothermal chemical reaction at the particle surface in a laminar flow,* Int. J. Heat Mass Transfer, Vol. 25, No. 7, pp. 1031–1042, 1982.

352. **Polyanin, A. D.,** *Qualitative features of internal problems of transient convective mass and heat exchange at large Peclet numbers,* Theor. Found. Chem. Eng., Vol. 18, No. 3, pp. 171–181, 1984.

353. **Polyanin, A. D.,** *Three-dimensional problems of the diffusion boundary layer,* J. Appl. Mech. Techn. Phys., No. 4, 1984.

354. **Polyanin, A. D.,** *Three-dimensional problems of unsteady diffusion boundary layer,* Int. J. Heat Mass Transfer, Vol. 33, No. 7, pp. 1375–1386, 1990.

355. **Polyanin, A. D.,** *Unsteady-state extraction from a falling droplet with nonlinear dependence of distribution coefficient on concentration,* Int. J. Heat Mass Transfer, Vol. 27, No. 8, pp. 1261–1276, 1984.

356. **Polyanin, A. D. and Dilman, V. V.,** *An algebraic method for heat and mass transfer problems,* Int. J. Heat Mass Transfer, Vol. 33, No. 1, pp. 183–201, 1990.

357. **Polyanin, A. D. and Dilman, V. V.,** *New methods of the mass and heat transfer theory. – 1. The method of asymptotic correction and the method of model equations and analogies,* Int. J. Heat Mass Transfer, Vol. 28, No. 1, pp. 25–43, 1985.

358. **Polyanin, A. D. and Dilman, V. V.,** *New methods of the mass and heat transfer theory. – 2. Methods of asymptotic interpolation and extrapolation,* Int. J. Heat Mass Transfer, Vol. 28, No. 1, pp. 45–58, 1985.

359. **Polyanin, A. D. and Dilman, V. V.,** *Methods of Modeling Equations and Analogies in Chemical Engineering,* CRC Press – Begell House, Boca Raton, 1994.

360. **Polyanin, A. D. and Dilman, V. V.,** *The method of asymptotic analogies in the mass and heat transfer theory and chemical engineering science,* Int. J. Heat Mass Transfer, Vol. 33, No. 6, pp. 1057–1072, 1990.

361. **Polyanin, A. D. and Dilman, V. V.,** *The method of the "carry over" of integral transforms in non-linear mass and heat transfer problems,* Int. J. Heat Mass Transfer, Vol. 33, No. 1, pp. 175–181, 1990.

362. **Polyanin, A. D. and Erokhin, L. Yu.,** *Heat transfer to bodies of complex shape,* Theor. Found. Chem. Eng., Vol. 24, No. 1, pp. 9–16, 1990.

363. **Polyanin, A. D. and Shevtsova, V. M.,** *Mass transfer between particles and a flow in a presence of a volume chemical reaction,* Fluid Dynamics, Vol. 22, No. 6, pp. 916–919, 1987.

364. **Polyanin, A. D. and Sergeev, Yu. A.,** *Convective diffusion to a reacting particle in a fluid. Nonlinear surface reaction kinetics,* Int. J. Heat Mass Transfer, Vol. 23, No. 9, pp. 1171–1182, 1980.

365. **Polyanin, A. D. and Sergeev, Yu. A.,** *On the concentration field of an orderly system of reacting plates distributed along a stream,* J. Appl. Math. Mech. (PMM), Vol. 45, No. 1, pp. 77–87, 1981.

366. **Polyanin, A. D. and Vyazmin, A. V.,** *Complicated mass and heat transfer between particles, drops and bubbles and a flow,* Theor. Found. Chem. Eng., Vol. 30, No. 6, pp. 542–550, 1996.

367. **Polyanin, A. D. and Vyazmin, A. V.,** *Mass and heat transfer to particles in a flow,* Theor. Found. Chem. Eng., Vol. 29, No. 2, pp. 128–139, 1995.

368. **Polyanin, A. D. and Vyazmin, A. V.,** *Mass and heat transfer between a drop or bubble and a flow,* Theor. Found. Chem. Eng., Vol. 29, No. 3, pp. 229–240, 1995.

369. **Polyanin, A. D., Kurdyumov, V. N., and Dilman, V. V.,** *Asymptotic correction method in chemical engineering problems,* Theor. Found. Chem. Eng., Vol. 26, No. 3, pp. 404–418, 1992.

370. **Polyanin, A. D., Vyazmin, A. V., Zhurov, A. I., and Kazenin, D. A.,** *Handbook on Exact Solutions of Heat and Mass Transfer Equations,* Faktorial, Moscow, 1998 [in Russian].

371. **Porter, M. R.,** *Handbook of Surfactants,* Chapman & Hall, London, 1994.

372. **Potapov, E. D., Serebryakova, N. G., and Troshin, V. G.,** *Interaction of porous spherical bodies in a slow viscous flow,* Fluid Dynamics, Vol. 27, No. 3, pp. 445–447, 1992.

373. **Povitskii, A. S. and Lyubin, L. Ya.,** *Foundations of Dynamics and Heat and Mass Transfer in Liquids and Gases in Weightless State,* Mashinostroenie, Moscow, 1972 [in Russian].

374. **Prandtl, L.,** *The mechanics of viscous fluids,* In: Aerodynamics Theory, Vol. 3, pp. 34–208; Springer-Verlag, Berlin, 1935.

375. **Prigogine, I.,** *Introduction to Thermodynamics of Irreversible Processes,* Charles C. Thomas, Springfild, 1955.

376. **Priimak, V. G.,** *Results and potentialities of direct numerical simulation of turbulent viscous fluid flows in circular pipe,* Physics Doklady, Vol. 36, No. 1, 1991.

377. **Princen, H. M.,** *Gravitational syneresis in foams and concentrated emulsions,* J. Colloid Interface Sci., Vol. 134, No. 1, pp. 188–197, 1990.

378. **Princen, H. M.,** *Osmotic pressure of foams and highly concentrated emulsions. 1. Theoretical consideration,* Langmuir, Vol. 2, No. 4, pp. 519–534, 1986.

379. **Princen, H. M.,** *Rheology of foams and highly concentrated emulsion. 1. Elastic properties and yield stress of a cylindrical model system,* J. Colloid Interface Sci., Vol. 91, No. 1, pp. 60–75, 1983.

380. **Princen, H. M. and Kiss, A. D.,** *Rheology of foams and highly concentrated emulsions,* J. Colloid Interface Sci., Vol. 112, No. 2, pp. 427–438, 1986.

381. **Princen, H. M. and Levinson, P.,** *The surface area of Kelvin's minimal tetrakaidecahedron: the ideal foam cell?,* J. Colloid Interface Sci., Vol. 120, No. 1, pp. 172–175, 1987.

382. **Proudman, I. and Pearson, J. R. A.,** *Expansions at small Reynolds number for the flow past a sphere and circular cylinder,* J. Fluid Mech., Vol. 2, No. 3, pp. 237–262, 1957.

383. **Prud'homme, R. K. and Khan, S. A. (editors),** *Foams: Fundamentals and Applications,* Marcel Dekker Inc., New York, 1995.

384. **Pugh, R. J.,** *Foaming, foam films, antifoaming and defoaming,* Adv. Colloid Interface Sci., Vol. 64, pp. 67–142, 1996.

385. **Pukhnachev, V. V.,** *Motion of a Viscous Fluid With Free Boundaries,* Izd. Novosib. Univ., Novosibirsk, 1989 [in Russian].

386. **Pushkarev, V. V. and Trofimov, D. I.,** *Physical and Chemical Characteristics of Sewage Purification From Surface-Active Substances,* Khimiya, Moscow, 1975 [in Russian].

387. **Ranger, K. B.,** *The circular disk straddling the interface of a two-phase flow,* Int. J. Mult. Flow, Vol. 4, pp. 263–277, 1978.

388. **Rao, S. S. and Bennett, C. O.,** *Steady state technique for measuring fluxes and diffusivities in binary liquid systems,* AIChE J., Vol. 17, No. 1, pp. 75–81, 1971.

389. **Rednikov, A. E. and Ryazantsev, Yu. S.,** *On thermocapillary motion of a drop under the action of radiation,* J. Appl. Mech. Techn. Phys., No. 2, 1989.

390. **Rednikov, A. E. and Ryazantsev, Yu. S.,** *On the thermocapillary motion of a drop with homogeneous internal heat evolution,* J. Appl. Math. Mech. (PMM), Vol. 53, No. 2, pp. 212–216, 1989.

391. **Rednikov, A. Ye., Ryazantsev, Yu. S., and Velarde, M. G.,** *Active drops and drops motion due to nonequilibrium phenomena,* J. Non-Equilibr. Thermodyn., Vol. 19, No. 1, p. 95, 1994.

392. **Rednikov, A. Ye., Ryazantsev, Yu. S., and Velarde, M. G.,** *Drop motion with surfactant transfer in a homogeneous surrounding,* Phys. Fluids, Vol. 6, No. 2, pp. 451–468, 1994.

393. **Reid, R. C., Prausnitz, J. M., and Sherwood, T. K.,** *The Properties of Gases and Liquids,* McGraw-Hill Book Comp., New York, 1977.

394. **Reichardt, H.**, *Vollständige Darstellung der tubulent Geschwindigkeitsverteilung in glatten Leitungen*, Zs. angew Math. Mech. (ZAMM), Bd. 31, Ht. 7, S. 208–219, 1951.

395. **Reichardt, H.**, *Gesetzmässingkeiten der freien Turbulenz*, VDI-Forrschungsheft, 1951.

396. **Reiner, M.**, *Rheology*, Springer-Verlag, Berlin, 1958.

397. **Reynolds, A. J.**, *The prediction of turbulent Prandtl and Schmidt numbers*, Int. J. Heat Mass Transfer, Vol. 18, No. 9, pp. 1055–1069, 1975.

398. **Reynolds, A. J.**, *Turbulent Flows in Engineering*, Wiley, London, 1974.

399. **Richard, J.-G.**, *Etude des profils de température dans un écoulement turbulent établi dans un tube cilindrique lisse*, Bull. Dir. étud. et rech., Ser. A, No. 2. pp. 1–173, 1972.

400. **Richardson, E. G. and Tyler, E.**, *The transverse velocity gradient near the mouths of pipes in which an alternating or continuous flow of air is established*, Proc. Phys. Soc. London, Vol. 42, pp. 1–15, 1929.

401. **Rimmer, P. L.**, *Heat transfer from a sphere in a stream of small Reynolds number*, J. Fluid Mech., Vol. 32, No. 1, pp. 1–7, 1968; Corrigenda: J. Fluid Mech., Vol. 35, No. 4, pp. 827–829, 1969.

402. **Rimon, J. and Cheng, S. I.**, *Numerical solution of a uniform flow over a sphere at intermediate Reynolds numbers*, Phys. Fluid, Vol. 12, No. 5, pp. 949–959, 1969.

403. **Rivkind, V. Ya. and Sigovtsev, G. S.**, *Motion of a drop with allowance for thermocapillary forces*, Fluid Dynamics, Vol. 17, No. 4, pp. 554–559, 1982.

404. **Roach, P. J.**, *Computational Fluid Dynamics*, Hermosa Publ., Albuquerque, 1972.

405. **Robertson, C. R. and Acrivos, A.**, *Low Reynolds number shear flow past a rotating circular cylinder. Part 2. Heat transfer*, J. Fluid Mech., Vol. 40, No. 4, pp. 705–718, 1970.

406. **Rohsenow, W. M., Hartnett, J. P., and Ganic, E. N.**, *Handbook of Heat Transfer Fundamentals*, McGraw-Hill, New York, 1985.

407. **Ross, S. and Prest, H. F.**, *On the morphology of bubble clusters and polyhedral foams*, Colloids and Surfaces, Vol. 21, Special issue, pp. 179–192, 1986.

408. **Rotem, Z. and Neilson, J. E.**, *Exact solution for diffusion to flow down an incline*, Can. J. Chem. Eng., Vol. 47, pp. 341–346, 1966.

409. **Rubinstein, R. and Barton, J. M.**, *Nonlinear Reynolds stress models and the renormalization group*, Phys. Fluids A, Vol. 2, pp. 1472–1476, 1990.

410. **Ruckenstein, E.**, *Mass transfer between a single drop and continuous phase*, Int. J. Heat Mass Transfer, Vol. 10, No. 12, pp. 1785–1792, 1967.

411. **Ruckenstein, E. and Jain, R. K.**, *Spontaneous rupture of thin liquid films*, Chem. Soc. London, Faraday Trans. II, Vol. 70, pp. 132–147, 1974.

412. **Ruckenstein, E. and Prieve, D. C.**, *Rate of deposition of Brownian particles under the action of London and double-layer forces*, Chem. Soc. London, Faraday Trans. II, Vol. 69, No. 10, pp. 1523–1536, 1973.

413. **Rusanov, A. I.**, *Michelle Generation in Solutions of Surface Active Substances*, Khimiya, St.-Peterburg, 1992 [in Russian].

414. **Rusanov, A. I.**, *Phase Equilibria and Surface Phenomena*, Khimiya, Leningrad, 1967 [in Russian].

415. **Rusanov, A. I., Levichev, S. A., and Zharov, V. T.**, *Surface Separation of Substances*, Khimiya, Leningrad, 1981 [in Russian].

416. **Rushton, E. and Davies, G. A.**, *Settling of encapsulated droplets at low Reynolds numbers*, Int. J. Mult. Flow, Vol. 9, No. 3, pp. 337–342, 1983.

417. **Rushton, E. and Davies, G. A.**, *The slow unsteady settling of two fluid spheres along their line of centers*, Appl. Sci. Res., Vol. 28, No. 1–2, pp. 37–61, 1973.

418. **Rvachev, V. L. and Slesarenko, A. P.,** *Algebra of Logic and Integral Transforms in Boundary Value Problems,* Naukova Dumka, Kiev, 1976 [in Russian].

419. **Ryazantsev, Yu. S.,** *Thermocapillary motion of a reacting droplet in a chemically active medium,* Fluid Dynamics, Vol. 20, No. 3, pp. 491–495, 1985.

420. **Rybczynski M. W.,** *Über die fortschreitende Bewegung einer flüssigen Kugel in einem zähen Medium,* Bull. Acad. Sci. Cracovie, Ser. A, Sci. Math., Bd. 1, S. 40–46, 1911.

421. **Ryskin, G. M.,** *Synopsis of PhD Thesis,* Leningrad Pedagogical Institute, Leningrad, 1976 [in Russian].

422. **Saffman, P. G.,** *On the boundary condition at the surface of a porous medium,* Stud. Appl. Math., Vol. 50, No. 2, pp. 93–101, 1971.

423. **Sahimi, M.,** *Flow in Porous Media and Fractured Rock,* VCH, Weinheim, 1995.

424. **Sakiadis, B. C.,** *Boundary-layer behavior on continuous solid surfaces. 2. Boundary layer on a continuous flat surface,* AIChE J., Vol. 7, No. 2, pp. 221–225, 1961.

425. **Samarskii, A. A., Galaktionov, V. A., Kurdyumov, S. P., and Mikhailov, A. P.,** *Peaking Regimes in Problems for Quasilinear Parabolic Equations,* Nauka, Moscow, 1987 [in Russian].

426. **Sanghi, S. and Aubry, N.,** *Mode interaction models for near-wall turbulence,* Fluid Mech., Vol. 247, pp. 455–488, 1993.

427. **Schlichting, H.,** *Boundary Layer Theory,* McGraw-Hill, New York, 1981.

428. **Schwartz, L. W. and Princen, H. M.,** *A theory of extensional viscosity for flowing foams and concentrated emulsions,* J. Colloid Interface Sci., Vol. 118, No. 1, pp. 201–211, 1987.

429. **Sebba, F.,** *Foams and Biliquid Foams-Aphrons,* Wiley, Chichester, 1987.

430. **Sedov, L. I.,** *Mechanics of Continuous Media, Vol. 1,* Nauka, Moscow, 1973 [in Russian].

431. **Sedov, L. I.,** *Plane Problems in Hydrodynamics and Aerodynamics,* Nauka, Moscow, 1966 [in Russian].

432. **Sedov, L. I.,** *Similarity and Dimensional Methods in Mechanics,* CRC Press, Boca Raton, 1993.

433. **Sehlin, R. C.,** *Forced-Convection Heat and Mass Transfer at Large Peclet Numbers From Axisymmetric Body in Laminar Flow: Prolate and Oblate Spheroids,* M.S. Thesis (Chem. Eng.), Carnegie Inst. Thechn., Pittsburgh, 1969.

434. **Sekimoto, K.,** *An exact non-stationary solution of simple shear flow in a Bingham fluid,* J. Non-Newtonian Fluid Mech., Vol. 39, No. 1, pp. 107–113, 1991.

435. **Sell, G. R., Foias, C., and Temam, R. (editors),** *Turbulence in Fluid Flows: a Dynamical Systems Approach,* Springer-Verlag, New York, 1993.

436. **Sellers, J. R., Tribus, M., and Klin, J. S.,** *Heat transfer to laminar flow in a round tube or flat conduit–the Graetz problem extended,* Trans. ASME, Vol. 78, No. 2, pp. 441–448, 1956.

437. **Sharma, A., Kishore, C. S., Salaniwal, S., and Ruckenstein, E.,** *Nonlinear stability and rupture of ultrathin free film,* Phys. Fluids, Vol. 7, No. 8, pp. 1832–1840, 1995.

438. **Sheludko, A. D.,** *Colloid Chemistry,* Mir, Moscow, 1984 [in Russian].

439. **Sherwood, T. K., Pigford, R. L., and Wilke, C. R.,** *Mass Transfer,* McGraw-Hill, New York, 1975.

440. **Shivamoggi, B.,** *Theoretical Fluid Dynamics,* Wiley VCH, Chichester, New York, 1998.

441. **Shkadov, V. Ya. and Zapryanov, Z. D.,** *Flow of Viscous Fluids,* Izd. Moscow Univ., Moscow, 1984 [in Russian].

442. **Shugai, G. A. and Yakubenko, P. A.**, *Spatio-temporal instability in free ultra-thin films*, Eur. J. Mech. B / Fluids, Vol. 17, No. 3, pp. 371–384, 1998.

443. **Shul'man, Z. P.**, *Convective Heat and Mass Transfer in Rheologically Complicated Fluids*, Energiya, Moscow, 1975 [in Russian].

444. **Shul'man, Z. P. and Baikov, V. I.**, *Rheodynamics and Heat and Mass Exchange in Film Flows*, Nauka i Tekhnika, Minsk, 1979 [in Russian].

445. **Shul'man, Z. P. and Berkovskii, B. M.**, *The Boundary Layer of Non-Newtonian Fluids*, Nauka i Tekhnika, Minsk, 1966 [in Russian].

446. **Sih, P. H. and Newman, J.**, *Mass transfer to the rear of sphere in Stokes flow*, Int. J. Heat Mass Transfer, Vol. 10, No. 12, pp. 1749–1756, 1967.

447. **Sinaisky, E. G.**, *Hydrodynamics of Physico-Chemical Processes*, Nedra, Moscow, 1997 [in Russian].

448. **Schowalter, W. R.**, *Mechanics of Non-Newtonian Fluids*, Pergamon Press, Oxford, 1978.

449. **Slezkin, N. A.**, *Dynamics of Viscous Incompressible Fluid*, Gostekhizdat, Moscow, 1955 [in Russian].

450. **Slobodov, E. B. and Chepura, I. V.**, *A cellular model of biphasal media*, Theor. Found. Chem. Eng., Vol. 16, No. 3, pp. 235–239, 1982.

451. **Smith, L. M. and Woodruff, L.**, *Renormalization-group analysis of turbulence*, Annual Rev. Fluid Mech., Vol. 30, 1988.

452. **Smol'skii, B. M., Shul'man, Z. P., and Gorislavets, V. M.**, *Rheodynamics and Heat Transfer in Nonlinear Viscoplastic Materials*, Nauka i Tekhnika, Minsk, 1970 [in Russian].

453. **So, R. M. C., Speciale, C. G., and Launder, B. E.** (editors), *Near-Wall Turbulent Flows*, Elsevier, Amsterdam, 1993.

454. **Soo, S. L.**, *Fluid Dynamics of Multiphase Systems*, Blaisdell Publ. Comp., Waltham, 1968.

455. **Spalding, D. B.**, *A single formula for the "law of wall"*, Trans. ASME, J. Appl. Mech., Vol. 28, No. 3, pp. 455–458, 1961.

456. **Spalding, D. B.**, *Kolmogorov's two-equation model of turbulence*, Proc. Royal Soc. London, Vol. A434, No. 1890, pp. 211–216, 1991.

457. **Sparrow, E. M. and Ohadi, M. M.**, *Numerical and experimental studies of turbulent heat transfer in a tube*, Numerical Heat Transfer, Vol. 11, No. 4, pp. 461–476, 1987.

458. **Speziale, C. G.**, *On nonlinear $K - l$ and $K - \epsilon$ models in turbulence*, Fluid Mech., Vol. 178, pp. 459–475, 1987.

459. **Sreenivasan, K. R.**, *Fractals and multifractals in fluid turbulence*, Annual Rev. Fluid Mech., Vol. 23, p. 539, 1991.

460. **Stanišić, M. M.**, *The Mathematical Theory of Turbulence*, Springer-Verlag, New York, 1993.

461. **Stechkina, I. B.**, *Diffusion settling of aerosols in fibrous filters*, Doklady Phys. Chem., Vol. 167, No. 6, 1966.

462. **Stewartson, K.**, *On the steady flow past a sphere at high Reynolds number using Oseen's approximation*, Philos. Mag., Vol. 1, No. 8, pp. 345–354, 1956.

463. **Stimson, M. and Jeffrey, G. B.**, *The motion of two spheres in a viscous flow*, Proc. Roy. Soc. London, Vol. A111, No. 757, p. 110, 1926.

464. **Stokes, G. G.**, *On the effect of the internal friction of fluids on the motion of pendulums*, Trans. Camb. Phil. Soc., Vol. 9, No. 2, pp. 8–106, 1851.

465. **Stuke, B.**, *Dynamische oberflächenspannuny polarer flüssigkeiten*, Zeitschrift für Electrochemishe, Bd. 63, S. 140–148, 1959.

466. **Subramanian, R. S.**, *Slow migration of a gas bubble in a thermal gradient*, AIChE J., Vol. 27, No. 4, pp. 646–654, 1981.

467. **Subramanian, R. S.**, *The motion of bubbles and drop in reduced gravity.* In: Transport Processes in Bubbles, Drop and Particle, pp. 1–42, Hemisphere, New York, 1992.

468. **Subramanian, R. S.**, *The Stokes force in a droplet in an unbounded fluid medium due to capillary effects*, J. Fluid Mech., Vol. 153, pp. 389–400, 1985.

469. **Subramanian, R. S.**, *Thermocapillary migration of bubbles and droplets*, Advances in Space Research, Pergamon Press, Vol. 3, No. 5, p. 145, 1983.

470. **Sykes, J. A. and Marchello, J. M.**, *Laminar flow of two immersible liquid falling films*, AIChE J., Vol. 15, No. 2, pp. 305–306, 1969.

471. **Szymczyk, J. and Siekmann, J.**, *Numerical calculation of the thermocapillary motion of a bubble under microgravity*, Chem. Eng. Comm., Vol. 69, pp. 129–147, 1988.

472. **Takagi, S., Prosperiti, A., and Matsumoto, Y.**, *Drag coefficient of a gas bubble in an axisymmetric shear flow*, Phys. Fluids, Vol. 6, No. 9, pp. 3186–3188, 1994.

473. **Tam, C. K. W.**, *The drag on a cloud of spherical particles in low Reynolds number flow*, J. Fluid Mech., Vol. 38, No. 3, pp. 537–546, 1969.

474. **Taylor, G. I.**, *Formation of emulsion in refinable film of flow*, Proc. Roy. Soc. London, Vol. A146, No. 858, pp. 501–523, 1934.

475. **Taylor, G. I.**, *Viscosity of a fluid, containing small drops of another fluid*, Proc. Roy. Soc. London, Vol. A138, No. 834, pp. 41–48, 1932.

476. **Taylor, T. and Acrivos, A.**, *On a deformation and drag of a falling viscous drop at low Reynolds number*, J. Fluid Mech., Vol. 18, No. 3, pp. 466–476, 1964.

477. **Temam, R.**, *Navier—Stokes equations*, North-Holland Publ. Comp., Amsterdam–New York, 1979.

478. **Ternovskii, I. G. and Kutepov, A. M.**, *Hydrocyclonization*, Nauka, Moscow, 1994 [in Russian].

479. **Thompson D'Arcy W.**, *On growth and form*, Cambridge Univ. Press, Cambridge, 1961.

480. **Tikhomirov, V. K.**, *Foams*, Khimiya, Moscow, 1983 [in Russian].

481. **Tikhomirov, V. K. and Vetoshkin, A. G.**, *Computation of cross-section area for Plato–Gibbs channels in polyhedral foams*, Colloid Journal, Vol. 54, No. 4, 1992.

482. **Tikhonov, A. N. and Samarskii, A. A.**, *Equations of Mathematical Physics*, Nauka, Moscow, 1972 [in Russian].

483. **Townsend, A. A.**, *The Structure of Turbulent Shear Flow*, Cambridge Univ. Press, London, 1976.

484. **Tripathi, A., Chhabra, R. P., and Sundarajan, T.**, *Power law fluid flow over spheroidal particles*, Ind. Eng. Chem. Res., Vol. 33, No. 2, pp. 403–410, 1994.

485. **Van Dyke, M. D.**, *Perturbation Methods in Fluid Mechanics*, Parabolic Press, Stanford California, 1975.

486. **Van Dyke, M. D.**, *An Album of Fluid Motion*, Parabolic Press, Stanford California, 1982.

487. **Vargaftic, N. V., Vinogradov, Y. K., and Yargin, V. S.**, *Handbook of Physical Properties of Liquids and Gases*, Begell House Inc., New York, 1996.

488. **Vetoshkin, A. G.**, *Analysis of model for hydraulic conductance of a foamy structure*, Theor. Found. Chem. Eng., Vol. 29, No. 5, pp. 423–426, 1995.

489. **Vetoshkin, A. G., Kazenin, D. A., and Kutepov, A. M.**, *Hydrodynamics of flows in a centrifugal foam suppressor*, Rus. J. Appl. Chem., Vol. 57, No. 1, 1984.

490. **Vetoshkin, A. G., Kazenin, D. A., Kutepov, A. M., and Makeev, A. A.**, *Theory of action of a centrifugal plate foam breaker*, Theor. Found. Chem. Eng., Vol. 20, No. 4, pp. 319–323, 1986.

491. **Vignes, A.,** *Diffusion in binary solutions,* Ind. Eng. Chem. Fundam., Vol. 5, No. 2, pp. 189–199, 1966.

492. **Vishik, M. J. and Fursikov, A. V.,** *Mathematical Problems of Statistical Hydromechanics,* Kluwer Acad. Publ., Dordrecht, 1988.

493. **Vochten, R. and Petre, G.,** *Study of the heat of reversible adsorption at the air-solution interface. 2. Experimental determination of the heat of reversible adsorption of some alcohols,* J. Colloid Interface Sci., Vol. 42, No. 2, pp. 320–327, 1973.

494. **Vochten, R., Petre, G., and Defay, R.,** *Study of the heat of reversible adsorption at the air-solution interface. 1. Thermodynamical calculation of the heat of reversible adsorption of nonionic surfactants,* J. Colloid Interface Sci., Vol. 42, No. 2, pp. 310–319, 1973.

495. **Voinov, O. V. and Petrov, A. G.,** *Motion of bubbles in fluids,* Itogi Nauki i Tekhniki (Mekhanika Zhidkosti i Gaza), Vol. 10, pp. 86–147, 1976 [in Russian].

496. **Voinov, O. V. and Petrov, A. G.,** *Flows with closed lines of flow and motion of droplets at high Reynolds numbers,* Fluid Dynamics, Vol. 22, No. 5, pp. 708–717, 1987.

497. **Voinov, O. V., Golovin, A. M., and Petrov, A. G.,** *Motion of an ellipsoidal bubble in a fluid of small viscosity,* J. Appl. Mech. Techn. Phys., No. 3, 1970.

498. **Voinov, O. V., Petrov, A. G., and Shrager, G. R.,** *Model of the flow internal circulation of a liquid drop in a gas stream,* Fluid Dynamics, Vol. 24, No. 6, pp. 966–968, 1989.

499. **Voloshchuk, V. M. and Sedunov, Yu. S.,** *Coagulation Processes in Disperse Systems,* Hydrometeoizdat, Leningrad, 1975 [in Russian].

500. **Vorontsov, E. G. and Tananaiko, Yu. M.,** *Heat Transfer in Fluid Films,* Tekhnika, Kiev, 1972 [in Russian].

501. **Vulis, L. A. and Kashkarov, V. P.,** *Theory of Jets of a Viscous Fluid,* Nauka, Moscow, 1965 [in Russian].

502. **Warsi, Z. U. A.,** *Fluid Dynamics,* CRC Press, Boca Raton, 1993.

503. **Waslo, S. and Gal-Or, B.,** *Boundary layer theory for mass and heat transfer in clouds of moving drops, bubbles or solid particles,* Chem. Eng. Sci., Vol. 26, No. 6, pp. 829–838, 1971.

504. **Weber, M. E.,** *Mass transfer from spherical drops at high Reynolds numbers,* Ind. Eng. Chem. Fundam., Vol. 14, No. 4, pp. 365–366, 1975.

505. **Westerterp, K. R., Van Swaajj, W. P. M., and Beenackers, A. A. C.,** *Chemical Reactor Design and Operation,* Wiley, New York, 1987.

506. **Wei, H. L. and Subramanian, R. S.,** *Migration of a pair of bubbles under the combined action of gravity and thermocapillarity,* J. Colloid Interface Sci., Vol. 172, No. 2, pp. 395–406, 1995.

507. **Weinberger, H. F.,** *Variational properties of steady fall in Stokes flow,* J. Fluid Mech., Vol. 52, No. 2, pp. 321–344, 1972.

508. **Wilkinson, L.,** *Non-Newtonian Fluids,* Pergamon Press, London, 1960.

509. **Wilson, A. J. (editor),** *Foams: Physics, Chemistry and Structure,* Springer-Verlag, New York, 1989.

510. **Winnikow, S.,** *Letter to the Editors,* Chem. Eng. Sci., Vol. 22, No. 3, p. 477, 1967.

511. **Yablonskii, G. S., Bykov, V. I., and Gorban', A. N.,** *Kinetic Models of Catalytic Reactions,* Nauka, Novosibirsk, 1983 [in Russian].

512. **Young, N. O., Goldstein, J. S., and Block, M. G.,** *The motion of bubbles in a vertical temperature gradient,* J. Fluid Mech., Vol. 6, No. 3, pp. 350–356, 1959.

513. **Zaitsev, V. F. and Polyanin, A. D.,** *Dynamics of spherical bubbles in non-Newtonian liquids,* Theor. Found. Chem. Eng., Vol. 26, No. 2, pp. 185–190, 1992.

514. **Zaitsev, V. F. and Polyanin, A. D.**, *Exact solutions of the boundary layer equations for power-law fluids*, Fluid Dynamics, Vol. 24, No. 5, pp. 686–690, 1989.

515. **Zaitsev, V. F. and Polyanin, A. D.**, *Discrete-Group Methods for Integrating Equations of Nonlinear Mechanics*, CRC Press, Boca Raton, 1994.

516. **Zaitsev, V. F. and Polyanin, A. D.**, *Handbook of Partial Differential Equations. Exact Solutions*, MP Obrazovaniya, Moscow, 1996 [in Russian].

517. **Zapryanov, Z. and Tabakova, S.**, *Dynamics of Bubbles, Drops and Rigid Particles*, Kluwer Acad. Publ., Dordrecht, 1999.

518. **Zapryanov, Z., Polyanin, A. D., and Ryazantsev, Yu. S.**, *Mathematical Modelling of Transport Phenomena*, St. Kliment Ohridski Univ. Press, Sofia, 1994 [in Bulgarian].

519. **Zdravkovich, M. M.**, *Flow Around Circular Cylinders: Vol. 1. Fundamentals*, Oxford Sci. Publ., Oxford, 1997.

520. **Zhizhin, G. V.**, *Laminar boundary layer of non-Newtonian fluid (qualitative investigation)*, J. Appl. Mech. Techn. Phys., No. 3, 1987.

521. **Zhizhin, G. V. and Ufimtsev, A. A.**, *Flow in the plane laminar boundary layer of dilatant liquids*, Fluid Dynamics, Vol. 12, No. 5, pp. 780–784, 1977.

522. **Žukauskas A. and Žiugžda J.**, *Heat Transfer at Cylinder in Cross Flow of Fluid*, Mokslas, Vilnius, 1979 [in Russian].

523. **Zhurov, A. I.**, *Shear flow around a porous cylinder*, Theor. Found. Chem. Eng., Vol. 29, No. 2, pp. 196–199, 1995.

524. **Zhurov, A. I., Polyanin, A. D., and Potapov, E. D.**, *Shear flow over a porous particle*, Fluid Dynamics, Vol. 30, No. 3, pp. 428–434, 1995.

525. **Zinchenko A. Z.**, *The slow axisymmetric motion of two drops in a viscous medium*, J. Appl. Math. Mech. (PMM), Vol. 44, No. 1, pp. 30–37, 1980.

# Index

# 386

turbulent boundary layer, 9, 40, 41
turbulent diffusion coefficient, 118
turbulent diffusion tensor, 118
turbulent flow, 8–11, 14, 23, 24, 32–36, etc.
  buffer layer, 33, 40
  core, 33, 34
  heat transfer, 123–125, 143–145, 147, 148
  jet, 23, 24
  logarithmic layer, 33, 40, 124, 143
  mass and heat transfer, 117, 118, 121
  nonisothermal, 243, 244
  plane channel, 36
  viscous sublayer, 33, 40
  wall region, 33, 40
  tube, 32–35
turbulent flux of admixture, 118
turbulent Prandtl number, 118, 124
turbulent Schmidt number, 118
turbulent viscosity, 9, 10, 23, 24, 32, 33, etc.
  Prandtl's relation, 10, 33
  Karman's relation, 10, 33
two-dimensional foam, 323, 324

# V

vacancy instability, 321
Van der Waals attraction, 308
Van Laar equation, 310
vapor phases, 236
velocity, mean flow rate,17, 26–28, 32, 86,
  126, 141, etc.
viscoelastic fluid, 264–266
viscoplastic media, 264, 265, 269
viscoplastic Shvedov–Bingham fluid, 269,
  272, 273, 277
viscosity
  apparent, 261, 284, 288
  Bingham, 265, 302
  dynamic, 2, 58, 93
  effective, 103, 261, 325

force, 236
kinematic, 2, 32, 94, 115, 116, 144, etc.
plastic, 272
structural, 265
transverse, 265
turbulent, 9, 10, 23, 24, 32, 33, 40, etc.
viscoplastic (foam), 328
viscous stress, 247
viscous sublayer, 33, 40
Voigt body, 326
volume rate, 268, 277
volume rate of flow, 15, 17, 19, 26–30, 49,
  126, etc.
volume reaction, 107, 111, 220–229, 336
  first-order, 222–230
  $n$th-order, 112, 221, 225, 230
vorticity region, 67
V-shaped body, 42, 43, 289, 290

# W

wake
  behind moving body, 24
  diffusion, 203, 206–211
  oscillating, 67
wall region, 33, 40
Weber number, 70, 93, 97, 188
Weissenberg effect, 264, 325
wetting angle for wall material, 15

# Y

yield stress, 264, 265, 269, 271, 279, 280,
  288, etc.
Young's relation, 16

# Z

zero-order volume reaction, 227, 228

**Other titles in the Topics in Chemical Engineering series**

This book is part of a series. The publisher will accept continuation orders which may be cancelled at any time and which provide for automatic billing and shipping of each title in the series upon publication. Please write for details.